GEOTECHNICAL, GEOLOGICAL AND EARTHQUAKE ENGINEERING

Volume 37

For further volumes:
http://www.springer.com/series/6011

Atilla Ansal • Mohamed Sakr
Editors

Perspectives on Earthquake Geotechnical Engineering

In Honour of Prof. Kenji Ishihara

Springer

Editors
Atilla Ansal
School of Engineering
Özyeğin University
Istanbul
Turkey

Mohamed Sakr
Faculty of Engineering
Geotechnical Engineering Research Group
Tanta University
Tanta
Egypt

ISSN 1573-6059 ISSN 1872-4671 (electronic)
Geotechnical, Geological and Earthquake Engineering
ISBN 978-3-319-10785-1 ISBN 978-3-319-10786-8 (eBook)
DOI 10.1007/978-3-319-10786-8

Library of Congress Control Number: 2015933626

Springer Cham Heidelberg New York Dordrecht London

Printed on acid-free paper

Springer is part of Springer Science+Business Media (www.springer.com)

Foreword

It was a privilege to organize the International Conference on Earthquake Geotechnical Engineering from Case History to Practice in the honour of Prof. Kenji Ishihara in Istanbul, Turkey during 17–19 June 2013.

A very prominent group of specialists in the field of earthquake geotechnical engineering delivered lectures of significant importance reflecting the recent developments in their field of interest during this conference. It was decided to compile most of these presentations in one book so that it may be available to specialists and those that are involved with research and application in the field of geotechnical engineering. The book is composed of 18 chapters.

The first chapter by Kenji Ishihara and Toshiyuki Kamata is concerned with features of the 2011 earthquake in Japan, characterized by predominance of the ground failure due to liquefaction and scour of the ground caused by tsunami. Unprecedented long duration of the shaking combined with large aftershocks have generated the worst situations resulting in the extensive damage due to liquefaction over the Tokyo Bay and the downstream plain areas of the Tone River 300~400 km distant away from the epicentral area. This chapter focus on the characteristic features in the occurrence of liquefaction and consequent damage in the area of the downstream reaches of the Tone River.

The second chapter by Liam Finn and Francisco Ruz is concerned with the amplification effect of soft surface layers. The effect on the seismic response of shallow soft surface layers is evaluated for developing the seismic section of the National Building Code of Canada for 2015 (NBCC 2015). The response of such layers is studied using recorded data from the 2011 Tohoku Earthquake.

The third chapter by Kyriazis Pitilakis, Evi Riga, and Anastasios Anastasiadis is about the proposal of a new design spectra and its application to the seismic risk assessment. Based on a worldwide database of strong ground motion records in different well constrained site conditions, a new soil classification is proposed for EC8 with the associated amplification factors and the normalized response spectra. Amplification factors and normalized response spectra are also proposed for the current EC8 classification. A detailed seismic risk assessment was performed for the city of Thessaloniki using (a) the capacity spectrum method, (b) the Uniform Hazard Spectrum (UHS) for rock conditions compiled applying the SHARE approach, (c) the current and the new site amplification factors to evaluate the site specific demand spectra, (d) the detailed inventory of the Thessaloniki building stock to select the appropriate capacity and fragility curves for each building typology, and (e) available functions to estimate the causalities and the economic losses from the physical building damages.

The fourth chapter by Ellen Rathje, Menzer Pehlivan, Robert Gilbert, and Adrian Rodriguez-Marek is on the topic of probabilistic approach related to site response analysis and for seismic hazard assessments. This chapter describes the convolution approach used to incorporate site response into PSHA. The main components used in the convolution approach are provided, and the required site characterization and site response analyses are discussed.

The fifth chapter by K. Önder Çetin and H. Tolga Bilge is on the stress scaling factors for seismic soil liquefaction engineering problems. On the basis of the compiled experimental data, a unified stress correction scheme, which enables to assess the combined effects of overburden (σ'_{v0}) and static (τ_{st}) shear stresses along with the degree of shear stress reversal, was introduced.

The sixth chapter by Takaji Kokusho is about the development of site amplification formula for virtual surface arrays to be consistent with vertical arrays using a number of KiK-net records during recent eight destructive earthquakes. A correlation between peak spectrum amplification and S-wave velocity ratio (base V_s/ surface V_s) improved much better if the surface V_s was evaluated from fundamental mode frequency combined with a thickness of equivalent surface layer in which peak amplification is exerted, rather than using the conventional V_s30.

The seventh chapter by Aslı Kurtuluş, Atilla Ansal, Gökçe Tönük, and Barbaros Çetiner is based on the observations recorded by the Istanbul vertical arrays during small earthquakes. Same events were also recorded by Istanbul Rapid Response Network (IRRN) which comprise 55 surface strong motion stations in the European side of Istanbul. An effort is made to model the recorded response at the downhole array sites as well as the at the IRRN stations using the acceleration records obtained by the deepest sensors on the engineering bedrock. Observations from the recorded response and results from 1D modelling of ground response have yielded in general good agreement between the observed and recorded soil response at the station sites.

The eighth chapter by Susumu Iai is about the centrifuge model tests and effective stress analyses performed on a breakwater subjected to tsunami, such as those that are seriously damaged during 2011 East Japan Earthquake. The centrifuge model tests at a scale of 1/200 are performed to simulate the failure of a breakwater subject to tsunami. With the effective stress analyses, this study demonstrates the importance of the mechanism of failure in the rubble mound due to seepage flow of pore water in addition to the force of tsunami action.

The ninth chapter by Sêco e Pinto, Pedro Simão points out the importance to analyze embankment dam behavior from the past experiences. Even though modern embankment dams may withstand the design earthquake without significant damages, it is important to prevent the occurrence of incidents and accidents of embankment dams under earthquakes and so a deep understanding of the triggering factors is needed. The structural components of potential risk depend mostly on storage capacity and on the height of the dam, as the potential downstream consequences are proportional to the mentioned values. Socio-economic risks can be expressed by a number of persons who need to be evacuated in case of danger and by potential downstream damage. The risks associated with dam projects are discussed.

The tenth chapter by Ross W. Boulanger, Jack Montgomery, and Katerina Ziotopoulou is concerned with the nonlinear deformation analyses (NDAs) used for evaluating the potential effects of liquefaction on the seismic performance of embankment dams. Confidence in these applications of NDAs may be significantly improved by establishing calibration and analysis protocols that reduce undesirable sources of variability between analysis results obtained by different analysts or different models. A key protocol for improving the quality of NDAs is the calibration and validation of a constitutive model using single-element simulations with the types of loading paths potentially important to the structure's response. Examples of single-element simulations are presented for several constitutive models and loading paths to illustrate some of the important features for NDAs of embankment dams affected by liquefaction.

Chapter 11 by Steven L. Kramer, Bita Astaneh Asl, Pelin Özener, and Samuel S. Sideras is about ground surface motions during liquefaction. This chapter reviews the process of liquefaction, the hydraulic conditions required for the manifestation of surface effects, and limitations in the inference of liquefaction triggering from the presence or absence of such effects are discussed. The effects of liquefaction triggering on soil stiffness and procedures for identification of the timing of liquefaction triggering are reviewed. The abilities of four advanced site response programs to compute ground motions in liquefiable soils are evaluated from the standpoints of pre- and post-triggering response. The implications of post-triggering response for the evaluation of ground surface response spectra are discussed.

Chapter 12 by Ikuo Towhata, Masahide Otsubo, Taro Uchimura, Masato Shimura, Bangan Liu, Toshihiko Hayashida, Damoun Taeseri, and Bertrand Cauvin is about the liquefaction mitigation of embedded lifelines. The gigantic earthquake in 2011 caused significant damage in lifeline in the Tokyo Metropolitan Area. In particular, the damage was significant in recent artificial islands where liquefaction affected embedded sewage pipelines profoundly. The present chapter addresses the ongoing model tests by which a variety of mitigation measures for sewage pipelines are examined. In the regions where future earthquake is expected, it is not possible to excavate pipes and reconstruct backfills now because of financial limitations. To cope with this situation, less expensive measures such as mechanical constraint, partial injection of grout, or limited installation of drainage measures are studied.

Chapter 13 by Susumu Yasuda, Iwao Suetomi, and Keisuke Ishikawa is about the effects of long duration of shaking during the 2011 Great East Japan Earthquake. The main shock was followed by big aftershocks because the earthquake was a "megathrust earthquake" with extremely large magnitude; $M\mathrm{w}=9.0$. The unique ground shaking caused serious liquefaction in a wide area of reclaimed land along Tokyo Bay though seismic intensities in the liquefied zones were not high. The occurrence of liquefaction, the settlement and the inclination of houses must have been affected by the aftershock. Large horizontal displacements were induced and caused roads to thrust and large horizontal displacement of liquefied ground had caused the severance of pipe joints and the shear failure of manholes, allowing an influx of muddy water into the pipes and manholes.

Chapter 14 by Ramon Verdugo is about the liquefaction caused during the 2010 Chile Earthquake of Magnitude 8.8. Several of the available records show a ground motion that exceeds 2 min of duration, which may explain the significant amount of liquefied sites. Field observations have shown that the earthquake triggered liquefaction in more than 170 different sites, covering a north-south distance of about 950 km, which approximately corresponds to twice of the length of the rupture zone. Liquefaction phenomenon induced damages to the road infrastructure, railroads system, buildings and houses. Liquefaction-induced ground failure displaced and distorted pile foundations of piers impacting seriously the operation of several ports. Especially interesting is the case of Juan Pablo II Bridge, where significant differential settlements were observed.

Chapter 15 by Wei F. Lee, C. C. Chen, M. H. Chang, and Louis Y. N. Ge presents a case study on non-plastic silty sand liquefaction. A new sampling technique, Gel-Push sampler, that could retrieve undisturbed non-plastic silty sand and results of series dynamic triaxial tests those investigated influences of fines contents, void ratio, as well as sampling disturbance are reported. Research progress presented here is hoped to be helpful in understanding mechanism as well as consequences of non-plastic silty sand liquefaction.

Chapter 16 by Michele Maugeri and Francesco Castelli is about the case history of a seismic retrofitting of a piled foundation. The approach adopted for the structural upgrading of a six storey reinforced concrete frame building in Italy, damaged by the 13 December 1990 Sicilian earthquake, is described and discussed. For evaluating the possibility to repair the building, an investigation on soil, structure and foundation was carried out. As building was founded on piles, for the seismic retrofitting of the structure, the response of the piles subjected to earthquake loading has been studied. Loading tests showed that the seismic event of December 13th 1990 have not damaged the effectiveness of the soil-pile system. The numerical analysis showed that the existing foundation was unable to carry on design seismic actions, the structural upgrading and seismic retrofitting of the piled foundation required the enlargement of the foundation and new bored piles.

Chapter 17 by George D. Bouckovalas and Yannis K. Chaloulos summarizes the main findings of a systematic research effort regarding the response of pile foundations in laterally spreading soils. The incentive of the research were findings from properly scaled (with regard to the pore fluid) centrifuge experiments suggesting that severe soil dilation may occur at the upper part of the pile, as a result of large soil-pile relative movement, causing soil pressures to significantly increase. A 3D nonlinear numerical methodology was developed and tested against the aforementioned experiments. Comparative analyses, with the old and the new boundaries, revealed that the former (which also reflect the kinematic response of the laminar box containers employed in model tests) can significantly underestimate soil pressures imposed to the foundation. In the sequel, the numerical methodology was applied parametrically (for various soil, pile and excitation characteristics) and a new set of multivariable relationships was statistically established for the practical estimation of ultimate soil pressures applied to the pile.

Chapter 18 by Yoshimichi Tsukamoto is about the use of Swedish weight sounding tests for earthquake reconnaissance investigations is integrated into one framework for analysing a wide range of stability problems of natural and reclaimed soil deposits during earthquakes, based on the past studies conducted by the author. It is highlighted that Swedish weight sounding tests are useful for subsurface soil profiling identifying thin weak soil layers, and can be used for determining the undrained shear strength, liquefaction strength and post-liquefaction settlement of natural and reclaimed soil deposits during earthquakes.

The editors would like to express their gratitude to all authors for their interest and efforts in preparing their manuscripts. Special thanks to Professor Kenji Ishihara for being so creative in generating major research in the area of earthquake geotechnical engineering and for his support in the organization of this conference.

Atilla Ansal and Mohamed Sakr

Contents

Contributors

Anastasios Anastasiadis Aristotle University of Thessaloniki, Thessaloniki, Greece

Atilla Ansal School of Engineering, Özyeğin University, Istanbul, Turkey

Bita Astaneh Asl University of Washington, Seattle, WA, USA

H. Tolga Bilge GeoDestek GeoEng. & Consultancy Services Ltd. Co., Ankara, Turkey

George D. Bouckovalas National Technical University of Athens, Athens, Greece

Ross W. Boulanger Department of Civil and Environmental Engineering, University of California at Davis, California, USA

Francesco Castelli University of Enna "Kore", Enna, Italy

Bertrand Cauvin Department of Civil Engineering, Engineering School ENTPE, Vaulx-en-Velin, France

K. Önder Çetin Middle East Technical University, Ankara, Turkey

Barbaros Çetiner Kandilli Observatory & Earthquake Research Institute, Boğazici University, Istanbul, Turkey

Yannis K. Chaloulos National Technical University of Athens, Athens, Greece

M. H. Chang MICE Engineering Consultants, Taipei Taiwan, Republic of China

C. C. Chen MICE Engineering Consultants, Taipei Taiwan, Republic of China

W.D. Liam Finn The University of British Columbia, Vancouver, BC, Canada

Louis Y. N. Ge MICE Engineering Consultants, Taipei Taiwan, Republic of China

Robert Gilbert University of Texas at Austin, Austin, TX, USA

Toshihiko Hayashida Fudo Tetra Corporation Japan, Tokyo, Japan

Kenji Ishihara Research and Development Initiative, Chuo University, Tokyo, Japan

Toshiyuki Kamata Chemical Grouting Co. Ltd, Tokyo, Japan

Takaji Kokusho Dept. Civil & Environment Eng., Chuo University, Tokyo, Japan

Steven L. Kramer University of Washington, Seattle, WA, USA

Asli Kurtuluş School of Engineering, Özyeğin University, Istanbul, Turkey

Wei F. Lee MICE Engineering Consultants, Taipei Taiwan, Republic of China

Bangan Liu Department of Civil Engineering, University of Tokyo, Tokyo, Japan

Michele Maugeri University of Catania, Catania, Italy

Jack Montgomery Department of Civil and Environmental Engineering, University of California at Davis, California, USA

Masahide Otsubo Department of Civil Engineering, University of Tokyo, Tokyo, Japan

Pelin Ozener Yildiz Technical University, Istanbul, Turkey

Menzer Pehlivan Mueser Rutledge Consulting Engineers, New York, NY, USA

Kyriazis Pitilakis Aristotle University of Thessaloniki, Thessaloniki, Greece

Ellen Rathje University of Texas at Austin, Austin, TX, USA

Evi Riga Aristotle University of Thessaloniki, Thessaloniki, Greece

Adrian Rodriguez-Marek Virginia Polytechnic Institute, Blacksburg, VA, USA

Francisco Ruz R y V Ingenieros, Santiago, Chile

Pedro Simão Sêco e Pinto National Laboratory of Civil Engineering (LNEC), Geotechnical Engineering, University of Coimbra, Coimbra, Portugal

Masato Shimura Department of Civil Engineering, University of Tokyo, Tokyo, Japan

Samuel S. Sideras University of Washington, Seattle, WA, USA

Iwao Suetomi Eight-Japan Engineering, Matsue, Japan

Iai Susumu Disaster Prevention Research Institute, Kyoto University, Kyoto, Japan

Damoun Taeseri Department of Civil Engineering, ETH-Zurich, Zurich, Switzerland

Gökçe Tönük School of Engineering, MEF University, Ayazağa, Istanbul, Turkey

Ikuo Towhata Department of Civil Engineering, University of Tokyo, Tokyo, Japan

Yoshimichi Tsukamoto Department of Civil Engineering, Tokyo University of Science, Chiba, Japan

Taro Uchimura Department of Civil Engineering, University of Tokyo, Tokyo, Japan

Ramón Verdugo CMGI, Santiago, Chile

Susumu Yasuda Tokyo Denki University, Tokyo, Japan

Katerina Ziotopoulou Department of Civil and Environmental Engineering, University of California at Davis, California, USA

Geotechnical Damage in the Downstream Reaches of the Tone River in the 2011 East Japan Earthquake

Kenji Ishihara and Toshiyuki Kamata

Abstract Features of the 2011 earthquake in Japan are characterized by predominance of the ground failure due to liquefaction and scour of the ground caused by Tsunami. Unprecedented long duration of the shaking combined with large aftershocks have generated the worst situations resulting in the extensive damage due to liquefaction over the Tokyo Bay and the downstream plain areas of the Tone River 300~400 km distant away from the epicentral area. In this paper, focus is placed on the characteristic features in the occurrence of liquefaction and consequent damage in the area of the downstream reaches of the Tone River.

Keywords Lateral flow · Liquefaction · Mega earthquake · Settlements

Introduction

An extremely large earthquake rocked the widespread area in the northern part of Japan Mainland at 14:46:33 (JST) on March 11, 2011. The magnitude of the quake was M=9.0, an event accompanied with an unprecedentedly large amount of energy release at a centroid depth of 24 km where the Pacific tectonic plate subsides into the North Pacific plate underneath the Japan Mainland. This quake was the largest ever recorded during the last 150 years since the start of seismic observation in Japan. Subsequently, the two big-scale aftershocks occurred, one at 15:09 with M=7.4 in the north and the other at 15:15 with M=7.7 in the south on the same day as displayed in the map of Fig. 1.

The fault zone covered a wide area about 500 km long and 200 km wide. It is noted that the biggest first quake with whopping M=9.0 occurred at an epicentre under the sea 130 km off the coast of Sendai, followed by the second largest event

K. Ishihara (✉)
Research and Development Initiative, Chuo University, Tokyo, Japan
e-mail: kenji-ishihara@e-mail.jp

T. Kamata
Chemical Grouting Co. Ltd, Tokyo, Japan
e-mail: t-kamata@chemicalgrout.co.jp

A. Ansal, M. Sakr (eds.), *Perspectives on Earthquake Geotechnical Engineering*,
Geotechnical, Geological and Earthquake Engineering 37,
DOI 10.1007/978-3-319-10786-8_1

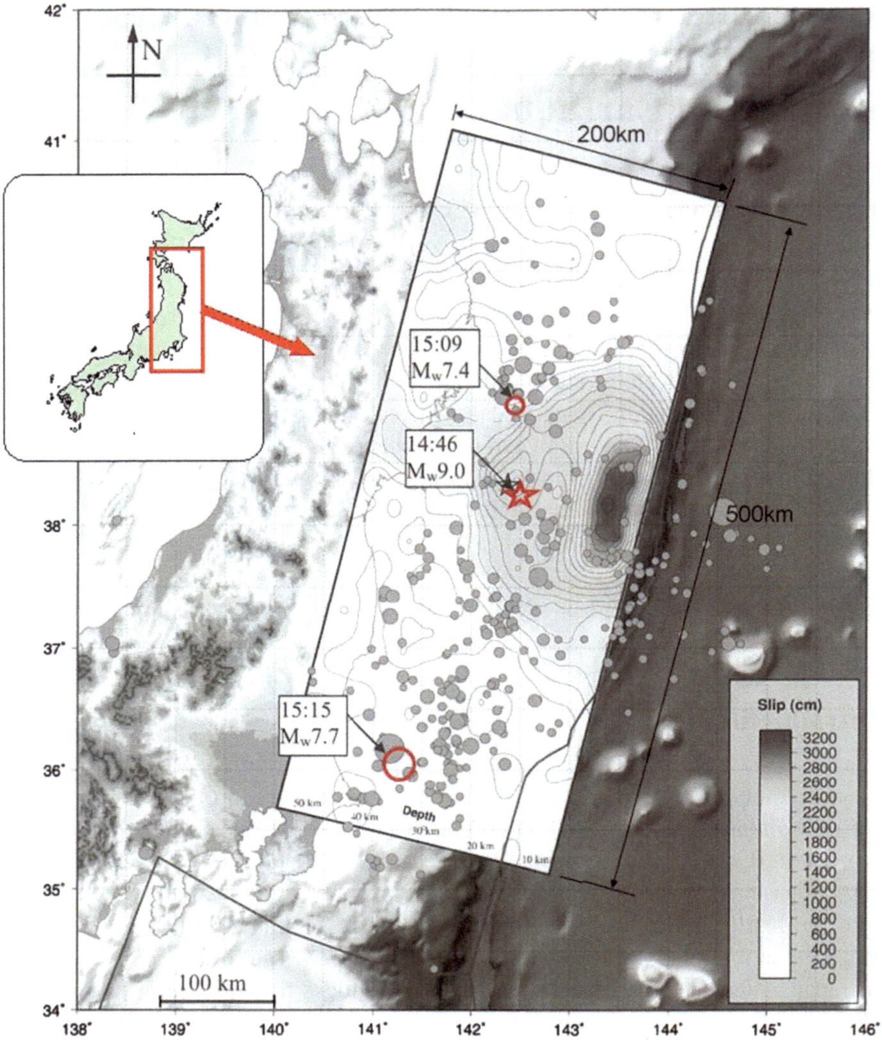

Fig. 1 Epicentral zone of the 2011 Great East Japan Earthquake. (Hayes et al. 2011)

at an epicentre about 150 km off the coast in the north. The fault ruptures moved southward, generating the third largest aftershock at 15:15 at a southern site 30 km off-shore from the coast of Fukushima prefecture. Strong shaking was felt and recorded at a number of sites over a widespread area from the northern prefectures, Iwate, Miyagi and Fukushima, down to the southern region such as Ibaragi, Chiba and Tokyo metropolis. Extensive liquefaction developed in the area of Tokyo Bay as well as over the flat lowlands surrounding lakes in the lower reaches of the Tone

River north of Tokyo. Occurrence of liquefaction and its consequence in the area of Tokyo Bay was described briefly in the previous papers by Ishihara K (2012), Ishihara and Araki (2012) and Ishikawa and Yasuda (2012). The characteristic features of liquefaction and associated ground movements incurred in the water-rich vast area downstream of the Tone River will be described in some details in the following pages of this paper.

Motion Characteristics in the Downstream Area of the Tone River

Amongst several seismic stations in this area, the K-NET named CHB004at Sawara in Katori indicated in Fig. 2 was the nearest to the city of Itako, Kamisu and Katori being considered in this paper. It was located at stiff soil site free from liquefaction. The major acceleration records in EW, NS and UD directions are shown in Fig. 3 and its trajectory in plan is displayed in Fig. 4. Time histories of velocity obtained by integrating the acceleration records are presented in Fig. 5. Spectral accelerations and velocities are shown in Figs. 6 and 7, respectively. It can be seen in these figures that the peak acceleration is 300 gal in E-W direction and the predominant period is 0.1 ~ 0.14 s in spectral acceleration, and 1.0 ~ 1.7 s in spectral velocity.

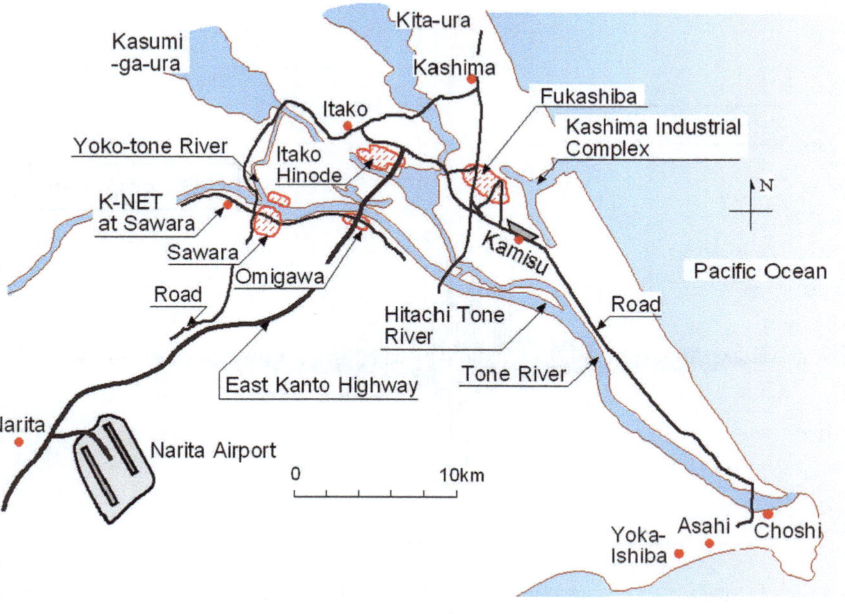

Fig. 2 A map of the downstream reaches of the Tone River

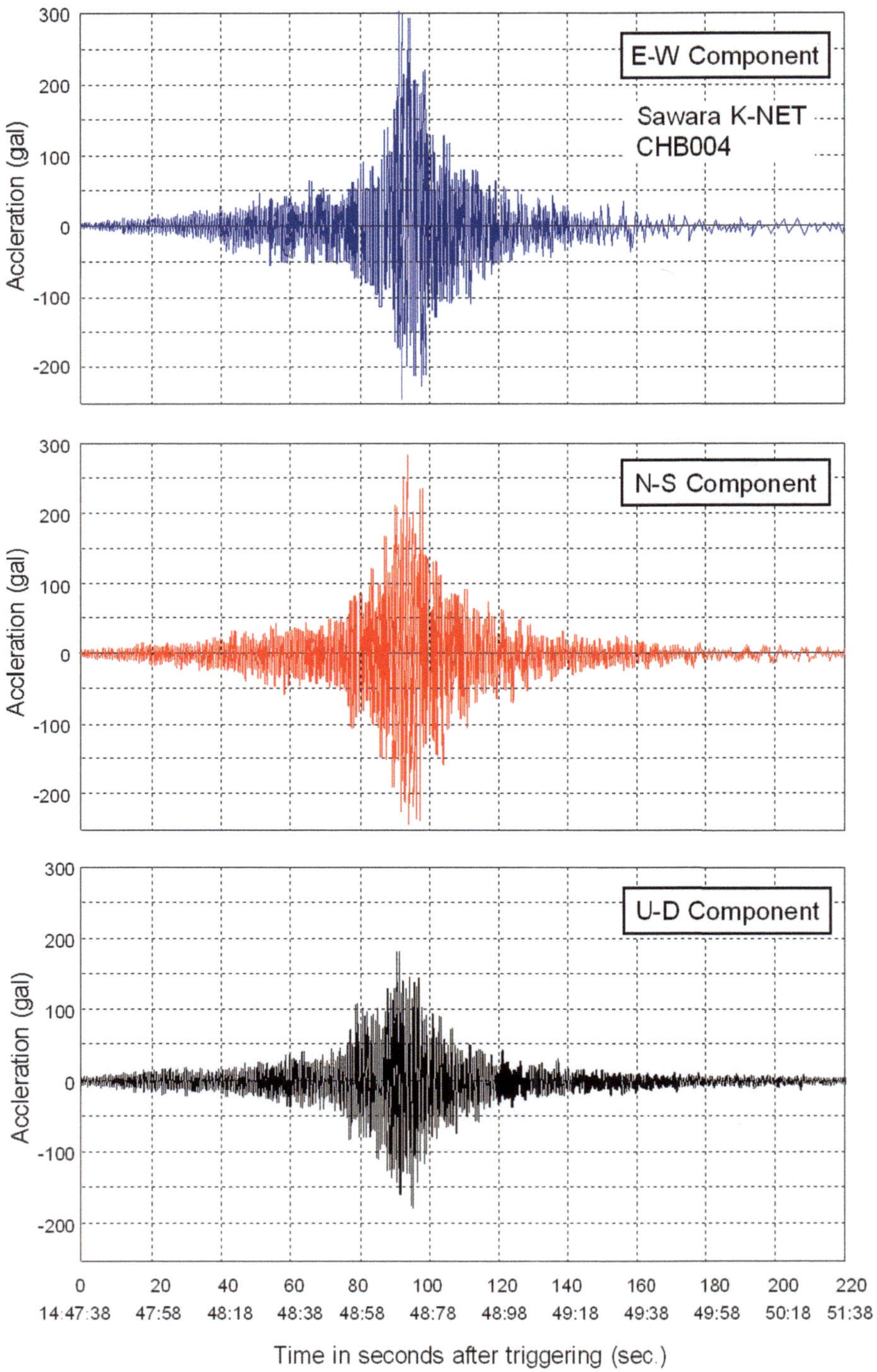

Fig. 3 Recorded motions at Sawara K-NET station, CHB004

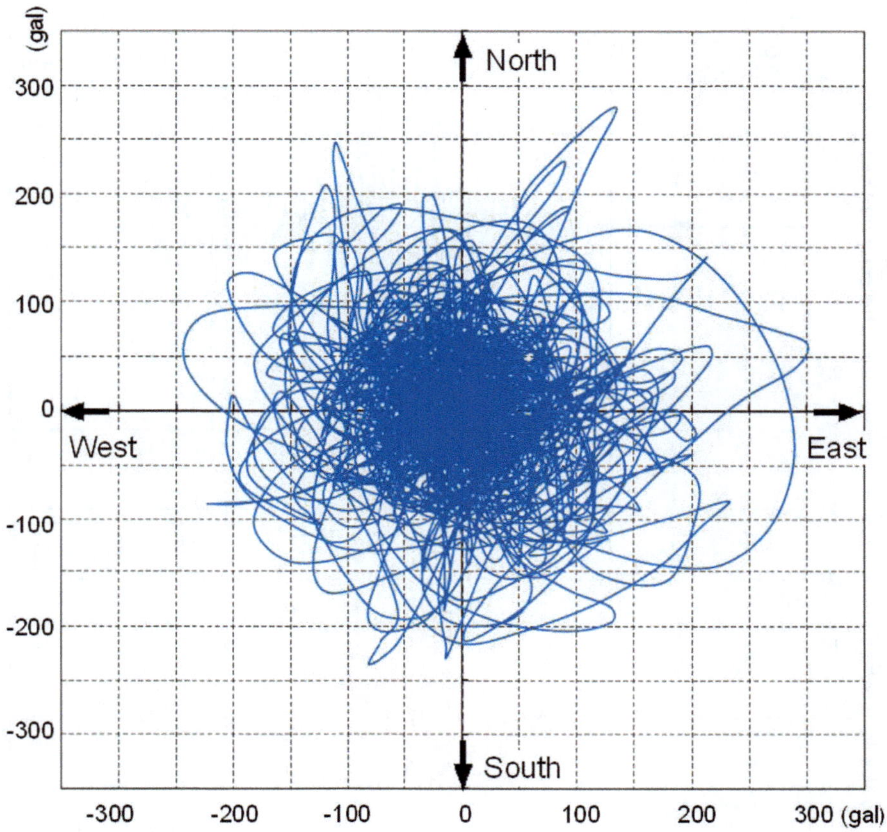

Fig. 4 Trajectory in plan of the recorded accelerations

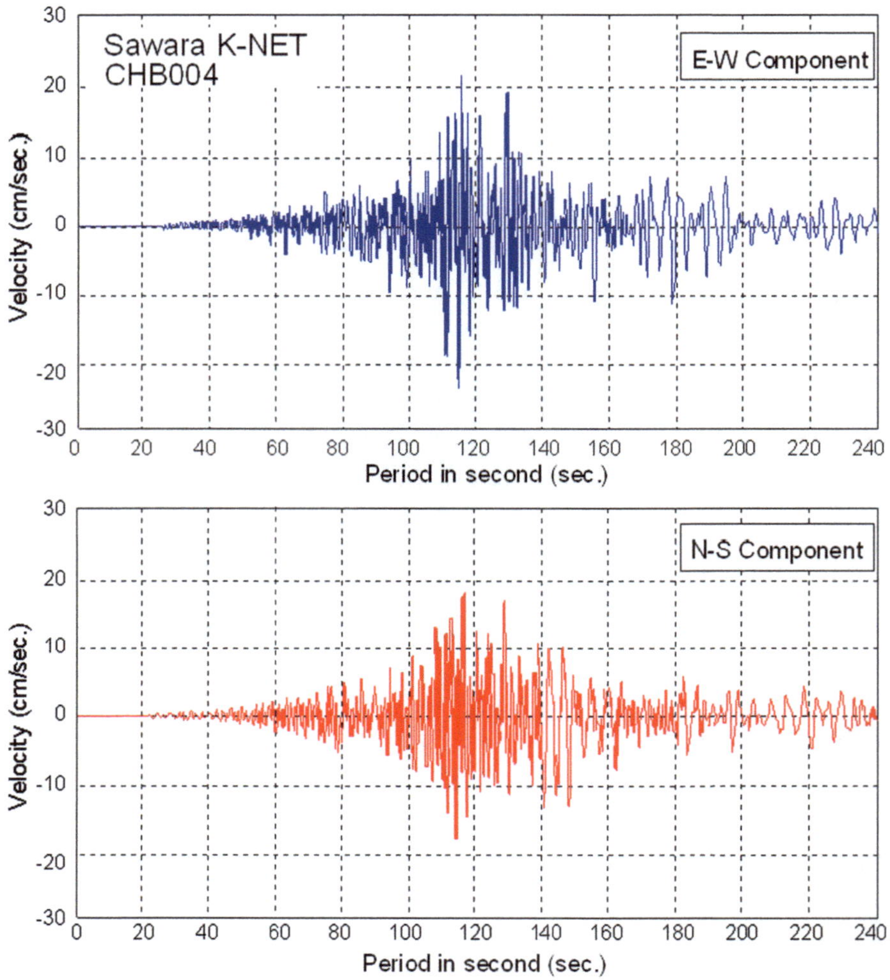

Fig. 5 Time histories of velocities obtained from recorded accelerationat the Sawara K-NET station

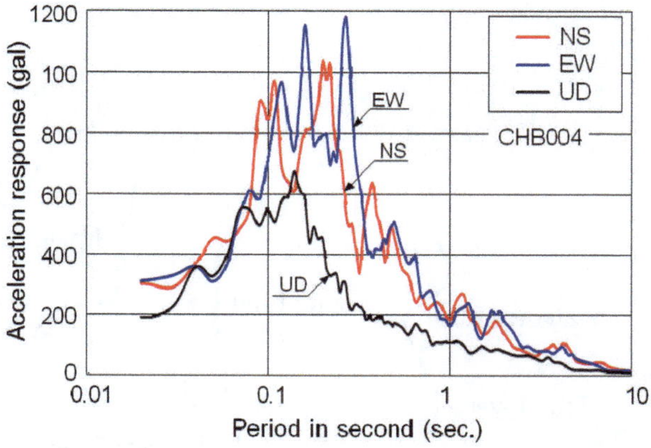

Fig. 6 Acceleration response spectra ofthe motions at Sawara K-NET station

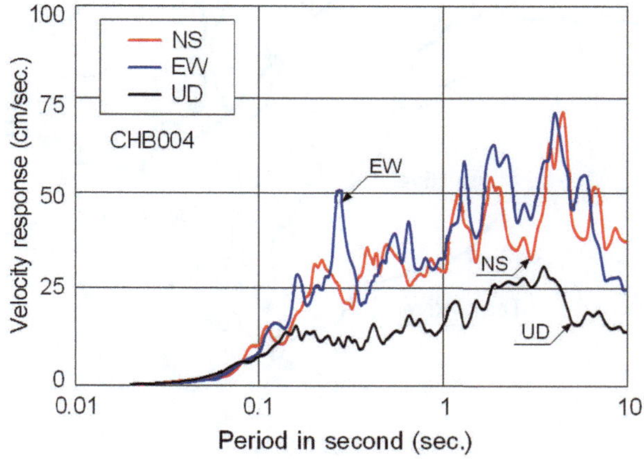

Fig. 7 Velocity response spectra ofthe motions at Sawara K-NET station

Ground Damage in the Downstream Area of the Tone River

Liquefaction in Kanto Region at Large

Extensive liquefaction was induced in reclaimed and alluvial deposits along rivers and bayshore areas in the western part of Kanto plain including Ibaragi, Chiba prefecture and Tokyo which are located as far as 300~400 km southwest of the epicentre of the main shock. The total land area in which signs of liquefaction were observed is purported to have been of the order of 70 km^2. In terms of the long distance from the epicentre and also in terms of the large expanse net area, the liquefaction in theKanto region by the 2011 earthquake was unprecedented and truly record-breaking. Shown in Fig. 8 is the distribution of places where apparent signs of liquefaction were observed such as sand oozing, boiling, ground cracking and

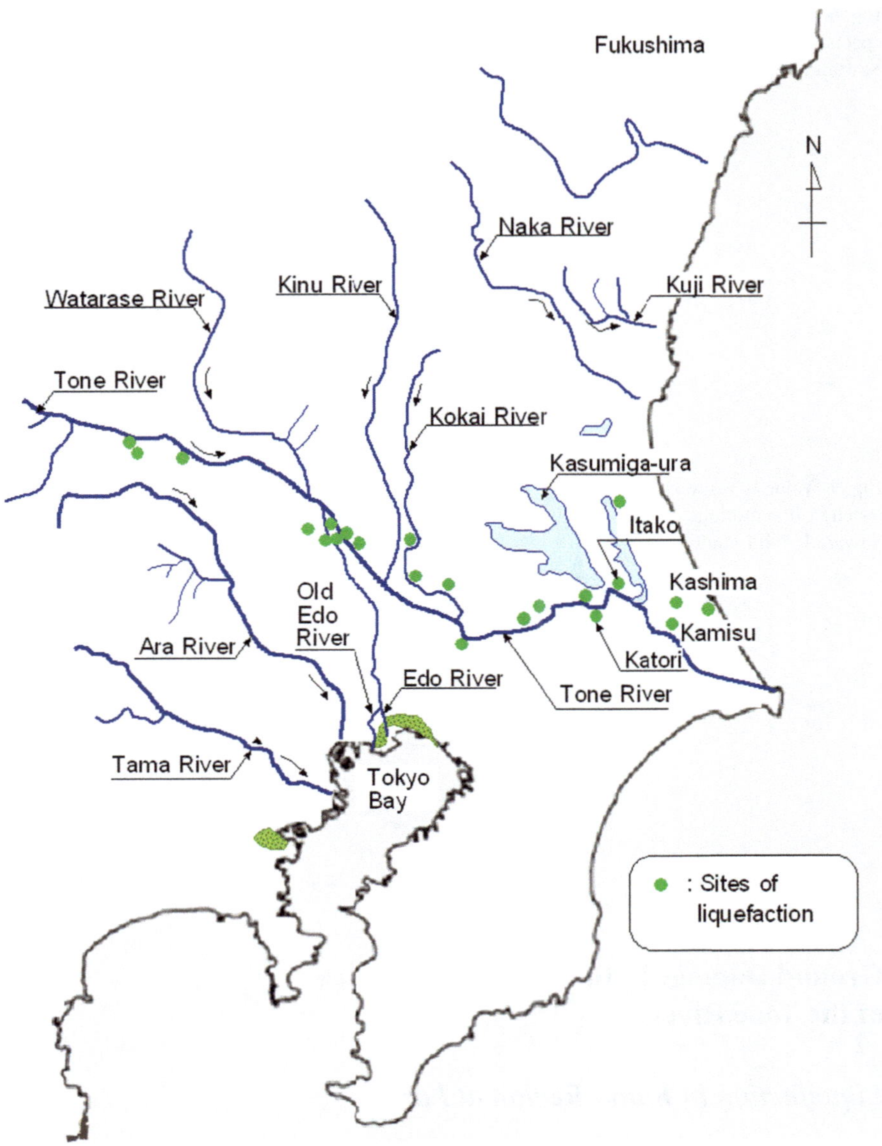

Fig. 8 Liquefaction—affected area in Kanto Region. (Yasuda 2011)

associated ground settlements and distorsions. Geomorphologically these may be classified as (1) those which developed in the reclaimed waterfront area along the Tokyo Bay, (2) the liquefaction in the lakeshore districts south of Kasumiga-ura in the lower reach of the Tone River, and (3) the spotwise occurrence of liquefaction at many locales along the Tone River and its tributaries where landfills had been conducted over the years in originally marshy flat lands. Features of the damage are also presented in the papres by Koseki et al. (2012), Senna et al. (2012) and Wakamatsu (2012).

Ground Damage in Itako City

1. *General*

As seen in an enlarged map of Fig. 2, Iatko city is located in the area surrounded by Kasumigaura Lake on the northwest, Kitaura Lake on the north and the large flat land in the lower reaches of the Tone River on the south. Figure 2 also shows the particular areas of severe damage due to liquefaction at Hinode district in Itako city. A map of Itako further enlarged is demonstrated in Fig. 9.

While places of liquefaction were scattered widely over the Itako city precinct, severe damage was concentrated predominantly in the district of Hinode about 25 km long and 15 km wide as indicated in Fig. 9. This area used to be an inlet-like lagoon as visualized in an old map of Fig. 10 as of 1881–1885. The lagoon was drained off later on by enclosing it with low-height levee on the east and by draining water out. The lagoon area as of 1893 prior to the drain-off is shown in Fig. 11. The drain-off was set out partly in 1939 and completed around 1952. The land was transformed to farmland mainly for rice production.

In early 1960's, the large project for developing Kashima near-shore industrial complex was inaugurated in the widespread area along the coast of the Pacific

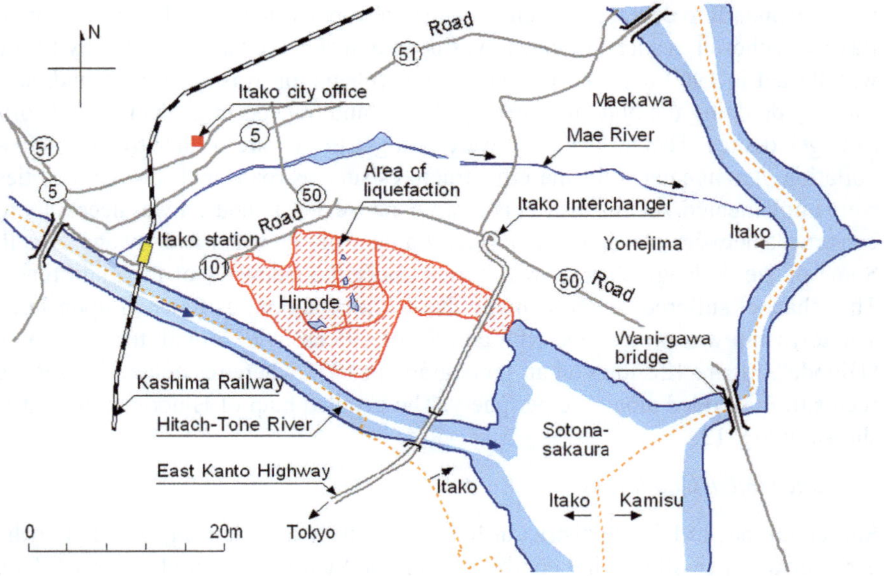

Fig. 9 Places of liquefaction, Hinode, in Itako city

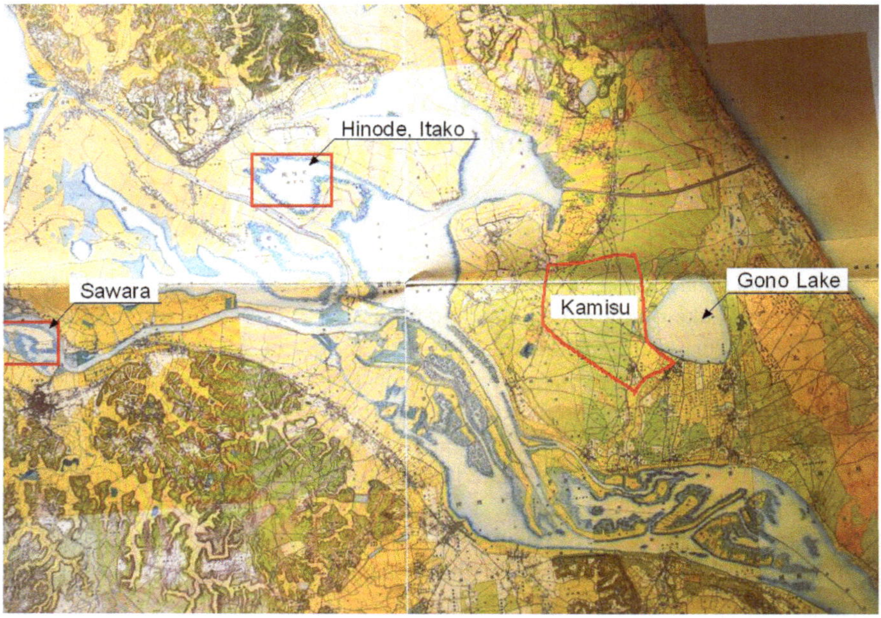

Fig. 10 An old map of the downstream area of Tone River (1881–1885)

Ocean east of Itako. This project involved near-beach inland excavation to construct ports and harbours to accommodate to large vessels for marine transportation. The location of the cut-in harbours is shown in Fig. 2. With increased population, there arouse demands to develop residential lands. In response to these, Hinode section of Itako was chosen as a land of new township and raising the land elevation by filling was started in 1969 over the farmland. The reclamation was performed hydraulically by dredging the bottom of nearby lagoon and transporting water-mixed soils to targeted sites. The surface soils were brought to the site further by land transportation. Arrangements for the infrastructures such as roads and sewage facilities were implemented. As this district is a vast expanse of flat lands, it was necessary to construct a network of water channels for drainage of rain water and sewage as well. Some of the drainage channel was buried under the sideways of the main roads. This channel suffered severe damage due to liquefaction, as touched upon later. The township was proclaimed with an area of 196 ha in 1974 with the new name "Hinode". Figure 12 shows an air photograph in 1971 where residential houses are seen sitting studded along the periphery. The detailed map of Hinode at present is shown in Fig. 13.

2. *Geotechnical Conditions*

Soil conditions in Hinode district at Itako were investigated, prior to and after the earthquake, by in-situ boring combined with the Standard Penetration Test (SPT). The soil profiles along the N-S line, ①, ②, ③, ④ in Fig. 13, is demonstrated in Fig. 14

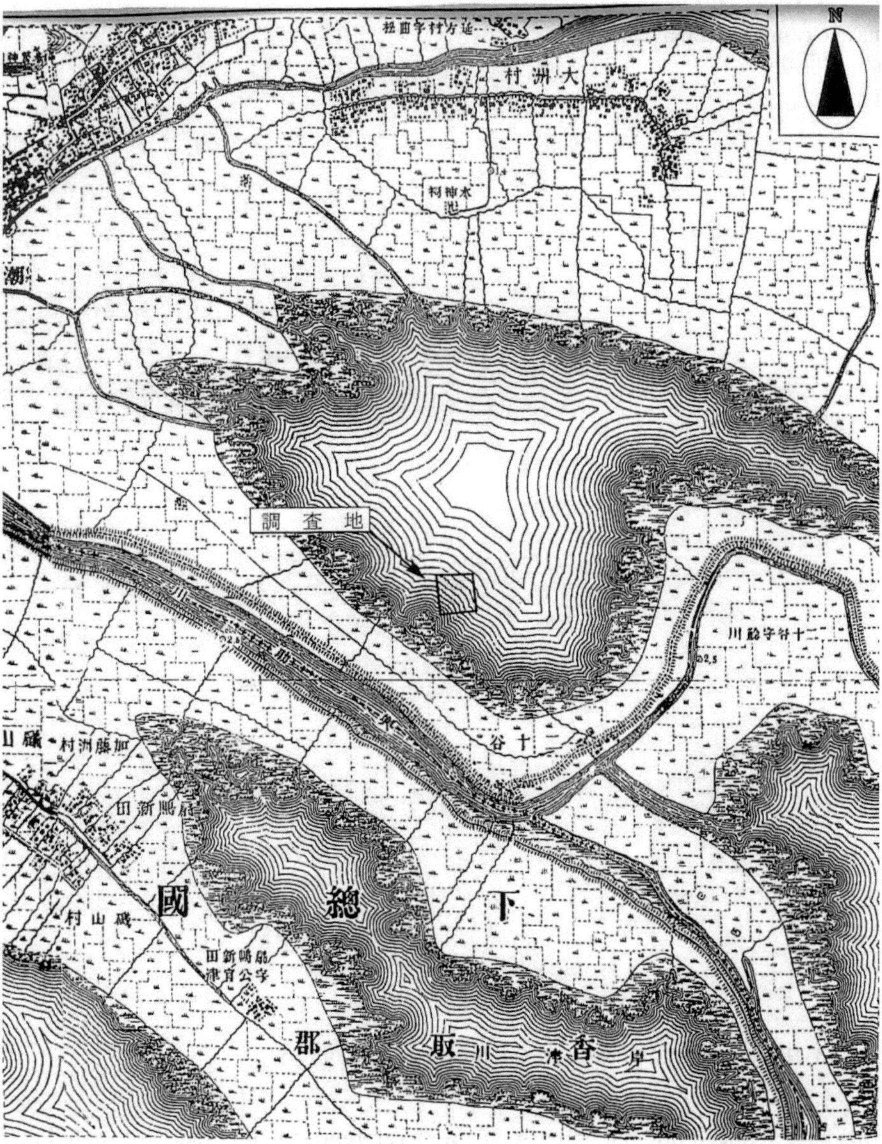

Fig. 11 Lagoon area as of 1893 prior to the drain-off (from Itako city office)

where it can be seen that below the surface fills there exists a loose deposit of sandy soil having a low SPT N-value of the order of 5–15. Saturated with a high ground water table of 1 ~ 2 m below the surface, the near-surface sand fill is deemed to be a new sand deposit which had been placed hydraulically at the time of the recent reclamation in 1965–1981. It is obvious that this deposit had developed liquefaction during the earthquake. It is also likely that the loose alluvial sand deposit below the

Fig. 12 Hinode district of Itako city in 1971 (from Itako city office)

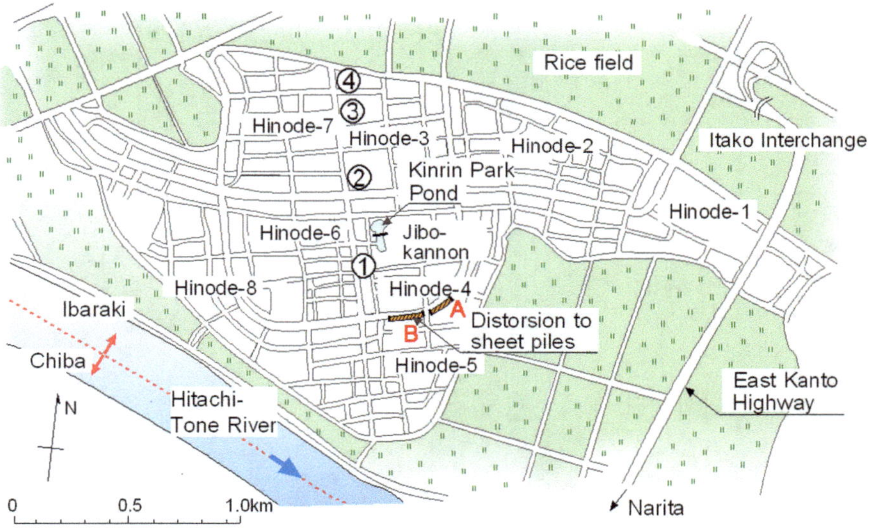

Fig. 13 Hinode district of Itako at present (from Itako city office)

fills have induced liquefaction as well. With reference to other data, it is noted that, below the artificial and alluvial sand deposits near the surface there exists a thick stratum of sandy silt or silt of alluvial origin with a SPT N-value of 1 ~ 2. Thus, the Hinode district in Itako city is considered to lie over the soft soil deposit which rests on a huge bowl-shaped basin lying over stiff soil formation. It is thus highly conceivable that a kind of sloshing or seiche motions must have been induced in the bowl shaped soft silty deposits masked by the liquefiable sands.

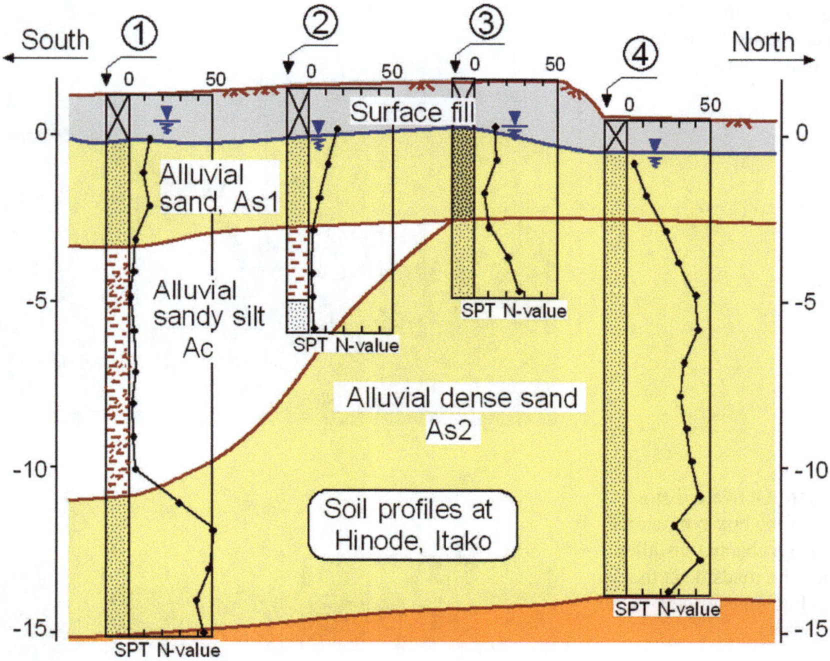

Fig. 14 Soil profile in Hinode, Itako

3. *Damage features at Hinode district in Itako*

As a result of strong shaking lasting for 2 min, liquefaction did develop extensively over the area in Hinode. Particularly notable was the tilting and settlements of private houses and uplifting of box-type water-draining channels buried beneath the sideway, accompanied by sand boiling and some lateral displacements. Figure 15 shows the settlement of about 50 cm with 1 ~ 2° tilt of residential houses located in Hinode-5 in Fig. 13. The damage to the road at point A in Hinode-5 (see Fig. 13) is displayed in Fig. 16 where the cover of concrete plates on the south sideway is seen being displaced up outward, accompanied by the settlement of the road pavement. Figure 17 shows the damage of the sideway at a place, B, indicated in Fig. 13 where the row of buried sheet piles along the road on the south side is seen being pushed up about 50 cm sidewards. The sheet piles appear to have been installed to a shallow depth of about 5.0 m as an outside wall of the box-type underground drainage channel about 2.0 m wide and 1.5 m in depth. A rough sketch for this damaged channel is displayed in Fig. 18.

There is a man-made pond at Hinode 4-chome as indicated in Fig. 13 in the middle of the Hinode district. The fills surrounding the pond moved 3 ~ 5 m towards the waters due to lateral spreading. A photo in Fig. 19 shows a distant view of the bridge from the south and Fig. 20a shows the damage to the east abutment of the

Fig. 15 Tilting of houses accompanied by settlements at Hinode, Itako

Fig. 16 Outward distortion of the box-type buried drainage channel installed along the roadside at the site A in Fig. 13—A view to the northeast

Fig. 17 Lift up and outward distorsion of the buried sheat pile row accompanied by lateral displacement at location B in Fig. 13—A view to the eastsouth

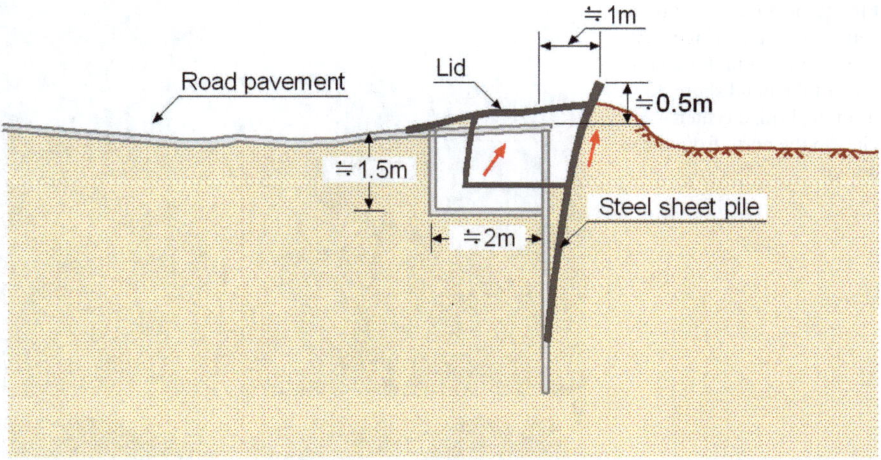

Fig. 18 Conceptual feature of distorsion of sheet-pile supported box-type drainage channel beneath the sideway. Hinode 5-chome, Itako

Fig. 19 General view of the bridge from the south side of the Kinrin Park at Hinode, Itako (see Fig. 13)

walkway bridge across the pond. A photo in Fig. 20b is the feature of lateral spread viewed down to the north from the bridge. As described by Tsukamoto et al. (2012), the soils in the fills are deemed to have developed liquefaction. The feature of the lateral spread is conceptually described in Fig. 21.

Fig. 20 **a** Breakage of the abutment of the bridge—A view to the north. **b** Lateral flow of the pond shore down from the bridge center—A view to the north from the bridge

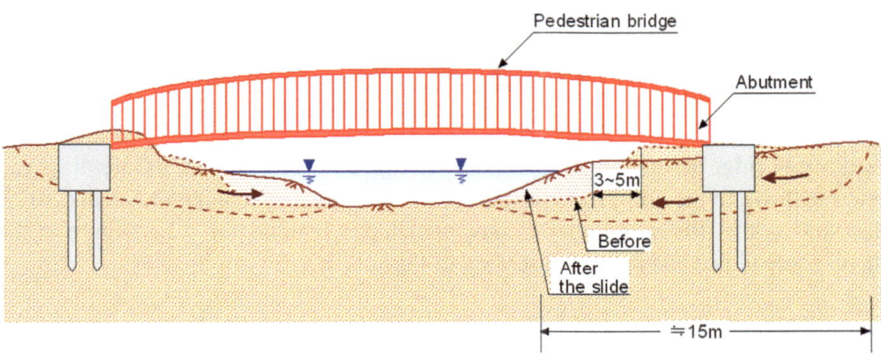

Fig. 21 Conceptional picture of lateral spread along the man-made pondin Kinrin Park, Itako

Ground Damage in Katori City

4. *General*

Katori city is located in the south and north sides of the Tone River as indicated in Fig. 2. The city precinct covers a wide area as shown in Fig. 22 including hilly area in the south and flat land north of the Tone River. The urban district traditionally called Sawara is a small area, but it had developed as a commercial centre for rice products taking advantage of river transportation. The exact location of Sawara is also indicated in Fig. 22. Today's urban section is shown in Fig. 23 superimposed to the old map. The urban part of the city south of the river had been developed over the years by reclaiming low-lying marshy lands which used to be flood plain of the River. The land reclamation was carried out sometimes by drain-off but many times by filling sandy soils dredged from nearby river bed. Chronological change of the land formation is shown in a suite of old maps shown in Fig. 24 where it can be seen that the area of the city center used to be pond-like small river around 1900 in marshy land in the flood plain.

From around 1950 onwards, the channel of the Tone River was reformed with construction of large levees on both sides with an aim of flood control. Then, in the era of rapid economic growth after the World War II, the central city portion was rapidly renovated through construction of modern infrastructures such as tall buildings, bridges, highway overpaths and buried lifelines. During this period, the

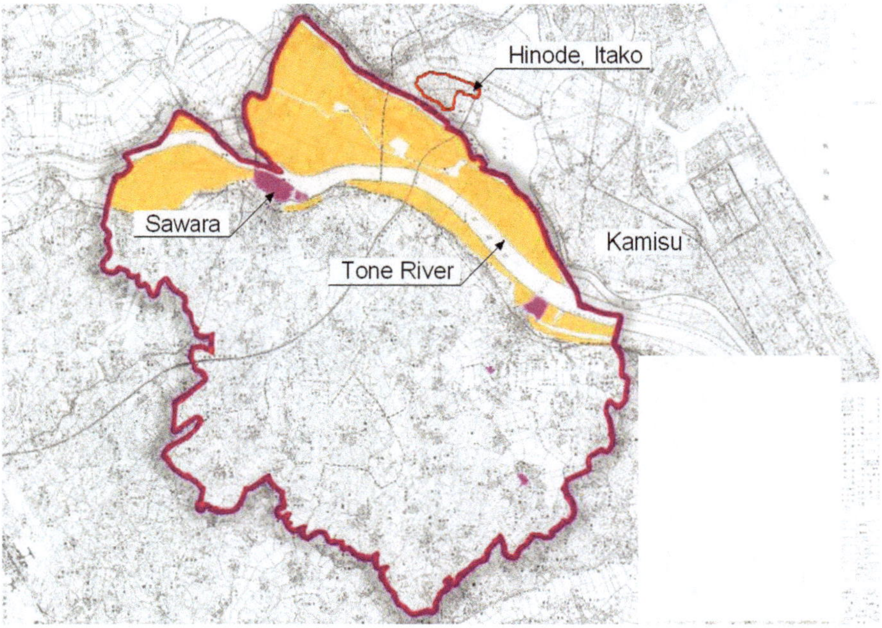

Fig. 22 District of Katori city

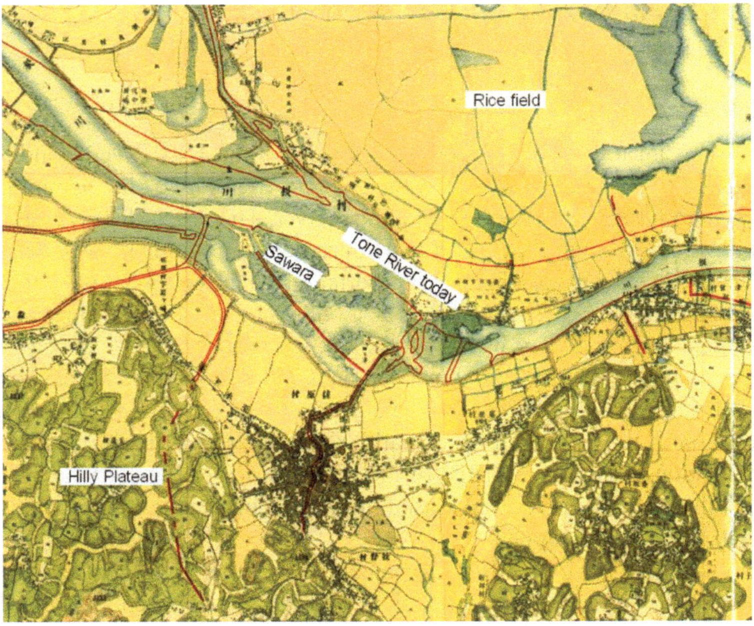

Fig. 23 Urban area in Katori city precinct superimposed to the old map of 1881–1885

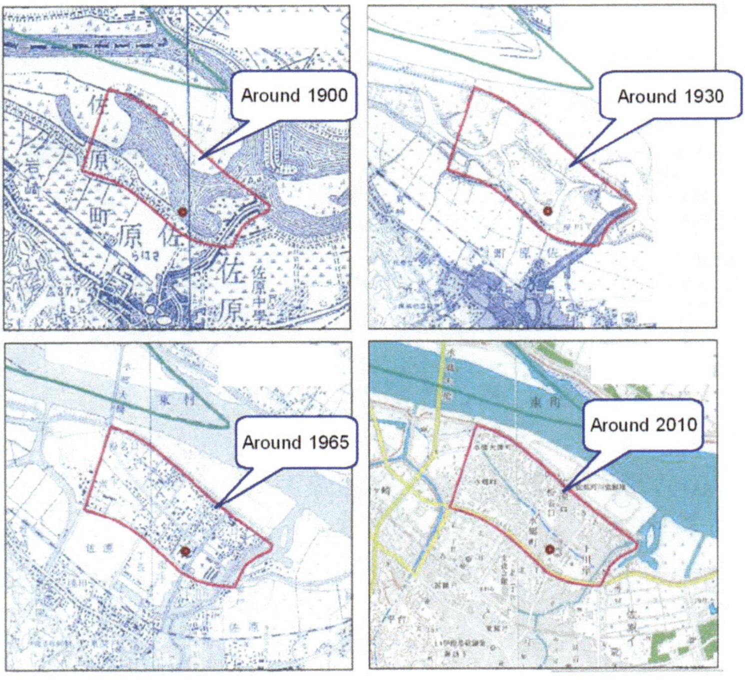

Fig. 24 Chronological change in land formation in Sawara in Katori city (from Katori city office)

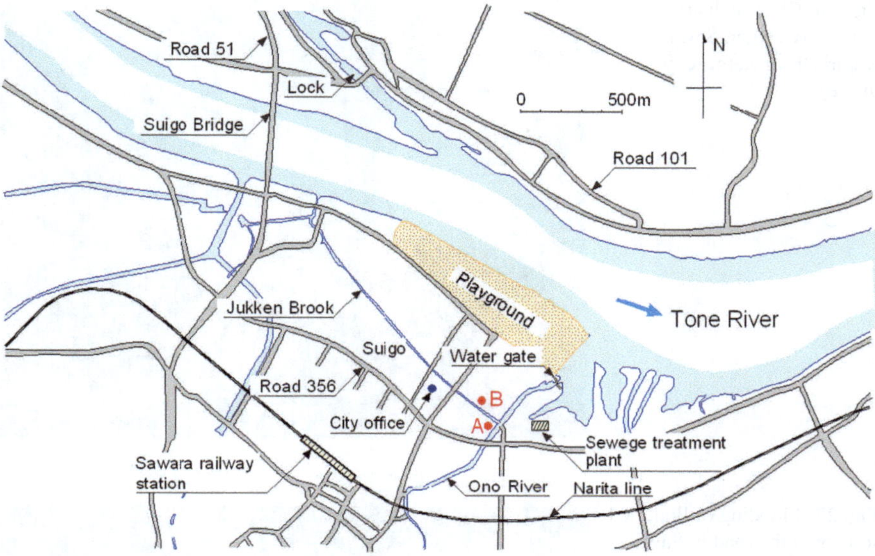

Fig. 25 Urban area in Sawara, Katori city

downstream portion of the old Ono River flowing into the Tone River from the south was widened and arranged to discharge excess flood water from the backland. A watergate was also constructed at the mouth to prevent backflow of excess water from the Tone River during flooding. The sewage treatment plant was constructed on the east to accommodate the sewage discharge from the newly developed urban area. The city area, as it is today, is shown in Fig. 25.

5. *Damage features in Sawara, Katori city*

During the 2011 earthquake, liquefaction developed extensively over the city areas, involving soil ejecta, inundation of roads with muddy soil-laden water, sinking and tilting of private houses. The havoc was dreadful particularly in the reclaimed area in the vicinity of Jukken Brook near the city office building which is located north of the main road No. 356 (see Fig. 25). The chaotic features of the liquefaction-related distress on the roads in the city are displayed in the photos in Figs. 26 and 27, where the muddy water is seen flooding over the road accompanied by tilting of houses and poles of electric power transmission. These damages were concentrated in the section along the Jukken Brook which had been a mashy inlet in the south of the Tone River. Lower stream portion of the Ono River was devastated awfully over the distance of about 400 m by lateral spread and upheaval of the river bottom as shown in the photo of Fig. 28. The ravage as above was also brought about in the Jukken Brook, as shown in Fig. 29, which was created artificially around 1960 through the city centre over the distance of 500 m. As can be seen in Fig. 29, the bankside of the brook had moved inwards accompanied by the upheaval of the riverbed. The damage shown in Fig. 29 is a view to the west from the point A in

Fig. 26 Chaotic distress in the city center, Sawara, Katori (from Katori city office)

Fig. 27 Flooding of liquefied soils over the road in Sawara, Katori (from Katori city office)

Fig. 28 Devastation in the downstream reach of the Ono River (from Katori city office)

Fig. 29 Upheaval of the river bottom in Jukken Brook, Sawara, Katori—A view view to the west from point A in Fig. 25 (from Katori city office)

Fig. 30 Lateral spreading in the Jukken Brook in Sawara—A view to the west from point B in Fig. 25

Fig. 25, near the entrance to the Ono River. Another view to the west from point B is shown in Fig. 30, where lateral displacement is seen on the bank on both sides.

6. *Soil Conditions in Sawara, Katori city*

As indicated in a suit of old maps in Fig. 24, the central part of the city is known used to be marshy lands corresponding to concavely meandering inlet-like section of the large Tone River. In the era around the turn to the nineteenth century, the land was dried off first and turned to the paddy field, but more recently it was reclaimed, around the beginning of the twentieth century, together with the construction of southern levee of the Tone River. Then, this area gradually developed into the town, as a center of commercial activities. Needless to say, all earthworks of civil engineering in the old era were conducted without knowledge of soil mechanics and thus carried out in simple and endemic ways leaving soils uncompacted and thus vulnerable to liquefaction during earthquakes.

There are many boring data obtained prior to and after the quake. Among these data, those archived by Chiba GeoEnvironment Information Center are quoted here. The locations of the boring are indicated in Fig. 31. A typical soil profile in the city center is shown in Fig. 32a. The exact location is at the northern corner of the Sawara middle school as indicated by the number 31780 in Fig. 31. At this site, evidence of liquefaction was apparently manifested on the ground surface. It can

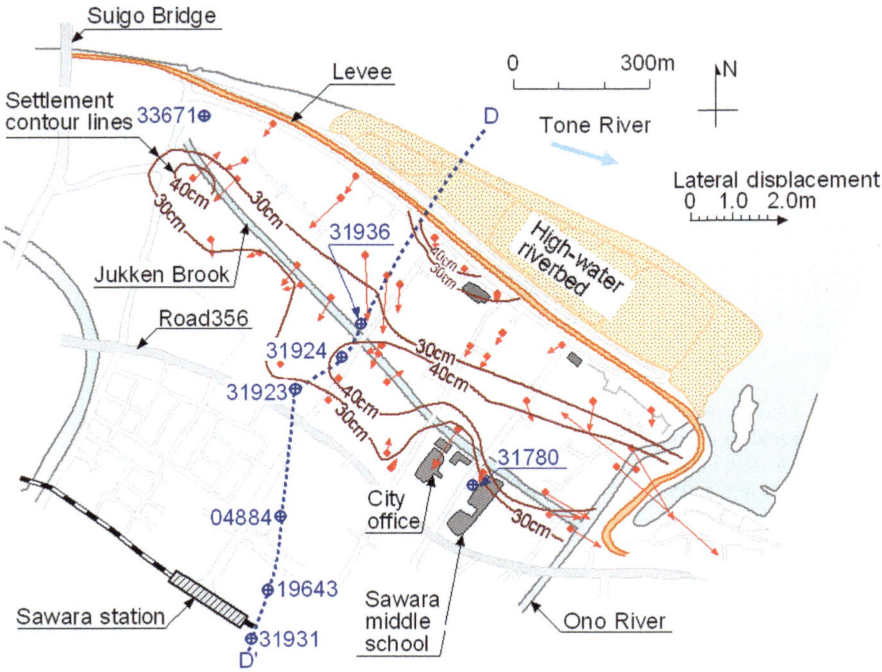

Fig. 31 Sites of boring and ground distorsions in Sawara city (from Katori city office)

be seen in Fig. 32a that underneath the loose sand deposit to a depth of 2.5 m, there exist a relatively stiff sand stratum with SPT N-value of 15 to 35, underlain by a layer of silty sand with medium density. As these three layers are identified as being composed of the sands of alluvial (Holocene) origin, they are denoted by As1, As2 and As3 in the order of depositional age. To be worthy of notice is the presence of underlying soft silty clay deposit to the depth 52 m. Because of such deep seated soft soils, the buildings in the Sawara middle school are supported by steel piles driven to the stiff sand as deep as 52 m. Another representative profile of subsurface soils is presented in Fig. 32b for the site No. 33671 situated in the northwest part of Sawara city as indicated in Fig. 31. It may be seen that the loose sand deposit to a depth of 9.5 m is underlaid by relatively stiff sand with SPT N-value of 7 to 20. It is to be noticed that a thick layer of silty fine sand with a medium density extends to a depth of 40 m until it reaches a stiff sand deposit. The soil profile at another point 31936, a site of severe liquefaction, is shown in Fig. 32c. It indicates that a man-made fill consisting of sandy soils exists near the surface underlain by silty sands As1 of alluvial origin. There are alternate layer of silts and sands (As1, Ac2) of alluvium below. It is likely that liquefaction occurred in the shallow-depth strata to a depth of about 10 m. The soil profile at point 19643 at Kita 3-chome is shown in Fig. 32d. In this area, no evident sign of liquefaction was observed at the time of the earthquake. Somewhat lower elevation of the ground water table must have prevented the liquefaction from occurring.

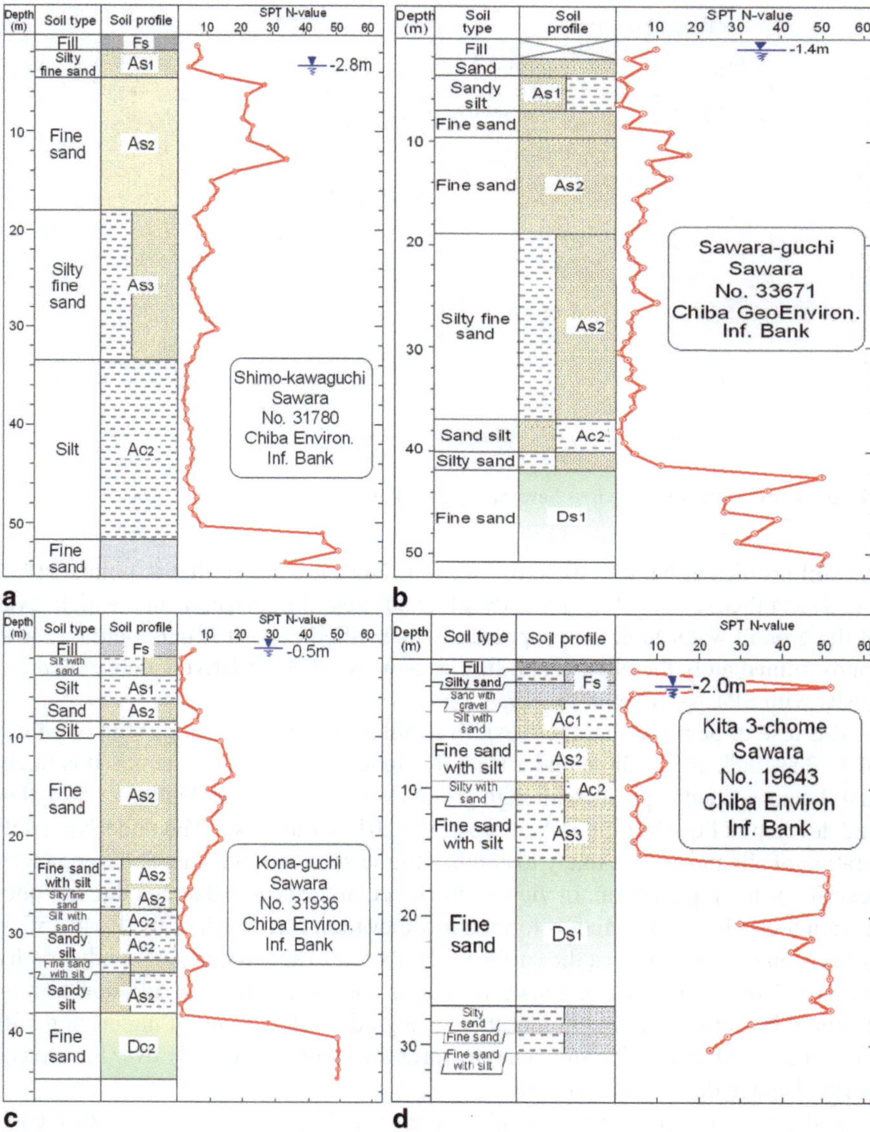

Fig. 32 **a** Soil profile in the premise of Sawara Middle school. **b** Soil profile at the northwest site of Sawara city. **c** Soil profile at Kona-guchi, Sawara. **d** Soil profile at Kita 3-chome, Sawara, Katori-city

To see overall soil conditions, the cross section D-D' in Fig. 31 across the city from north to south is displayed in Fig. 33, where it is noticed that whilst the northern part is flanked by the high embankment of the Tone River, the residential area along the Jukken Brook is lowest in elevation with the ground water table near the surface, and therefore suffered severe devastation due to liquefaction. Looking over

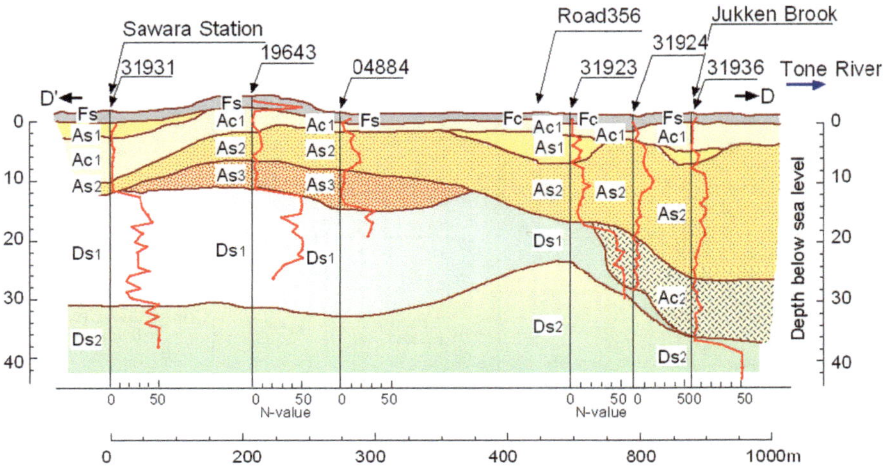

Fig. 33 Cross section D-D' across Sawara city (See Fig. 31)

the soil profiles at large in the cross section D-D', it is noted that at points 31936 and 31924 there exist soft and muddy silt-sands near the surface with the high level of the ground water table. This apparently man-filled deposit is underlaid by loose sands immediately therebelow and further downwards by relatively dense sand deposits with SPT N-value of 10~15.

It is also to be noted that there exist a stratum of soft silt or sands with N-values of 2~3 as thick as 20~30 m underneath the liquefied near-surface strata. It is likely that the man-filled deposits with high ground water table to the depth of 1.5~2.0 m had developed liquefaction at the time of the 2011 earthquake. The underlying soft stratum of silt and sand is likely to amplify the seismic motion which also enhanced severity of the liquefaction. In view of the fact that the severe damage due to liquefaction took place in the man-reformed low elevation area north of the Road 356, it is envisioned in Fig. 33 that the surface fills (Fs or Fc) and underlying sandy or silty soils (As1 or Ac1) were the causes to trigger the liquefaction. In addition to this, it should be noticed in Fig. 33 that there is a gradual sloping up of underlying stiff soils of diluvial origin (Pleistoncene) towards the south particularly over the area of heavy damage due to liquefaction.

Soil conditions in Sawara were also investigated by means of the Swedish cone penetration tests. The outcome of these studies is reported by Tsukamoto et al. (2012).

4. *Characteristic features of the ground damage*

As a consequence of the extensive liquefaction in the city of Sawara, the ground suffered a considerable amount of distorsions such as settlements and lateral movements. Shown in Fig. 31 are the distributions of the settlement in terms of contour lines and the lateral displacement at many spots in terms of vector. It can be seen in Fig. 31 that the settlement as much as 40 cm or even more had taken place in the area due northeast of the city office extending southeast towards the mouth of the Ono River. If this map is compared with the old map in Figs. 23 and 24, it is evidently known that this area used to be a deepest portion of water-stagnant inlet. It is of interest to notice that the area enclosed by the contour line of 30 cm in Fig. 31 does coincide with the portion of the old lagoon shown in Fig. 24. With respect to the lateral displacement, the values in term of vector are directed by and large towards the area of the largest settlement, except for the one on the right bank of the Ono River which is directed largely in the opposite direction. This indicates a large distorsion induced in the eastern bank towards the centre of the Ono River.

5. *Local effects of uneven soil stratification upon seismic motions*

It is conceivable that the seismic wave coming up supposedly vertically must have been amplified differently and distorted in the area of spacially altering underground stiff soils or rock formations. Thus, the secondary waves such as surface waves must have been generated as a result of reflection or refraction due to the inclination of underlying stiff soil formation or near-surface soft soil stratification. This fact appears to have been the case in the area of large settlement in Sawara particularly in the belt-shaped zone along the Jukken Brook. It can be seen in Fig. 33 that the stiff sand layers denoted by Ds1 and Ds2 become shallower towards the south with the underground slopes of 1/15. Thus, it is likely that subordinate types of waves had been generated due to the presence of inclined stiff soil layers under the ground. If any kind of the secondary waves is generated, the shear stresses near the surface are non-zero and therefore, the stress ratio defined as the seismically induced shear stress divided by the confining stress due to overburden becomes substantially large. It should be recalled that, according to the basic principle in soils mechanics, the soil behaviour is dominated not by the shear stress itself but by the stress ratio. This aspect of the concern related to accelerated triggering of liquefaction was addressed in the previous paper by Ishihara and Araki (2012).

Ground Damage in Kamisu City

1. *General*

The city is located in the downstream plain area of the Tone River as shown in the map of Figs. 2 and 8. An enlarged map in Fig. 34 indicates that the Kamisu city is about 7 km east of Itako city and about 5 km from the coast of the Pacific Ocean. The district of severe liquefaction in Kamisu is at Horiwari and Fukashiba and sporadically distributed spots in their vicinity.

According to an old map, there existed a fairly large lake called Gono Lake as shown in Fig. 10 having a triangular shape in the north of the today's Gono Lake. Its location on the present map is also shown in Fig. 34 with a dashed line. This lake was buried around 1960 when the Kashima near-shore industrial complex was developed, but the belt zone about 1.5 km long and 200 m wide was left and reshaped

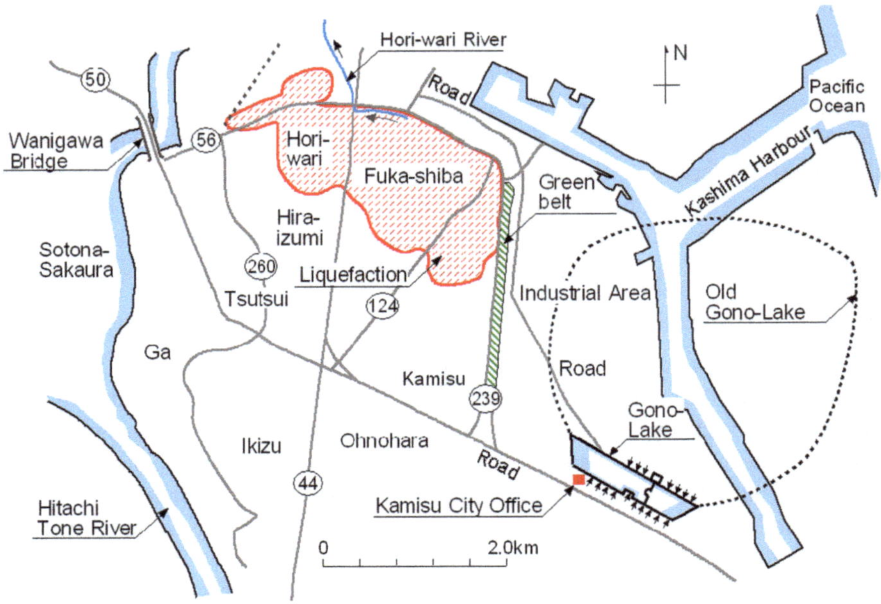

Fig. 34 Areas of liquefaction-induced damage in Kamisu city

as the new Gono Lake as accordingly indicated in Fig. 34. As the surrounding area is composed of freshly deposited sandy soils, the shore line of this lake was also damaged by the lateral flow of liquefied materials.

2. *Damage features*

In Fukabori district, where the ground damage was most severe, the deposits near the surface were composed of freshly filled sandy soils. Over this area, the gravel deposit were excavated in 1950 ~ 1970 to a maximum of 10 m depth even below the ground water table to take gravels for construction purposes. Later on, the large pits were buried by borrow sands. Then, this area was transformed into residential district. When houses, facilities and buried lifelines were built in 1980's, there was no compaction efforts expended to stabilize loose sand deposits for providing competent bases and foundations. The consequence of liquefaction was awfully disastrous as seen in the photos shown in Fig. 35a where houses look as if they were afloat on lake water. Figure 35b shows another example of houses sunken and tilted on the liquefied ground. Infrastructures such as roads and buried pipelines were horribly destroyed in Kamisu. Figure 36a shows the destruction to the road in Horiwari area involving 50 cm overall sinking of the paved surface, accompanied by the lateral movement of the sideways towards the centre. Figure 36b shows about 50 cm floating up of the box-type water drainage buried channel about 2 m wide and 1.5 m deep. Private houses in the vicinity were all damaged by sinking and tilting.

Fig. 35 Sinking and tilting of houses in the Fukashiba district of Kamisu. **a** Mud water around private houses. **b** Private houses tilted and subsided

3. *Geotechnical conditions*

The soil conditions in Kamisu were investigated by means of the Standard Penetration Test (SPT) combined with Swedish cone penetration tests. The results of the investigations by the Swedish cone are reported in the paper by Kawabe et al. (2012). Soi1 investigations were performed at numerous sites at Fukashiba and Horiwari as shown in Fig. 37. Some of the results of these borings are displayed in Fig. 38 in the form of sideviews where it is apparently seen that the sand fills (Bn, Fl) or alluvial sand (As1) developed liquefaction. Particularly to be noticed is the existence of the thick deposit of man-made fill, denoted by Fl, towards the north-east, as typically demonstrated by the boring B4-1 and K-21 in Fig. 38a. This is the typical deposit formed by backfilling the water-pooled open pits by hurling sands after taking gravels out in recent years. It is to be noted that in the area typically shown in Fig. 38b, thick strata of gravels, Ag, remain at depths 4 ~ 10 m, as they existed before excavation. This is the gravel layer excavated at the time of the construction of the Kashima Harbour nearby.

Fig. 36 Damage to the road in Horiwari, Kamisu. **a** Settlement of the road with the walkway that had moved toward the center. **b** Floating up of the buried water-drainage channel on both sides of the road

a

b

12. *Lateral flow along the shore of Gono Lake*

As mentioned above, the present rectangular shape of the Gono Lake was formed recently in the period 1960 ~ 1965 when the construction of the nearshore Kashima industrial complex was put into execution as a large national project. The shore line of the new size-curtailed lake consists all the way of sand deposits not stabilized enough to cope with the risk of liquefaction. Thus, the lateral flow amounting to 2 ~ 3 m was induced towards the lake all the way along the bank, destroying the concrete-block-paved gentle slopes. Figures 39a and 40a show pictures of the slides on the south and north sides of the bank, respectively. The conceivable features of the soil movement are schematically depicted in Figs. 39b and 40b. As indicated, the slide in the northern shore extended almost 30 m backwards from the lakeshore with the ejecta of sand and ground cracking visible even at places far beyond the shore line. On the southern side, detailed soil investigations were carried out using the Swedish cone. The results of the study reported by Tsukamoto et al. (2012) indicate that the near-surface sand fills had caused liquefaction and the lateral flow of the shore.

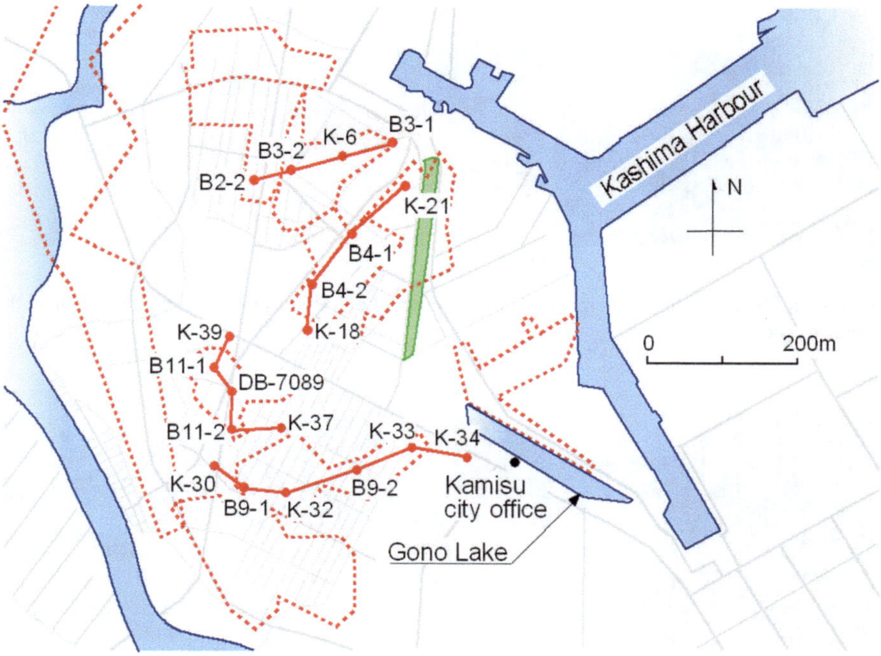

Fig. 37 Places of soil borings at Fukashiba district in Kamisu city

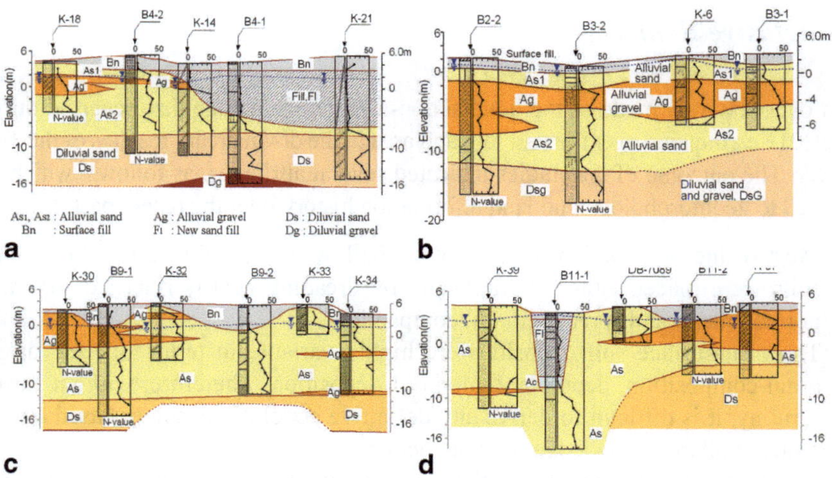

Fig. 38 Side views of soil profiles at Fukashiba, Kamisu (from Kamisu city office). **a** Soil profiles in Fukashiba, Kamisu. **b** Soil profile in Fukashiba, Kamisu. **c** Soil Profile in Fukashiba, Kamisu. **d** Soil profile in Fukashiba, Kamisu

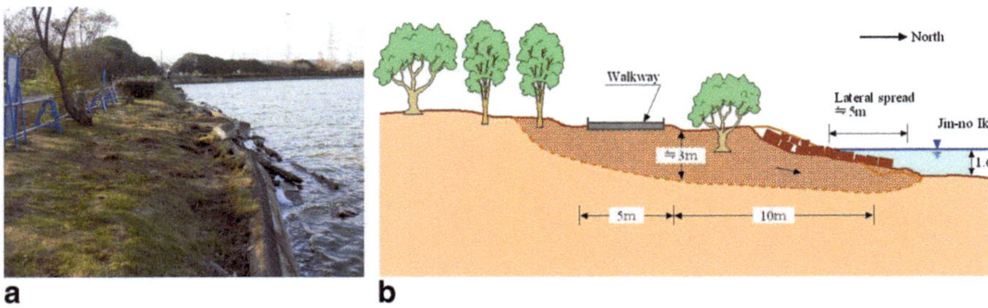

Fig. 39 **a** Sliding in the southern shore of the Gono Lake. **b** Conceptual picture of the slide in the southern shore

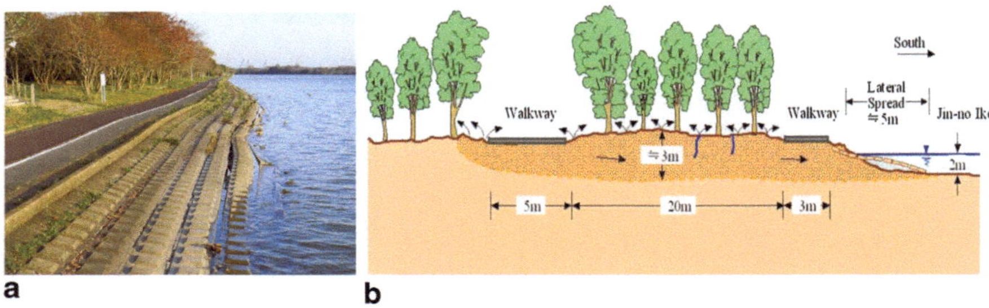

Fig. 40 **a** Sliding in the northern shore of the Gono Lake. **b** Conceptual picture of the slide in the north shore

Conclusive Remarks

An overview was made of the liquefaction-associated damage that occurred during the 2011 earthquake over the vast lowland area in the downstream reaches of the Tone River. The outcome of the studies indicated several attributes, as follows, which are related to geomorphology and land reformation history over the recent past.

1. Most of the liquefaction took place at shallow depths of 1~5 m where sandy soils were placed either hydraulically by dreading nearby riverbeds or transported onland and spread without compaction.
2. The near-surface soil deposits were highly variable in terms of their origin, grain-composition, depths and mode of deposition. Therefore, without further scrutiny, it is difficult to derive any definitive conclusions solely based on currently available knowledge on liquefaction.
3. Amongst several evidence of liquefaction, fissuring with sand ejection along linear lines through a distance of 10~50 m was one of the characteristic features manifested on the ground surface. This is suggestive of occurrence of tensile stresses near the surface during the shaking of the earthquake, which in turn, implies that reflected or refracted, or surface waves were generated, because of complex underground topography.

Acknowledgements In preparing the draft of this paper, a number of data on soil profiles and photos were kindly offered by officers of regional cities. The authors wish to extend their sincere thanks to Ms. M. Murata of Itako city, Mr. T. Ando of Kamisu city and Mr. T. Takemoto of Katori city.

References

Ishihara K (2012) Liquefaction in Tokyo Bay and Kanto Region in the 2011 Great Eat Japan Earthquake. International Symposium on engineering Lessons Learned from the Giant Earthquake, key note presentation, Tokyo

Ishihara K, Araki K (2012) Liquefaction in Tokyo Bay and Kanto Regions in the 2011 Great Eat Japan Earthquake. Second International Conference on Performanced-Based Design in Earthquake Geotechnical Engineering, Taormina

Ishikawa K, Yasuda S (2012) Liquefaction along Tokyo Bay during the 2011 Tohoku-Pacific Ocean Earthquake in Japan. 2nd International Conference on Performance-Based Design in Earthquake Geotchnical Engineering, Taormina, Italy, Paper No. 610

Kawabe S, Tsukamoto Y, Kokusho T, Takahashi R (2012) Soil liquefaction observed at Katori City located along the Lower Stream of Tonegawa river during 2011, Great East Japan Earthquake

Koseki J, Miyashita Y, Deng J, Araki H (2012) Liquefaction along Tone River Basin in Japan, caused by the 2011 off the Pacific Coast of Tohoku Earthquake. 2nd International Conference on Performance-Based Design in Earthquake Geotechnical Engineering, Taormina, Italy, Paper No. 1.14

Senna S, Hasegawa N, Maeda T, Fujiwara H (2012) Liquefaction damage of the Tonerawa Basin caused by the 2011 Off the Pacific Coast of Tohoku Earthquake. International Symposium on Engineering Lessons Learned from the 2011 Great East Japan Earthquake, Tokyo, Japan, pp 719–730

Tsukamoto Y, Kawabe S, Kokusho T, Araki K (2012) Soil liquefaction observed at areas located along the Lower Stream of Tonegawa River during 2011 Great East Japan Earthquake. International Symposium on Engineering Lessons Learned from the 2011 Great East Japan Earthquake

Wakamatsu K (2012) Recurrent liquefaction induced by the 2011 Great East Japan Earthquake Conference with the 1987 Earthquake. International Symposium on Engineering Lessons Learned from the 2011 Great East Japan Earthquake, Tokyo, Japan, pp 675–686

Yasuda S (2011) Liquefaction-induced damage in Kanto Region. Report of Investigation for the Tohoku Off-Ocean Earthquake, Japanese Geotechnical Society (in Japanese)

Amplification Effects of Thin Soft Surface Layers: A Study for NBCC 2015

W.D. Liam Finn and Francisco Ruz

Abstract The seismic response of shallow soft surface layers is of concern in developing the seismic section of the National Building Code of Canada for 2015. The response of such layers is studied using recorded data from the 2011 Tohoku earthquake. Fourteen sites have been studied in detail so far that had soft, shallow surface layers. At each site two records were available, one at the bottom of the borehole at a depth of the order of 100–500 m and one on the surface. Site response analyses were conducted to determine the ground motion at the top of the rock underlying the surface layer, so that the amplification of the soil layer itself relative to the rock surface could be determined. These analyses were conducted using the program SHAKE. The properties in SHAKE were calibrated to get a good match between the measured and computed surface motions for the recorded input at the bottom of the rock.

Keywords Site response · Shallow site amplification factors · Amplification correlations with V_{s30} · Tohoku 2011 data

Introduction

The soil conditions at a site are a major factor in determining the damage potential of incoming seismic waves from bedrock. The softer surface soils amplify the motions from bedrock as they pass to the surface. Damage patterns in Mexico City from the 1985 Michoacan earthquake provide a vivid example of this effect. Peak accelerations in the incoming rock motions were generally less than 0.04 g and had predominant periods of around 2 s. Many clay sites in the dried lakebed on which the original city was founded had site periods of around 2 s also and were excited into resonant response by the rock motions. As a result, the bedrock motions were

W. D. L. Finn (✉)
The University of British Columbia, Vancouver, BC, Canada
e-mail: finn@civil.ubc.ca

F. Ruz
R y V Ingenieros, Santiago, Chile
e-mail: francisco@ruzvukasovic.cl

A. Ansal, M. Sakr (eds.), *Perspectives on Earthquake Geotechnical Engineering*,
Geotechnical, Geological and Earthquake Engineering 37,
DOI 10.1007/978-3-319-10786-8_2

amplified about 5 times as they passed through the clay layers. The amplified motions in turn excited resonant response in buildings with periods of about 2 s with devastating effects. In the 1989 Loma Prieta earthquake, major damage occurred on soft soil sites in the San Francisco—Oakland region where strong motion spectral accelerations were amplified 2–4 times compared to adjacent rock sites and caused severe damage, (Housner 1989; Idriss 1990).

In the current Canadian building code (NBCC 2010), the amplification potential of site conditions are quantified by the use of foundation factors. All site conditions are classified into one of five classes (A–E) based on the average shear wave velocity in the top 30 m, V_{s30}, and a set of short and long period amplification factors are assigned to each site class. Problems arise with this approach, when the surface soils on the rock are less than 30 m thick. Prior to NBCC 2010, sites with soft soil thicknesses less than 30 m were classified using V_{s30} even if that meant incorporating some rock to make up the thickness, with the caveat that this process was not allowed to move the site into one of the rock categories, A, B. This was a matter of practice and was not incorporated in the code. NBCC 2010 directed that the appropriate Site Class for the shallow surface deposits should be based on the average properties of the softer surface materials. There was little evidence to support this development. The 2011 Tohoku earthquake in Japan has provided many recordings on soft shallow sites and this data is being studied to get an insight into the ground motion amplification potential of these sites and to provide a sound basis for classifying these sites for NBCC 2015.

Theoretical Basis of Amplification

The basic mechanism of site amplification (Finn and Wightman 2003) is best illustrated by examining the effect of an uniform surface layer on incoming bedrock motions. Consider the elastic undamped layer shown in Fig. 1 which is characterized by a thickness, H. a shear wave velocity, Vs, and a density, ϱ.

Okamoto (1973) has shown that, if bedrock motion is a harmonic wave with a period equal to the period of the elastic surface layer ($T=4\ H/V_s$), then the amplification factor $AF=a_t/a_r$ for undamped motion at the surface is

$$AF = 2/\kappa \tag{1}$$

Fig. 1 Elastic layer on elastic half space

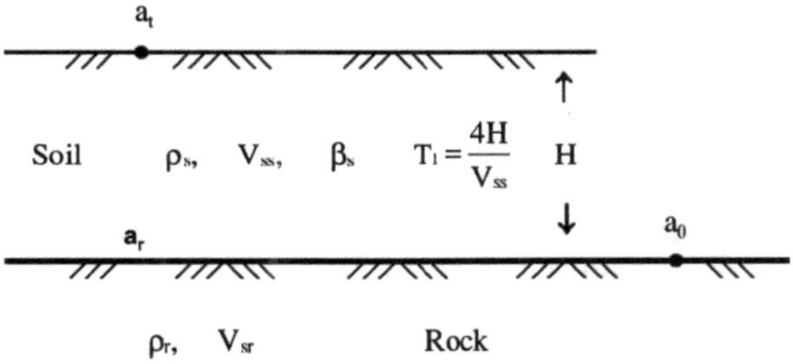

where, the impedance ratio, is $\varrho_s Vss/\varrho_r Vsr$, a_t is the surface acceleration, a_r is the incoming bedrock acceleration to the upper layer and $\varrho_{s,}$ ϱ_r are the soil and rock densities respectively. The factor 2 in Eq. 1 is due to wave reflection from the surface of the soil layer. If damping is considered

$$AF = 2/(\kappa + \beta_s \pi/2)$$

where β_s is the equivalent viscous damping ratio in the soil.

The acceleration, $a_{r,}$ is rarely available. Therefore ground motions for seismic design are determined by first selecting appropriate rock or stiff soil outcrop motions, a_o, and using these as input to a nonlinear or equivalent linear site response analysis that accommodates the split between up-going and down-going component motions at the interface between soil and rock or stiff soil. In this case the amplification is estimated as $AF = a_t/a_0$ where a_0 is the outcrop motion. Since a_0 is also a surface motion, the amplification factor is now given by

$$AF = 1/(\kappa + \beta_s \pi/2) \quad (2)$$

These theoretical results show that the important parameters controlling ground motion amplification are (i) the relationship between the periods of the incoming motions from bedrock and the predominant period of the surface layer, (ii) the impedance ratio between the bedrock and the surface layer and (iii) soil damping (Finn and Wightman 2003).

Analyses of Tohoku Borehole Sites

The amplification of ground motions passing through shallow soft surface layers is of concern in developing design motions for the National Building Code of Canada for 2015 (NBCC 2015). The response of such layers is being studied using recorded data from the 2011 Tohoku earthquake obtained from KiK-net at http://www.kyoshin.bosai.go.jp/. Data from a total of 30 Tohoku sites will be studied ultimately. Fourteen sites have been studied so far with soft surface layers less than 18 m in thickness with 2 exceptions. Details of these sites are shown in Table 1. The PGA at the top of rock was calculated using SHAKE analysis as will be described later.

Ghofrani et al. (2012) made a careful, detailed study of all Tohoku borehole data and studied the amplification effects of the surface layers in a very interesting paper. They defined amplification with respect to the motions recorded at the bottom of the borehole, assuming that negligible amplification would occur between the bottom and the top of the rock column. In practice, amplification is usually related to motions on a stiff soil or rock layer directly underlying the surface layer. Therefore in the present study the ground motions at the top of the rock column were calculated and the amplification of the surface motions was defined relative to the top of rock motions.

Table 1 Details of instrumented boreholes

Site	Name	PGA surface	PGA top rock	PGA base rock	Downhole depth (m)	H soil (m)	Vs30 (m/s)
1	MYGH04	0.70	0.41	0.12	100	4	850
2	MYGH12	0.53	0.31	0.16	102	6	748
3	IWTH27	0.75	0.38	0.11	100	4	670
4	IBRH16	0.51	0.28	0.12	300	12	626
5	FKSH09	0.43	0.25	0.10	200	10	585
6	TCGH13	0.56	0.41	0.14	140	4	584
7	IBRH18	0.45	0.29	0.16	507	15	559
8	FKSH17	0.28	0.18	0.07	100	6	544
9	FKSH10	1.08	0.48	0.18	200	16	487
10	IBRH15	0.60	0.25	0.10	90	5	450
11	SITH11	0.20	0.06	0.02	106	14	372
12	SITH06	0.23	0.10	0.06	200	50	369
13	FKSH19	0.61	0.31	0.13	100	18	360
14	FKSH18	0.58	0.19	0.10	100	30	307

At each site two records are available, a rock motion at depths of the order of 100–500 m and on the surface. The objective of this study was to estimate the amplification of motions between the top of bedrock and the surface of the soft layer. This entailed site response analyses to determine the motions at top of rock. These response analyses were conducted using the program SHAKE (Schnabel et al. 1972). The properties in SHAKE were calibrated to get a good match between the measured and computed surface motions for the recorded input at the bottom of the rock. Typical results are shown in Figs. 2, 4 and 6. Comparisons between recorded and calculated spectra suggest that the SHAKE analyses are good enough to give

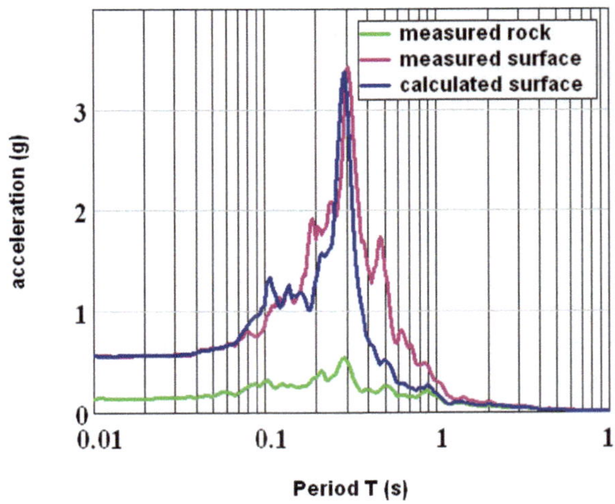

Fig. 2 Recorded and computed response spectra at site TCGH13

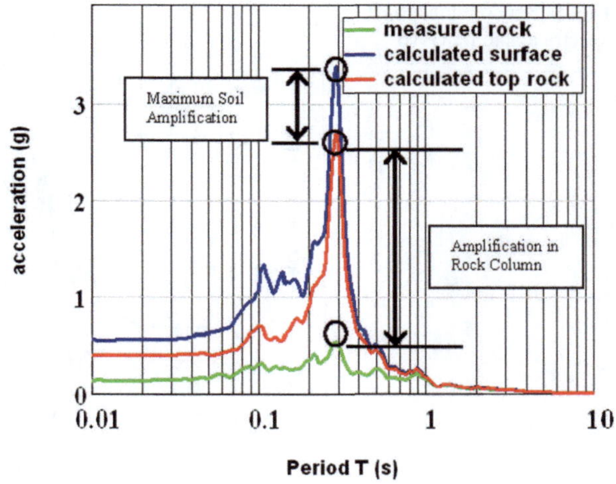

Fig. 3 Recorded and computed response spectra at site TCGH13

useful estimates of the amplification of motions by the soft surface layers. The amplification of spectral acceleration at the top of the rock column with respect to that at the base of the rock column is shown in Figs. 3, 5 and 7 which also shows the spectral amplification at the top of the soil layer with respect to the top of the rock. It appears that variable but significant amplification can occur in the rock column. The amplification in the rock column is also indicated by a comparison of PGA values at the top and bottom of the rock column in Table 1.

The average amplification factor for the response spectrum at the surface relative to the top of rock was calculated for each site and is plotted against V_{s30} in Fig. 8. There is a good correlation ($R^2=0.80$) between average amplification of spectral acceleration and V_{s30}. The standard deviation is $\sigma=0.5$. The mean correlation is given by

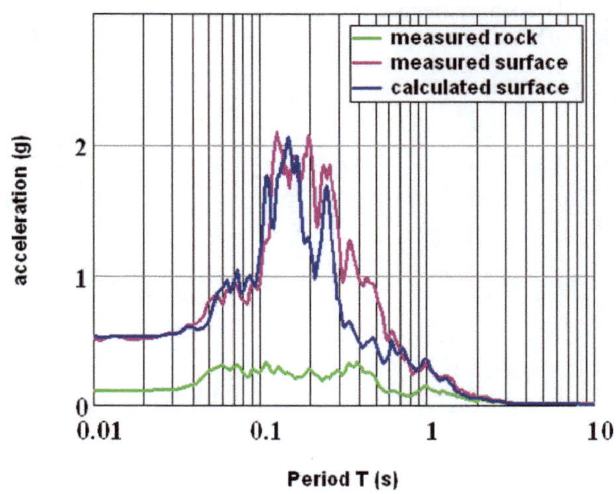

Fig. 4 Recorded and computed response spectra at site IBRH16

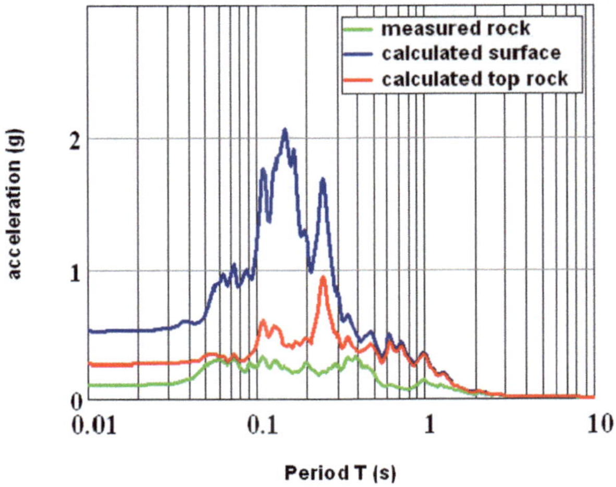

Fig. 5 Recorded response spectra at site IBRH16

$$\text{Mean AF} = 167.66\, V_{s30}^{0.719} \tag{3}$$

where AF is the amplification factor.

NBCC 2010 recommends that the classification of shallow sites for site amplifications should be based on the properties of the sites in particular Vs. In Fig. 9 average spectral amplifications are plotted against Vs. There seems to be very little correlation between the average spectral amplification and the velocity of the surface layer in contrast to the strong correlation between mean spectral amplification and V_{s30}. It would seem that the recommendation in NBCC 2010 regarding classification of the surface layer needs to be reviewed critically.

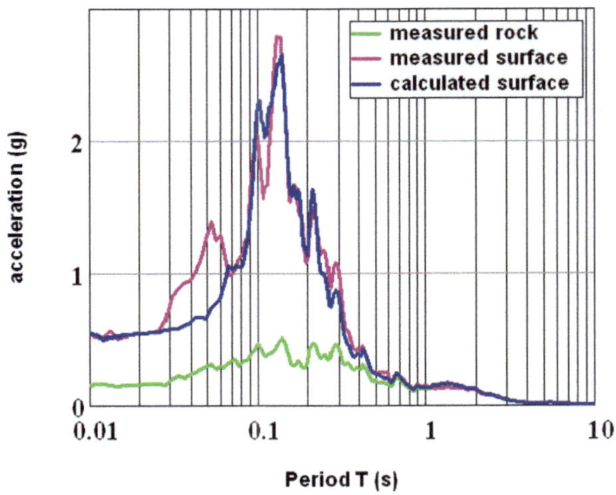

Fig. 6 Recorded and computed response spectra at site MYGH12

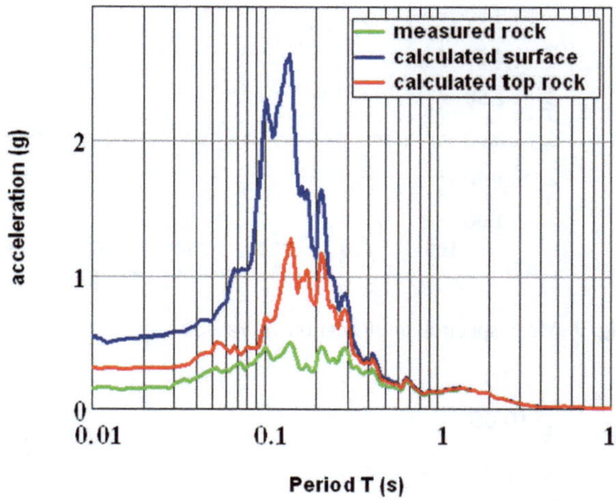

Fig. 7 Recorded response spectra at site MYGH12

It is clear from examination of Figs. 2–7 that the seismic response of the surface layers is a narrow band short period response, typically in the period range T=0.1–0.3 s. Therefore the maximum spectral amplification may be more important than the mean. The maximum peak spectral amplification is shown in Fig. 10 as a function of $V_{s30.}$ There is a considerable scatter in the data and the correlation V_{s30} with R^2=0.36. The defining equation of the correlation is

$$Max\ AF = -0.0091\,V_{s30} + 8.6893 \tag{4}$$

This type of scatter is typical of field response data as shown in Fig. 11, where Borcherdt (1994) shows field data for the period range T=0.4–2.0 s. The data shows a scatter band of 2 standard deviations about the mean.

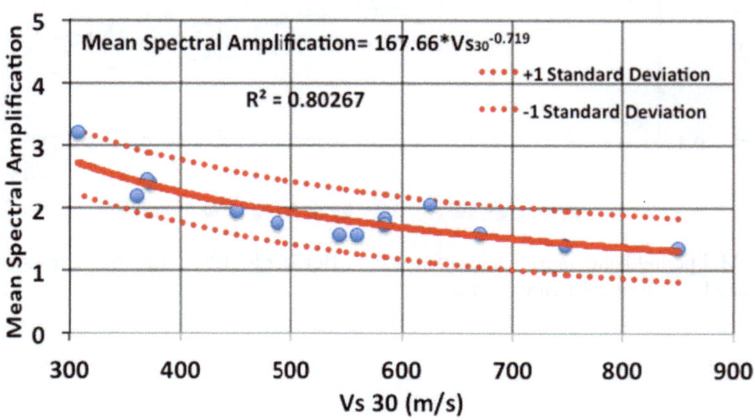

Fig. 8 Average spectral acceleration as a function of V_{s30} with ± one standard deviation (0.5)

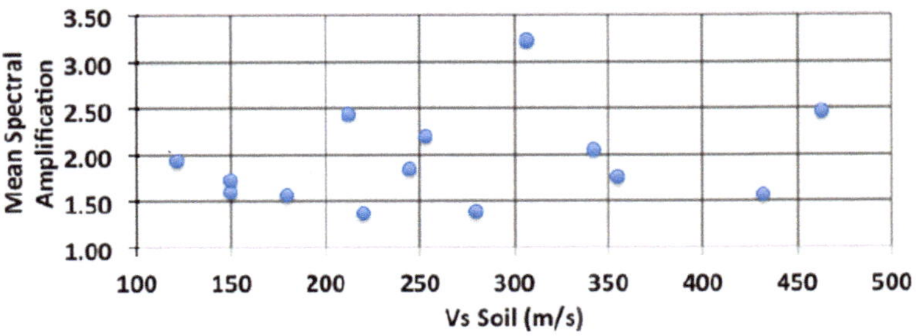

Fig. 9 Mean spectral amplification vs Vs_soil

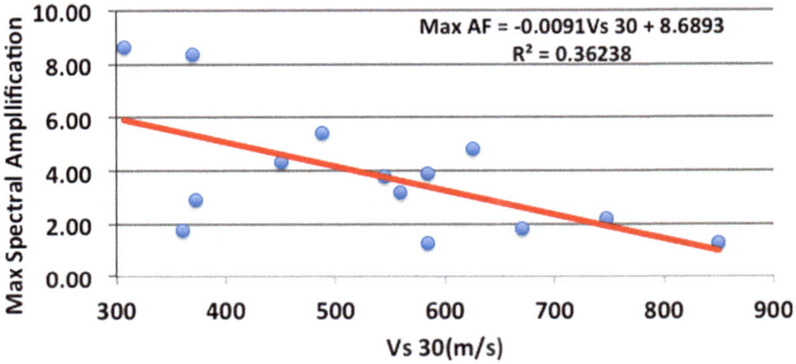

Fig. 10 Maximum spectral acceleration as a function of V_{s30}.

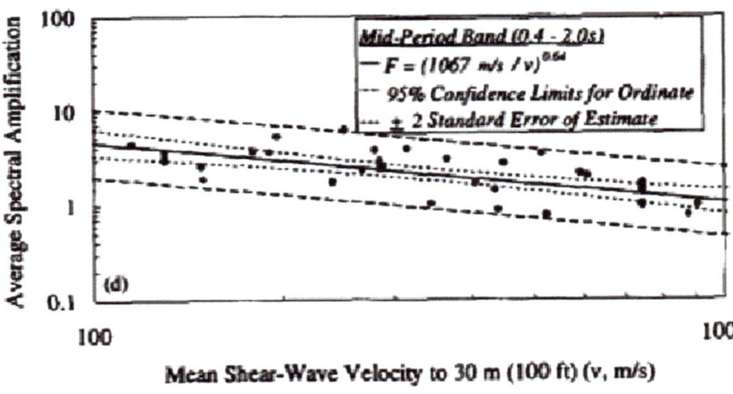

Fig. 11 Mid period band amplification-(0.4–2.0 s), Borcherdt (1994) used by permission of the Earthquake Engineering Research Institute

Fourier Spectra Amplification Factors Based on Site Period

The amplification factors currently used in US and Canadian codes are based on the pioneering work of Borcherdt (1994). He developed continuous amplification factors based on Fourier amplitude ratios for specific period ranges as a function of V_{s30}. His findings are summarized in Fig. 12. For periods greater than T=0.5 s, the amplification factors are not very sensitive to period but are much greater than the short period amplification factors for periods less than T=0.5. These results led to the recommendation to adopt 2 sets of amplification factors; one for short periods and one set for longer periods.

Period dependent Fourier amplitude spectra were developed in the present study for the same period ranges as used by Borcherdt (1994). The results are shown in Fig. 13. For the Tohoku borehole sites, the amplification factors in the short period range T=0.1–0.5 s are much higher than the longer period amplification factors in contrast to the findings by Borcherdt (1994) for Loma Prieta data which was recorded on much thicker deposits. The curves representing the three longer period ranges (Fig. 13) are not as close as in Fig. 12 but the amplifications given by all three curves could be reasonably represented by the mid-period curve, indicating a certain lack of sensitivity to differences in the longer period ranges.

The role of site period is investigated further by relating the fundamental period of the soft soil surface layer, T=4 H/V_s, where V_s is the average shear wave velocity in the soil layer, to V_{s30}. A remarkable correlation between the period of the surface soil layer and V_{s30} is evident in Fig. 14 with $R^2=0.95$. The defining equation of the correlation is:

$$Log\left(f_0\right) = 1.62Log\left(V_{s30}\right) - 3.62 \quad (5)$$

Ghofrani et al. (2012) conducted a similar study on all available Tohoku data but based their site period on the total depth of the borehole. Their equation is:

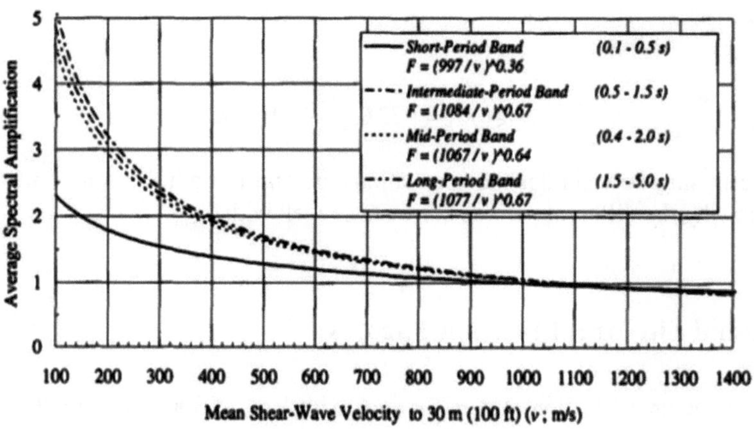

Fig. 12 Short and long period amplification factors for Loma Prieta response data (Borcherdt 1994), used by permission of the Earthquake Engineering Research Institute

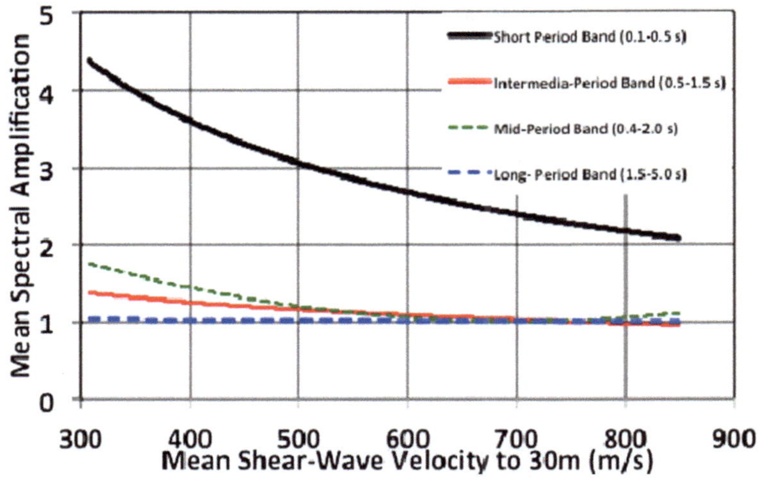

Fig. 13 Fourier spectra amplification factors as a function of shear wave velocity

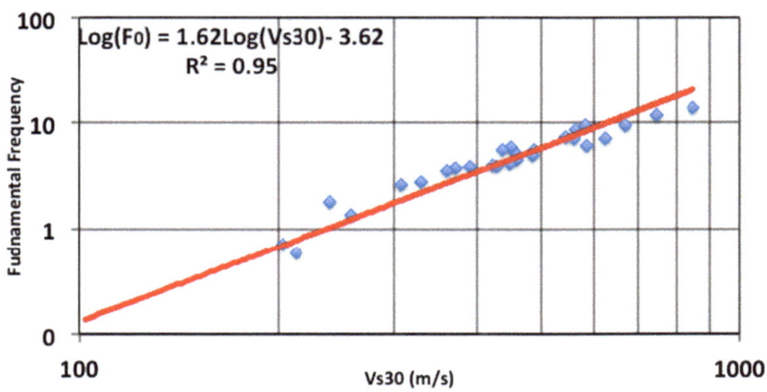

Fig. 14 Fundamental site frequency vs mean shear wave velocity in the top 30 m

$$Log(f_0) = 1.33Log(V_{s30}) - 2.92 \tag{6}$$

It is surprising how similar the correlations are and in both cases it is difficult to explain why the different periods correlate so well with $V_{s30.}$

Regional Site Amplification Factors

The average amplification factors for sites with Vs_{30} of 300, 400 and 800 m/s respectively are shown in Fig. 15. The average spectral amplification was calculated for each site characterized by shear wave velocity and the resulting spectral

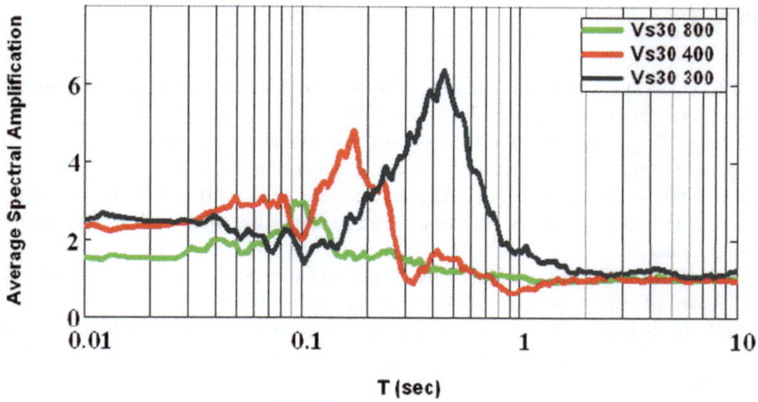

Fig. 15 Average spectral amplification factors for sites with different V_{s30} and periods T

amplifications are shown in Fig. 15 as a function of period. The peak spectral amplification increases with decreasing shear wave velocity and the period at which the peak occurs increases. Most of the spectral peaks occur in the range of 0.1–0.2 s if the shear wave velocity is greater than 400 m/s. Below this velocity the period reaches T=0.45 s.

Concluding Remarks

An important finding from the present preliminary study is that the average shear wave velocity in a soft surface layer, significantly less than 30 m thick, may not a reliable indicator of amplification potential. Therefore the current recommendation for classifying such soil layers in NBCC 2010 may need reconsideration for NBCC 2015. A strong correlation was found between V_{s30} and average amplification factors for the thin, soft surface layers which suggests a return to practice prior to 2010 may be appropriate. A surprising finding was the very strong correlation between the period of the surface layer and $V_{s30.}$ Ghofrani et al. (2012) found a similar correlation with the entire borehole period. There appears to be no fundamental theoretical basis for the strong correlations between T and V_{s30}.

For the Tohoku borehole sites, the amplification factors in the short period range T=0.1–0.5 s are much higher than the longer period amplification factors in contrast to the findings by Borcherdt (1994) for Loma Prieta data which were recorded on much thicker deposits. The peak spectral amplification increases with decreasing shear wave velocity and the period at which the peak occurs increases.

Acknowledgements Ground motion data and site conditions for the borehole studies described in this report were obtained from KiK-net at http://www.kyoshin.bosai.go.jp, the NIED strong motion seismograph networks. The high quality data from many deep boreholes provides unique resources for evaluating many of the procedures used in earthquake engineering, especially the procedures for site response analysis.

References

Borcherdt RD (1994) Estimates of site-dependent response spectra for design (methodology and justification). Earthquake Spectra 10(4):617–653

Finn WD Liam, Wigthman Adrian (2003) Ground motion amplification factors for the proposed 2005 edition of the National Building Code of Canada. Can J Civ Eng 30:272–278

Ghofrani H, Atkinson GM, Goda K (2012) Implications of the 2011 M9.0 Tohoku japan earthquake for the treatment of site effects in large earthquakes, submitted to: Bulletin of Earthquake Engineering

Housner GW (1989) Competing against time, Report to Governor Deukmejian of California. Governor's board of inquiry on the 1989 LomaPrieta Earthquake

Idriss IM (1990) Response of soft soil sites during earthquakes. In: Duncan JM (ed) Proceedings, H. Bolton seed memorial symposium. BiTech Publishers, Vancouver, B.C., Berkeley, Vol 2, pp 273–289

NBCC (2010, 2015) National building code of Canada. National Research Council of Canada, Ottawa

Okamoto S (1973) Introduction to earthquake engineering. University of Tokyo, Tokyo

Schnabel PB, Lysmer J, Seed HB (1972) SHAKE, a computer program for earthquake response analysis of horizontally layered sites, *Report EERC 72–12,* Earthquake Engineering Research Center, University of California, Berkeley

New Design Spectra in Eurocode 8 and Preliminary Application to the Seismic Risk of Thessaloniki, Greece

Kyriazis Pitilakis, Evi Riga and Anastasios Anastasiadis

Abstract Based on a worldwide database of strong ground motion records in different well constrained site conditions, a new soil classification is proposed for EC8 with the associated amplification factors and the normalized response spectra for 5% damping. Amplification factors and normalized response spectra are also proposed for the current EC8 classification. Then we perform a detailed seismic risk assessment for the city of Thessaloniki using (a) the capacity spectrum method, (b) the Uniform Hazard Spectrum (UHS) for rock conditions compiled applying the SHARE approach, (c) the current and the new site amplification factors to evaluate the site specific demand spectra, (d) the detailed inventory of the Thessaloniki building stock to select the appropriate capacity and fragility curves for each building typology, and (e) available functions to estimate from the physical building damages the causalities and the economic losses. The main conclusions are summarized as following: (a) EC8 site classification, amplification factors and normalized response spectra should be revised. (b) The current methods for the seismic risk assessment are not robust enough; they need serious upgrade in all their components, and yet the uncertainties will remain quite high.

Keywords EC8 · Soil amplification factors · Soil classification system · Seismic risk

K. Pitilakis (✉) · E. Riga · A. Anastasiadis
Aristotle University of Thessaloniki, P.O. Box 424, 54124 Thessaloniki, Greece
e-mail: kpitilak@civil.auth.gr

E. Riga
e-mail: eviriga@civil.auth.gr

A. Anastasiadis
e-mail: anas@civil.auth.gr

A. Ansal, M. Sakr (eds.), *Perspectives on Earthquake Geotechnical Engineering,*
Geotechnical, Geological and Earthquake Engineering 37,
DOI 10.1007/978-3-319-10786-8_3

Introduction

Eurocode 8-Part 1 accounts for site effects through appropriate elastic design spectra and amplification factors based on different soil classes. The influence of local site conditions is reflected with the shape of the PGA-normalized response spectra and the "soil factor" S, which represents ground motion amplification with respect to outcrop conditions, while the main parameter for soil classification is the average shear wave velocity of the upper 30 m of the soil profile ($V_{s,\,30}$). Damages of buildings and in general the risk assessment at city scale is evaluated using these period-dependent amplification factors applied on the computed ground motion estimated at rock conditions for given hazard return periods and using appropriate risk assessment methods.

The present work is divided in two parts. In the first one we use a global strong-motion database, which was compiled in the framework of SHARE FP7 Collaborative Project of the European Union (www.share-eu.org), to evaluate the present EC8 normalized spectra and associated amplification factors for different soil conditions. Improved soil factors are proposed for the most common EC8 soil classes B and C. An additional database consisting of strong motion records only from sites, which dispose a well-documented soil profile concerning dynamic properties and depth up to the 'seismic' bedrock ($V_s > 800$ m/s) is also compiled. This database is used to propose a new site classification system, which, besides the classical geotechnical parameters N_{SPT}, S_u and PI, uses also the fundamental period of the site, the thickness of soil deposits and the average shear wave velocity of the whole soil column to the seismic bedrock $V_{s,\,av}$, instead of $V_{s,\,30}$. Amplification factors and normalized response spectra for the new soil classes and for the two seismicity levels defined in EC8, i.e. $M_s \leq 5.5$ and $M_s > 5.5$, are also proposed. We also present the estimated period-dependent amplification factors to be used in case that a Uniform Hazard Spectrum (UHS) for rock conditions is available. Uncertainties in the estimation of soil amplification factors are captured using a logic-tree approach, which allows the use of alternative models and methods in an effective way.

In the second part the improved and the new period-dependent soil amplification factors for EC8 and the new soil classification system respectively are used to perform the risk assessment of the city of Thessaloniki in Greece. We use the UHS estimated for Thessaloniki for rock site conditions and for a return period of 475 years applying the most recent methodology developed in SHARE; then we estimate in a first stage the ground response at surface using the current soil amplification factors of EC8, the improved soil amplification factors for EC8 and the new site classification system. In a second stage we estimate the expected damages to the building stock of the city, using the Capacity Spectrum Method and appropriate fragility curves for all building typologies. Economic and social losses are also estimated.

The aim of this second part is to discuss the differences of the various hazard models on the risk assessment in a real city scale when combined with other involved parameters, namely, among others, the design demand spectra, the capacity curves of the building typologies and their fragility functions. We found that the differences in the final risk assessment could be very important and the uncertainties involved very high, which leads to the conclusion that the whole methodology and all its components should be seriously improved to get more reliability and safety.

Part 1: Improved Amplification Factors for EC8

Introduction

Based on the results of extended research studies and valuable records after several recent earthquakes like 1989 Loma Prieta, 1994 Northridge, 1995 Kobe and many others all over the world, current seismic code provisions have largely accepted the significant role of local site conditions. Their influence is described through appropriate elastic design spectra based on different soil categories and levels of shaking intensity. The main and almost universally adopted parameter for site classification is $V_{s,30}$, i.e. the average shear wave velocity of the upper 30 m of the soil profile, calculated from the total time needed for a shear wave to travel these 30 m. This parameter, initially introduced by Borcherdt and Glassmoyer (1992) and Borcherdt (1994), is supposed to provide a quantitative parameterization of the site.

In EC8 the $V_{s,30}$ parameter is used along with N_{SPT} blow count, plasticity index PI and undrained shear strength C_u to define five soil types (A to E), while two extra special ground types (S1 and S2) are also proposed for special soils (Table 1). Elastic response spectra are proposed for two different levels of seismic action, Type 1 and Type 2. Type 1 spectra have more energy in long-period motions and are proposed for use in regions having high seismic activity and stronger earthquakes; it is estimated that Type 1 spectra should be used when the earthquakes that contribute to the seismic hazard have a surface wave magnitude M_s greater than 5.5. Type 2 spectra are recommended for $M_s \leq 5.5$, having larger normalized spectral amplitudes at short periods. Ground motion amplification, accounting for local soil and site effects is expressed through a constant soil factor S, which increases uniformly the normalized elastic response spectra in all periods.

Even if $V_{s,30}$ is not the fundamental parameter to describe site response, it has the advantage that it can be obtained easily and at relatively low cost, since the depth of 30 m is a typical depth of geotechnical investigations and sampling borings, and has definitely provided engineers with an unambiguous and quantitative parameter for site classification. The main weakness is that the single knowledge of the V_s profile at the upper 30 m cannot quantify properly the effects of the real impedance contrast, which is one of the main sources of the soil amplification, as for example in case of shallow (i.e. <15–20 m) loose soils on rock or deep soil profiles with variable stiffness and contrast. As a result, the use of $V_{s,30}$ as the basic criterion for site classification in engineering practice, disregarding the soil type, geotechnical criteria and the thickness of soil deposits, combined with the limited number of code site classes can be misleading in many cases (Pitilakis 2004). The use of $V_{s,30}$ as a proxy to seismic amplification has been questioned by several recent works

Table 1 Ground types defined in EC8 (CEN 2004)

Ground type	Description of stratigraphic profile	Parameters		
		$V_{s,30}$ (m/s)	N_{SPT} (blows/30 cm)	C_u (kPa)
A	Rock or other rock-like geological formation, including at most 5 m of weaker material at the surface	>800	–	–
B	Deposits of very dense sand, gravel, or very stiff clay, at least several tens of meters in thickness, characterized by a gradual increase of mechanical properties with depth	360–800	>50	>250
C	Deep deposits of dense or medium-dense sand, gravel or stiff clay with thickness from several tens to many hundreds of meters	180–360	15–50	70–250
D	Deposits of loose-to-medium cohesionless soil (with or without some soft cohesive layers), or of predominantly soft-to-firm cohesive soil	<180	<15	<70
E	A soil profile consisting of a surface alluvium layer with V_s values of type C or D and thickness varying between about 5 and 20 m, underlain by stiffer material with V_s>800 m/s			
S_1	Deposits consisting, or containing a layer at least 10 m thick, of soft clays/silts with a high plasticity index (PI>40) and high water content	<100 (indicative)	–	10–20
S_2	Deposits of liquefiable soils, of sensitive clays, or any other soil profile not included in types A—E or S_1			

(e.g. Castellaro et al. 2008; Kokusho and Sato 2008; Lee and Trifunac 2010), a summary of which can be found in Pitilakis et al. (2013). In addition, several alternative or supplementary indicators for soil amplification have been suggested, such as depth-to-basement (Steidl 2000, DIN 4149:2005), average shear wave velocity over depths other than 30 m (Boore and Joyner 1997; Gallipoli and Mucciarelli 2009; Luzi et al. 2011) or predominant site period/frequency (e.g. Japan Road Association 1980, 1990; Cadet et al. 2008, 2012), resulting accordingly in a variety of proposed classification schemes with different parameters defining the different soil classes. Two indicative examples of such classification systems are the ones proposed by Rodriguez-Marek et al. (2001) and Pitilakis et al. (2004a, 2006a).

In this first part of the work we have multiple goals. First, the adequacy of EC8 normalized response spectra and soil amplification S factors for the different soil categories proposed in EC8 is examined using a very large set of ground motion records from a worldwide strong motion database recently compiled in the framework of the European research project called 'Seismic Hazard Harmonization in Europe' (SHARE) (Yenier et al. 2010). Improved parameters and normalized shapes for EC8 elastic response spectra are proposed, where necessary. Secondly, a new improved soil and site classification system is presented, derived from a comprehensive analysis of strong ground motion records of the SHARE database, focusing on sites which dispose a well-documented soil profile concerning dynamic properties and depth up to the 'seismic' bedrock (V_s > 800 m/s). This classification system and the accompanying elastic response spectra are appropriate for seismic code provisions and in particular for EC8. Finally we propose period-dependent amplification factors for the improved EC8 spectra using the current soil classifcation, as well as for the new classication scheme. Both are compared to the current EC8 period-dependent amplification factors.

Data Selection

In the framework of SHARE, an extended strong motion database was compiled, by unifying existing national and international databases, such as the European strong motion database, the Turkish national strong-motion database, the Next Generation Attenuation database, the Kik-Net database, the global worldwide database compiled by Cauzzi and Faccioli, the Internet Site for European Strong-motion Data and the Italian Accelerometric Archive database (Yenier et al. 2010). The database covers earthquakes dating back to 1930s and contains a total of 14193 recordings from 2448 earthquakes recorded at 3708 stations. The soil and site documentation of the stations included in the database is restricted to the $V_{s,\,30}$ values and site classification according to EC8. Further information on the compilation and characteristics of the database can be found in SHARE (http://www.share-eu.org/). The version of the database used in the present study is v3.1 (March 2010) and contains 13500 records.

For the validation of EC8 elastic response spectra, the analyses were conducted using three different sub-sets of data with different levels of PGA, all extracted from the main SHARE database. Only recordings corresponding to sites with available classification according to EC8 classification scheme were used. An event magnitude criterion was set, which allowed the use of records with surface wave magnitude $M_s \geq 4$, excluding in this way small magnitude events. Strong motion recordings with usable spectral period less than 2.5 s were also discarded, leading to a dataset of 7161 3-component accelerograms, hereinafter called DS1. Magnitude M and peak ground acceleration PGA distributions with respect to the epicentral distance R, for the records in DS1 are shown in Fig. 1a, b respectively. It is observed that there are significantly fewer records with PGA values exceeding 200 cm/s^2 and many weak motion records with peak values less than 20 cm/s^2. For this reason, and considering the fact that design spectra should be derived basically from records of strong earthquakes, all the analyses were performed not only for the whole dataset (DS1), but also for two smaller datasets, one with records with PGA greater or equal to 20 cm/s^2 (hereafter DS2), and one with records with PGA greater or equal to 150 cm/s^2 (hereafter DS3), which may be considered as representative for high seismicity regions.

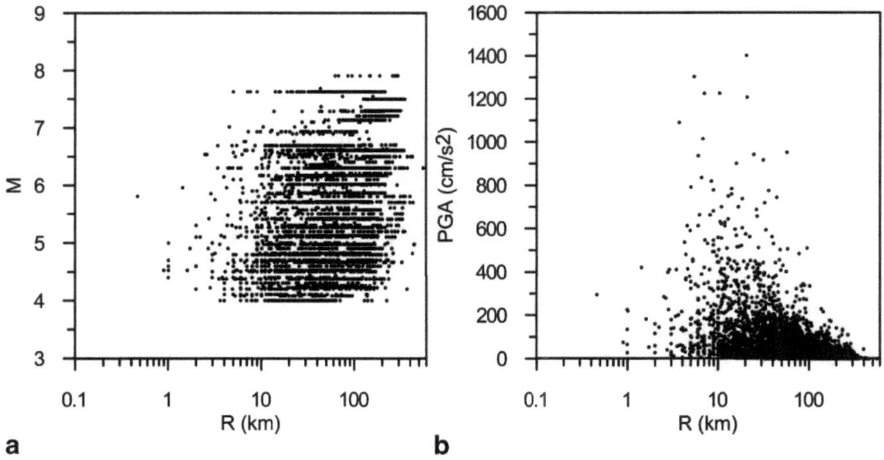

Fig. 1 Data coverage of the DS1 database records (7161 accelerograms) in terms of **a** Magnitude M—Epicentral distance R and **b** Peak Ground Acceleration (PGA)—Epicentral distance R

For the proposal of the new classification system, a subset of the SHARE database was compiled, containing records from sites which dispose a well-documented soil profile concerning dynamic properties and depth up to the 'seismic' bedrock (V_s>800 m/s). For this reason, the raw data of shear wave propagation velocity V_s with depth were collected and parameters such as $V_{s,\,30}$ or $V_{s,\,av}$ were calculated from scratch for as many stations included in the SHARE database as possible. This updated database, hereafter called SHARE-AUTH database, contains 3666 strong motion records from 536 stations from Greece, Italy, Turkey, Japan and USA and constitutes a reliable set of empirical measurements for estimation of influence of local site conditions. The geographic distribution of the selected stations and records is presented in Fig. 2. A more detailed description of the SHARE-AUTH database, as well as a list of all the stations it includes, can be found in Pitilakis et al. (2013).

For the 536 stations of the database, new site parameters, not included in the original database, were calculated. These include the thickness of the soil deposits H (i.e. depth to "seismic" bedrock—V_s>800 m/s), the time-based average shear wave velocity $V_{s,\,av}$ of the entire soil deposit (calculated with Eq. 1) and the fundamental period T_0 of the soil deposit (calculated with Eq. 2):

$$V_{s,av} = H \,/ \sum_{i=1,N_b} h_i / V_i \tag{1}$$

$$T_0 = 4H \,/\, V_{s,av} \tag{2}$$

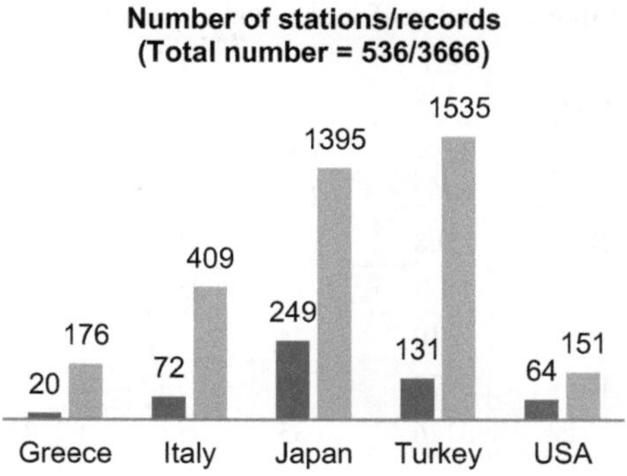

Fig. 2 Geographic distribution of the stations (in *dark gray*) and records (in *light gray*) of the SHARE-AUTH database

where h_i and V_i denote respectively the thickness and shear-wave velocity of the i-th layer, in a total of N_b layers from the surface to the "seismic" bedrock.

$V_{s,30}$ value for each site was recalculated using Eq. (3):

$$V_{s,30} = 30 / \sum_{i=1,N} h_i / V_i \tag{3}$$

where N is the number of layers existing in the top 30 m. However, for some sites (mostly K-net sites) available shear-wave velocity models did not extend down to 30 m. For these cases, the $V_{s,30}$ values were estimated using Eq. (4), which assumes a constant shear wave velocity between the bottom of the velocity model and the depth of 30 m:

$$V_{s,30} = 30 / \left[tt(d) + (30 - d) / V_b \right] \tag{4}$$

where d is the depth to the bottom of the velocity model, tt(d) is the travel time to depth d and V_b is the velocity at the bottom of the model (Boore 2004). Equation (4) was applied to 83 sites.

For the assessment of the elastic response spectra for the soil classes of the new classification system, the same criteria were applied as for the validation of EC8 spectra. As a result, a subset of the SHARE-AUTH database was used (referred to as dataset DS4), consisting of 715 strong motion records with surface wave magnitude $M_s \geq 4$, PGA$\geq$20 cm/s^2 and usable spectral period T$\geq$2.5 s.

The number of available strong motion records per EC8 soil class and seismicity type in each one of the four datasets is shown in Table 2. It should be noted that the majority of the records come from sites, which are classified as B and C, while few data are actually available for soil classes D and E.

Table 2 Number of strong motion records for each dataset

Soil class	DS1		DS2		DS3		DS4	
	Type 2	Type 1	Type 2	Type 1	Type 2	Type 1	Type 2	Type 1
A	402	264	105	125	9	23	15	19
B	1508	1896	419	1151	38	214	172	163
C	1133	1775	353	1261	44	219	128	132
D	10	4	3	1	–	–	3	1
E	73	96	33	49	5	7	33	49
Total	3126	4035	913	2587	96	463	351	364

Validation of EC8 Elastic Response Spectra

Normalized Acceleration Response Spectra

Datasets DS1, DS2 and DS3 were at first used to validate the spectral shapes proposed in EC8. Datasets DS1 and DS2 were utilized for both seismicity types, while DS3 was considered as more representative for Type 1 seismicity. The 5 % damped spectral values of the two horizontal components were included in the original database. In order to combine the two horizontal components into one single earthquake intensity measure, the geometric mean (GM) of the response spectra for the two orthogonal horizontal components of motion was calculated for each strong motion record. Each GM spectrum was then normalized to the GM PGA of the record. All available records were grouped for the EC8 soil classes and seismicity type. For each soil class and seismicity type the median of these normalized spectra was calculated, along with the 16th and 84th percentiles, which represent the average minus and plus one standard deviation respectively. The normalized empirical response spectra were then compared with the normalized EC8 design spectra.

Figure 3 illustrates the median spectra derived from DS1 dataset with the 16th and 84th percentiles and the normalized acceleration spectra recommended in EC8 provisions for soil class A. Figure 4 illustrates the normalized spectra for soil classes B and C derived from DS2 and DS3 datasets for Type 1 seismicity. For soil classes A, B and C, EC8 spectra match the empirical data to a satisfactory extent for all datasets examined, lying between the median and the 84th percentile of the empirical normalized spectra.

Figure 5 illustrates the normalized spectra for soil class D records derived from DS1 dataset, which has only ten records for Type 2 and four records for Type 1 seismicity. Keeping in mind that the empirical spectra have been derived from such a poor sample, the ordinates of both Type 1 and 2 EC8 spectra do not provide a satisfactory fit to the median empirical spectra. For Type 2 seismicity, the empirical spectra are shifted towards much longer periods than the proposed ones, while for Type 1 seismicity the plateau of the empirical spectra reaches much higher values than the spectrum proposed by EC8. The lack of sufficient data for soil class D does not justify any change in spectral shapes. However, the need for more strong-motion records from class D sites is more than evident.

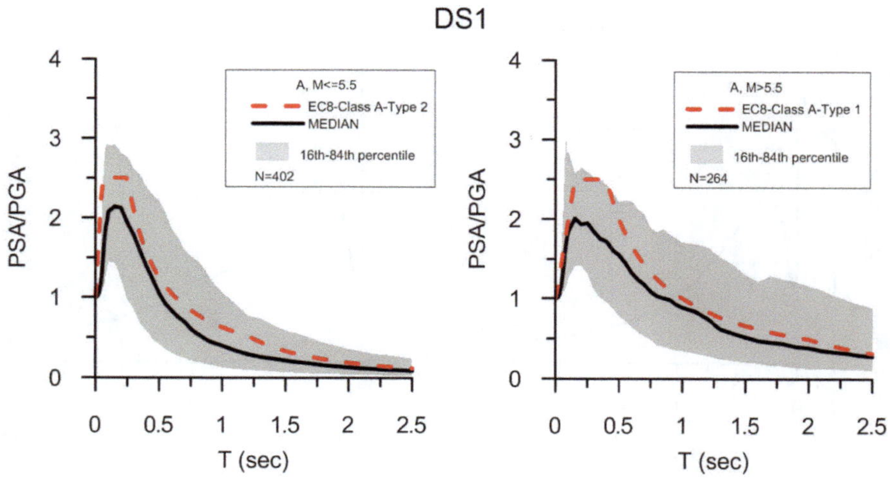

Fig. 3 Normalized elastic acceleration response spectra for soil class A, for Type 2 (*left*) and Type 1 seismicity (*right*), obtained from DS1 dataset. Dashed lines correspond to EC8 proposed spectra, shaded areas correspond to the region between the 16th and 84th percentile and solid lines correspond to median empirical spectra

Figure 6 illustrates the normalized spectra for soil class E records derived only from DS2 dataset. EC8 spectra are on the conservative side for periods greater than 0.3 s, lying close to or even above the 84th percentiles. For smaller periods EC8 spectra are unable to capture the wide range of normalized spectral acceleration values, which is observed in the empirical spectra. A potential need to increase the plateau for both Type 1 and Type 2 spectra is identified.

To summarize, for soil classes A, B and C the spectral shapes provided by EC8 are in good agreement with the derived empirical data for both seismicity types prescribed in EC8. For soil classes D and E the sample of data is not as rich as for the other soil classes and hence the results may not be as convincing as for soil classes A, B and C. However, we found a clear tendency in soil class D spectral shapes to differ substantially from the EC8 shapes. Equally important differences are found in soil class E where the EC8 spectra seem to be conservative enough for medium and high periods, but probably in short period the plateau should be somehow increased. At this stage, no improvement is proposed for the normalized response spectra of the current version of EC8. Another observation is that EC8 elastic spectra do not seem to have been derived based on a common rationale for all soil classes. For example, in some cases (e.g. soil classes A, B and C) EC8 spectra lie between the median and the 84th percentile, while in other cases (e.g. soil class E) EC8 spectra lie closer to or even above the 84th percentile of the empirical normalized spectra.

Soil Amplification Factors

To account for the epistemic uncertainties, amplification factors were estimated using a logic tree approach, which is illustrated schematically in Fig. 7. This approach

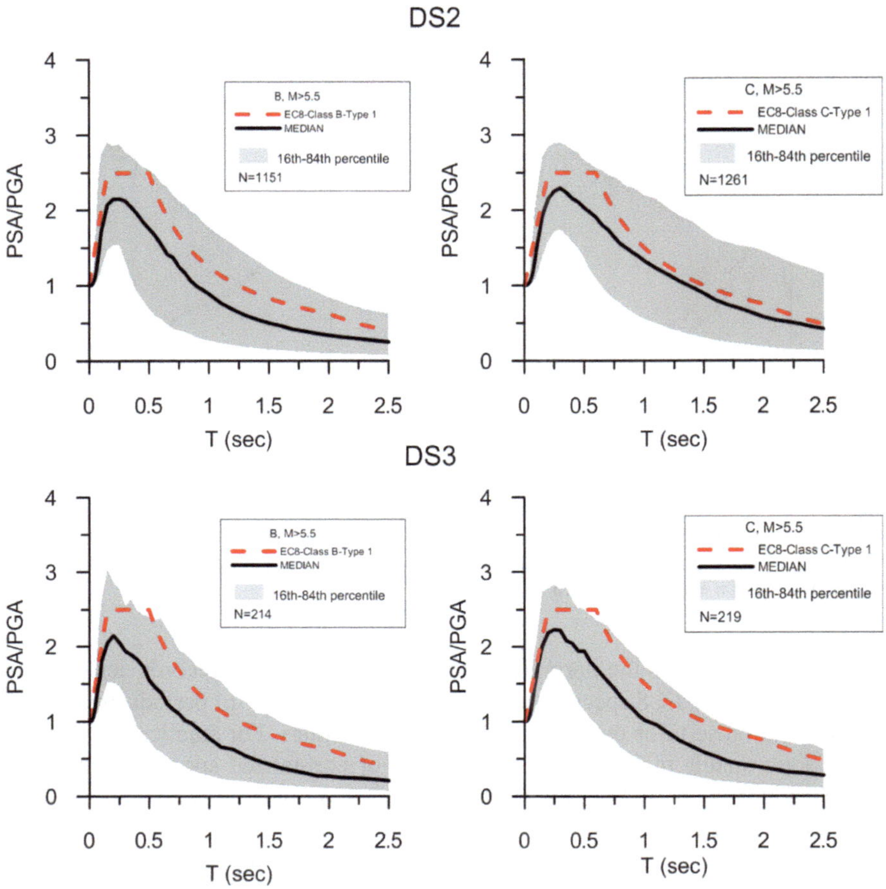

Fig. 4 Normalized elastic acceleration response spectra for Type 1 seismicity obtained from DS2 and DS3 datasets for soil class B (*left*) and soil class C (*right*). See Fig. 3 caption for further description

combines two state-of-the-art methods with equal weights. A detailed description of both approaches is available in Pitilakis et al. (2012).

The first method (Approach 1, Choi and Stewart 2005) constitutes a non-reference site approach, which calculates period-dependent amplification factors using Ground Motion Prediction Equations (GMPEs) for the estimation of reference acceleration spectral values in rock sites. The amplification factor for ground motion j within site class i, S_{ij}, at spectral period T, is evaluated from the geometric mean of 5 % damped acceleration response spectra for the two horizontal components of shaking, GM_{ij}, and the reference ground motion for the site, $(GM_r)_{ij}$, using Eq. (5) (Choi and Stewart 2005):

$$S_{ij}(T) = GM_{ij} / (GM_r)_{ij} \tag{5}$$

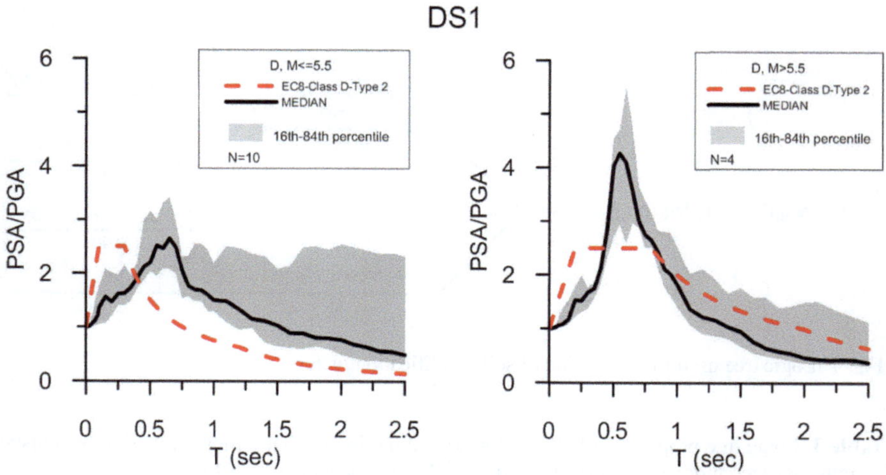

Fig. 5 Normalized elastic acceleration response spectra for soil class D, for Type 2 (*left*) and Type 1 seismicity (*right*), obtained from DS1 dataset. See Fig. 3 caption for further description

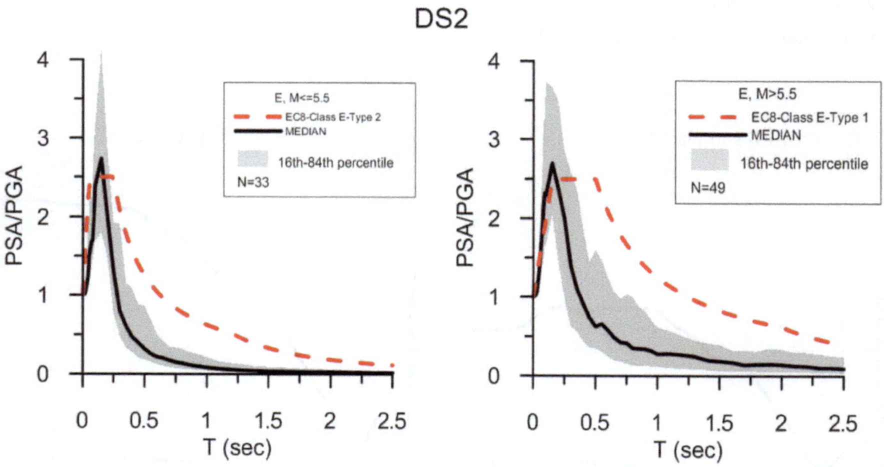

Fig. 6 Normalized elastic acceleration response spectra for soil class E, for Type 2 (*left*) and Type 1 seismicity (*right*), obtained from DS2 dataset. See Fig. 3 caption for further description

In Eq. (5), GM_{ij} and $(GM_r)_{ij}$ were computed at the same spectral period, ranging from 0 to 2.5 s. The reference motion parameter $(GM_r)_{ij}$ was estimated as the weighted average of the rock predictions of the four GMPEs proposed in SHARE for active shallow crustal regions (Delavaud et al. 2012), since the majority of the SHARE database stations are in active regions. The proposed GMPEs and corresponding weights, as well as the abbreviations used for the GMPEs hereafter are included in Table 3. The metadata necessary for the implementation of the GMPEs

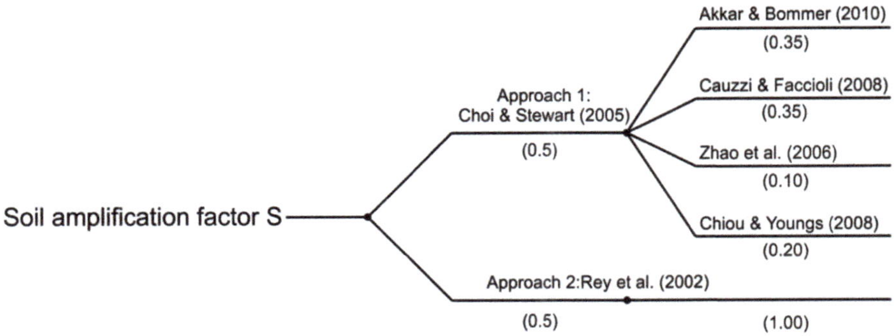

Fig. 7 Logic tree used for estimation of soil amplification factors

Table 3 Logic tree proposed in SHARE for ground motion prediction for active shallow crustal regions. Selected models and corresponding weights. (Delavaud et al. 2012)

Model	Abbreviation	Proposed weight
Akkar and Bommer (2010)	A&B	0.35
Cauzzi and Faccioli (2008)	C&F	0.35
Zhao et al. (2006)	Zh	0.10
Chiou and Youngs (2008)	C&Y	0.20

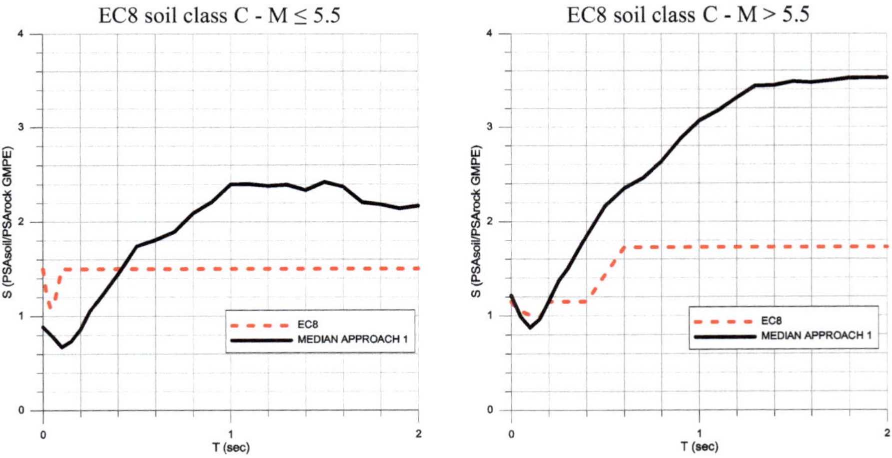

Fig. 8 Amplification factors estimated with Approach 1 for EC8 soil class C using DS1 dataset, for Type 2 (*left*) and Type 1 seismicity (*right*)

were derived from SHARE strong motion database. Figure 8 illustrates representative results for soil class C using DS1dataset.

In order to estimate with Approach 1 a single period-independent amplification factor for each soil class and each seismicity level, similar to the soil factor "S" proposed in EC8, the amplification factors were averaged over a range of periods from T=0 to T=2.0 s. The resulting amplification factors, however, do not only reflect the amplification related to the increase of ordinates of soil elastic spectra with respect to rock elastic spectra, but also the amplification due to the change in shape of PGA-normalized response spectra, since average spectra of softer soils tend to have a larger plateau, which is shifted towards longer periods, compared to rock spectra (Rey et al. 2002). In order to isolate the amplification related to the increase of ordinates of soil spectra with respect to rock spectra, the period-averaged amplification factors were divided by the spectral shape ratio values (SR), which reflect the difference in the shape of soil spectra with respect to rock spectra, when all spectra are constrained to have the same ordinate at zero period spectra, e.g. by normalizing them to PGA (Rey et al. 2002).

In the second method (Approach 2, Rey et al. 2002) period-independent amplification factors are calculated for each soil class and for different magnitude intervals (M.I.) of $M_s=0.5$, with respect to the rock sites of the database using Eq. (6):

$$S = (I_{soil} / I_{rock}) \cdot (1 / SR) \tag{6}$$

where SR is the spectral shape ratio, I_{soil} and I_{rock} are the spectrum intensities for soil and rock respectively, originally defined by Housner (1952) for spectral velocities and here adapted for spectral accelerations (Eq. 7):

$$I = \int_{0.05}^{2.5} \overline{R \cdot S_a(T)}\, dt \tag{7}$$

where $\overline{R \cdot S_a(T)}$ denotes the log-average of distance-normalized 5% spectral ordinates $R \cdot S_a(T)$ for each soil class and magnitude interval. I_{soil}/I_{rock} ratios provide a scaling factor for site effect that represents an average amplification globally affecting the whole spectrum (Rey et al. 2002).

A sample plot for DS2 dataset, showing the log-average of distance-normalized response spectra for soil class C with respect to soil class A, for the M.I. covering a magnitude range from 4 to 5.5, is illustrated in Fig. 9.

Tables 4 and 5 summarize the soil factors obtained for EC8 soil classes with the different approaches and datasets, for Type 2 and Type 1 seismicity, respectively, as well as the weighted average, current EC8 S values and proposed S values. The average S factors derived from DS1 dataset, which contains all records regardless of PGA, generally leads to higher values compared to DS2 dataset, which makes use only of records with PGA≥20 cm/s^2. For Type 1 seismicity, even lower S factors are obtained when using only records with large values of PGA (DS3 dataset with PGA≥150 cm/s^2). However, when examining the two approaches separately, we observe that the removal of weaker ground motion records from the database leads to a clear reduction of soil factors S as far as Approach 2 is concerned, while in Approach 1 there is no apparent trend.

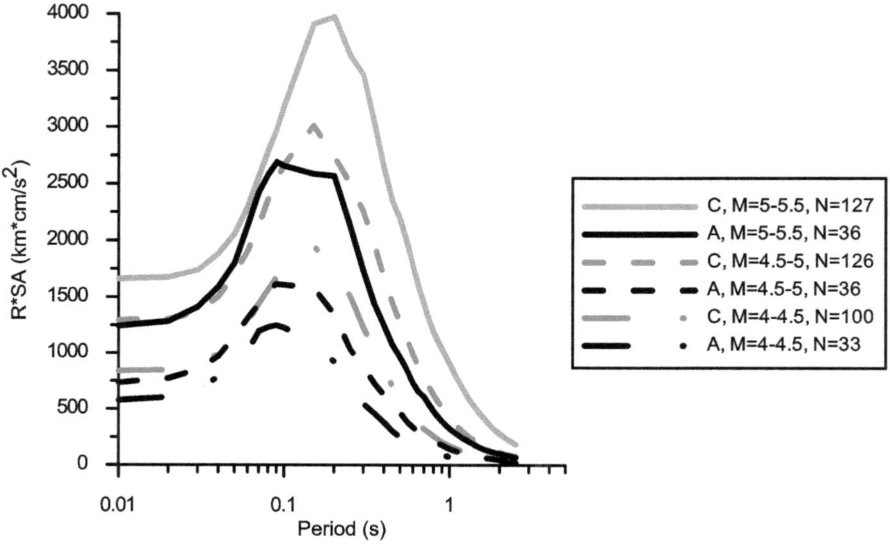

Fig. 9 DS2: Log-average, distance-normalized acceleration response spectra for sites of soil class C (*grey lines*) and rock sites (*black lines*) for Type 2 magnitude intervals

Comparing the determined weighted average S factors with those indicated in EC8 provisions, EC8 factors for class B are in good comparison to the empirical data for both seismicity types, while for soil class C the S factors derived from all datasets are higher compared to EC8. For soil class D the estimated soil factors, which were derived from a limited dataset, are also higher than the EC8 factors for Type 2 seismicity but relatively close to the EC8 factors for Type 1 seismicity. For soil class E, the weighted average soil factors are found quite low for Type 1 seismicity, a result, which is attributed to both the limited data and the averaging process.

Based on the results derived from this comprehensive study, improved soil factors S are proposed for potential use in an EC8 update, supposing that no further changes are made in the definition of soil classes, seismicity types and the shape of normalized design spectra. The PGA ranges of DS2 and DS3 datasets can be considered as more representative for Type 2 and Type 1 seismicity, respectively. However, since the soil factors obtained from DS2 are more conservative than the ones obtained from DS3, we decided to keep the soil factors derived from DS2 as more appropriate for both seismicity types. For soil classes D and E, due to the insufficient datasets, it was decided to keep the same S period independent amplification factors as in EC8, knowing however from the few available records that most probably the EC8 amplification factors might be inadequate. Again, site-specific ground response analyses should be recommended by EC8 for the estimation of soil amplification for soil class D and in some cases for soil class E sites. The proposed soil factors S for EC8 soil classes are given in Tables 4 and 5.

Table 4 Soil factors for EC8 soil classes—Type 2 seismicity

$M_s \leq 5.5$											
Soil class	SHARE-DS1			SHARE-DS2			SHARE-DS3			EC8	Proposed
	Ap.1	Ap.2	W.A.	Ap.1	Ap.2	W.A.	Ap.1	Ap.2	W.A.		
B	0.90	1.55	1.23	1.51	1.37	1.44	–	–	–	1.35	*1.40*
C	1.93	2.54	2.23	2.19	2.12	2.16	–	–	–	1.50	*2.10*
D	3.36	3.07	3.22	2.92	2.00	2.46	–	–	–	1.80	*1.80*[a]
E	0.98	1.79	1.39	1.30	1.96	1.63	–	–	–	1.60	*1.60*[a]

[a] site specific ground response analysis required

Table 5 Soil factors for EC8 soil classes—Type 1 seismicity

$M_s > 5.5$											
Soil class	SHARE-DS1			SHARE-DS2			SHARE-DS3			EC8	Proposed
	Ap.1	Ap.2	W.A.	Ap.1	Ap.2	W.A.	Ap.1	Ap.2	W.A.		
B	1.47	1.34	1.41	1.53	1.08	1.31	1.49	0.94	1.22	1.20	*1.30*
C	2.09	2.24	2.16	2.06	1.46	1.76	1.82	1.15	1.48	1.15	*1.70*
D	1.74	1.42	1.58	1.56	0.92	1.24	–	–		1.35	*1.35*[a]
E	0.91	1.07	0.99	0.97	0.83	0.90	0.93	0.78	0.85	1.40	*1.40*[a]

[a] site specific ground response analysis required

New Soil Classification Scheme

Description of the New Classification Scheme

A new soil classification system that includes soil type, stratigraphy, thickness and stiffness, fundamental period of soil deposit (T_0) and average shear wave velocity of the entire soil deposit ($V_{s,\,av}$) as key parameters was initially proposed by Pitilakis et al. (2004a, 2006a). According to this site classification scheme, the sites are classified into six basic categories (A, B, C, D, E and X). The classification system proposed herein uses exclusively experimental data from the SHARE-AUTH database. The proposed improvements concern mainly the limits of values of the parameters describing each soil class. The new improved soil classification system is more precise, convenient and practical from a geotechnical perspective. At the same time it introduces as primary classification parameters the predominant period of the site, which is a fundamental factor for site amplification, and the depth to bedrock or to a layer with significant impedance contrast. The secondary parameters are soil type and stratigraphy, as well as the standard penetration test blow count N_{SPT}, plasticity index PI and undrained shear strength S_u. The proposed classification system is described in detail in Table 6, where $V_{s,\,av}$, N_{SPT} and S_u are average values over the whole soil column until the bedrock. This refinement of the proposed by EC8 classification scheme comprises the main factors affecting site response while minimizing the amount of data required for site characterization. The estimation of T_0 and $V_{s,\,av}$ is nowadays quite easy using various instrumental techniques and empirical methods (Pitilakis 2004; Chávez-García 2007). In addition, the disaggregation of the effect of soil stratigraphy and/or the soil type on seismic response with different subclasses improves the estimation of ground motion characteristics in many cases and in particular for deep soil profiles, for soft soils and for the case of a thin layer of rather soft soil on very stiff or rock type basement.

The proposed classification system is in general compatible with EC8, introducing at the same time some extra subclasses, which allows users to consider the influence of the depth to seismic bedrock and some important geotechnical properties. It should be stressed, however, that there is no complete correspondence between the proposed soil classes and the main EC8 soil classes. For instance, a site classified as C1 in the new classification, may be a B site according to EC8. This is a reasonable outcome as the soil class B in EC8 is too broad. This refinement of site classification is authorized under EC8, leaving to the national code committees to include more detailed parameters of the local soil and site conditions, as well as spectral shapes and corner periods, in their National Annexes. Unfortunately, this rarely takes place and the site classes' criteria and spectral shapes proposed in EC8 are simply adopted without any further refinement (Trifunac 2012).

Figure 10 illustrates the classification of the sites of the SHARE-AUTH database based on both EC8 and the proposed classification scheme. Due to lack of necessary data, subclasses D1, D2 and D3 were treated as a unified soil class (D).

Table 6 Proposed soil and site characterization

Soil class	Description	T_0 (s)	Remarks
A1	Rock formations		$V_s \geq 1500$ m/s
A2	Slightly weathered/segmented rock formations (thickness of weathered layer <5.0 m)	≤0.2	Surface weathered layer: $V_{s,av} \geq 200$ m/s Rock formations: $V_s \geq 800$ m/s
	Geologic formations resembling rock formations in their mechanical properties and their composition (e.g. conglomerates)		$V_s \geq 800$ m/s
B1	Highly weathered rock formations whose weathered layer has a considerable thickness (5.0–30.0 m)	≤0.5	Weathered layer $V_{s,av} \geq 300$ m/s
	Soft rock formations of great thickness or formations which resemble these in their mechanical properties (e.g. stiff marls)		$V_{s,av}$: 400–800 m/s $N_{SPT} > 50$ $S_u > 200$ kPa
	Soil formations of very dense sand—sand gravel and/or very stiff/to hard clay, of homogenous nature and small thickness (up to 30.0 m)		$V_{s,av}$: 400–800 m/s $N_{SPT} > 50$ $S_u > 200$ kPa
B2	Soil formations of very dense sand—sand gravel and/or very stiff/to hard clay, of homogenous nature and medium thickness (30.0–60.0 m), whose mechanical properties increase with depth	≤0.8	$V_{s,av}$: 400–800 m/s $N_{SPT} > 50$ $S_u > 200$ kPa
C1	Soil formations of dense to very dense sand—sand gravel and/or stiff to very stiff clay, of great thickness (>60.0 m), whose mechanical properties and strength are constant and/or increase with depth	≤1.5	$V_{s,av}$: 400–800 m/s $N_{SPT} > 50$ $S_u > 200$ kPa
C2	Soil formations of medium dense sand—sand gravel and/or medium stiffness clay (PI > 15, fines percentage > 30 %) of medium thickness (20.0–60.0 m)	≤1.5	$V_{s,av}$: 200–450 m/s $N_{SPT} > 20$ $S_u > 70$ kPa
C3	Category C2 soil formations of great thickness (>60.0 m), homogenous or stratified that are not interrupted by any other soil formation with a thickness of more than 5.0 m and of lower strength and $V_{s,av}$ velocity	≤1.8	$V_{s,av}$: 200–450 m/s $N_{SPT} > 20$ $S_u > 70$ kPa

Table 6 (continued)

Soil class	Description	T_0 (s)	Remarks
D1	Recent soil deposits of substantial thickness (up to 60.0 m), with the prevailing formations being soft clays of high plasticity index (PI > 40), high water content and low values of strength parameters	≤2.0	$V_{s,av} \leq 300$ m/s $N_{SPT} < 25$ $S_u < 70$ kPa
D2	Recent soil deposits of substantial thickness (up to 60.0 m), with prevailing fairly loose sandy to sandy-silty formations with a substantial fines percentage (not to be considered susceptible to liquefaction)	≤2.0	$V_{s,av} \leq 300$ m/s $N_{SPT} < 25$
D3	Soil formations of great overall thickness (> 60.0 m), interrupted by layers of category D1 or D2 soils of a small thickness (5.0–15.0 m), up to the depth of ~40 m, within soils (sandy and/or clayey, category C) of evidently greater strength, with $V_{s,av} \geq 300$ m/s	≤3.0	$V_{s,av}$: 150–600 m/s
E	Surface soil formations of small thickness (5.0–20.0 m), small strength and stiffness, likely to be classified as category C and D according to its geotechnical properties, which overlie category A formations ($V_{s,av} \geq 800$ m/s)	≤0.7	Surface soil layers $V_{s,av} \leq 400$ m/s
X	Loose fine sandy-silty soils beneath the water table, susceptible to liquefaction (unless a special study proves no such danger, or if the soil's mechanical properties are improved), Soils near obvious tectonic faults, Steep slopes covered with loose lateral deposits, Loose granular or soft silty-clayey soils, provided they have been proven to be hazardous in terms of dynamic compaction or loss of strength. Recent loose landfills, Soils with a very high percentage in organic material, Soils requiring site-specific evaluations		

$V_{s,av}$ is average shear wave velocity over the whole soil column until the bedrock

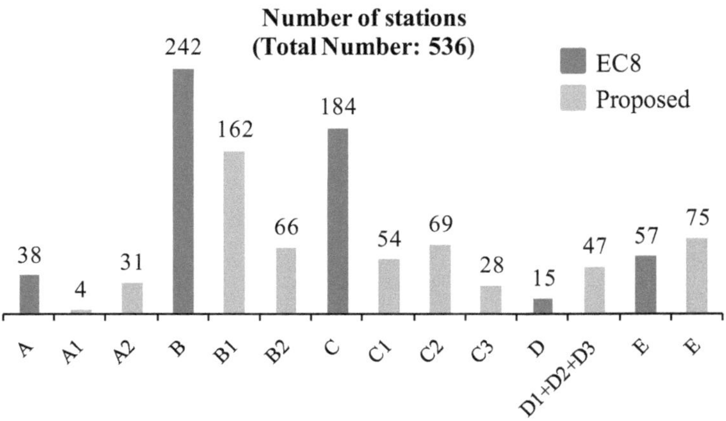

Fig. 10 Classification of the sites of SHARE-AUTH database according to EC8 (in *dark gray*) and the proposed classification scheme (in *light gray*)

Shape of Horizontal Elastic Response Spectra

For the equations describing the proposed normalized response spectra for the soil classes of the new classification system, the general form of the equations proposed by EC8 was adopted, allowing however for a differentiation of spectral amplification parameter β, which in EC8 is constant and equal to 2.5. The equations describing the elastic acceleration spectra for 5 % damping are the following:

$$0 \le T \le T_B : \frac{S_a(T)}{PGA_{rock}} = S \cdot \left[1 + \frac{T}{T_B} \cdot (\beta - 1)\right] \tag{8}$$

$$T_B \le T \le T_C : \frac{S_a(T)}{PGA_{rock}} = S \cdot \beta \tag{9}$$

$$T_C \le T \le T_D : \frac{S_a(T)}{PGA_{rock}} = S \cdot \beta \cdot \frac{T_C}{T} \tag{10}$$

$$T_D \le T : \frac{S_a(T)}{PGA_{rock}} = S \cdot \beta \cdot T_C \cdot \left(\frac{T_D}{T^2}\right) \tag{11}$$

PGA_{rock} is the design ground acceleration at rock-site conditions, S is the soil amplification factor, T_B and T_C are the limits of the constant spectral acceleration branch, T_D is the value defining the beginning of the constant spectral displacement range of the spectrum and β is the spectral amplification parameter. Parameters S, T_B, T_C,

T_D and β depend on soil class and on the level of seismicity, which is described by the magnitude of the most significant earthquake contributing to the local hazard. For reasons of homogeneity, the magnitude-dependent criterion used in this study follows the EC8 scheme. The adoption of these two different types of spectra helps for a better representation of the shape of the response spectra, which differentiates substantially between records from small and large magnitude events and also allows for a simplified introduction of non-linear site response. Parameters T_B, T_C, T_D and β, which define the shape of the response spectra, were determined by fitting the general spectral equations (Eqs. 8–11 divided by soil amplification factor S) to the empirical data. For this purpose, subset DS4 of the SHARE-AUTH database was used.

The normalized acceleration response spectra from the selected strong-motion records, i.e. the acceleration spectra divided by PGA, were plotted for each soil class and for the two levels of seismicity. For each soil class and level of seismicity, the median normalized acceleration spectra, as well as the 16th and 84th percentiles were calculated. The proposed normalized acceleration response spectra were evaluated through fitting the general spectral equations as close as possible to the 84th percentile, in order to account as much as possible for the uncertainties associated with the nature of the problem. Moreover, it was considered important that normalized acceleration response spectra were derived based on a common rationale, and that the same degree of conservatism was followed for all soil classes, and spectral periods, which is not the case for the normalized response spectra of EC8 (Pitilakis et al. 2012).

Figure 11 is a representative plot, illustrating the median and the range between 16 and 84 % percentiles, and the proposed design normalized acceleration spectra for soil class B1. Similar plots for all the new soil classes are available in Pitilakis et al. (2013). Type 2 spectra generally exhibit a narrower range of constant spectral

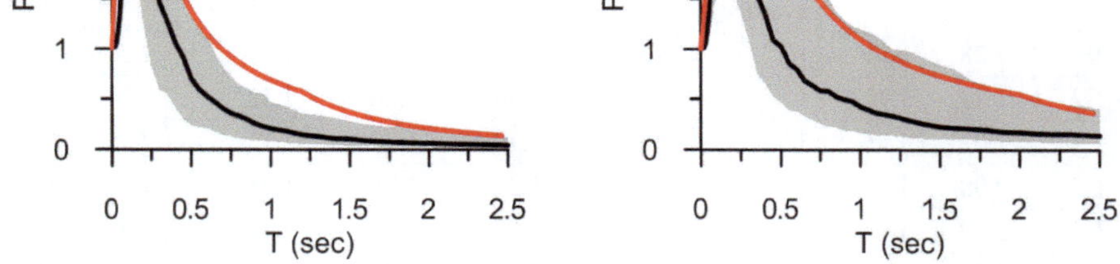

Fig. 11 Normalized elastic acceleration response spectra for soil class B1 of the new classification system for Type 2 seismicity (*left*) and Type 1 seismicity (*right*). *Red lines* represent the proposed spectra. The range of the 16th to 84th percentile is illustrated as a *gray area*

acceleration compared to Type 1. This justifies the decision to select two different spectrum types, since, in this way, overestimation of spectral ordinates in areas affected only by low-to-moderate magnitude earthquakes can be avoided.

The proposed values for the coefficients T_B, T_C, T_D and β required to define the normalized response spectra are given in earlier, along with the corresponding soil amplification factors.

New Soil Amplification Factors

Soil amplification factors S for the proposed classification system were estimated using the logic tree approach described earlier, using dataset DS4. Table 7 summarizes the soil amplification factors obtained for all soil classes with the two different approaches, as well as the weighted average and proposed S values. For soil classes B1, B2, C1, C2 and C3, the new soil amplification factors resulted from rounding the weighted average of the soil factors estimated with approaches 1 and 2. For soil class D, where actually limited data are available, we slightly reduced the soil amplification factors calculated from the logic tree approach, since we judged them as over-conservative. In contrast, for soil class E, again due to the relatively small dataset, we decided to maintain the soil factor values proposed by EC8, since the definition of soil classes E in EC8 and the new classification system are similar. Based on the presently available data, site-specific ground response analyses for both soil classes D and E should be recommended by the seismic code provisions.

New Elastic Response Spectra

The proposed coefficients T_B, T_C, T_D, β and S defining the elastic response spectra for the new soil classes are summarized in Table 8. The proposed new elastic accel-

Table 7 Soil factors for the new classification system

Soil class	Type 2 ($M_s \leq 5.5$)				Type 1 ($M_s > 5.5$)			
	Ap. 1	Ap. 2	Weighted average	Proposed	Ap. 1	Ap. 2	Weighted average	Proposed
B1	1.28	0.99	1.13	*1.2*	1.03	1.03	1.03	*1.1*
B2	1.89	1.17	1.53	*1.5*	1.36	1.28	1.32	*1.3*
C1	2.02	1.46	1.74	*1.8*	2.19	1.27	1.73	*1.7*
C2	2.08	1.39	1.74	*1.7*	1.35	1.15	1.25	*1.3*
C3	2.59	1.61	2.10	*2.1*	1.57	1.07	1.32	*1.3*
D	2.19	2.26	2.23	*2.0*[a]	2.03	1.79	1.91	*1.8*[a]
E	1.54	1.30	1.42	*1.6*[a]	1.10	0.94	1.02	*1.4*[a]

[a] site specific ground response analysis required

Table 8 Parameters of the proposed acceleration response spectra

Soil class	Type 2 ($M_s \leq 5.5$)					Type 1 ($M_s > 5.5$)				
	T_B (s)	T_C (s)	T_D (s)	S	β	T_B (s)	T_C (s)	T_D (s)	S	β
A	0.05	0.30	1.2	1.0	2.5	0.10	0.40	2	1.0	2.5
B1	0.05	0.25	1.2	1.2	2.75	0.10	0.40	2	1.1	2.75
B2	0.05	0.30	1.2	1.5	2.5	0.10	0.50	2	1.3	2.5
C1	0.10	0.25	1.2	1.8	2.5	0.10	0.60	2	1.7	2.5
C2	0.10	0.40	1.2	1.7	2.5	0.10	0.60	2	1.3	2.5
C3	0.10	0.50	1.2	2.1	2.5	0.10	0.90	2	1.3	2.5
D[a]	0.10	0.70	1.2	2.0	2.5	0.10	0.70	2	1.8	2.5
E[a]	0.05	0.20	1.2	1.6	2.75	0.10	0.35	2	1.4	2.75

[a] site specific ground response analysis required

eration response spectra, normalized to the design ground acceleration at rock-site conditions PGA_{rock}, are illustrated in Fig. 12.

Period-Dependent Amplification Factors

In Fig. 13 we present the period-dependent amplification factors for the current version of EC8, the improved EC8 and the new classification scheme, which result by dividing the elastic response spectrum of each soil class with the corresponding response spectrum for rock.

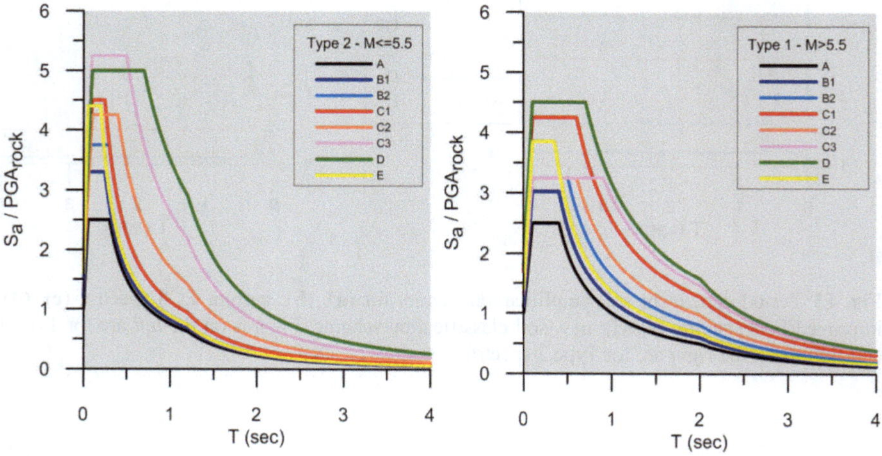

Fig. 12 Type 2 (*left*) and Type 1 (*right*) elastic acceleration response spectra for the proposed classification system

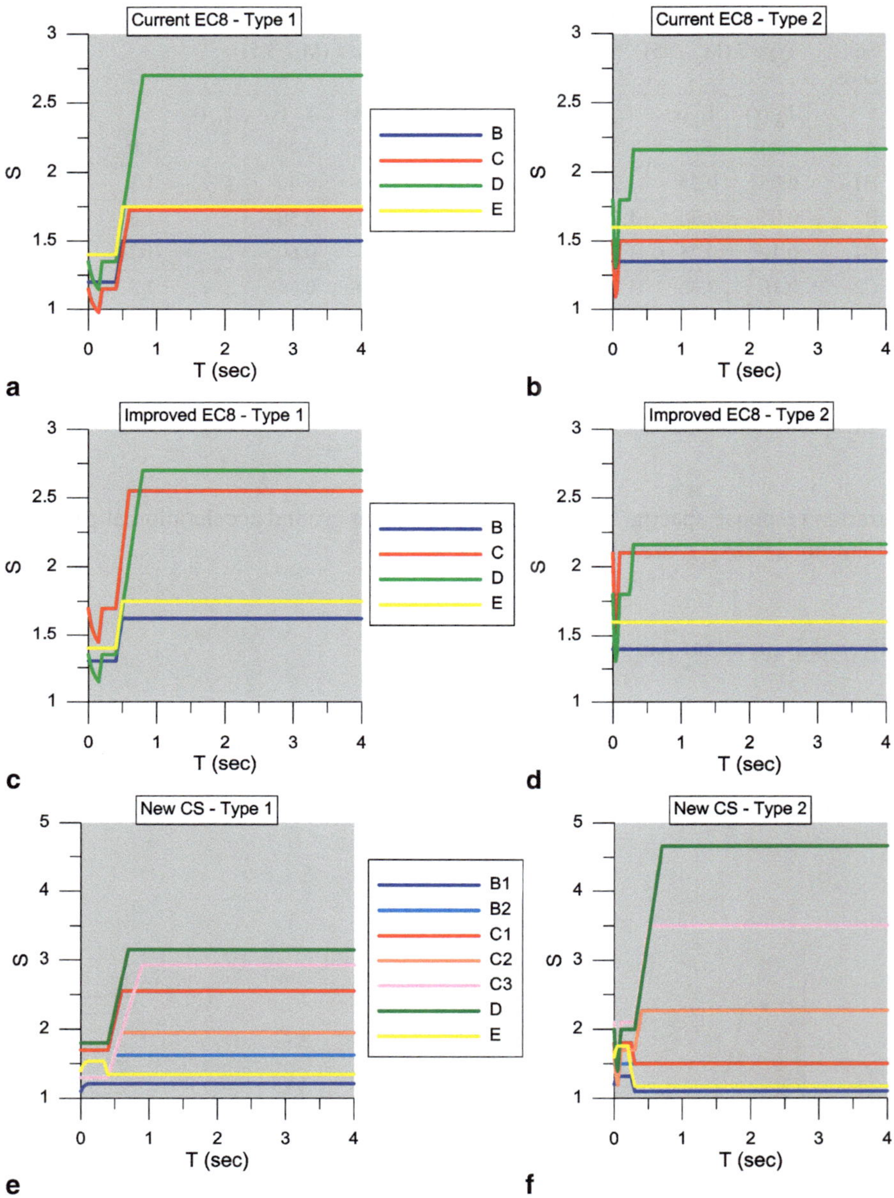

Fig. 13 Period-dependent soil amplification factors for (**a**), (**b**): current EC8 spectra, (**c**), (**d**): improved EC8 spectra, (**e**), (**f**): new soil classification scheme. Figures on the *left* are for Type 1 and figures on the *right* are for Type 2 spectra

Part 2: Risk Assessment of Thessaloniki

Introduction

Thessaloniki is the second largest city in Greece with about one million inhabitants. It has a long seismic history of devastating earthquakes. The most recent one occurred in 1978 (M6.5, R=25 km). Since then a lot of studies have been performed in the city to estimate the seismic hazard and to assess the seismic risk, including the present one. Due to the very good knowledge of the different parameters it has been selected as pilot study in several major EU research projects, namely SYNER-G, RISK-UE, LESS-LOSS and SHARE. In the present work for the estimation of seismic risk and loss assessment, we selected a smaller part of the whole agglomeration shown in Fig. 14. This is decided with respect to the better availability of a detailed inventory of the building stock. The selected area is divided in twenty Sub-City Districts (SCD) according to the Urban Audit database of EUROSTAT (http://www.urbanaudit.org).

The site specific seismic hazard has been evaluated using the Uniform Hazard Spectrum estimated for rock site conditions applying the methodology developed in SHARE in combination with the current and improved EC8 spectra, as well as the spectra and amplification factors of the new soil classification system. The three hazard models have been then used to assess the building damages using the capacity spectrum method and available fragility functions for the building typologies in Thessaloniki. The associated socioeconomic losses have been estimated using available empirical functions.

Site Zonation

An extensive program of geotechnical and geophysical studies undertaken the past 20 years, provided a solid and detailed information of the geotechnical conditions in the city (Pitilakis et al. 1992, 2001, 2003, 2004b, c, 2006b, c, 2007a, b; Pitilakis and Anastasiadis 1998; Raptakis et al. 1994a, b; Raptakis 1995; Anastasiadis and Pitilakis 1996, 1997; Anastasiadis et al. 2001, 2002). Detailed geotechnical maps are available; in summary the geotechnical zonation of the city is very well known. The average zonation of the city, based exclusively on the EC8 and the new classification system, is illustrated in Figs. 15 and 16 respectively.

Seismic Hazard

The seismic risk assessment was carried out using the Uniform Hazard Spectra (UHS) for rock-site conditions provided for all Sub-City Districts of Thessaloniki from the new hazard model developed in SHARE for a mean return period $T_m=475$

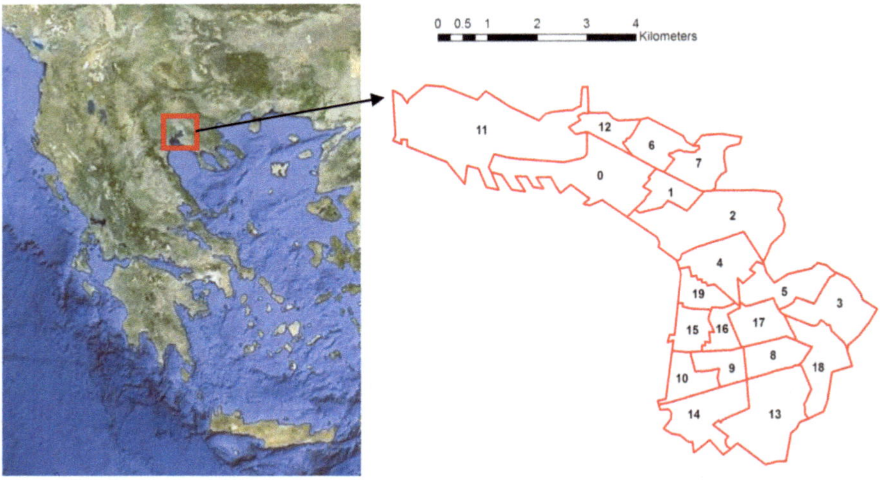

Fig. 14 Study area of Thessaloniki case study. Red lines illustrate Urban Audit SCDs boundaries

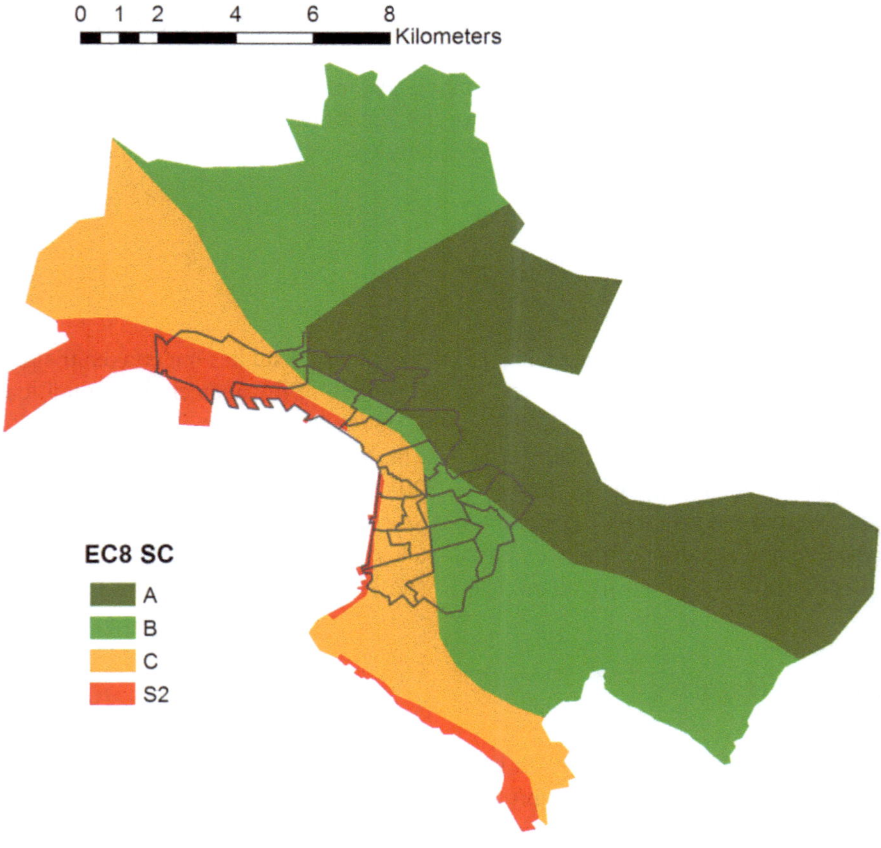

Fig. 15 Map of EC8 soil classes (based on $V_{s,30}$) for Thessaloniki, compiled utilizing all available geological, geophysical and geotechnical information

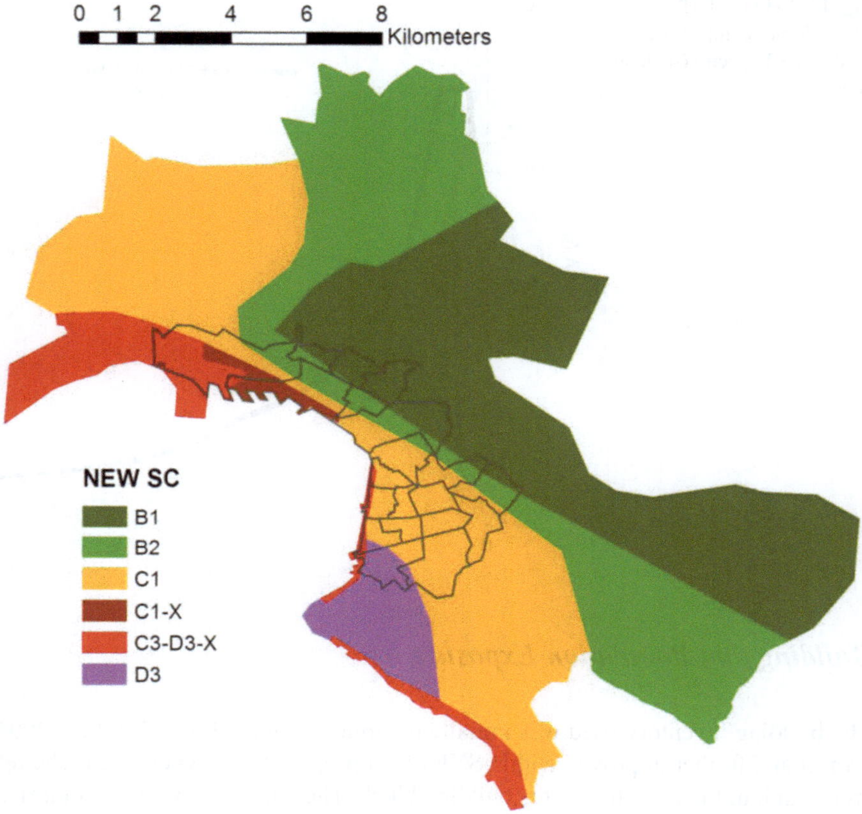

Fig. 16 Map of the soil classes of the proposed new soil classification system for Thessaloniki, compiled utilizing all available geological, geophysical and geotechnical information

years. Figure 17 illustrates the UHS for SCD 19, along with the design spectrum EC8 Type 1 spectra for PGA values equal to 0.2 g, all for rock-site conditions. SHARE UHS presents lower amplitudes and a narrower plateau compared to the EC8 code spectrum.

Peak ground acceleration for rock-site conditions (PGArock) for the twenty SCDs, as defined in the corresponding UHS, ranges from 0.195 g to 0.202 g.

Based on the computed UHS for rock conditions and on the detailed geotechnical mapping (see Figs. 15 and 16), the site dependent hazard and the demand spectra in each Sub-City District is estimated by multiplying the UHS ordinates with the period dependent amplification factors given in Fig. 13 for the three hazard models: the current EC8 (Hazard1), the improved EC8 (Hazard 2) and finally the new classification scheme and corresponding amplification factors (Hazard 3).

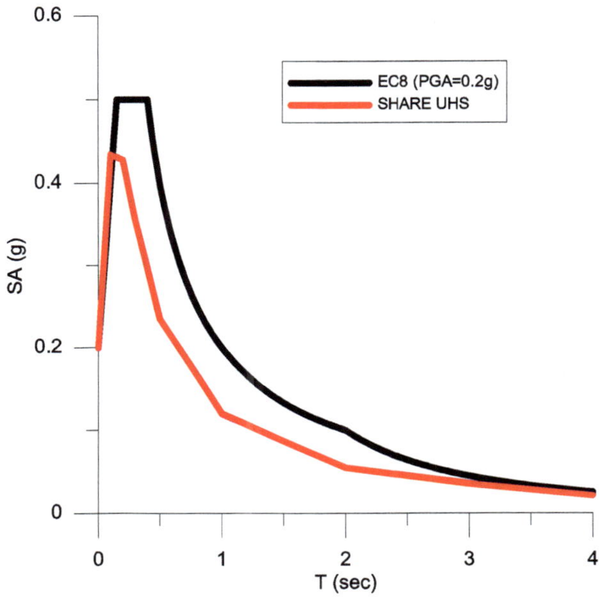

Fig. 17 SHARE UHS for Thessaloniki compared to EC8-Type 1 spectra (rock site conditions)

Building and Population Exposure

The building inventory used was initially compiled within RISK-UE (2001–2004) project and further improved within SYNER-G project (www.syner-g.eu). The reference unit of the inventory is the building block. The building inventory comprises 2893 building blocks with 27738 buildings, the majority of which (25639) are reinforced concrete (RC) buildings. The application presented herein was limited only to the RC buildings of the building stock of the city, built in different periods mainly after the World War II, which covers more than 90 % of the buildings of the city.

The classification of RC buildings is based on a Building Typologies Matrix representing practically all common RC building types in Greece (Kappos et al. 2006) (Table 9). Regarding the structural system, both frames and frame-with-shear walls (dual) systems are included, with a further distinction based on the configuration of the infill walls. Regarding the height, three subclasses are considered (low-, medium- and high-rise). Finally, as far as the level of seismic design is concerned, four different levels are considered:

- *No code (or pre-code):* R/C buildings with very low level of seismic design or no seismic design at all, and poor quality of detailing of critical elements.
- *Low code*: R/C buildings with low level of seismic design (roughly corresponding to pre-1980 codes in S. Europe, e.g., the 1959 Code for Greece).
- *Moderate code*: R/C buildings with medium level of seismic design (roughly corresponding to post-1980 codes in S. Europe, e.g., the 1985 Supplementary Clauses of the Greek Seismic Codes) and reasonable seismic detailing of R/C members.
- *High code*: R/C buildings with enhanced level of seismic design and ductile seismic detailing of R/C members according to the new generation of seismic codes (similar to Eurocode 8).

Table 9 RC Building Typology Matrix (BTM) for Thessaloniki (Kappos et al. 2006)

Type	Structural system	Height	Code level
RC1	Concrete moment frames	(L) Low-rise (1–3) (M) Mid-rise (4–7) (H) High-rise (8+)	(N)o/pre code (L)ow code (M)edium code (H)igh code
RC3	Concrete frames with unreinforced masonry infill walls		
3.1	Regularly infilled frames	(L) Low-rise (1–3) (M) Mid-rise (4–7) (H) High-rise (8+)	(N)o/pre code (L)ow code (M)edium code (H)igh code
3.2	Irregularly infilled frames (pilotis)	(L) Low-rise (1–3) (M) Mid-rise (4–7) (H) High-rise (8+)	(N)o/pre code (L)ow code (M)edium code (H)igh code
RC4	RC dual systems (RC frames and walls)		
4.1	Bare Systems (no infill walls)	(L) Low-rise (1–3) (M) Mid-rise (4–7) (H) High-rise (8+)	(N)o/pre code (L)ow code (M)edium code (H)igh code
4.2	Regularly infilled dual systems	(L) Low-rise (1–3) (M) Mid-rise (4–7) (H) High-rise (8+)	(N)o/pre code (L)ow code (M)edium code (H)igh code
4.3	Irregularly infilled dual systems (pilotis)	(L) Low-rise (1–3) (M) Mid-rise (4–7) (H) High-rise (8+)	(N)o/pre code (L)ow code (M)edium code (H)igh code

The classification of the RC buildings of the study area based on the BTM of Table 9 is illustrated in Fig. 18. Most of the buildings are either regularly or irregularly infilled dual systems (building types RC4.2 and RC4.3), while the majority of RC buildings are pre-1980 constructions and thus have been designed with low level of seismic code.

Based on the Urban Audit Data for Thessaloniki, the population of the study area is 376589. Since there were no available data for the exact number of inhabitants of each building of the inventory, we assumed a constant number of inhabitants per square meter of floor area for all buildings independently of their use.

Building Damage Model

The Capacity Spectrum Method (CSM) was used to determine the seismic performance of the building stock comparing the capacity of the structures to the seismic demand (Freeman 1998, Fajfar and Gapsersic 1996). An iterative process is applied to estimate the so-called "performance point", which is the intersection between the capacity curve of the structure and the demand spectrum, properly modified to ac-

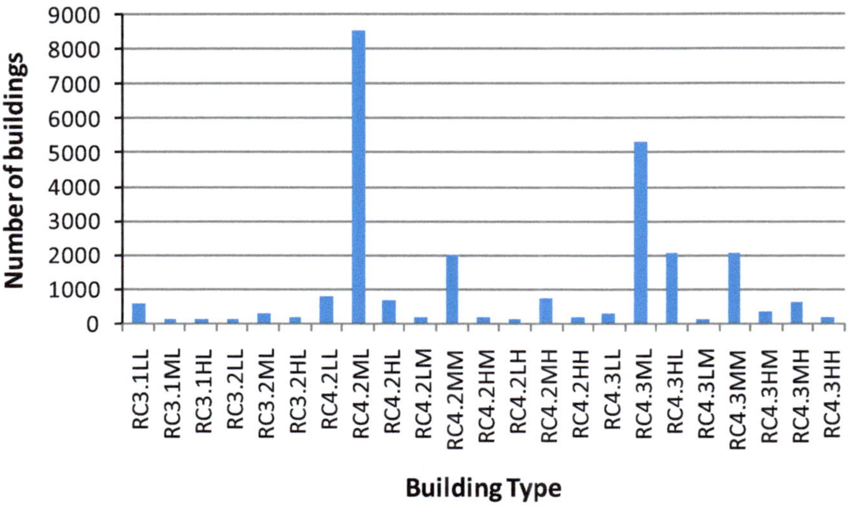

Fig. 18 Classification of the RC buildings of the study area based on the BTM of Table 9 (Kappos et al. 2006). The first letter of each building type refers to the height of the building (*L* low, *M* medium, *H* high), while the second letter refers to the seismic code level of the building (*N* no, *L* low, *M* medium, *H* high)

count for the elastic and hysteretic damping of the structure. A detailed description of the CSM method is available in ATC-40 (1996). The estimated spectral displacement is then overlaid with the available fragility curves in order to compute the damage probability in each of the different damage states.

Already available capacity and fragility curves for the Thessaloniki building typologies were used (D' Ayala et al. 2012). The capacity and the fragility curves have been estimated analytically, mainly with non-linear static (pushover) analysis; several uncertainties are inherently involved in these procedures, which will be only shortly at the end of the chapter. Five damage states are defined as follows: DS1 (slight), DS2 (moderate), DS3 (substantial to heavy), DS4 (very heavy) and DS5 (collapse). The definition of the damage states thresholds in terms of spectral displacement is given in Table 10. With this definition, the threshold between slight (DS1) and moderate (DS2) damage state for the most frequent building classes in Thessaloniki is for a spectral displacement slightly greater than the yield displacement (Fig. 19). The yield displacement itself in a bilinear capacity curve is normally defined as the point where the building response becomes strongly nonlinear and the strength level is higher than the design strength. However, it is difficult to define this threshold, since it depends on the way that the actual pushover curve is "bilinearized" (D' Ayala et al. 2012). A proposed methodology for this procedure is given in ATC-40 (1996).

The EarthQuake Risk Model code (EQRM—http://sourceforge.net/projects/eqrm) (Geoscience Australia) was used for the estimation of expected physical damages. The EQRM code has been designed for four damage states (slight,

Table 10 Damage state spectral displacement thresholds (D' Ayala et al. 2012)

Damage state	Bare frames Infilled frames with $S_{du,\,bare} < 1.1S_{du}$	Infilled frames with $S_{du,\,bare} \geq 1.1S_{du}$	Bare dual Infilled dual- shear wall drop strength	Infilled dual-infill walls failure
DS1	$0.7S_{dy}$		$0.7S_{dy}$	
DS2	$S_{dy}+0.05\,(S_{du}-S_{dy})$		$S_{dy}+0.05\,(S_{du}-S_{dy})$	
DS3	$S_{dy}+(1/3)\,(S_{du}-S_{dy})$	$S_{dy}+(1/2)\,(S_{du}-S_{dy})$	$S_{dy}+(1/2)\,(S_{du}-S_{dy})$	$0.9S_{du}$
DS4	$S_{dy}+(2/3)\,(S_{du}-S_{dy})$	S_{du}	S_{du}	$S_{du,\,bare}$
DS5	S_{du}	$S_{du,\,bare}$	$1.3S_{du}$	$1.3S_{du,\,bare}$

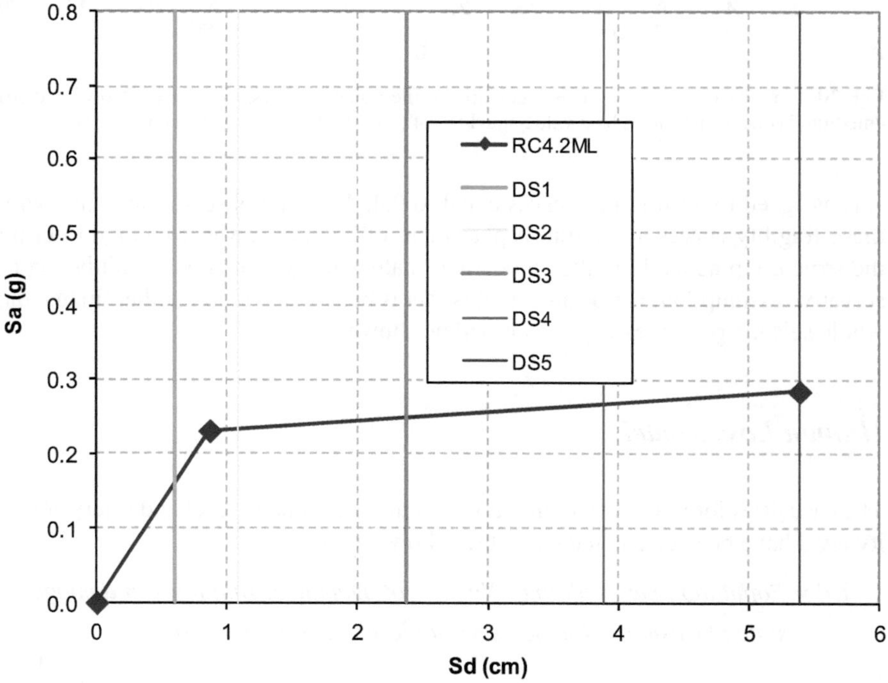

Fig. 19 Definition of damage states along the capacity curve for the most frequent building type (D' Ayala et al. 2012)

moderate, extensive and complete). Consequently the damage states have been modified as follows: DS1 for slight damage, DS2 for moderate, DS3 for extensive, while DS4 and DS5 damage states have been combined to express complete damage state. Representative plots of the used fragility curves as a function of spectral displacement (S_d) referring to mid-rise regularly infilled dual systems designed with "low" (RC4.2ML, most frequent typology) and "medium" seismic code provisions (RC4.2MM) are illustrated in Fig. 20. The lognormal standard deviation beta value

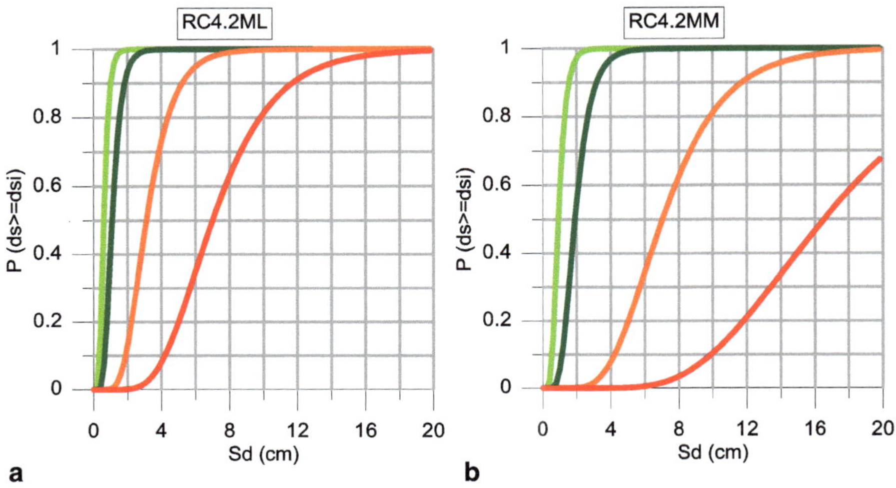

Fig. 20 Fragility curves for mid-rise regularly infilled dual R/C systems, for (**a**) "low" and (**b**) "medium" code design, for a beta value equal to 0.4 (modified after D' Ayala et al. 2012)

was assigned to all fragility curves equal to 0.4. A crucial observation is that with these fragility curves a spectral displacement of 2 cm or lower is enough to bring the structure practically in the "moderate" state damage, which is certainly a conservative assumption. Unfortunately this fact is penalizing, among other things, the whole seismic risk assessment, as it will be shown later.

Human Loss Model

The casualties for any given building type, building damage level and injury severity level have been calculated using the following equation:

$$Kij = Population\ per\ Building * Number\ of\ Damaged\ Buildings\ in\ damage\ state\ j * Casualty\ Rate\ for\ severity\ level\ i\ and\ damage\ state\ j \quad (12)$$

The injury severity levels are defined according to HAZUS-MH (FEMA 2003) by a four level injury scale, as described in Table 11. The casualty rates used to estimate the social losses are those defined by HAZUS-MH (FEMA 2003) for reinforced concrete building (Table 12).

Moreover we assumed that all buildings have only residential use, which is not entirely correct and the number of inhabitants of each building is proportional to the floor area of the building (constant number of inhabitants per square meter of floor area for all buildings). Finally, regarding the casualty rates for the complete damage states, we reasonably assumed that 5 % of the "complete" damaged buildings fully collapse, a percentage, which is close to the actual records form past earthquakes in Greece.

Table 11 Description of the injury severity levels. (FEMA 2003)

Injury severity	Injury description
Level 1	Injuries requiring basic medical aid without requiring hospitalization
Level 2	Injuries requiring medical care and hospitalization, but not expected to progress into a life threatening status
Level 3	Injuries that pose an immediate life threatening condition if not treated adequately and expeditiously. The majority of these injuries result because of structural collapse and subsequent collapse or impairment of the occupants
Level 4	Instantaneously killed or mortally injured

Table 12 Casualty rates for reinforced concrete moment frame structures. (FEMA 2003)

Injury severity	Slight damage	Moderate damage	Extensive damage	Complete damage
Level 1	0.05	0.25	1	5[a]—40[b]
Level 2	–	0.03	0.1	1[a]—20[b]
Level 3	–	–	0.001	0.01[a]—5[b]
Level 4	–	–	0.001	0.01[a]—10[b]

[a] Smaller values are related with partial collapse of the buildings
[b] Larger values are given for total collapse

Economic Loss Model

Economic losses are expressed though the mean damage ratio (MDR), which represents the total cost of repair divided by the cost of reconstruction. The final economic loss is then calculated by multiplying the total replacement cost with the mean damage ratio. For the estimation of MDR, the loss indices (repair cost/replacement cost) used in Kappos et al. (2003) were applied (Table 13). MDR is calculated using the following equation:

$$MDR = [D1] \cdot 0.005 + [D2] \cdot 0.05 + [D3] \cdot 0.2 + [D5] \cdot 0.8 \quad (13)$$

where [D1], [D2], [D3] and [D5] are the percentages of surface area in damage grades slight, moderate, extensive and complete, respectively.

Table 13 Loss indices for R/C structures (Kappos et al. 2003)

Damage State	Central loss index (%)
DS1	0.5
DS2	5
DS3	20
DS5	80

Damage and Loss Assessment

Damaged Floor Area

We calculated for each building type belonging to each building block the probabilities for slight, moderate, extensive and complete damage. We then multiplied these probabilities with the total floor area of the buildings of the specific building block that are classified to the specific building type and estimated for this building class the floor area, which will suffer each damage state. Repeating this for all building blocks which belong to the same Sub-City District (SCD) and for all building types, we can calculate for the specific SCD the total floor area of each building type that will suffer each damage state. The percentages of damaged floor area per damage state for each SCD and for the three hazard models described are illustrated in Figs. 21, 22 and 23. The total percentages of damaged floor area per damage state for the whole study area are given in Table 14.

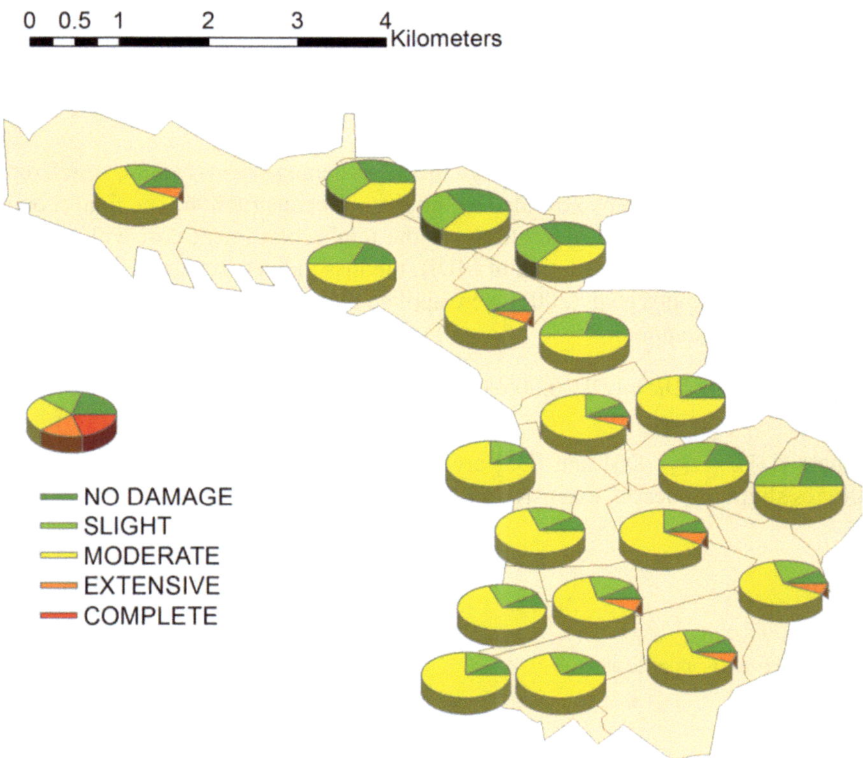

Fig. 21 Hazard 1: Seismic risk per Sub-City District for a return period of 475 years in terms of the percentage of damaged floor area per damage state

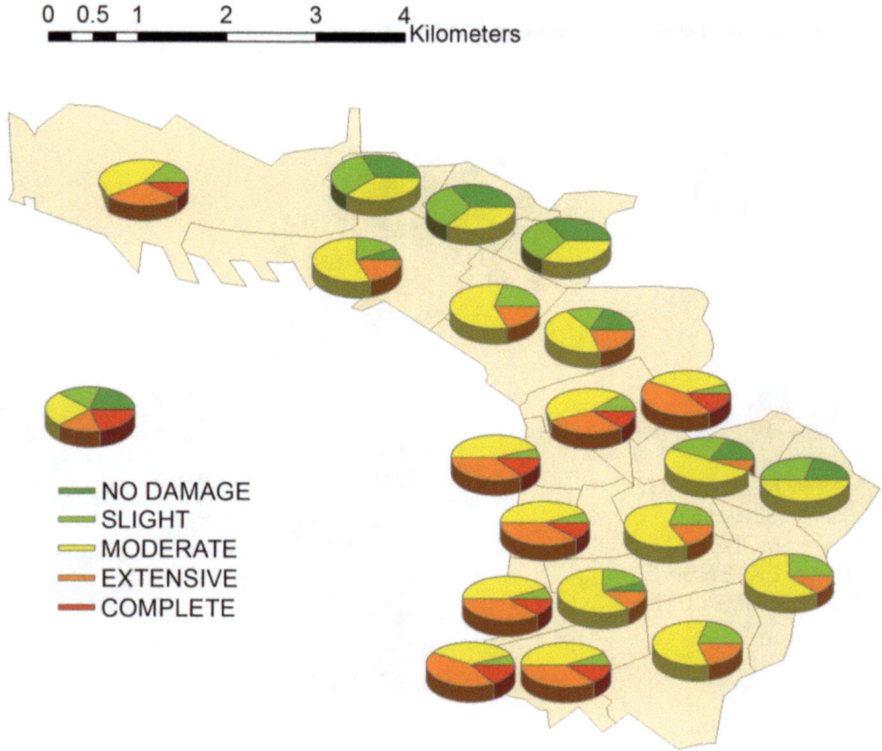

Fig. 22 Hazard 2: Seismic risk per Sub-City District for a return period of 475 years in terms of the percentage of damaged floor area per damage state

We dispose a quite accurate damage assessment after the 1978 earthquake for almost 50 % of the present study area. Of course the density of the buildings, the typology and code design level differ from the present situation. Moreover the direct comparison with the damages observed in the 1978 earthquake is not straightforward, on the one hand because the 1978 event is not of identical intensity with the herein studied hazard models, and on the other hand because the damages of the 1978 event have been classified only in three categories (green, yellow and red). Actually, as it shown in Fig. 24, where we compare the response spectra of the two components of the only available record with the computed ones for soil category C or C1 according to the two classification schemes (EC8 and new respectively), the 1978 earthquake spectrum is close to the spectrum of hazard model 1, but is much less severe than the spectra of hazard models 2 and 3.

Due to the different definitions of damage states, the present four damage states (five with the "no-damage" state) should be split into three (green, yellow and red), in order to be able to make a rough comparison between the estimated damages and the actual damages from the 1978 earthquake. We subjectively decided to hold in the "green" tag the "no", the "slight" and 30 % of the "moderate" damages. Then, in

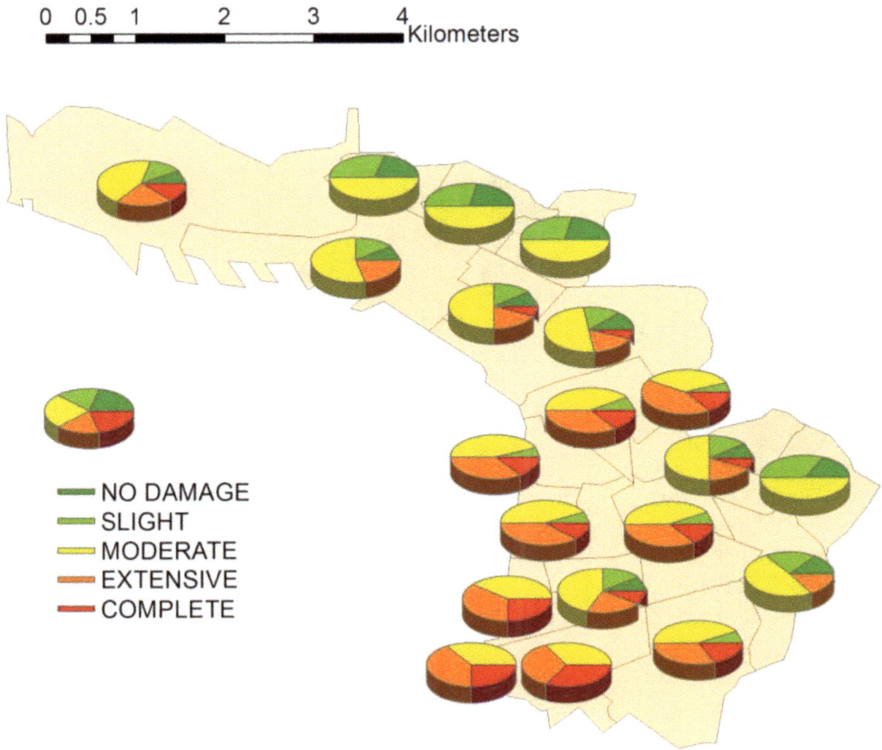

Fig. 23 Hazard 3: Seismic risk per Sub-City District for a return period of 475 years in terms of the percentage of damaged floor area per damage state

Table 14 Percentages of damaged floor area per damage state for the three hazard models, for the whole city district and for a return period of 475 years

	Hazard 1 (%)	Hazard 2 (%)	Hazard 3 (%)
No	15.36	11.28	8.56
Slight [D1]	23.88	17.34	15.31
Moderate [D2]	56.54	47.81	45.03
Extensive [D3]	4.17	18.94	21.76
Complete [D5]	0.06	4.64	9.34

the "yellow" tag we put the rest of the "moderate" and the "extensive", and finally in the "red" tag the "complete". In Table 15 we give the comparison of the damages estimated herein with the observed damages during the 1978 earthquake. We observe that the percentage of the "red" buildings from Hazard 2 is in good agreement with that of the 1978 event, while the methodology adopted herein in general leads to more "yellow" and less "green" buildings compared to the damages observed in 1978.

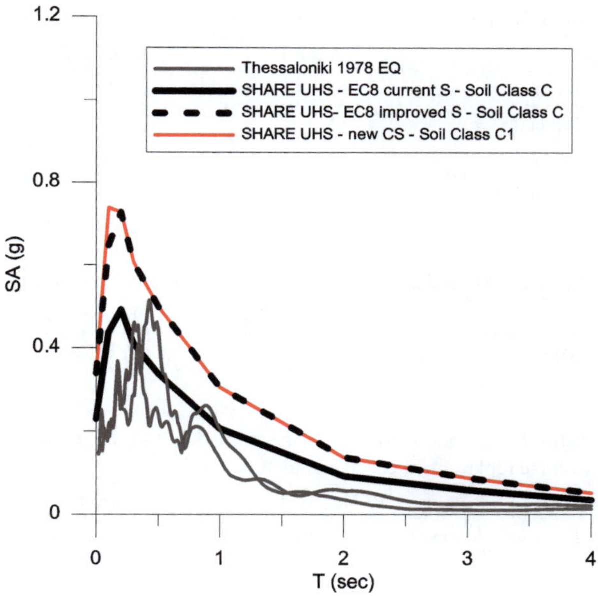

Fig. 24 Elastic response spectra (5 %) of the two components of the only record available from Thessaloniki 1978 earthquake compared to SHARE UHS for EC8 soil class C and C1

Table 15 Comparison of the herein computed damage percentages with the observed percentages of damages during the M6.5 1978 earthquake (Penelis 2008)

	Thessaloniki 1978 EQ (%)	Hazard 1 (%)	Hazard 2 (%)	Hazard 3 (%)
Green	74.50	56.20	42.96	37.38
Yellow	21.0	43.75	52.41	53.28
Red	4.50	0.06	4.64	9.34

Human Losses

Casualties were estimated using Eq. (12). The total casualties for the whole study area and the injury severity levels described in Table 11 are given in Table 16.

Economic Losses

Economic losses, which are expressed by the mean damage ratio (MDR), were calculated by inserting the percentages [D1], [D2], [D3] and [D5] given in Table 14 into Eq. (13). The resulting total MDR values for the whole study area are given in Table 17.

By multiplying the mean damage ratio with the replacement cost, the economic losses can be evaluated. Assuming an average replacement cost equal to 1000 €/m^2, economic losses are given in Table 18.

Table 16 Level 1—Level 4 casualties for hazard models 1–3 for a return period of 475 years

	Hazard 1	Hazard 2	Hazard 3
Level 1	762	2619	4094
Level 2	85	520	932
Level 3	1	52	106
Level 4	1	102	208

Table 17 Mean damage ratios for hazard models 1–3 for a return period of 475 years

	Hazard 1 (%)	Hazard 2 (%)	Hazard 3 (%)
MDR	3.83	9.98	14.15

Table 18 Economic losses for hazard models 1–3, for a return period of 475 years, assuming an average replacement cost equal to 1000 €/m²

	Hazard 1	Hazard 2	Hazard 3
Economic losses (€)	1.3 billions €	3.4 billions €	4.8 billions €

Discussion

The observed differences in the damage assessment and losses are primarily attributed to the hazard models. The improved or new S factors for EC8 (Hazards 2 and 3) increase the building damages and consequently the social and economic losses, since they both result in higher ordinates of the elastic, and therefore the inelastic demand spectra for the same structure. In Fig. 25a we compare the elastic demand spectra for the two EC8 soil classes B and C when using the current S factors to the corresponding spectra when using the improved S factors. The available capacity curves of two of the most-frequent building classes (RC4.2ML and RC4.2MM) (D' Ayala et al. 2012) are also illustrated. The significant discrepancies (especially for soil class C) result in performance points with higher spectral displacement values when using the improved S factors instead of the current ones, which inevitably leads to higher vulnerability for Hazard 2 compared to Hazard 1. In Fig. 25b we compare the elastic demand spectra for EC8 soil class B (using both current and improved S factors) with the elastic demand spectrum for soil class C1, since the majority of the building blocks which are classified as B according to EC8 are classified as C1 according to the new classification system. We again observe that the ordinates of C1 spectrum are much higher than those of B spectra, a fact which leads to higher vulnerability for Hazard 3 compared to Hazards 1 and 2.

At this point we should emphasize that the methodology of the damage assessment, applying the capacity spectrum method, involves significant uncertainties, and the previously presented results should be interpreted with caution. A thorough discussion on the epistemic and aleatory uncertainties related to all possible parameters involved in the procedure is far beyond the scope of the present work. Never-

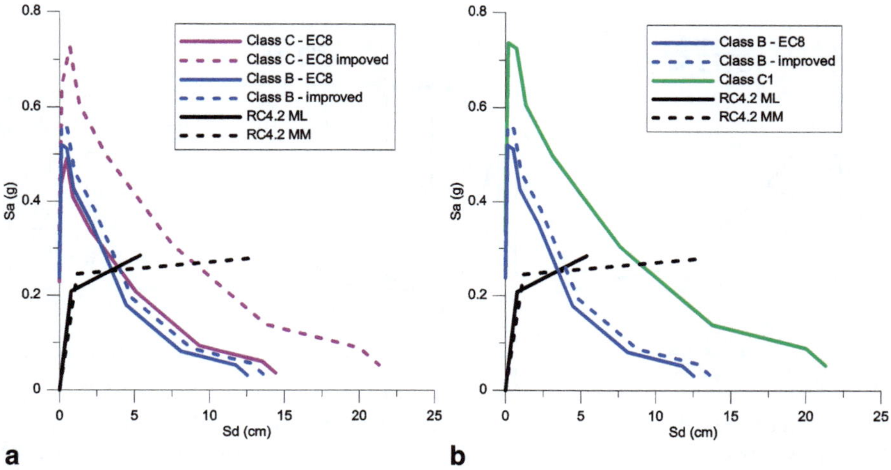

Fig. 25 SHARE UHS (elastic) amplified with (**a**) the current and the improved S factors for EC8 soil classes B and C and (**b**) the current and the improved S factors for EC8 soil class B and the S factors for new soil class C1. The capacity curves of two of the most-frequent building classes (RC4.2ML and RC4.2MM) are also illustrated

theless, we will highlight in the following some of them, which are to our point of view the most important.

The estimation of the percentage of the buildings classified to each damage level for each building type is made with the following steps: (a) Estimation of the elastic demand spectra considering the seismic hazard and the local site conditions; (b) estimation of the inelastic demand spectra to account for the hysteretic damping associated with the inelastic behaviour of structures; (c) evaluation of the capacity curve of each building typology; (d) estimation of the performance point, which provides the spectral acceleration (S_a) and spectral displacement (S_d) values for the damage assessment; (e) selection of the fragility curves for the specific building type; (f) use of the estimated S_d to estimate from the fragility curves the probabilities of a building of the specific typology to suffer each damage state for the given seismic scenario. Figures 26a, b, 27a and b illustrate the essentials of the methodology for the most frequent building type in Thessaloniki (RC4.2ML) for soil type C according to the present and the improved EC8 code amplification factors respectively. In all the above described steps various uncertainties are involved. We will concentrate ourselves on some comments regarding the definition and construction of (i) the inelastic demand spectra, (ii) the capacity curves and (iii) the fragility curves.

i. *Elastic and Inelastic Demand Spectra*

Reduction of elastic spectra to account for the hysteretic damping associated with the inelastic behaviour of structures is carried out using either over-damped elastic or constant-ductility inelastic spectra. Over the last years the use of reduction factors to evaluate the inelastic spectral values has become the subject of many studies and debates (Lin et al. 2005, Gencturk et al. 2007). In the present work,

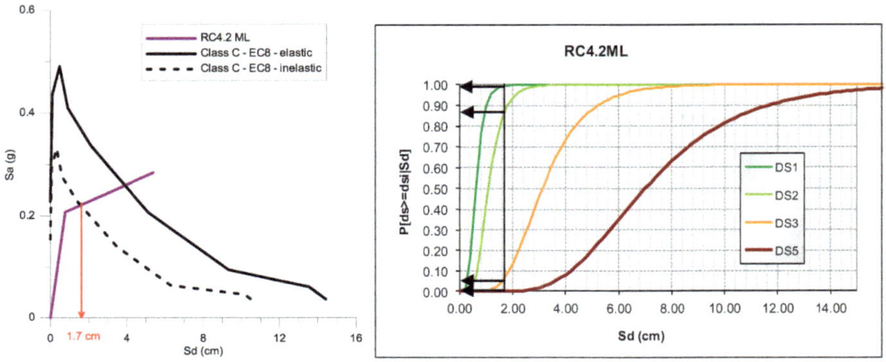

a b

Fig. 26 Procedure of building damage assessment for building type RC4.2ML on EC8 soil class C using the present amplification factors: (**a**) Estimation of the performance point ($S_{d'p}$, $S_{a'p}$) as the intersection of the inelastic demand spectrum with the capacity curve. (**b**) Estimation of the probabilities for each damage state using $S_{d'p}$ as input in the appropriate fragility curves

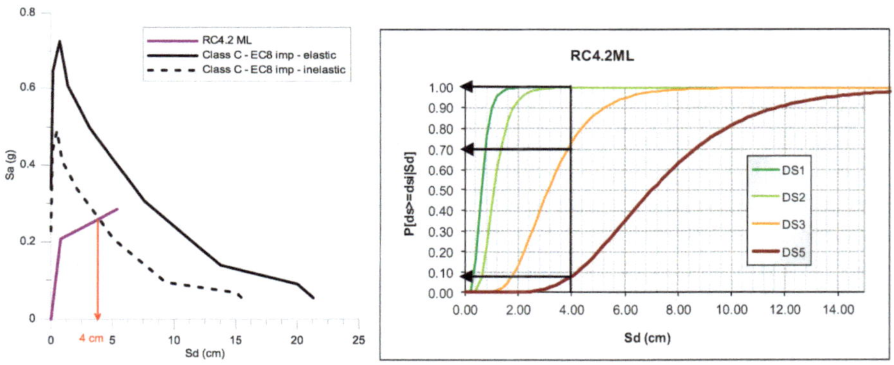

a b

Fig. 27 Procedure of building damage assessment for building type RC4.2ML on EC8 soil class C using the improved amplification factors: **a** Estimation of the performance point ($S_{d,p}$, $S_{a,p}$) as the intersection of the inelastic demand spectrum with the capacity curve. **b** Estimation of the probabilities for each damage state using $S_{d,p}$, as input in the appropriate fragility curves

the damping-based spectral reduction factors by Newmark and Hall (1982) were used, which have been adopted by ATC-40 (ATC 1996) and HAZUS (FEMA 2003). Damping-based reduction factors have also been proposed by other researchers (e.g. Lin and Chang 2003 and Priestley et al. 2007). The method adopted herein is probably resulting in higher reduction factors (Cassarotti et al. 2009) and consequently damages.

ii. *Capacity curves and Performance Point*

Conventional force-based static (pushover) methods have been widely used to estimate the capacity curves. The method presents several weaknesses, some of them already reported in FEMA 440 (2005). Among the most important are the following: inappropriate prediction of deformations when higher modes are important and/or the structure is highly pushed into its nonlinear post-yield range; inappropriate prediction of local damage concentrations; difficulty in the reproduction of important dynamic effects and to incorporate three-dimensional and cyclic earthquake loading effects; definition of the global yielding and collapse damage states; effects of building irregularity. Besides, as an analytical approach, it encompasses the challenges of the accuracy of the building modeling and the consideration of the age of the building and the design code applied in the construction period. Due to all these problems, it is not rare that the rather limited large-scale experimental data often provide higher ordinates of the yielding acceleration than the analytical curves; higher values of yield acceleration would lead to smaller displacement values at the performance points, and thus to lower vulnerability.

In the present work we used the bilinear curves proposed by D'Ayala et al. (2012). Figure 28 illustrates the capacity curves of the most frequent building classes in Thessaloniki along with the elastic and inelastic response spectra for soil class B (Hazard 2). The analytically estimated yielding acceleration for these cases varies between 0.20 g and 0.25 g, while the performance spectral displacement ranges between 1.7 and 2.5 cm approximately.

iii. *Fragility curves*

The proposal of analytical fragility curves for typical buildings is subject of numerous research efforts the past 15 years. Most of the analytical functions proposed so far are based on pushover analyses, while the incremental dynamic analysis (Vamvatsikos and Cornell 2002) applied recently seems to have several advantages. The construction of the curves involves several subjective decisions, which may change

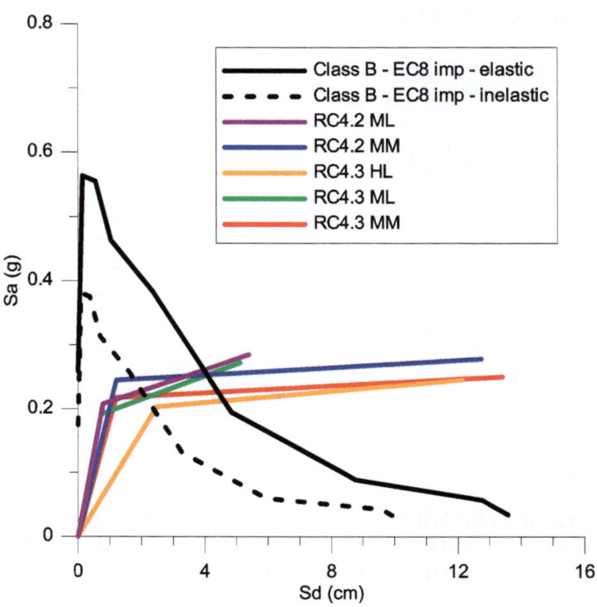

Fig. 28 Elastic and inelastic response spectra for EC8 soil class B using the improved S factors together with the capacity curves of the most-frequent building classes

seriously their form. Among the most important is the definition of the damage states. Even the same researcher in different periods may propose different curves for the same type of building. To illustrate the important scatter of the proposed curves we present in the following figures two examples elaborated in the frame of SYNER-G research project (http://www.syner-g.eu, coordinator K. Pitilakis). Figure 29 presents fragility curves for yielding and collapse proposed by different researchers for the most frequent building type in Thessaloniki, while in Fig. 30 we present the whole spectrum of fragility curves for mid-rise buildings with bare

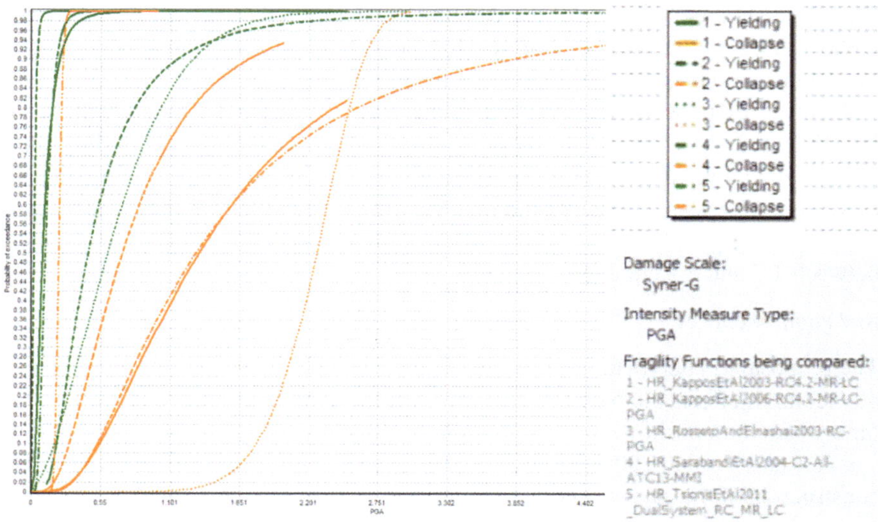

Fig. 29 Comparison of fragility curves for the most frequent building type in Thessaloniki (dual systems, medium rise, low code) by different researchers

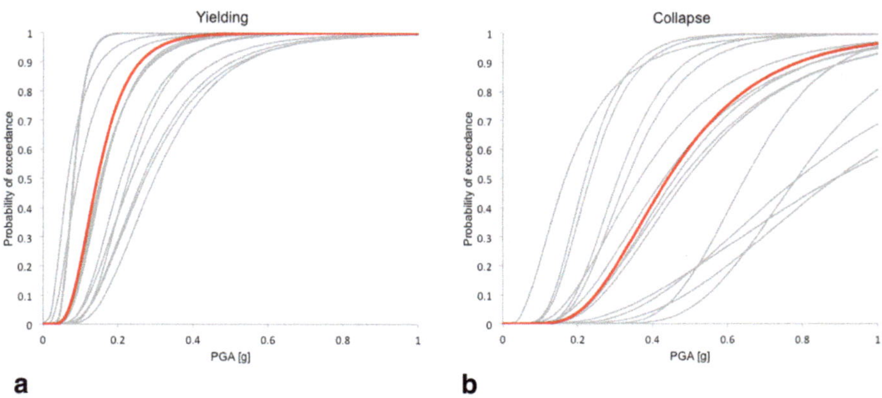

Fig. 30 Individual fragility curves by different researchers (*grey*) and mean fragility curves (*red*) for mid-rise buildings with bare moment resisting frame with lateral load design for damage states (**a**) yielding and (**b**) collapse. (Crowley et al. 2011)

moment resisting frame with lateral load design. It is obvious that the impact that an inadequate selection might have on the final evaluation of damages is very important.

Conclusions

The validity and accuracy of the elastic response spectra defined in Eurocode 8 provisions was checked against a large worldwide dataset of strong motion records in terms of both PGA-normalized spectra and soil factors. The EC8 spectral shapes, for the given definition of soil categories, are on average in good agreement with the empirical data. Within the scope of a potential EC8 update, improved values were proposed for the soil amplifications factors S for EC8 soil classes B and C, while for soil classes D and E, due to the lack of data we proposed to keep for the moment the EC8 amplification factors, recommending however the need for site-specific ground response analyses in case of important projects.

In the framework of the improvement and refinement of EC8 regarding the selection of a more accurate design input motion, a new soil and site classification system was also developed, aiming at eliminating the different limitations of the present soil classification. The new classification system uses parameters such as the thickness of soil deposits, the average shear wave velocity to the seismic bedrock instead of $V_{s,30}$, and the fundamental period of the site, along with traditionally used parameters, such as S_u, N_{SPT}, PI, and soil description, which is more refined than the present one. Site amplification factors, as well as normalized acceleration response spectra were proposed for each soil class of the new classification system and the two seismicity contexts used in EC8 provisions. The new classification system is more precise, convenient and practical from a geotechnical engineering point of view.

The improved soil amplification factors for EC8 and the new soil classification system were finally used to perform the risk assessment of the city of Thessaloniki in Greece, applying the Capacity Spectrum Method. The seismic hazard (outcrop conditions), for 475 years of return period, has been estimated using the new SHARE approach. The site-specific demand spectra have been estimated applying the period-dependent amplification factors to the computed UHS. As it was expected, the improved S factors for EC8 or the use of the new classification system and corresponding S factors increases the expected physical damages to buildings, as well as the social and economic losses compared to the use of the current EC8 S factors. Critical points to which special attention should be paid when implementing the Capacity Spectrum Method were finally highlighted.

Data and Resources

All records used in this study were selected from the worldwide strong motion database compiled in the framework of SHARE (Yenier et al. 2010). For the compilation of SHARE-AUTH database, shear wave velocity profiles were gathered from the Italian Accelerometric Archive (Working Group ITACA 2010, http://itaca.mi.ingv.it), Kik-Net (http://www.kik.bosai.go.jp/), K-Net (http://www.k-net.bosai.go.jp/), the Turkish national strong-motion database (Akkar et al. 2010, in DVD format), ROSRINE program website (http://gees.usc.edu/ROSRINE/), Dr. David Boore's personal webpage (http://www.daveboore.com/), selected USGS open-file reports and the shear wave velocity archives of the Research Unit of Soil Dynamics and Geotechnical Earthquake Engineering of AUTH.

Acknowledgements This research has been mainly funded by the European Community's Seventh Framework Program [FP7/2007–2013] under grant agreement n° 226967 (Seismic Hazard Harmonization in Europe, http://www.share-eu.org/). The authors would like to thank Anna Karatzetzou, MSc, for calculating the inelastic demand spectra and helping with the implementation of the Capacity Spectrum Method.

References

Akkar S, Bommer JJ (2010) Empirical equations for the prediction of PGA, PGV and spectral accelerations in Europe, the Mediterranean region and the Middle East. Seismol Res Lett 81:195–206. doi:10.1785/gssrl.81.2.195

Akkar S, Çağnan Z, Yenier E, Erdoğan Ö, Sandikkaya A, Gülkan P (2010) The recently compiled Turkish strong motion database: preliminary investigation for the seismological parameters. J Seismolog 14:457–479. doi:10.1007/s10950-009-9176-9

Anastasiadis A, Pitilakis K (1996) Shear modulus Go and damping of typical Greek soils at low strain amplitudes. Technika Chronika. Sci J Tech Chamber Greece 16(3):9–18

Anastasiadis A, Pitilakis K (1997) Variation of shear modulus G and damping of typical Greek soils with strain amplitude. Technika Chronika. Sci J Tech Chamber Greece 17:37–47

Anastasiadis A, Raptakis D, Pitilakis K, (2001). Thessaloniki's detailed microzoning: subsurface structure as basis for site response analysis. Pure Appl Geophy 158:2597–2633

Anastasiadis A, Apessou M, Pitilakis K, (2002). Earthquake hazard assessment in Thessaloniki, Greece: Level II—Site response analysis. International Conference: earthquake loss estimation risk ruction, Bucharest, Romania, 24–26 October

ATC-40 (1996) Seismic evaluation and retrofit of concrete buildings. Applied Technology Council, Redwood City, California

Boore DM (2004) Estimating (or NEHRP Site Classes) from shallow velocity models (Depths < 30 m). B Seismol Soc Am 94:591–597. doi:10.1785/0120030105

Boore DM, Joyner WB (1997) Site amplifications for generic rock sites. B Seismol Soc Am 87:327–341

Borcherdt RD (1994) Estimates of site-dependent response spectra for design (methodology and justification). Earthquake Spectra 10:617–653. doi:10.1193/1.1585791

Borcherdt RD, Glassmoyer G (1992) On the characteristics of local geology and their influence on ground motions generated by the Loma Prieta earthquake in the San Francisco Bay region, California. B Seismol Soc Am 82:603–641

Cadet H, Bard PY, Duval AM (2008) A new proposal for site classification based on ambient vibration measurements and the Kiknet strong motion data set. In: Proceedings of the 14th world conference on earthquake engineering, Beijing

Cadet H, Bard PY, Rodriguez-Marek A (2012) Site effect assessment using KiK-net data: Part 2—site amplification prediction equation based on f_0 and V_{sz}. Bull Earthquake Eng 10:451–489. doi:10.1007/s10518-011-9298-7

Castellaro S, Mulargia F, Rossi PL (2008) V_{s30}: proxy for seismic amplification? Seismol Res Lett 79:540–543. doi:10.1785/gssrl.79.4.540

Casarotti C, Monteiro R, Pinho R (2009) Verification of spectral reduction factors for seismic assessment of bridges. B New Zeal Soc Earthquake Eng 42:111–121

Cauzzi C, Faccioli E (2008) Broadband (0.05 to 20 s) prediction of displacement response spectra based on worldwide digital records. J Seismolog 12:453–475. doi:10.1007/s10950-008-9098-y

CEN (European Committee for Standardization) (2004) Eurocode 8: Design of structures for earthquake resistance, Part 1: General rules, seismic actions and rules for buildings. EN 1998-1:2004. Brussels, Belgium

Chávez-García FJ (2007) Site effects: from observation and modeling to accounting for them in building code. In: Pitilakis KD (ed) Earthquake Geotechnical Engineering, 4th International Conference of Earthquake Geotechnical Engineering—Invited lectures, Springer, pp 53–72

Chiou BSJ, Youngs RR (2008) An NGA model for the average horizontal component of peak ground motion and response spectra. Earthquake Spectra 24:173–215. doi:10.1193/1.2894832

Choi Y, Stewart JP (2005) Nonlinear site amplification as function of 30 m shear wave velocity. Earthquake Spectra 21:1–30. doi:10.1193/1.1856535

Crowley H, Colombi M, Silva V, Ahmad N, Fardis M, Tsionis G, Papailia A, Taucer F, Hancilar U, Yakut A, Erberik MA (2011) Fragility functions for common RC building types in Europe, Deliverable 3.1, Syner-G Project 2009/2012

D'Ayala D, Kappos A, Crowley H, Antoniadis P, Colombi M, Kishali E, Panagopoulos G, Silva V (2012) Providing building vulnerability data and analytical fragility functions for PAGER, Final Technical Report

Delavaud E, Cotton F, Akkar S, Scherbaum F, Danciu L, Beauval C, Drouet S, Douglas J, Basili R, Sandikkaya MA, Segou M, Faccioli E, Theodoulidis N (2012) Toward a ground-motion logic tree for probabilistic seismic hazard assessment in Europe. J Seismolog. doi:10.1007/s10950-012-9281-z

DIN 4149:2005 (2005) Buildings in German earthquake regions—design loads, analysis, and structural design of buildings. Normenausschuss Bauwesen (NABau) im Deutschen Institut für Normung e.V. (DIN), April 2005, 84 p

Fajfar P, Gaspersic P (1996) The N2 method for the seismic damage analysis for RC buildings. Earthquake Eng Struct Dynam 25:23–67

FEMA (2003) HAZUS-MH technical manual, federal emergency management agency, Washington DC

FEMA (2005) FEMA 440 improvement of nonlinear static seismic analysis procedures, ATC-55, Washington DC

Freeman SA (1998) The capacity spectrum method as a tool for seismic design, Proceedings of the 11th European Conference on Earthquake Engineering, Paris, France

Gallipoli MR, Mucciarelli M (2009) Comparison of site classification from V_{S30}, V_{S10}, and HVSR in Italy. B Seismol Soc Am 99:340–351. doi:10.1785/0120080083

Gencturk B, Elnashai AS, Song J (2007) Fragility relationships for populations of buildings based on inelastic response. Department of Civil and Environmental Engineering. University of Illinois at Urbana-Champaign, Urbana

Housner GW (1952) Spectrum intensities of strong-motion earthquakes. In: Proceedings of Symposium on Earthquakes and Blast Effects on Structures, Earthquake Engineering Research Institute, pp 20–36

Japan Road Association (1980) Specifications for Highway Bridges Part V, Seismic Design, Maruzen Co., LTD

Japan Road Association (1990) Specifications for Highway Bridges Part V, Seismic Design, Maruzen Co., LTD

Kappos AJ, Panagiotopoulos C, Panagopoulos G, Papadopoulos E (2003) RISK-UE WP4—Reinforced Concrete Buildings (Level I and II analysis)

Kappos AJ, Panagopoulos G, Panagiotopoulos C, Penelis G (2006) A hybrid method for the vulnerability assessment of R/C and URM buildings. Bull Earthquake Eng 4(4):391–413. DOI 10.1007/s10518-006-9023-0

Kokusho T, Sato K (2008) Surface-to-base amplification evaluated from KiK-net vertical array strong motion records. Soil Dyn Earthquake Eng 28:707–716. doi:10.1016/j.soildyn.2007.10.016

Lee VW, Trifunac MD (2010) Should average shear-wave velocity in the top 30 m of soil be used to describe seismic amplification? Soil Dyn Earthquake Eng 30:1250–1258. doi:10.1016/j.soildyn.2010.05.007

Lin YY, Chang KC (2003) An improved capacity spectrum method for ATC-40. Earthquake Eng Struct Dynam 32:2013–2025

Lin YY, Miranda E, Chang KC (2005) Evaluation of damping reduction factors for estimating elastic response of structures with high damping. Earthquake Eng Struct Dynam 34:1427–1443. doi: 10.1002/eqe.499

Luzi L, Puglia R, Pacor F, Gallipoli MR, Bindi D, Mucciarelli M (2011) Proposal for a soil classification based on parameters alternative or complementary to $V_{s,\,30}$. Bull Earthquake Eng 9:1877–1898. doi:10.1007/s10518-011-9274-2

Newmark NM, Hall WJ (1982) Earthquake spectra and design, Earthquake Engineering Research Institute

Penelis G (2008) Thessaloniki 1978 earthquake. Turning point in seismic protection of Greece. Proceedings of 30 years after the Thessaloniki earthquake. Memoirs and perspectives, Polytechnic School of Aristotle University of Thessaloniki, May 2008 (in Greek)

Pitilakis K (2004) Site effects. In: Ansal A (ed) Recent advances in earthquake geotechnical engineering and microzonation. Kluwer Academic Publishers, pp 139–197

Pitilakis K, Anastasiadis A, (1998) Soil and site characterization for seismic response analysis. Proceedings of the XI ECEE, Paris 6–11 Sept. 1998, Inv. Lectures, pp 65–90

Pitilakis K, Anastasiadis A, Raptakis D (1992) Field and laboratory determination of dynamic properties of natural soil deposits. Proceedings of the 10th World Conference on Earthquake Engineering. Madrid 5:1275–1280

Pitilakis K, Makra K, Raptakis D (2001) 2D vs. 1D site effects with potential applications to seismic norms: The cases of EUROSEISTEST and Thessaloniki, Invited Lecture, Proc. XV ICSMGE Satellite Conference on Lessons Learned from Recent Strong Earthquakes, Istanbul, 123–133

Pitilakis K et al (2003) Microzonation study of Thessaloniki, Final report

Pitilakis K, Gazepis C, Anastasiadis A (2004a) Design response spectra and soil classification for seismic code provisions. In: Proceedings of 13th World Conference on Earthquake Engineering, paper n. 2904, Vancouver, B.C., Canada

Pitilakis K, Kappos A, Hatzigogos T, Anastasiadis A, Alexoudi M, Argyroudis S, Penelis G, Panagiotopoulos Ch, Panagopoulos G, Kakderi K, Papadopoulos I, Dikas N, (2004b) Synthesis of the application to Thessaloniki city, RISK-UE Report, Thessaloniki

Pitilakis K et al (2004c) RISK-UE "An Advanced Approach to Earthquake Risk Scenarios with Applications to Different European Towns" Research Project, European Commission, DG CII2001-2004, CEC Contract Number: EVK4-CT-2000-00014, (2001–2004)

Pitilakis K, Gazepis C, Anastasiadis A (2006a) Design response spectra and soil classification for seismic code provisions, in: Proceedings of geotechnical evaluation and application of the seismic Eurocode EC8 2003–2006, ETC-12 Workshop, NTUA Athens, pp 37–52

Pitilakis K et al (2006b) SRMLIFE "Development of a global methodology for the vulnerability assessment and risk management of lifelines, infrastructures and critical facilities. Application to the metropolitan area of Thessaloniki" Research Project, General Secretariat for Research and Technology, 2003–2006, Greece

Pitilakis K, Alexoudi M, Argyroudis S, Anastasiadis A (2006c) Seismic risk scenarios for an efficient risk management: the case of Thessaloniki (Greece), advances in earthquake engineering for Urban Risk Ruction, NATO Science Series, Ed. S.T. Wasti and G. Ozcebe, Springer Netherlands, pp 229–244

Pitilakis K, Anastasiadis A, Kakderi K, Argyroudis S, Alexoudi M (2007a) Seismic zonation, vulnerability assessment and loss scenarios in Thessaloniki, Proceedings of the 2nd Japan-Greece Workshop on Seismic Design, Observation, and Retrofit of Foundations, April 3–4, Tokyo, Japan

Pitilakis K, Anastasiadis A, Kakderi K, Argyroudis S, Alexoudi M (2007b) Vulnerability assessment and risk management of lifelines, infrastructures and critical facilities. The case of Thessaloniki' s Metropolitan area, Proceedings of the 4th International Conference on Earthquake Geotechnical Engineering, Thessaloniki, Greece, 25–28 June

Pitilakis K, Riga E, Anastasiadis A (2012) Design spectra and amplification factors for Eurocode 8. Bull Earthquake Eng 10:1377–1400. doi: 10.1007/s10518-012-9367-6

Pitilakis K, Riga E, Anastasiadis A (2013) New code site classification, amplification factors and normalized response spectra based on a worldwide ground-motion database. Bull Earthquake Eng. doi: 10.1007/s10518-013-9429-4

Priestley MJN, Calvi GM, Kowalsky MJ (2007) Displacement-based seismic design of structures, 658 IUSS Press, Pavia 721

Raptakis D (1995) Contribution to the determination of the geometry and the dynamic characteristics of soil formations and their seismic response, Ph. D. Thesis (in Greek), Dep. of Civil Engineering, Aristotle University of Thessaloniki

Raptakis D, Anastasiadis A, Pitilakis K, Lontzetidis K (1994a) Shear wave velocities and damping of Greek natural soils, Proceedings of the 10th European Conference on Earthquake Engineering, Vienna, vol 1, pp 477–482

Raptakis D, Karaolani E, Pitilakis K, Theodulidis N, (1994b) Horizontal to vertical spectral ratio and site effects: The case of a downhole array in Thessaloniki (Greece), Proceedings of XXIV General Assembly, ESC, Athens, Vol III, pp 1570–1578

Rey J, Faccioli E, Bommer JJ (2002) Derivation of design soil coefficients (S) and response spectral shapes for Eurocode 8 using the European strong-motion database. J Seismolog 6:547–555. doi:10.1023/A:1021169715992

RISK-UE (2001–2004) An advanced approach to earthquake risk scenarios with applications to different european towns. Research Project, European Commission, Project No: EVK4-CT-2000–00014

Rodriguez-Marek A, Bray JD, Abrahamson NA (2001) An empirical geotechnical seismic site response procedure. Earthquake Spectra 17:65–87. doi:10.1193/1.1586167

Steidl JH (2000) Site response in southern California for probabilistic seismic hazard analysis. B Seismol Soc Am 90:S149–S169. doi:10.1785/0120000504

SYNER-G (2009–2012) Systemic seismic vulnerability and risk analysis for buildings, lifeline networks and infrastructures safety gain. FP7 Research Project, European Commission, Project No: 244061, coordinator K. Pitilakis

Trifunac MD (2012) Earthquake response spectra for performance based design—a critical review. Soil Dyn Earthquake Eng 37:73–83.doi: 10.1016/j.soildyn.2012.01.019

Vamvatsikos D, Cornell CA (2002) Incremental dynamic analysis. Earthquake Eng Struct Dynam 31:494–514. doi:10.1002/eqe.141

Yenier E, Sandikkaya MA, Akkar S (2010) Report on the fundamental features of the extended strong motion databank prepared for the SHARE project (v1.0)

Zhao JX, Zhang J, Asano A, Ohno Y, Oouchi T, Takahashi T, Ogawa H, Irikura K, Thio HK, Somerville PG, Fukushima Y, Fukushima Y (2006) Attenuation relations of strong ground motion in Japan using site classification based on predominant period. B Seismol Soc Am 96:898–913. doi:10.1785/0120050122

Incorporating Site Response into Seismic Hazard Assessments for Critical Facilities: A Probabilistic Approach

Ellen Rathje, Menzer Pehlivan, Robert Gilbert and Adrian Rodriguez-Marek

Abstract Probabilistic seismic hazard analysis (PSHA) is now the standard approach for developing earthquake ground motions for seismic design because it takes into account the important sources of uncertainty and variability associated with ground motion prediction. PSHA typically does not account for local site response but rather provides a uniform hazard spectrum that is used as the input into site response analyses. This approach does not account rigorously for the uncertainties in the site response analysis, and therefore the resulting surface motion is associated with an unknown hazard level. Seismic hazard assessments for critical facilities, such as nuclear power plants, require that ground motions be defined for specified hazard levels and also include the effects of local site conditions. To achieve this objective, the uncertainty and variability in the site amplification must be quantified and incorporated into the PSHA. This paper describes the convolution approach used to incorporate site response into PSHA. The main components used in the convolution approach are provided, and the required site characterization and site response analyses are discussed. The convolution approach is illustrated through examples from recent projects.

Introduction

Probabilistic seismic hazard analysis (PSHA) has become the standard approach for developing earthquake ground motions for seismic design because it takes into account the important sources of uncertainty and variability associated with ground

E. Rathje (✉) · R. Gilbert
University of Texas at Austin, Austin, TX, USA
e-mail: e.rathje@mail.utexas.edu

M. Pehlivan
Mueser Rutledge Consulting Engineers, New York, NY, USA
e-mail: menzer.pehlivan@gmail.com

R. Gilbert
e-mail: bob_gilbert@mail.utexas.edu

A. Rodriguez-Marek
Virginia Polytechnic Institute, Blacksburg, VA, USA
e-mail: adrianrm@vt.edu

A. Ansal, M. Sakr (eds.), *Perspectives on Earthquake Geotechnical Engineering*,
Geotechnical, Geological and Earthquake Engineering 37,
DOI 10.1007/978-3-319-10786-8_4

motion prediction. PSHA accounts for all possible earthquake events as well as the variability associated with ground motion prediction to compute a hazard curve, which represents the annual rate of exceedance of different levels of ground shaking. Seismic source characterization defines the possible earthquake events (i.e., magnitudes and locations) and their recurrence, while ground motion characterization predicts the probability distribution of ground shaking for each earthquake event (Reiter 1990). This calculation generally is performed for rock conditions and hazard curves are computed for a range of response spectral periods. The spectral accelerations associated with a design hazard level (e.g., 2 % probability of exceedance in 50 years) are obtained for each spectral period from the corresponding hazard curve and used to generate a uniform hazard spectrum (UHS). This response spectrum commonly is used for design or to select time series for subsequent dynamic analyses.

The UHS derived from a PSHA represents ground shaking for rock conditions and thus is only appropriate for design if the overlying structure is founded on rock. To account for local soil conditions, site response analyses must be performed. The traditional approach for accounting for site response uses the rock UHS as input into site response analyses (Fig. 1). This response spectrum is used to select or spectrally match a suite of time series that, on average, match the UHS. These motions are then used as input into one-dimensional site response analyses to compute the site amplification as a function of period. Uncertainty in the material properties (e.g., shear wave velocity, nonlinear modulus reduction and damping curves) may be taken into account through sensitivity analyses in which two or three possible site profiles are analyzed. The multiple input motions and multiple site profiles result in variability in the computed site amplification. Using the median amplification

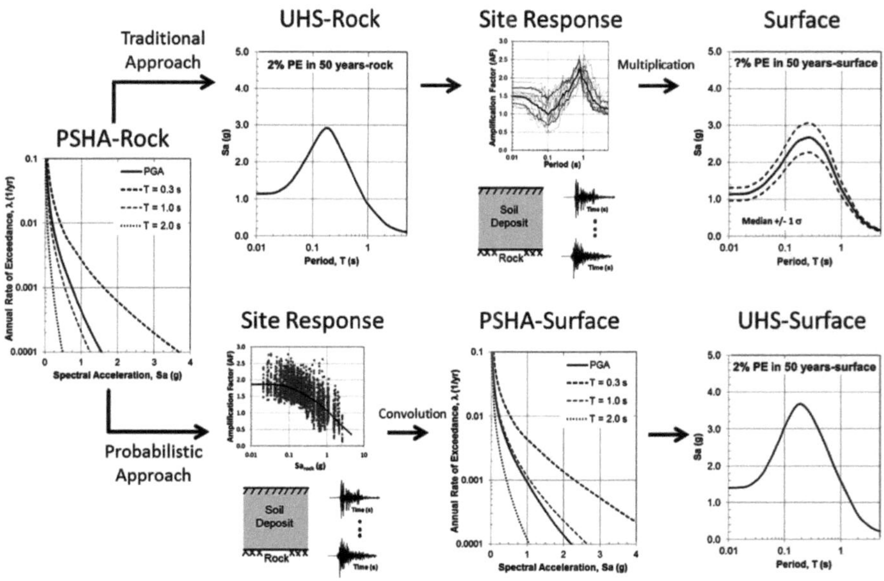

Fig. 1 Approaches to incorporate site response into seismic hazard studies

and its standard deviation, a median surface response spectrum and plus/minus one standard deviation spectra are computed. However, the hazard level associated with the computed surface motion is unknown because the uncertainty in the site amplification is not incorporated within the PSHA.

To generate a surface response spectrum with a known hazard level the site amplification and its uncertainty must be incorporated into the PSHA through a probabilistic approach (Fig. 1). This approach computes hazard curves at the ground surface by considering each rock motion amplitude, its annual rate of occurrence, and the probability that it is amplified to exceed a given surface motion amplitude. The role of site response analysis in this calculation is to define the probability distribution of site amplification given a rock motion amplitude. This site amplification relationship is developed from a large suite of site response analyses performed for a range of input rock motion intensities and multiple site property realizations defined through Monte Carlo simulation. The resulting hazard curves for shaking at the ground surface allow a UHS associated with a known hazard level to be defined for the ground surface of a soil site.

This paper describes the main components required to incorporate site response into PSHA. The site characterization and interpretation needed to develop the statistical models for the Monte Carlo simulation are outlined. The development of a site-specific statistical model for shear wave velocity is illustrated through an example from a recent seismic hazard assessment of a nuclear power plant site in South Africa. The computation of soil hazard curves is demonstrated through a hypothetical site in southern California.

Seismic Hazard Curves Incorporating Site Response

Seismic hazard curves from PSHA show the annual rate of exceedance for different levels of ground shaking for different spectral periods. The PSHA involves a seismic source characterization, which defines the locations and recurrence rates of earthquakes of different sizes, and a ground motion characterization, which defines the probability distribution of ground shaking given an earthquake scenario (i.e., magnitude and distance). This information is used to compute the annual rate of exceedance (λ) for a ground shaking amplitude (z) given all possible earthquake scenarios (Kramer 1996):

$$\lambda_r(z) = \lambda_o \int_m \int_r P[Sa_r > z \mid m, r] \cdot f(m,r)\, dr \cdot dm \tag{1}$$

where λ_r represents λ for rock conditions, Sa is spectral acceleration, $P[Sa_r > z \mid m, r]$ is the conditional probability that the spectral acceleration on rock exceeds z for a given magnitude (m) and distance (r), $f(m,r)$ is the probability density function for the earthquake scenarios, and λ_o is the activity rate (i.e., the rate of earthquakes).

Computed seismic hazard curves generally represent rock conditions because the ground motion prediction equations used in the ground motion characterization are best suited for rock. While more recent ground motion predictions equations (e.g., Boore and Atkinson 2008; Abrahamson and Silva 2008) can predict ground shaking at soil sites, these models do not incorporate comprehensive site information and thus do not account for the detailed, site-specific conditions (i.e. shear wave velocity profile). The most robust approach to computing a seismic hazard curve for the local site conditions involves developing a site-specific ground motion prediction equation that predicts the spectral acceleration at the surface of the soil site (Sa_s) and using this relationship in Eq. (1) to produce (McGuire et al. 2001):

$$\lambda_s(z) = \lambda_0 \int_m \int_r P[Sa_s > z|m,r] \cdot f(m,r) \cdot dr \cdot dm \tag{2}$$

where λ_s represents λ at the surface of a soil site. The development of a site-specific ground motion prediction equation is not trivial. An empirically based model requires a large number of motions to be recorded at the site of interest, which is not realistically possible. An alternative simulation approach involves stochastic ground motion simulations for a range of magnitudes and distances that include local site response and empirical regression on the computed surface motions (McGuire et al. 2001). Again, this approach is not realistic for typical projects.

An approximation to Eq. (2) that uses site amplification relationships was proposed by Bazzurro and Cornell (2004) and is called the convolution approach because it convolves the rock hazard curve with the probability density function for site amplification (AF). Assuming that AF is a function of the spectral acceleration on rock (Sa_r) and is independent of earthquake magnitude and distance, the soil hazard calculation becomes:

$$\lambda_s(z) = \int_{Sa_r} P[AF > \frac{z}{x}|x] \cdot f_{Sa_r}(x) \cdot dx \tag{3}$$

where $P[AF > \frac{z}{x}|x]$ is the probability that AF is greater than $\frac{z}{x}$ given $Sa_r = x$, and $f_{Sa_r}(x)$ is the probability density function for Sa_r normalized by the activity rate. In discrete form, Eq. (3) can be written as:

$$\lambda_s(z) = \sum_{x_j} P[AF > \frac{z}{x_j}|x_j] \cdot p_{Sa_r}(x_j) \tag{4}$$

where $p_{Sa_r}(x_j)$ is the annual probability of occurrence for $Sa_r = x_j$ This probability is obtained by differentiating a previously defined rock hazard curve. Assuming AF is lognormally distributed, $P\left[AF \geq \frac{z}{x}\middle|x\right]$ is computed as:

$$P\left[AF \geq \frac{z}{x}\middle| x\right] = \hat{\Phi}\left(\frac{ln\left[\frac{z}{x}\right] - \mu_{lnAF|x}}{\sigma_{\ln AF|x}}\right) \tag{5}$$

where $\mu_{lnAF|x}$ is the mean value of the natural logarithm of AF given $Sa_r = x$ and $\sigma_{\ln AF|x}$ is the standard deviation of the natural logarithm of AF given $Sa_r = x$. Note that $\widehat{\Phi}(\cdot)$ is the standard complementary Gaussian cumulative distribution function. Parameters $\mu_{lnAF|x}$ and $\sigma_{\ln AF|x}$ are obtained from an AF relationship for the site and are a function of bedrock amplitude x.

In the approach outlined above, the role of the site response analysis is to develop an AF relationship that predicts AF at each spectral period as a function of the spectral acceleration on rock (Sa_r) at the same period. A significant number of site response analyses are performed using a large suite of input motions that spans the range of Sa_r indicated by the rock hazard curves, and the properties of the site (i.e., shear wave velocity, layering, modulus reduction and damping curves) are varied via Monte Carlo simulation. The AF predictive relationship and its standard deviation are developed by statistical regression of the AF data generated by the site response analyses.

An attractive aspect of the convolution approach is that the calculation of the rock hazard curve, which is the domain of the engineering seismologist, is separated from the calculation of the site response, which is the domain of the geotechnical engineer. While the engineering seismologist and geotechnical engineer still need to be familiar with the analyses performed by the other, this approach allows them to focus on their domain expertise.

Input Requirements

The inputs needed for the site response analyses include appropriate input motions and the dynamic material characterization of the site. The specific site characterization and input motion requirements for the convolution approach are described below.

Site Characterization Requirements

The site characterization required for site response analyses performed as part of the convolution approach is significant. While the types of data required for the site characterization are the same as for any site response project (i.e., shear wave velocity versus depth, nonlinear shear modulus reduction and damping curves), the data must be collected such that statistical models can be developed that describe the variation of these properties across the site. These statistical models are used to generate between 60 and 100 realizations of the site properties.

Shear wave velocity is measured to a depth that coincides with the reference rock conditions used to develop the rock hazard curves. The analysis and interpretation of the shear wave velocity data must be informed by the statistical model that will be used to generate velocity profiles for the Monte Carlo simulations. Toro (1995) developed a statistical model that is commonly used to develop shear wave velocity profiles. This model consists of a layering model and a velocity model. The layering model describes the variation of layer thickness (or rate of layer transitions/boundaries) with depth and is used to generate the thicknesses of layers within a simulated velocity profile. The shear-wave velocity model describes the statistical variation of shear-wave velocity within each layer and is used to assign velocities to each of the generated layers.

The layering model developed by Toro (1995) is a non-homogenous Poisson model where the average layer transition rate (λ_t) is depth dependent. The depth dependence is parameterized using an exponential model given by:

$$\lambda_t(d) = c_3(d + c_1)^{c_2} \tag{6}$$

where λ_t is the layer transition rate in m^{-1} and d is depth in m. Toro (1995) analyzed over 500 shear wave velocity profiles from across the U.S. and determined the coefficients in (6) to be $c_1 = 10.86$, $c_2 = -0.89$, and $c_3 = 1.98$. The resulting depth dependence of λ_t is shown in Fig. 2a, along with the corresponding average layer thickness, simply defined as $1/\lambda_t$. The average layer thickness ranges from about 4 m at the surface to about 30 m at a depth of 90 m.

For each simulated shear wave velocity profile, the layer thicknesses are generated using the layering model and then velocities are assigned to these layers. Assuming a log-normal distribution for shear wave velocity, the shear wave velocity of layer i (V_i) is derived from the median Vs at mid-depth of layer i ($\hat{V}_i$), the standard deviation of the natural logarithm of Vs (σ_{lnVs}), and a random standard normal variable for layer i (Z_i) using (Toro 1995):

$$V_i = \exp(\ln[\widehat{V}_i] + Z_i \cdot \sigma_{lnVs}) \tag{7}$$

To model the interlayer correlation between velocities, a first-order, auto-regressive model is used (Toro 1995) to generate the values of Z_i:

$$Z_1 = \varepsilon_1$$

$$Z_i = \rho Z_{i-1} + \varepsilon_i \cdot \sqrt{1 - \rho^2} \quad \text{For } i > 1 \tag{8}$$

where ε_i are independent, normal random variables with zero mean and unit standard deviation and ρ is the interlayer correlation coefficient. To use (7) and (8) to simulate a velocity profile, a model for interlayer correlation is required, as well as a median velocity profile and its standard deviation in natural log units.

Toro (1995) developed a model for ρ that is dependent on depth (d) and the distance between the mid-point of adjacent layers, which is related to layer thickness (th):

$$\rho(d,th)=[1-\rho_d(d)]\cdot\rho_t(th)+\rho_d(d) \tag{9}$$

where $\rho_d(d)$ is the depth-dependent correlation coefficient and $\rho_t(th)$ is the thickness-dependent correlation coefficient. These correlation coefficients are defined as (Toro 1995):

$$\rho_d(d)=\begin{cases}\rho_{200}\cdot\left[\dfrac{d+d_o}{200+d_o}\right]^b & , d\le 200m\\ \rho_{200} & , d>200m\end{cases} \tag{10}$$

$$\rho_t(th)=\rho_o\cdot exp\left(\frac{-th}{\Delta}\right) \tag{11}$$

where ρ_{200}, d_o, ρ_0 and Δ are model parameters. Generally, this model incorporates larger interlayer correlation for deeper layers and thinner layers. Toro (1995) fit this model to data from measured shear wave velocity profiles and estimated the model parameters for generic site conditions defined by site class. Figure 2b plots ρ_d as a function of depth and ρ_t as a function of thickness for the Toro (1995) model parameters for sites with Vs30 = 180 to 360 m/s. Figure 2b shows that ρ_d increases with depth, ρ_t decreases with thickness, and ρ_t approaches zero at a thickness of about 20 m. The expression used to calculate ρ from ρ_d and ρ_t (Eq. 9) results in $\rho\to\rho_t$ as $\rho_t\to 1$ and $\rho\to\rho_d$ as $\rho_t\to 0$.

The median velocity profile and its standard deviation (Figs. 2c and 2d) are obtained from a large number of measured velocity profiles across a site. All of the velocity profiles are plotted together and within each main soil layer the natural logarithm of the velocities are averaged to obtain the median velocity. The standard deviation of the natural logarithm of the velocities (σ_{lnVs}) is also computed for each layer. Typically σ_{lnVs} takes on values between 0.05 and 0.4. Toro (1995) compiled σ_{lnVs} across velocity profiles measured in the same site class and reported values between 0.3 and 0.4. These values generally represent an upper-bound for a site-specific analysis because they were computed from velocity profiles measured at different sites.

Figure 3 shows example velocity profiles generated from Monte Carlo simulation using the site response program Strata (Kottke and Rathje 2008). These velocity profiles were generated from the baseline velocity profile in Fig. 2c, σ_{lnVs} equal to 0.2, and the Toro (1995) interlayer correlation model for sites with Vs30 = 180-360 m/s. The profiles in Fig. 3a do not include variation in the layer thickness, and thus each layer maintains the same thickness as specified in the baseline profile.

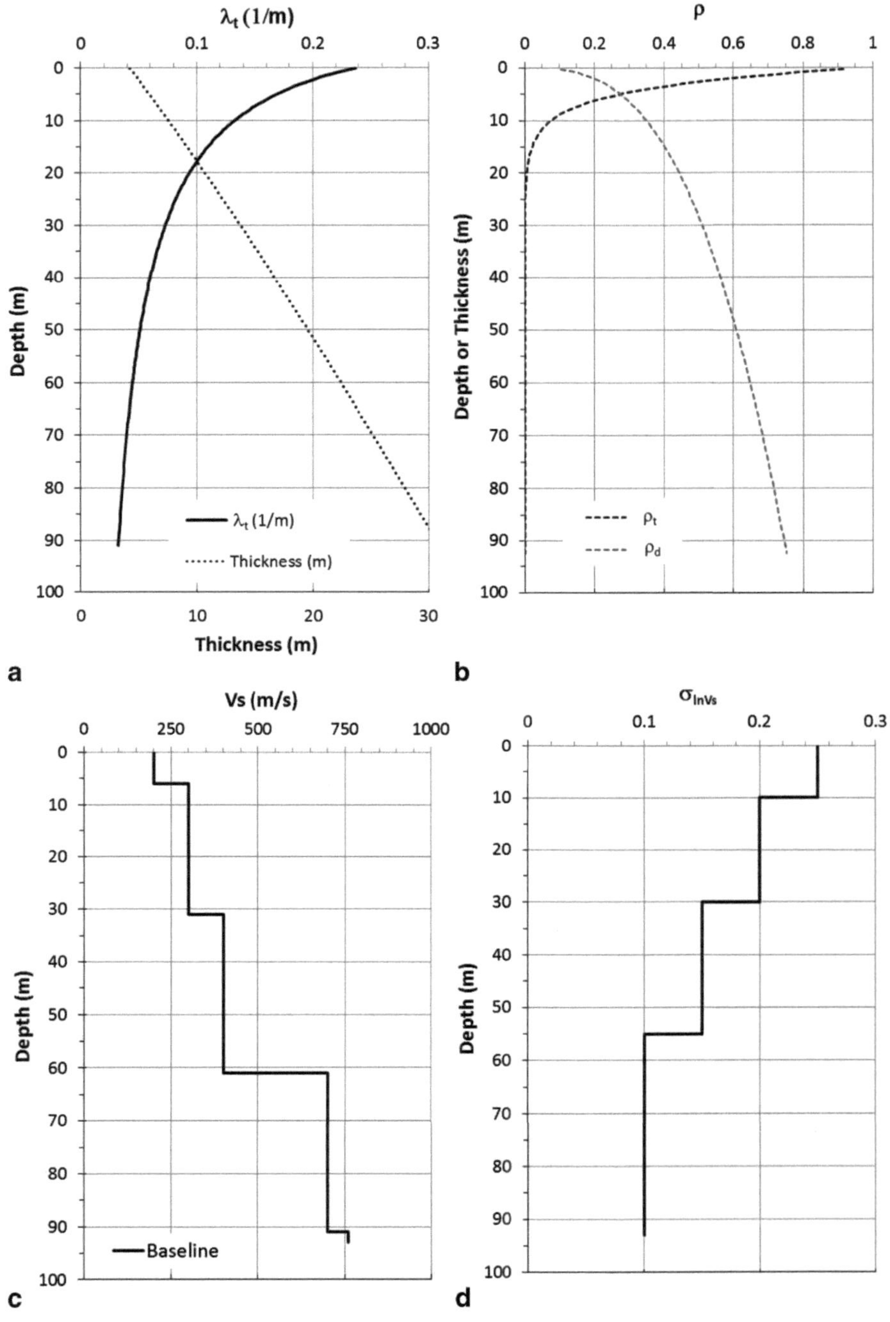

Fig. 2 Site characterization requirements for shear wave velocity. **a** Poisson model for layer thickness, **b** model for interlayer correlation, **c** median shear wave velocity profile, and **d** standard deviation of the natural log of Vs (σ_{lnVs})

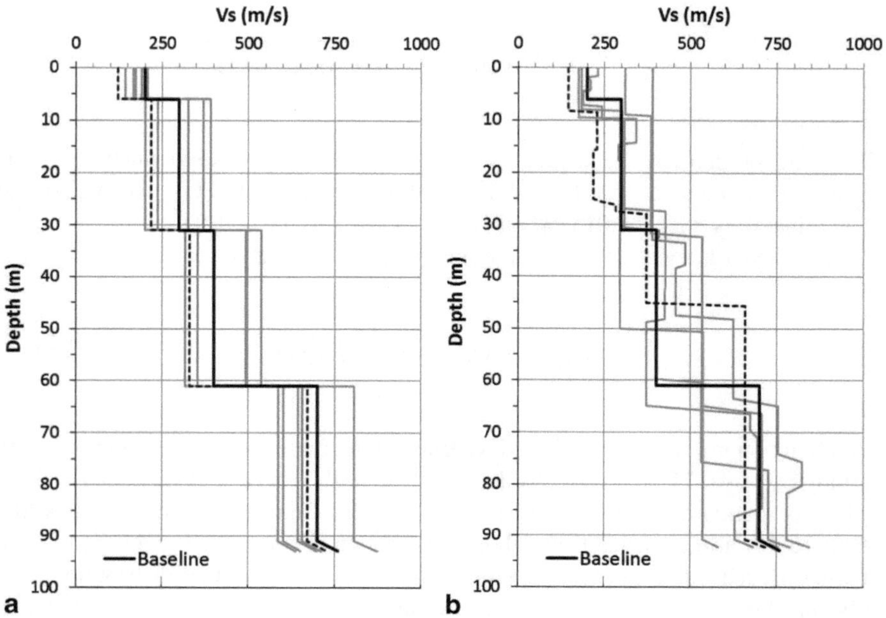

Fig. 3 Statistical generation of shear wave velocity profiles. **a** Profiles in which layer thickness is not varied and **b** profiles in which layer thickness is varied using the Toro (1995) layering model

Because of the interlayer correlation, the velocity profile for a single realization (as demonstrated by the dashed profile in Fig. 3a) has some layers close to the baseline while other layers are further from the baseline. The profiles in Fig. 3b include variation in the layer thickness. In this case the thicknesses of layers for each velocity realization are different from the baseline, as demonstrated by the dashed profile in Fig. 3b. To generate these profiles, the layer thicknesses first are generated and then the baseline velocity profile, standard deviation, and interlayer correlation are used to assign velocities to the layers.

The uncertainty in the nonlinear modulus reduction and damping curves also is taken into account through Monte Carlo simulation. The baseline modulus reduction (G/G_{max}) and damping (D) curves may be assigned from generic models based on soil type, plasticity index, confining pressure, etc. or from laboratory measurements on samples obtained from the site. The Monte Carlo simulation requires the baseline curves as well as an estimate of variability. Darendeli and Stokoe (2001) is one of the few studies that provide an estimate of the standard deviation associated with the nonlinear curves for a given set of generic soil conditions. The Darendeli and Stokoe (2001) model assumes that the parameters G/G_{max} and D are normally distributed at a given strain level, and that the standard deviation for each is a function of the magnitude of G/G_{max} and D, respectively. The standard deviation of the normalized shear modulus (σ_{NG}) is given as:

$$\sigma_{NG} = 0.015 + 0.16 \cdot \sqrt{0.25 - (G/G_{max} - 0.5)^2} \tag{12}$$

This function predicts the largest standard deviation (σ_{NG}=0.095) when G/G_{max} is equal to 0.5, and smaller values of standard deviation when G/G_{max} is smaller and larger than 0.5 ($\sigma_{NG} \sim 0.015$ when G/G_{max} is close to 0.0 or 1.0).

The standard deviation of the damping ratio ($\sigma_{D(\%)}$) is given as:

$$\sigma_{D(\%)} = 0.0067 + 0.78 \cdot \sqrt{D(\%)} \tag{13}$$

This function produces a standard deviation that increases with increasing damping ratio ($\sigma_{D(\%)} \sim 0.75\%$ at D=1% and $\sigma_{D(\%)} \sim 3.0\%$ at D=15%). Because these values of standard deviation are based on samples from many different sites, the standard deviation derived from laboratory tests on site-specific specimens may be smaller. Because the standard deviation model assumes a normal distribution, cut-offs must be applied to avoid unrealistic values (i.e., $G/G_{max}>1.0$, $G/G_{max}<0.0$, and $D<0\%$ are not allowed).

G/G_{max} and D curves are not independent of each other because a modulus curve that is more linear (i.e., above the baseline) is associated with a damping curve that is below the baseline. Thus, generation of G/G_{max} and D curves should include negative correlation between the curves ($\rho_{D,NG}<0.0$). A commonly used value of $\rho_{D,NG}$ is −0.5. The Fig. 4 displays pairs of modulus reduction and damping curves generated using $\rho_{D,NG}=-0.5$. Because of the negative correlation, a modulus reduction curve that is above the baseline curve is associated with a damping curve that is below the baseline (i.e. note the dashed curves in Fig. 4).

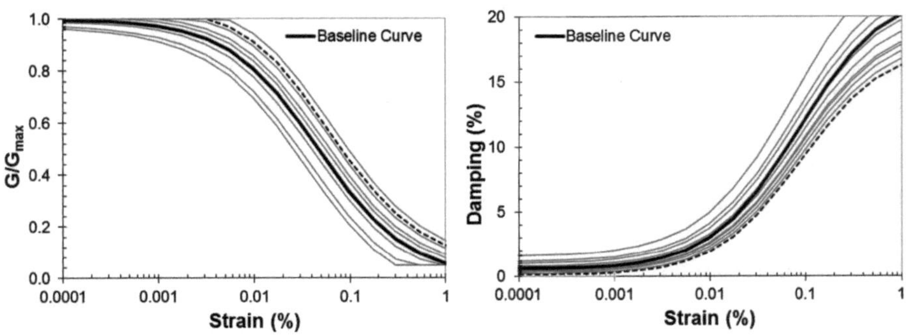

Fig. 4 Statistical generation of shear modulus reduction and damping curves using Darendeli and Stokoe. (2001)

Input Motion Requirements

Because the goal of the site response analyses is to define an amplification relationship that predicts AF as a function of Sa_r, the input motions must span the range of input intensities indicated by the rock hazard curves. For traditional equivalent-linear site response analyses or nonlinear analyses, the input motions are specified by a large suite of input acceleration-time histories. An alternative to traditional equivalent-linear site response analyses is random vibration theory (RVT)-based equivalent-linear site response analysis (Silva et al. 1997; Rathje and Ozbey 2006). This approach to site response analysis still computes the response using the one-dimensional wave equation and strain-compatible soil properties, but the input motions are described by only a response spectrum and duration. As a result, the input motions can be derived directly from ground motion prediction equations or from UHS obtained from the rock hazard curves.

Figure 5 shows the response spectra of the input motions selected for a suite of site response analyses with the spectra color-coded based on the input peak ground acceleration (PGA). Figure 5a shows 130 response spectra from individual acceleration-time histories selected from earthquakes with magnitudes between 6.1 and 7.9 and recorded at distances between 1.5 and 90 km. Figure 5b shows 146 acceleration response spectra generated from the Abrahamson and Silva (2008) ground motion prediction equation for Vs30 = 760 m/s and earthquakes with magnitudes between 5.0 and 8.0 and distances between 0 and 100 km. Some of the events use epsilon greater than zero to achieve the largest intensities. For both suites of input motions, the input Sa_r spans at least two orders of magnitude.

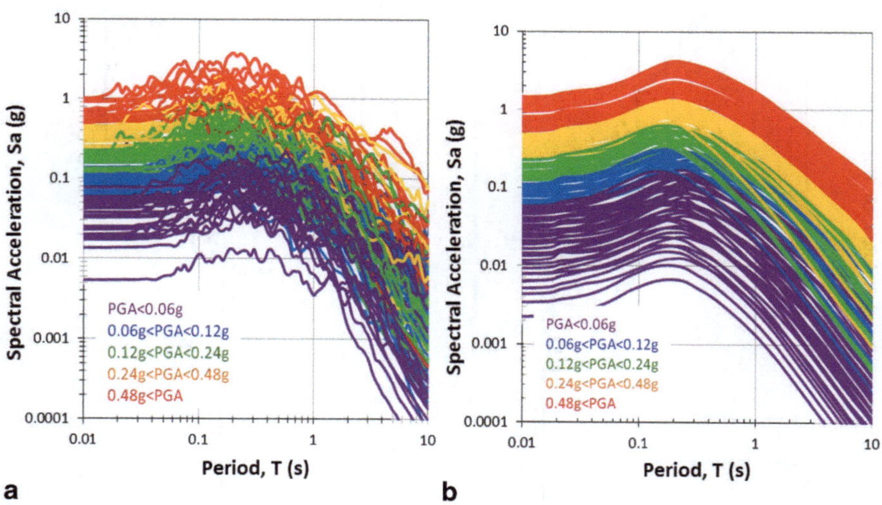

Fig. 5 Input motion selection for **a** time series site response analysis and **b** RVT site response analysis (Pehlivan 2013)

Example Applications

The convolution approach to incorporating site response into PSHA involves both the development of a site-specific statistical model for the site properties and the development of site amplification functions that predict AF as a function of Sa_r. An example of the development of a site-specific statistical model for shear wave velocity is presented from the Thyspunt Nuclear Siting Project in South Africa, while an example of developing site amplification functions and computing soil hazard curves is presented for a hypothetical site in southern California.

Development of a Site-Specific Model for Shear Wave Velocity: South Africa

The Thyspunt Nuclear Siting Project involved a probabilistic seismic hazard assessment for a proposed nuclear power plant site in Thyspunt, South Africa (Bommer et al. 2013). The site is located on the southern coast of South Africa and the geology consists of thinly bedded siltstone, mudstones, and quartzitic sandstone of Palaeozoic age. Shear wave velocities were measured by PS suspension logging within 26 boreholes across the site. The measured shear wave velocity profiles are plotted in Fig. 6a. The site is a rock site with an average shear wave velocity of

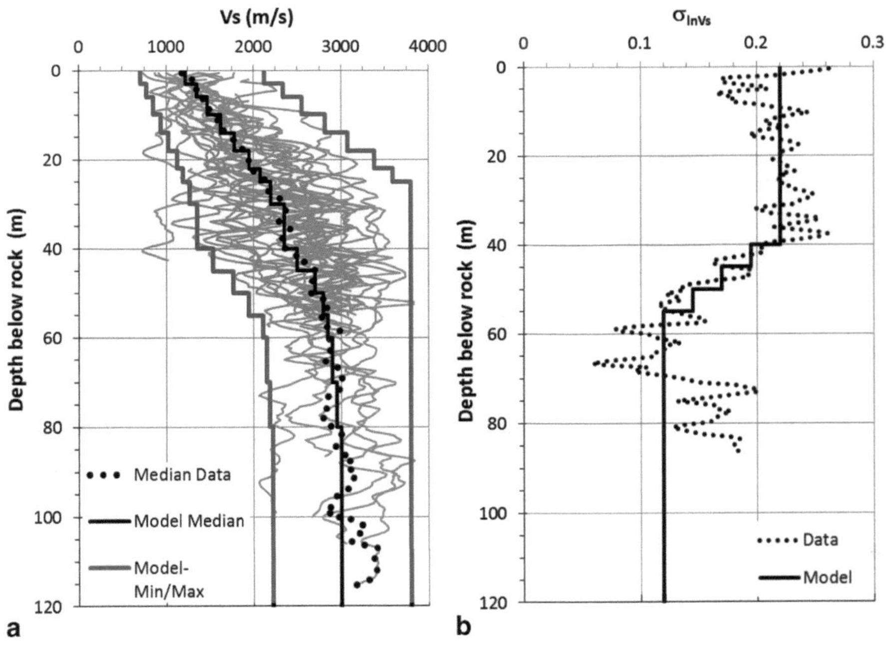

Fig. 6 Site-specific shear wave velocity model for the Thyspunt Nuclear Site in South Africa: **a** median shear wave velocity profile and **b** σ_{lnVs} profile

about 1200 m/s at the surface and the velocity increases to about 3000 m/s at a depth of 80 m. While this is a rock site, site response analyses were required to account for the effects of the materials with shear wave velocities less than the reference rock velocity of 3000 m/s.

The median velocity profile and its standard deviation (in natural log units) as a function of depth are developed based on the measured velocity profiles (Fig. 6). The median Vs profile from the velocity data is shown in Fig. 6a along with the layered model that fits the median. The computed σ_{lnVs} as a function of depth is shown in Fig. 6b for depths that have five or more profiles. σ_{lnVs} is about 0.22 over the top 40 m and decreases to about 0.12 at a depth of 60 m. The model of σ_{lnVs} for use in analysis also is shown in Fig. 6b.

The Monte Carlo simulations of Vs may generate unrealistically large or small velocities for a layer. The lognormal distribution is truncated to avoid these unrealistic values. The proposed maximum and minimum values of Vs for each layer are shown in Fig. 6a. These values were assigned as $\pm 2.5\ \sigma_{lnVs}$ values with an additional physical constraint of a maximum of 3800 m/s. Generally, these proposed maximum and minimum values bound the range of measured Vs values.

The measured shear wave velocity profiles were used to develop a site-specific layering model and a site-specific interlayer correlation model. To develop the layering model, each Vs profile was discretized into layers of constant slowness (i.e., reciprocal of Vs) based on visual inspection of the slowness profile. Slowness was used rather than shear-wave velocity because slowness emphasizes the lower velocity layers, which more strongly influence site amplification. The observed numbers of layer transitions were compiled within moving 10-m bins and plotted versus depth (Fig. 7a). The width of the bin did not significantly change the results. The parameters of the exponential model were obtained from maximum likelihood estimation ($c_1 = 52.0$, $c_2 = -1.19$, $c_3 = 37.7$). The final model is shown in Fig. 7a and indicates λ_t equal to 0.3 m^{-1} near the surface (i.e., average layer thickness of

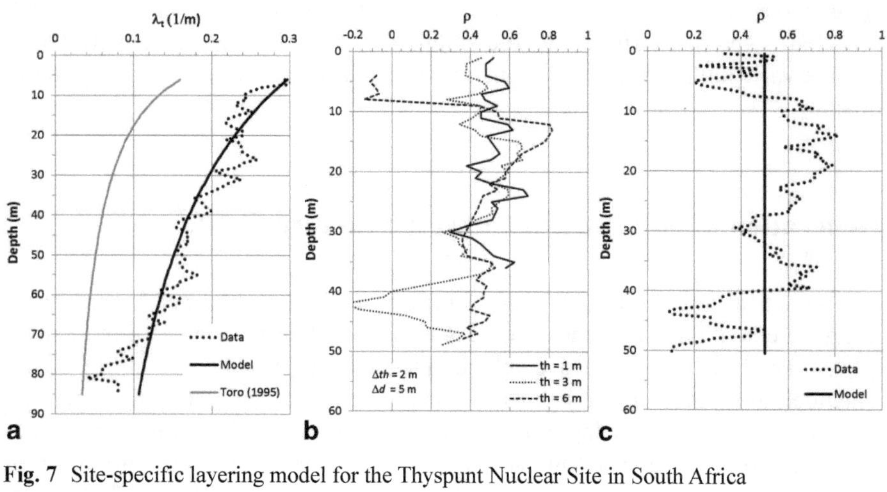

Fig. 7 Site-specific layering model for the Thyspunt Nuclear Site in South Africa

about 3 m) and λ_t equal to 0.11 m^{-1} at a depth of 80 m (i.e., thickness about 9 m). Also shown in Fig. 7a is the generic layering model developed by Toro (1995), which generally displays smaller λ_t (i.e., thicker layers) than the Thyspunt data at all depths. This difference is attributed to the fact that the Toro (1995) model is based on Vs profiles at predominantly soil sites, while the Thyspunt site is a weathered rock site. Weathering will lead to more variations in velocity with depth, which leads to thinner layers.

The interlayer correlation model was developed by first computing the normalized residual (ε) of the natural logarithm of Vs in each layer and then calculating the correlation coefficient between ε values from adjacent layers. To investigate the influence of depth (d) and interlayer separation distance (th) on the interlayer correlation coefficient, ρ was computed for ε values in different bins of d and th, and these bin had widths of Δd and Δth. Bins are required such that enough data are available with which to compute ρ. Values of Δd between 5 and 20 m and values of Δth between 1 m and 9 m were considered. ρ was computed for bins that contained at least 10 pairs of ε.

Figure 7b plots computed ρ values versus depth for different values of th using bin widths of $\Delta th = 2m$ and $\Delta d = 5m$. The influence of th is not strong and a clear depth dependence is not observed. These results support a correlation model that is independent of interlayer separation dependence. Figure 7c shows the computed ρ values versus depth where ρ is computed using all pairs of adjacent layers. Again there is no systematic variation of ρ with depth. Thus, an average ρ value of 0.5 was used.

To illustrate the velocity profiles generated using the models described, a subset of 10 shear-wave velocity profiles are plotted versus depth in Fig. 8. These 10 realizations show that the layer thicknesses vary from profile to profile and the thickness generally increases with depth. The simulated profiles are located both above and below the median Vs model and a single profile may be above the median at some depths and below the median at others. Ten simulated velocity profiles are not enough to ensure that the median and standard deviation of the shear-wave velocity model are captured. Using 60 to 100 simulations both the median velocity profile and the variation of σ_{lnVs} with depth can be captured by the Monte Carlo simulations.

Generation of Soil Hazard Curves: Southern California

A hypothetical soil site in southern California is used to generate soil hazard curves using the convolution approach. These analyses are described in detail in Pehlivan (2013). The baseline shear wave velocity profile of the site is given in Fig. 2c and Darendeli and Stokoe (2001) generic modulus reduction and damping curves are assigned to each velocity layer based on the mean effective confining pressure at the middle of each layer (Pehlivan 2013). Equivalent-linear (EQL) site response analyses are performed using the program Strata (Kottke and Rathje 2008). One

Fig. 8 Examples of generated shear wave velocity profiles for the Thyspunt Nuclear Site in South Africa

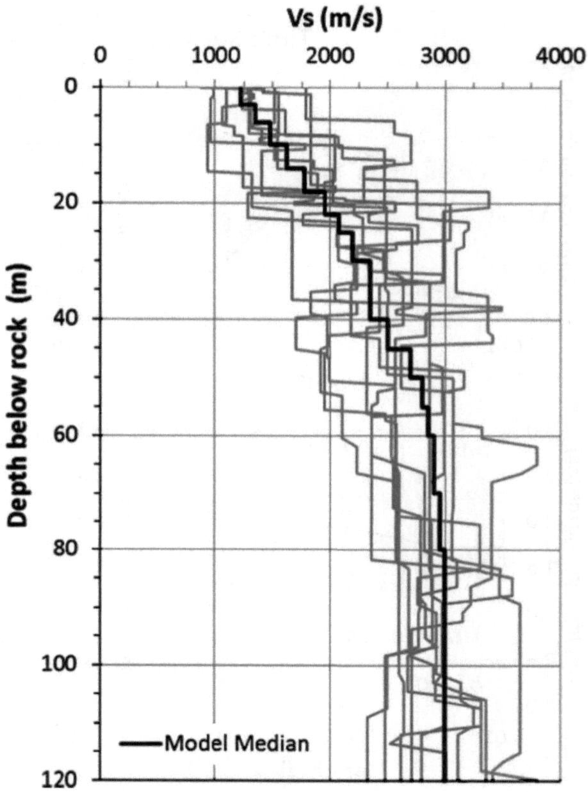

suite of analyses is performed using the time series from Fig. 5a, while a separate suite is performed using the RVT input motions from Fig. 5b.

Figure 9 shows the AF models developed from the site response analyses for two spectral periods. Figure 9a shows AF versus Sa_r for PGA from time series EQL analyses and Fig. 9b shows the PGA results from the RVT EQL analyses. In general the AF decreases with increasing Sa_r due to soil nonlinearity and the increased damping induced by larger intensity shaking. For the time series analyses, the AF data display variability both due to motion-to-motion variability and the shear wave velocity variability introduced by the Monte Carlo simulations. For the RVT analyses, the AF data display variability only due to shear wave velocity variability. The AF data are fit with a higher order polynomial in terms of lnAF and $\ln Sa_r$, and the relationships from time series and RVT are very similar. The standard deviation of the lnAF (σ_{lnAF}) is about 30 % larger for time series than for RVT due to time series analyses also including motion-to-motion variability. However, the difference between the σ_{lnAF} from time series and RVT is not always this large (Pehlivan 2013).

The shape of the AF relationship for T = 1.0 s (Figs. 9c and 9d) is different than for PGA. Amplification increases from about 1.7 to 2.0 as Sa_r increases from

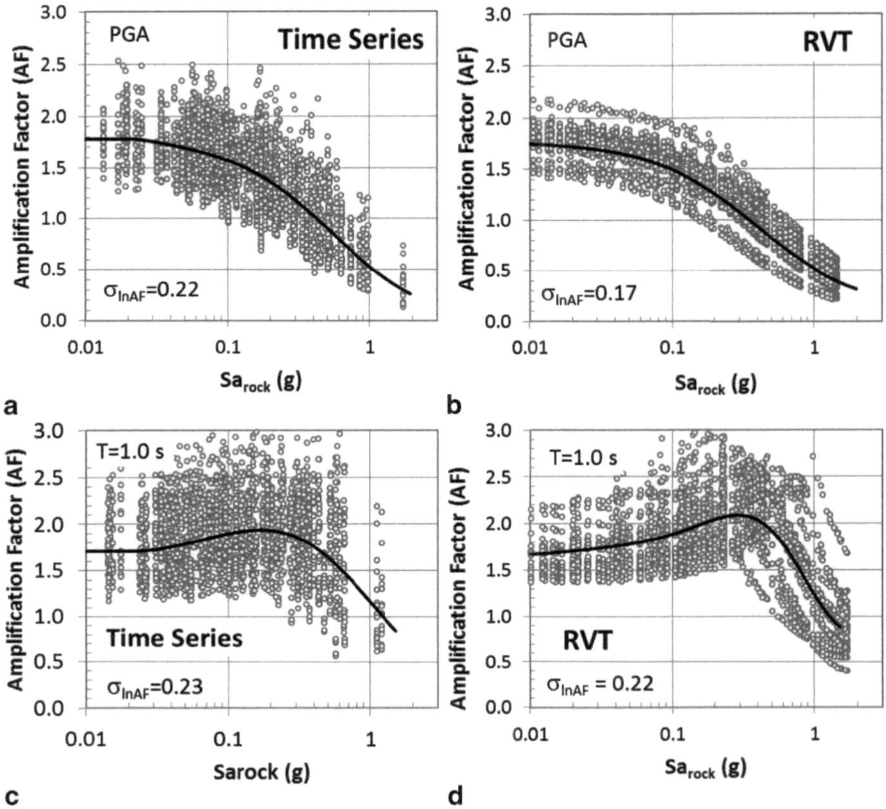

Fig. 9 Amplification models for PGA and T = 1.0 s developed for southern California site using time series and RVT site response analyses

0.01 to 0.3 g, and then amplification decreases as Sa_r increases further. This shape of the amplification relationship is a result of the spectral period of 1.0 s being slightly larger than the small-strain natural period of the site (~0.8 s for this site). At this spectral periods, the site period elongation associated with soil nonlinearity produces an increase in AF with increasing Sa_r. As Sa_r increases further, the site period elongates beyond the site period under consideration and AF decreases. The σ_{lnAF} values for T = 1.0 s are comparable with those for PGA. For this site the largest values of σ_{lnAF} are around 0.5 and occur for periods between 0.2 and 0.4 s (Pehlivan 2013).

For both the time series and RVT analyses, the scatter in the data varies with Sa_r. σ_{lnAF} are computed for different bins of Sa_r to quantify the change in σ_{lnAF} with Sa_r and these values are shown in Fig. 10 for PGA. For time series analysis, σ_{lnAF} increases from 0.15 to 0.33 as Sa_r increases. The overall σ_{lnAF} for the entire time series dataset is 0.22 (Fig. 9a). For RVT analysis, which has an overall σ_{lnAF} of 0.17 (Fig. 9b), σ_{lnAF} increases from 0.10 to 0.22 as Sa_r increases. For both time series and RVT, the σ_{lnAF} for the largest input intensity range is larger than the overall σ_{lnAF} computed for the entire dataset.

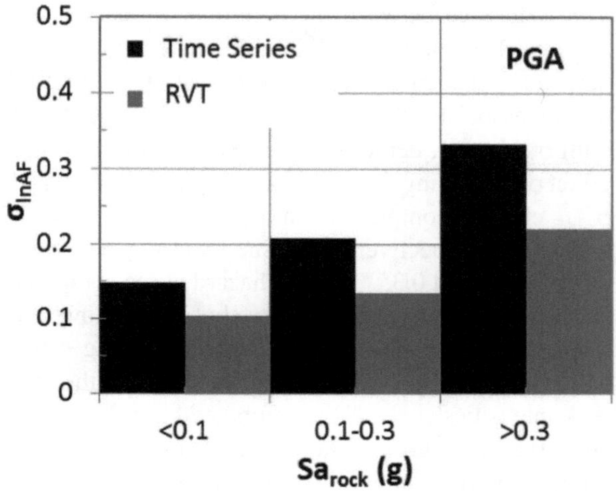

Fig. 10 σ_{lnAF} as a function of Sa_r for PGA models developed for southern California site using Time Series and RVT site response analyses

The AF relationship and its σ_{lnAF} both play a role in the calculation of the surface soil hazard curve. Figure 11 shows computed surface hazard curves for PGA using the time series AF relationship from Fig. 9a and different values of σ_{lnAF}. Figure 11a plots probabilistic soil hazard curves for σ_{lnAF} between 0.1 and 0.5 along with the deterministic soil hazard curve (i.e., σ_{lnAF}=0.0). All of the soil hazard curves are similar for λ greater than 0.01 1/year (i.e., smaller ground motion amplitudes) and they start to deviate at smaller λ (i.e., larger ground motion amplitudes).

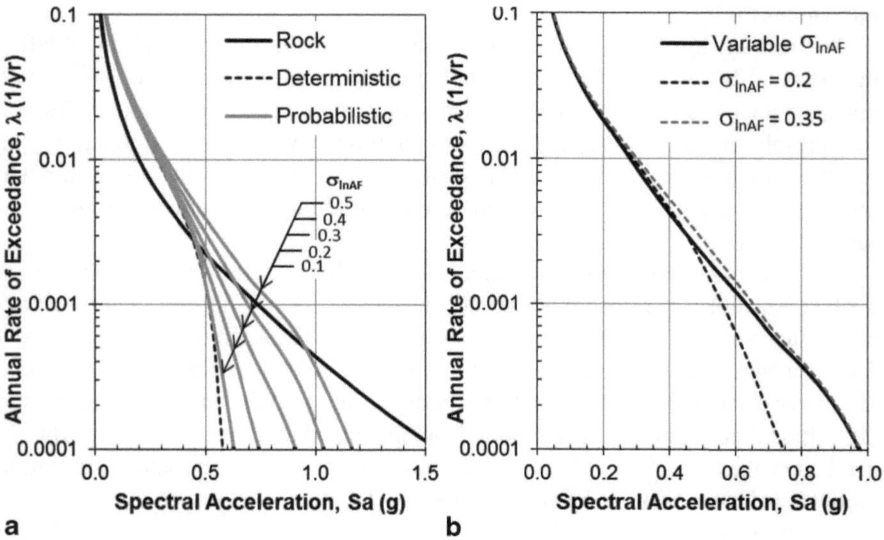

Fig. 11 Influence of σ_{lnAF} on the computed surface hazard curves for PGA for southern California site

The difference between the deterministic curve and the probabilistic curve with σ_{lnAF}=0.1 is minimal, but the difference is more significant for σ_{lnAF} greater than 0.1. At λ=0.0004 1/yr (i.e., 2% probability of exceedance in 50 years) the ground motion associated with σ_{lnAF}=0.5 is 75% larger than the ground motion associated with σ_{lnAF}=0.1, demonstrating the importance of properly quantifying σ_{lnAF}. The effect of modelling the variation of σ_{lnAF} with Sa_r is shown in Fig. 11b. The variable σ_{lnAF} curve is computed with σ_{lnAF}=0.15, 0.2, and 0.35 for Sa_r<0.1 g, 0.1–0.3 g, and>0.3 g, respectively. Also shown in Fig. 11b are curves computed for constant σ_{lnAF} of 0.2 and 0.35. The soil hazard curve computed with variable σ_{lnAF} initially is similar to the σ_{lnAF}=0.20 curve but then follows the σ_{lnAF}=0.35 curve. At larger intensities (i.e., smaller λ) the hazard is being driven predominantly by amplification of the larger rock intensities and thus the σ_{lnAF} associated with the larger rock intensities controls the computed hazard curve. Because engineering design is concerned with ground motion levels associated with smaller λ, it is important to properly model the variation of σ_{lnAF} with Sa_r.

Conclusions

This paper describes a probabilistic approach to incorporate the dynamic response of local soil conditions into probabilistic seismic hazard analysis. This convolution approach modifies the rock hazard curve at each spectral period to account for the site response and its uncertainty, such that a hazard curve for the soil surface is computed. Performing this calculation for a range of spectral periods allows a uniform hazard spectrum to be defined for the ground surface of a soil site for a known hazard level.

The role of the site response analysis in the convolution approach is to define the probability distribution of site amplification given a rock motion level. Site response analyses are performed for a range of input rock motion intensities and for multiple site property realizations defined through Monte Carlo simulation. The results from these analyses are used to develop a site amplification relationship that predicts the median site amplification and its variability as a function of the input rock intensity.

The site characterization requirements for the Monte Carlo simulations were described and illustrated through a model developed for a proposed nuclear power plant site in South Africa. The development of the site-specific statistical model for shear wave velocity demonstrated that generic models for layer thickness and interlayer correlation coefficient may not apply due to differences in geology and depositional environment. The model developed for the South Africa site had thinner layers than the Toro (1995) model, and the developed interlayer correlation model was independent of layer thickness and depth.

A hypothetical site in southern California was used to develop site amplification models, generate soil hazard curves, and demonstrate the influence of the site amplification model on the computed soil surface hazard curves. In particular, the standard deviation of the amplification factor (σ_{lnAF}) plays an important role in defining the soil surface hazard curve. Accurately defining the variation of σ_{lnAF} with input intensity is important to properly quantifying the ground motion at hazard levels associated with design levels.

Acknowlegments The research presented in this paper was partially support by funding from the Nuclear Regulatory Commission under grant NRC-04-09-134. This support is gratefully acknowledged. We would also like to acknowledge the contributions from the Ground Motion Characterization team from Thyspunt Nuclear Siting Project: Prof. Julian Bommer, Prof. Frank Scherbaum, and Dr. Peter Stafford.

Referencess

Abrahamson N, Silva W (2008) Summary of the Abrahamson and Silva NGA ground-motion relations. Earthq Spectra 24(1):67–97

Bazzurro P, Cornell CA (2004) Nonlinear soil-site effects in probabilistic seismic-hazard analysis. Bull Seismol Soc Am 94(6):2110–2123

Bommer JJ, Coppersmith KJ, Hattingh E, Nel AP (2013) An application of the SSHAC Level 3 process to the probabilistic seismic hazard assessment for the thyspunt nuclear site in South Africa. Transactions, 22nd conference on structural mechanics in reactor technology, San Francisco—August 18–23

Boore DM, Atkinson GM (2008) Ground-motion prediction equations for the average horizontal component of PGA, PGV, and 5 %-damped PSA at spectral periods between 0.01 s and 10.0 s. Earthquake Spectra 24(1), 99–138

Darendeli MB, Stokoe II KH (2001) Development of a new family of normalized modulus reduction and material damping curves. Geotechnical Engineering Report GD01-1, University of Texas

Kottke A, Rathje EM (2008) Technical manual for strata. Rep. No. 2008/10, Pacific Earthquake Engineering Research Center, Berkeley

Kramer SL (1996) Geotechnical earthquake engineering. Prentice–Hall, Upper Saddle River

McGuire R, Silva WJ, Costantino CJ (2001) Technical basis for revision of regulatory guidance on design ground motions: hazard-and risk consistent ground motion spectra guidelines. U.S. Nuclear Regulatory Commission Report NUREG/CR-6728

Pehlivan M (2013) Incorporating site response analysis and associated uncertainties into the seismic hazard assessment of nuclear facilities. PhD Dissertation, University of Texas, pp 254

Rathje EM, Ozbey MC (2006) Site-specific validation of random vibration theory-based seismic site response analysis. J Geotech Geoenviron Eng 132(7):911–922

Reiter L (1990) Earthquake hazard analysis: issues and insights. Columbia University Press, New York, pp 254

Silva WJ, Abrahamson NA, Toro G, Costantino C (1997) Description and validation of the stochastic ground motion model. Report for Brookhaven National Laboratory, Upton

Toro GR (1995) Probabilistic models of site velocity profiles for generic and site-specific ground-motion amplification studies. Technical Rep. No.779574, Brookhaven National Laboratory, Upton

Stress Scaling Factors for Seismic Soil Liquefaction Engineering Problems: A Performance-Based Approach

K. Önder Çetin and H. Tolga Bilge

Abstract Most of the widely used seismic soil liquefaction triggering methods propose cyclic resistance ratio (CRR) values valid at the reference normal effective stress ($\sigma'_{v,0}$) of one atmosphere and zero static shear stress ($\tau_{st,0}$) states. Then, a series of correction factors are applied on this reference CRR, for the purpose of assessing the variability due to normal effective and static shear stress states (i.e. K_σ and K_α corrections) acting on the horizontal plane. In the literature, a number of relationships suggested to be used as part of liquefaction triggering methodologies. However, the presence of a wide range of correction factors, some of which with even contradicting trends, suggests that more research needs to be performed to reduce this uncertainty. Additionally, these stress correction factors are treated as being strain-independent and are applied disjointedly to CSR or CRR. The main motivation of this on-going study is defined as to develop a strain-dependent semi-empirical framework to assess combined effects of i) $\sigma'_{v,0}$, ii) $\tau_{st,0}$ acting on the plane, where cyclic shear stresses either produce iii) shear stress reversal or not. For this purpose, cyclic simple shear tests were performed on laboratory reconstituted sand samples. Additionally, cyclic test data were compiled from the available literature. On the basis of probabilistic assessment of this data, a unified correction scheme, which incorporates the interdependent effects of both overburden and static shear stresses along with the degree of cyclic shear stress reversal, has been developed.

Introduction

Most of widely used seismic soil liquefaction triggering relationships propose cyclic resistance ratio (CRR) values valid at the reference stress state of one atmosphere vertical effective and zero static shear stresses. Then, a series of correction factors is applied on this reference cyclic resistance (or stress ratio) for the purpose

K. Ö. Çetin (✉)
Middle East Technical University, Ankara, Turkey
e-mail: ocetin@metu.edu.tr

H. T. Bilge
GeoDestek GeoEng. & Consultancy Services Ltd. Co., Ankara, Turkey
e-mail: htbilge@geodestek.com

A. Ansal, M. Sakr (eds.), *Perspectives on Earthquake Geotechnical Engineering*,
Geotechnical, Geological and Earthquake Engineering 37,
DOI 10.1007/978-3-319-10786-8_5

of assessing the variability due to vertical (or confining) effective and static shear stresses (i.e. K_σ and K_α corrections) acting on the horizontal plane.

Following the pioneer study of Seed (1983), a number of recommendations (e.g.: Harder and Boulanger (1997), Hynes and Olsen (1999), Vaid et al. (2001), Boulanger (2003a,b), Çetin et al. (2004) etc.) were presented aiming to clarify the use of these correction factors. However, the presence of wide range of correction factors, some of which exhibit contradicting trends, suggests that more needs to be done to close the gap, and hopefully to resolve existing conflicts. Inspired from this, the main motivation is defined as to develop a semi-empirical framework to assess combined effects of (i) normal effective overburden stress (σ'_{v0}), ii) static (τ_{st}) shear stress acting on the plane, where cyclic shear stresses may produce iii) shear stress reversals or not. For this purpose, an extensive laboratory testing program was executed. A total number of 150 cyclic laboratory simple shear tests were performed on Monterey, Nevada and Morro Bay very "loose" to very "dense" sands, subjected to a wide range of confining, static and cyclic shear stress conditions. In addition, after a careful literature review, 136 cyclic laboratory simple shear tests from Wu et al. (2003) and Boulanger et al. (1991) databases were compiled and studied. On the basis of the compiled experimental data, a unified stress correction scheme, which enables to assess the combined effects of overburden (σ'_{v0}) and static (τ_{st}) shear stresses along with the degree of shear stress reversal, will be introduced.

Following this introduction, a brief review of the available literature will be presented with special emphasis on the strength and limitations of currently available models.

Existing Stress Correction Factors

Since the pioneering study of Lee and Seed (1967a, b), which focused on the effects of confining and static shear stresses on cyclic response of saturated sandy soils, almost five decades have passed. Seed et al. (1984) introduced the first set of K_σ and K_α corrections to be used in liquefaction triggering assessments. Since then, as it will be presented later in this chapter, despite their strain- and inter-dependent behavior, these correction factors have continued to be treated strain-independently and disjointedly. Consistent with this invalid trend, the discussion of the literature regarding these correction factors, will also be presented separately.

K_σ *Corrections*

Monotonic undrained shear strength of a normally consolidated saturated soil increases with increasing consolidation (or confinement) stresses. Same response is also valid under cyclic loading; however, cyclic shear stress normalized with confining stress, referred to as cyclic stress ratio, tends to decrease as slope of failure envelope flattens out with increasing confining stresses.

Lee and Seed's (1967a) study is considered as one of the pioneering efforts, which addressed that the peak deviator stress to cause "failure" (defined as r_u of 100 and 20% double amplitude (DA) axial strain) increased with increasing effective consolidation stress. The basis of this conclusion is founded on cyclic triaxial tests performed on reconstituted Sacramento River sand specimens of a wide range of relative density (D_R). Seed and Peacock (1971)'s seismic soil liquefaction triggering chart, widely referred to as the pioneering study defining the liquefaction/non-liquefaction boundary as a function of relative density and cyclic stress ratio, had no reference to K_σ effects. Later, Castro and Poulos (1977) evaluated the shear strength of reconstituted sand samples as a function of cyclic mobility ratio (CMR), which was defined as the ratio of half of cyclic deviator stress to initial effective confinement stress (i.e., $\sigma_d/2\sigma'_{c,0}$). It was reported that the relationship between strength and σ'_{c0} was linear and function of D_R.

Seed and Idriss (1982), based on field performance data, developed a chart solution expressing the relationship between overburden corrected standard penetration test blow count, N_1, and cyclic stress ratio (CSR) triggering liquefaction (i.e: cyclic resistance ratio, CRR), for a moment magnitude (M_w) 7.5 event. As shown in Fig. 1, mostly with the intend of extending SPT-based cyclic resistance ratio values valid at the reference stress state, to significantly higher confining stresses, especially useful for assessing the liquefaction triggering performance of dam foundations, Seed (1983) introduced the first set of K_σ correction factors. In 1984, Seed et al. presented their widely used relationship between procedure and overburden-corrected SPT blow counts, $N_{1,60}$ and CSR triggering liquefaction during a $M_w = 7.5$ event. Consistent with Seed (1983) and Seed et al. (1984), with the argument that K_σ

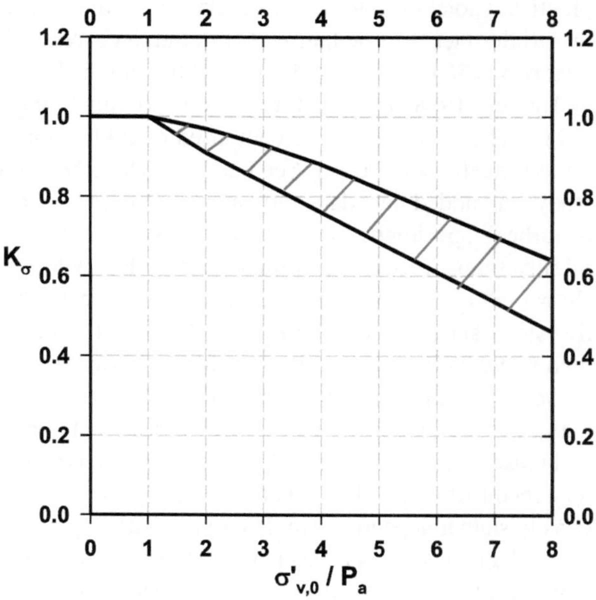

Fig. 1 Relationship between $\sigma'_{v,0}$ and K_σ (after Seed 1983)

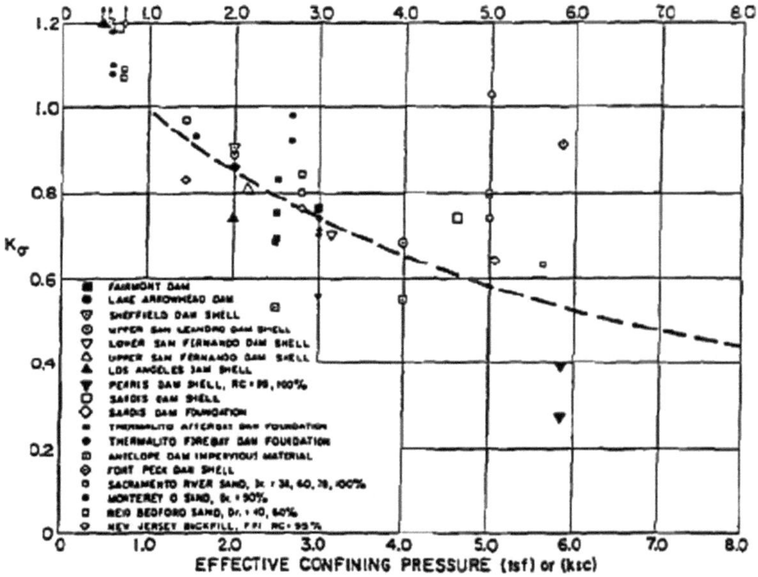

Fig. 2 Relationship between $\sigma'_{v,0}$ and K_σ (Seed and Harder 1990)

corrections were not applied when assessing liquefaction triggering case histories (i.e.: back analysis), which establish the basis of liquefaction triggering relationship, consistently, it was recommended not to apply K_σ corrections for liquefaction engineering assessment of soil layers (i.e.: forward analysis) with a vertical effective stress less than 1 atmosphere. Unfortunately, this—at first glance consistent and practical choice- produced unconservatively biased predictions for deep soil layers due to the fact that median vertical effective stress of liquefaction triggering case histories is 56 kPa (or 65 kPa if weighting applied, Çetin 2000) but not 100 kPa.

Harder (1988) proposed a relationship employing a similar approach to Seed (1983), and used a database composed of cyclic laboratory test data covering sandy to gravelly soils. Later, Seed and Harder (1990) re-evaluated the results of this study, and added new data corresponding to both laboratory reconstituted and "undisturbed" specimens. It was reported that "undisturbed" specimens were consist of silty sands, recovered in thin-walled tubes with stationary piston samplers. As shown in Fig. 2, K_σ test results, this time, were presented to cover the vertical effective stress state less than 1 atm., which suggested a K_σ value of 1.2 or higher at median vertical effective stress state of liquefaction triggering case history database.

Scatter in the test data, especially at larger stresses is concluded to be due to heterogeneity in: (i) soil type and density state ranging from silt to sand and gravel, and loose to dense, respectively, (ii) sample preparation technique including laboratory reconstituted and "undisturbed" samples. Despite such a scatter, proposing a density state independent single curve is listed as one of the drawbacks of the study.

Vaid and Thomas (1995) performed cyclic triaxial tests on reconstituted Fraser River sand at $D_R = 19$, 40 and 59%. It was concluded that at the loosest state, an

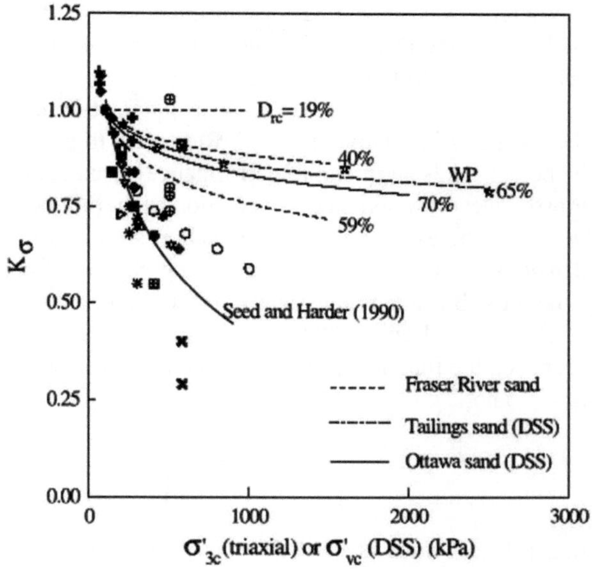

Fig. 3 Relationship between $\sigma'_{v,0}$ and K_σ (Vaid and Thomas 1995)

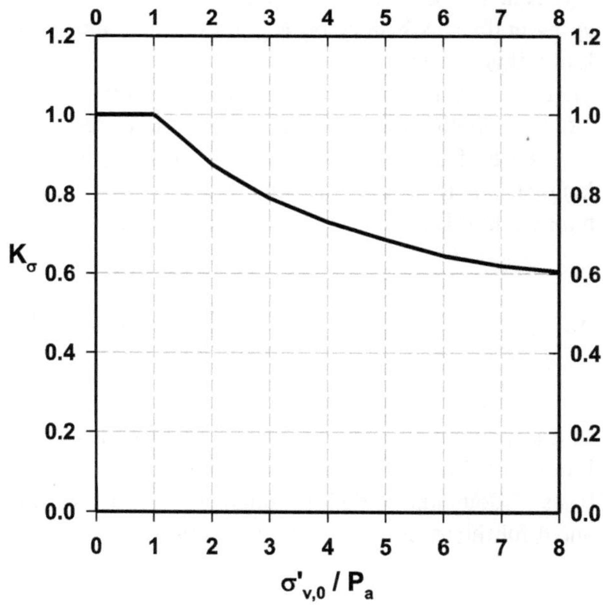

Fig. 4 Relationship between $\sigma'_{v,0}$ and K_σ (after Harder and Boulanger 1997)

increase in $\sigma'_{c,0}$ has almost no effect on CRR; whereas, this effect becomes more pronounced with increasing relative density state. The proposed K_σ correction factors by Vaid and Thomas (1995) are presented in Fig. 3, again suggesting K_σ corrections exceeding 1.0 for vertical effective stresses less than 1 atm.

As part of NCEER Workshop, Harder and Boulanger (1997) re-visited earlier efforts, and proposed the curve given in Fig. 4, relating K_σ and effective confinement pressure. It was stated that for confining stresses higher than 1 atm., K_σ decreases with increasing D_R. The argument for proposing a relative density independent K_σ scheme was reported as "the decrease in K_σ with increasing relative density is likely to be offset by the decrease in post liquefaction deformability that accompanies the increases in relative density". Reference to post liquefaction deformations can be perceived as an intend to performance-based approach; however, back then and mostly even now, existing liquefaction triggering boundary curves had no association with the expected post liquefaction strain levels. Hence, for varying relative density soils, a reliable quantitative comparison of post liquefaction performances is not possible, and this remained as the limitation of the study.

Based on Olsen's (1984) earlier experimental findings, Hynes and Olsen (1999) proposed the functional form presented in Eq. 1, where K_σ is defined as a function of $\sigma'_{v,0}$ and D_R.

$$K_\sigma = \left(\frac{\sigma'_{v,0}}{P_a}\right)^{f-1} \tag{1}$$

In Eq. 1, "f" varies as a function of D_R and with increasing D_R, f also increases, and consequently produces severe reductions in K_σ values at higher confining stresses. Hynes and Olsen also addressed the strong dependency of soil deposition method, aging and stress history on K_σ, but no quantitative resolution was recommended. Later, this correction scheme was adopted by Youd et al. (2001), in their state-of-the-art study, and "f" values of 0.8, 0.7 and 0.6 were recommended to be used for "loose", "medium-dense" and "dense" sands, respectively.

Inspired from Konrad's (1988) relative state parameter index term (ξ_R), Boulanger (2003a) proposed a critical state theory-based framework. ξ_R was defined as a function of D_R, mean effective confining stress (p') and grain type (Q) as follows:

$$\xi_R = \frac{1}{Q - \ln\left(\frac{100p'}{P_a}\right)} - D_R \tag{2}$$

In Eq. 2, consistent with Bolton (1986), Q values of 10 for quartz and feldspar, 8 for limestone, 7 for anthracite and 5.5 for chalk, are recommended to be used. On the basis of Boulanger (2003a), Boulanger and Idriss (2004) proposed the use of Eqs. 3 and 4 for the assessment of K_σ corrections.

$$K_\sigma = 1 - C_\sigma \cdot \ln\left(\frac{\sigma'_v}{P_a}\right) \tag{3}$$

$$C_\sigma = \frac{1}{18.9 - 17.3 \cdot D_R} \le 0.3 \tag{4}$$

Boulanger (2003a) reported that the proposed theoretical framework reduced the conservatism imposed at high overburden stresses.

Aside from these experimentally-based approaches, on the basis of statistical assessment of liquefaction triggering case histories, (e.g., Çetin et al. 2004; Moss et al. 2006) proposed alternative K_σ correction factors. In Eq. 5, K_σ correction factor by Çetin et al. (2004) is presented. Despite the efforts to incorporate D_R (or $N_{1,60}$) into K_σ limit state model, statistical assessment results refused the dependency.

$$K_\sigma = \exp\left(-0.278 \cdot \ln\left(\frac{\sigma'_v}{P_a}\right)\right) \tag{5}$$

Most recently, Bilge and Çetin (2011) restudied earlier efforts, and addressed strain dependent nature of K_σ corrections. On the basis of cyclic laboratory test results compared within a probabilistic framework, a set of strain (or r_u) compatible K_σ corrections was proposed. In Eq. 6, the K_σ correction factor recommended for a performance level expressed by 6% double amplitude cyclic shear strain level, is presented along with ± one standard deviation of model error.

$$\ln(K_\sigma) = \ln\left[\left(\frac{\sigma'_{v0}}{P_a}\right)^{-0.0047 \cdot D_R}\right] \pm 0.044 \tag{6}$$

Figure 5 comparatively summarizes major K_σ correlations as proposed by Youd et al. (2001), Boulanger and Idriss (2004), Çetin et al. (2004) along with Bilge and Çetin (2011). A consistent trend is observed among the K_σ corrections of Youd et al. (2001), Çetin et al. (2004) and Bilge and Çetin (2011).

As a summary, in the current state of practice, there exists a consensus (i) regarding the need to apply K_σ correction for stresses other than 1.0 atm., (ii) for stresses

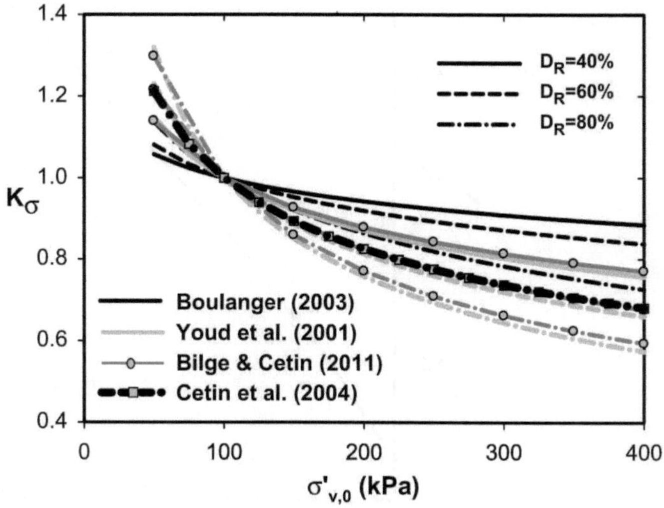

Fig. 5 Comparison among major K_σ correction factors

less than and greater than 1.0 atm., K_σ corrections are greater than and less than 1.0, respectively. However, quantification of K_σ correction still remains as a controversial issue, and for critical projects site specific solutions are recommended (Harder and Boulanger 1997; Youd et al. 2001).

K_α Corrections

Since mid-60's, owing to its significant importance on cyclic straining and corollary excess pore water pressure generation responses of cohesionless soils, a number of experimental studies have been performed to quantify the effects of presence of initial static shear stress (τ_{st}) on the plane, where later cyclic shear stresses will be applied. Despite this early awareness, it has still remained as one of the least understood and quantified topics of liquefaction engineering.

On the basis of anisotropically consolidated cyclic triaxial tests performed on sand specimens, Lee and Seed (1967b) stated that the strength of specimens increased with increasing i) D_R, effective consolidation stress, and effective principle stress ratio (i.e. $K_{c,0}=\sigma'_1/\sigma'_3$), and decreased if stress reversals occured during cyclic loading. The dependence of excess pore water pressure response on shear stress reversal was also referred to, which outlined the most of current state of knowledge.

Seed (1983) introduced a correction factor, K_α, which is recommended to be used to adjust cyclic resistance ratio, given for level or almost level sites. Seed (1983) also introduced a normalized static stress ratio term α, to quantify the level of initial static shear stress, as given in Eq. 7, and proposed a the graphical solution presented in Fig. 6 to assess K_α as a function of α.

Fig. 6 Relationship between α and K_α (after Seed 1983)

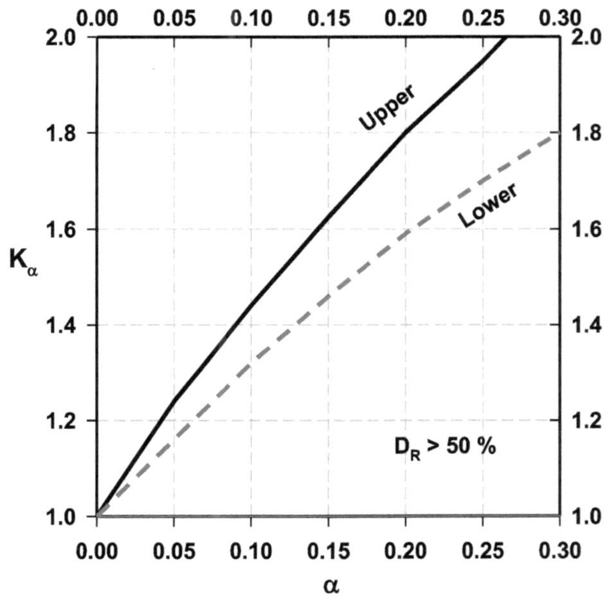

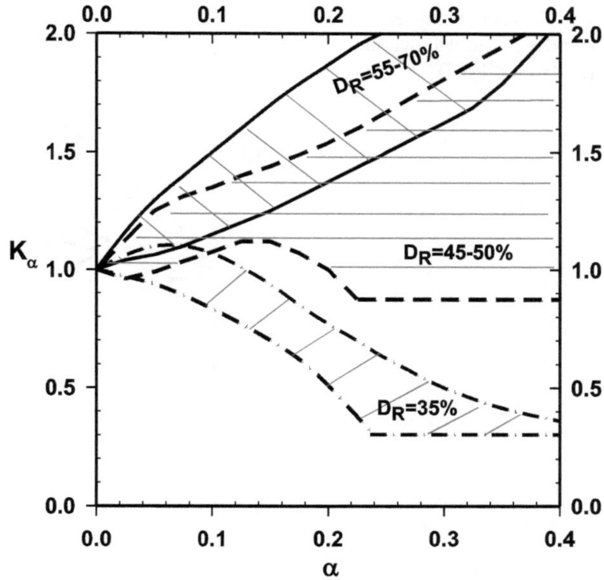

Fig. 7 Relationship between α and K_α (after Seed and Harder, 1990)

$$\alpha = \frac{\tau_{st,0}}{\sigma'_{v,0}} \tag{7}$$

For relatively "denser" soils, the presence of static shear stresses is concluded to increase cyclic resistance. However, starting with Vaid and Finn (1979), Vaid and Chern (1983), and Vaid et al. (1985), on the basis of cyclic simple shear tests performed on dry Ottowa sand, it was documented that cyclic loading resistance might increase or decrease depending on relative density state, the magnitude of initial static shear stress ($\tau_{st,0}$) and cyclic shear strain levels.

Later, on the basis of laboratory cyclic test data, for which a 7.5 % shear strain failure criterion was adopted as the basis of comparison, Seed and Harder (1990) recommended a set of K_α correction factors for a range of D_R values, given in Fig. 7. Although the recommended K_α values exhibit a wide range of uncertainty, this study was regarded as a milestone after presenting a jointed α and D_R dependent K_α correction framework.

Later, Boulanger et al. (1991), on the basis of an extensive experimental study aiming to investigate the effects of static shear stresses on both cyclic resistance and pore pressure generation responses of saturated cohesionless soils, concluded that the adopted failure (or resistance) criterion significantly affected the results. Based on a strength criterion, expressed in terms of CRR corresponding to 3 % single amplitude (SA) shear strain level, a set of correction factors was proposed. As part of the NCEER Workshop, Harder and Boulanger (1997) revisited these earlier curves, and proposed their final NCEER recommendations, as shown in Fig. 8.

Relatively more recently, Vaid et al. (2001) proposed a new set of correlations based on cyclic triaxial tests performed on Fraser River sand. In this study, the

Fig. 8 Relationship between α and K_α (after Harder and Boulanger, 1997)

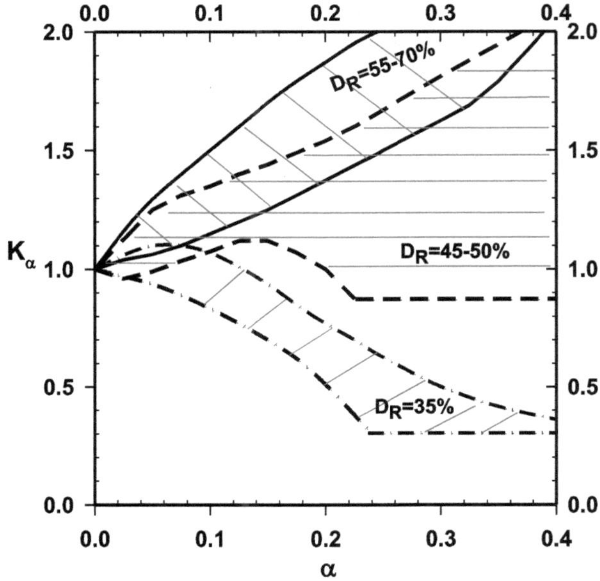

interdependency between K_α and K_σ correction factors was introduced along with a unified correction factor ($K_{\sigma\alpha}$), to take into account the combined effects of confining and static shear stress states. Additionally, Vaid et al. concluded that separately assessed K_α and K_σ correction factors of Harder and Boulanger resulted in over-conservative results, as illustrated in Fig. 9. Later, Wu et al. (2003) reported contrary observations based on cyclic simple shear test results. The discrepancy in

Fig. 9 Relationship between α and K_α (after Vaid et al., 2001)

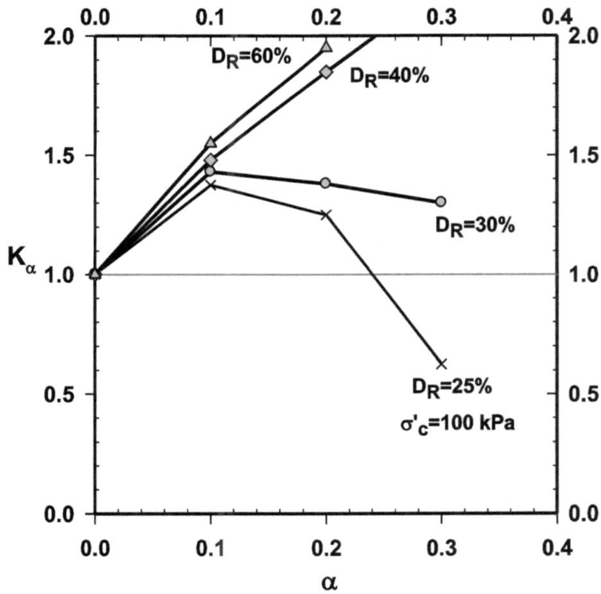

conclusions is attributed to the use of different types of cyclic laboratory tests: i.e.: triaxial vs. simple shear.

More recently, Boulanger (2003b) proposed a new set of K_α correction factors as a function of relative state parameter index (ξ_R), which intrinsically considers the effective confinement stress and relative density states. A closed-form K_α solution for $\alpha \leq 0.35$ is recommended as follows:

$$K_\alpha = a + b \cdot \exp\left(-\frac{\xi_R}{c}\right) \tag{8}$$

$$a = 1267 + 636 \cdot \alpha^2 - 634 \cdot \exp(\alpha) - 632 \cdot \exp(-\alpha) \tag{9}$$

$$b = \exp\left[-1.11 + 12.3 \cdot \alpha^2 + 1.31 \cdot \ln(\alpha + 0.0001)\right] \tag{10}$$

$$c = 0.138 + 0.126 \cdot \alpha + 2.52 \cdot \alpha^3 \tag{11}$$

As shown in Fig. 10, Boulanger (2003b) compared the proposed K_α values with cyclic laboratory test results compiled from effective confining stresses varying from 2 to 16 atmosphere. It is not clear if these comparisons were performed after eliminating the variability in effective confining stress states (i.e.: after application of K_σ corrections). Nevertheless, the model is an improvement over previous attempts,

Fig. 10 Relationship between α and K_α (after Boulanger 2003b)

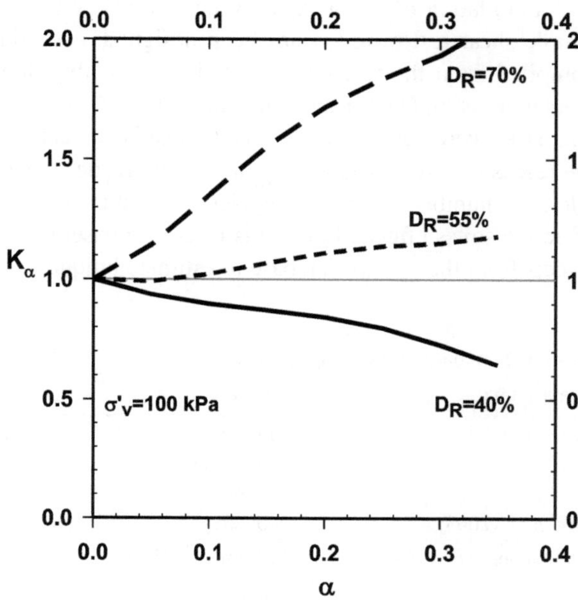

as it proposes a closed-form assessment framework, which considers combined effects of relative density and confinement (or vertical) stress, on K_α correction factor. However, still the K_σ correction is independent of α state, and the use of an independent $K_\sigma$correction scheme, as recommended by Boulanger and Idriss (2004), may result in biased conclusions.

Similar to current state of practice regarding K_σ corrections, there is a consensus regarding the need to correct for K_α effects. However, unlike K_σ corrections, there exists lack of agreement regarding trends on how initial static shear stresses may affect cyclic response of saturated cohesionless soils. To further clarify this, for a medium dense sand sample (e.g.: $D_R = 50\,\%$) subjected to static shear stresses expressed by a α value of 0.2, recommended K_α values may vary in the range of 0.9 to 1.8.

Considering their significance in liquefaction engineering assessments, and inspired by the lack of consensus on even major trends, the main motivation of this study is defined as to develop a unified, strain compatible approach to evaluate the effects of variations in effective overburden stress and static shear stress on cyclic response of saturated clean sands. For the purpose of compiling a high quality database, a detailed laboratory testing program has been executed, in addition to the assessment of available literature. Details of these efforts will be presented next.

Laboratory Testing Program and Database Compilation Efforts

Efforts aiming to develop a semi-empirical or an empirical model naturally require the compilation of a high quality database. For this purpose an extensive cyclic simple shear test program has been designed, and additional test data have been compiled from the well-documented, high quality simple shear test databases of Boulanger et al. (1991) and Wu et al. (2003). As presented in Table 1, two separate databases were consolidated.: (i) a K_σ test is characterized by $\alpha \leq 0.03$; whereas (ii) the rest is referred to as $K_{\alpha,\sigma,SRR}$ tests. The K_σ database is used to develop baseline CRR vs. number of uniform cycles, N relationships for assessing purely vertical effective stress effects. Later, this reference baseline will be used to "subtract" K_σ effects from the $_{K\alpha,\sigma,SRR}$ database to estimate unique $K_{\alpha,\ SRR}$ trends.

Table 1 Summary of the compiled data

Data source	Sand type	# of K_σ tests	# of $K_{\alpha,\sigma,SRR}$ tests
This study	Monterey	19	72
	Morro Bay	2	12
	Nevada	37	8
Wu et al. (2003)	Monterey	58	15
Boulanger et al. (1991)	Sacramento River	26	37
	Total	142	144

Fig. 11 Grain size distribution of sand specimens

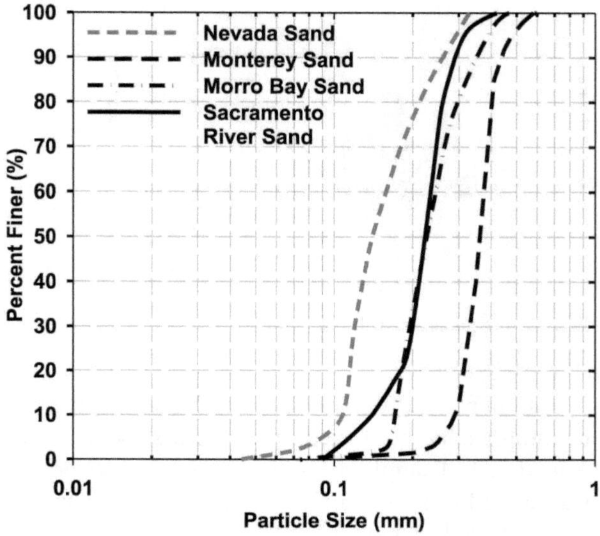

Before discussing the details of the testing program, two major data outsources will be reviewed. As summarized in Table 1, 26 out of 142, and 37 out of 144 K_σ and $K_{\alpha,\,SRR}$ test data, respectively, were compiled from Boulanger et al. (1991) database. As part of a comprehensive research study, cyclic responses of clean saturated sands subjected to uni- and bi-directional cyclic shear stresses, were investigated. Simple shear tests were performed on Sacramento River sand, index properties of which, were listed in Table 2. Grain size characteristics is also presented in Fig. 11. Specimens were reconstituted by using moist tamping method to various relative density states ranging from 35 to 55 %. Samples were consolidated under a vertical effective stress of 207 kPa. The frequency of cyclic loading was adopted as 0.2 Hz.

58 out of 142 of K_σ, and 15 out of 144 $K_{\alpha,Q,SRR}$ data were compiled from Wu et al. (2003) database. Stress-controlled simple shear tests were performed on wet pluviated Monterey No. 0/30 sand specimens, which were K_0 consolidated to a wide range of relative densities under vertical effective stresses of 40, 80 or 180 kPa. Frequency of cyclic loading was reported as 0.1 Hz. Index properties as well as grain size characteristics are summarized in Table 2 and Fig. 11, respectively.

Table 2 Index properties of specimens

Sand Type	G_S	e_{min}	e_{max}	D_{10}	D_{30}	D_{60}	C_u	C_c	USCS
Monterey	2.64	0.54	0.86	0.30	0.33	0.39	1.3	0.9	SP
Nevada	2.67	0.53	0.89	0.11	0.12	0.18	1.6	0.8	SP
Sacramento River	2.68	0.64	1.01	0.14	0.20	0.22	1.6	1.3	SP
Morro Bay	2.65	0.58	0.91	0.18	0.20	0.25	1.4	0.9	SP

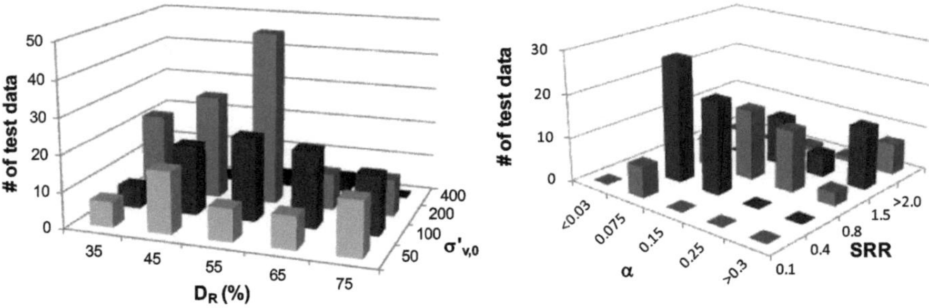

Fig. 12 Frequency distribution plots

Additionally, within the scope of this study, an extensive cyclic simple shear testing program has been executed to increase both the quantity and variability of data. A total of 150 simple shear tests have been performed on Monterey No. 0/30, Nevada and Morro Bay sands. Index properties and grain size characteristics were presented in Table 2 and Fig. 11, respectively. Specimens were reconstituted to density states ranging from 37 to 85 % by wet pluviation method, and then K_0-consolidated under a wide range of effective stresses varying from 35 to 400 kPa. Frequency of loading was adopted as 0.1 Hz.

Database frequency distributions are presented in the vertical effective stress ($\sigma'_{v,0}$) vs. relative density (D_R); and static stress ratio (α) vs. stress reversal ratio ($SRR = \tau_{st}/\tau_{cyc}$) domains, as presented in Fig. 12. The choices of $\sigma'_{v,0}$ vs. D_R and α vs. SRR are not arbitrary, where the first domain expresses dilatancy characteristics of samples; and the latter one presents static and cyclic shear loading scheme. In the α vs. SRR domain, K_σ tests (i.e.: $\alpha < 0.03$ tests) are intentionally excluded to provide a clear presentation of $K_{\alpha SRR}$ test conditions.

On the basis of laboratory tests performed at the reference confining and shear stress states (i.e. $\sigma'_{v,0} = 100$ kPa and $\alpha = 0.0$), $CRR_{\sigma'_v=100kPa,\alpha=0}$ vs. number of uniform loading cycle (N) relationships corresponding to performance levels expressed by shear strain of =0.5 %, 1.0 % and 3.0 % (single amplitude, SA) were established for a number of relative density states. Figure 13 schematically illustrates the methodology followed to estimate $K_{\sigma,\alpha,SRR}$ values.

For each test, number of cycles to reach to selected performance level (strain- or r_u-based) is determined, as illustrated in Fig. 13, and the corresponding reference cyclic resistance ratio ((i.e.: $CRR_{\sigma'_v=100kPa,\alpha=0}$) is evaluated by using the upper plot. The ratio of cyclic test CSR corresponding to the shear strain level γ (i.e.: $CRR_{\sigma'_v,\alpha,\gamma=3\%}$) to $CRR_{\sigma'_v=100kPa,\alpha=0}$ is then used to assess the proposed unified correction scheme, $K_{\sigma,\alpha,SRR}$, as given in Eq. 12.

$$K_{\sigma,\alpha,SRR,\gamma=3\%} = \frac{CRR_{\sigma'_v,\alpha,\gamma=3\%}}{CRR_{\sigma'_v=100kPa,\alpha=0,\gamma=3\%}} = K_\sigma \cdot K_\alpha \cdot K_{SRR} \qquad (12)$$

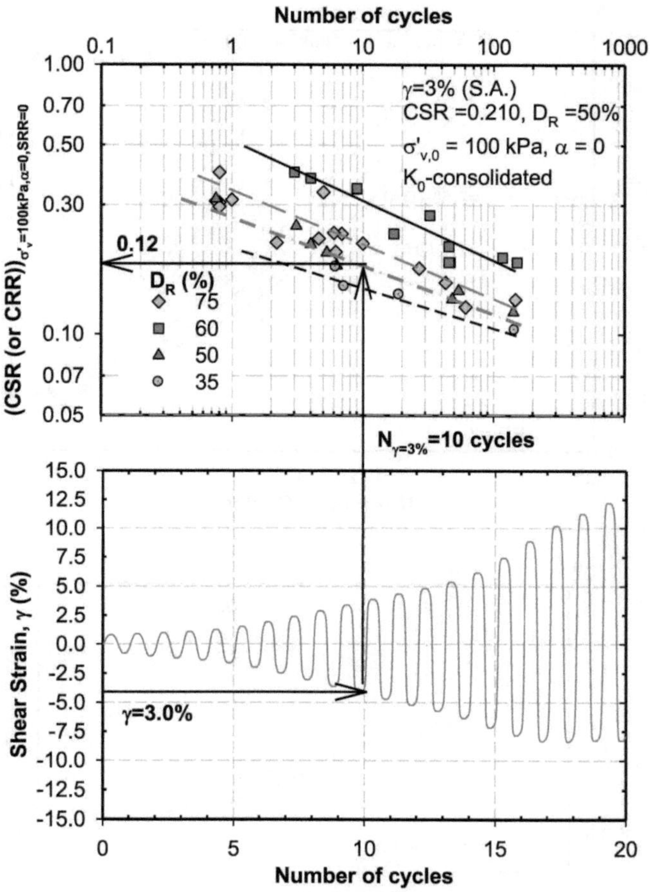

Fig. 13 Schematic illustration of stress correction factor assessment framework

This outlined approach was followed to consolidate cyclic laboratory test results, and develop a general conclusions regarding $K_{\sigma,\alpha,SRR}$ response. For the purpose of assessing potential strain dependency in K_σ, for a medium dense sample (i.e.: $D_R=50\,\%$) which was isotropically consolidated (i.e.: $\alpha=0.0$ and SRR$=0.0$) at confining stresses of 40, 100 and 200 kPa, CRR vs. N curves were developed for double amplitude $=1$, 3 and 7.5 %, as shown in Fig. 14. Later for an arbitrarily selected N value of 10, CRR values corresponding to 40 and 200 kPa confining stress states were evaluated, and the ratio of them to the CRR at the reference state of 100 kPa was reported as K_σ values. Estimated K_σ values are shown in lower half of Fig. 14. Note that N-dependent non-linear response of CRR is not within the scope of this study and readers are referred to Çetin and Bilge (2012) for a complete discussion. As clearly illustrated by this exercise and Fig. 14, K_σ exhibits a shear strain dependent response, and it increases with increasing shear strain performance levels for confining stress ranges less than 1 atm. A similar exercise was also performed

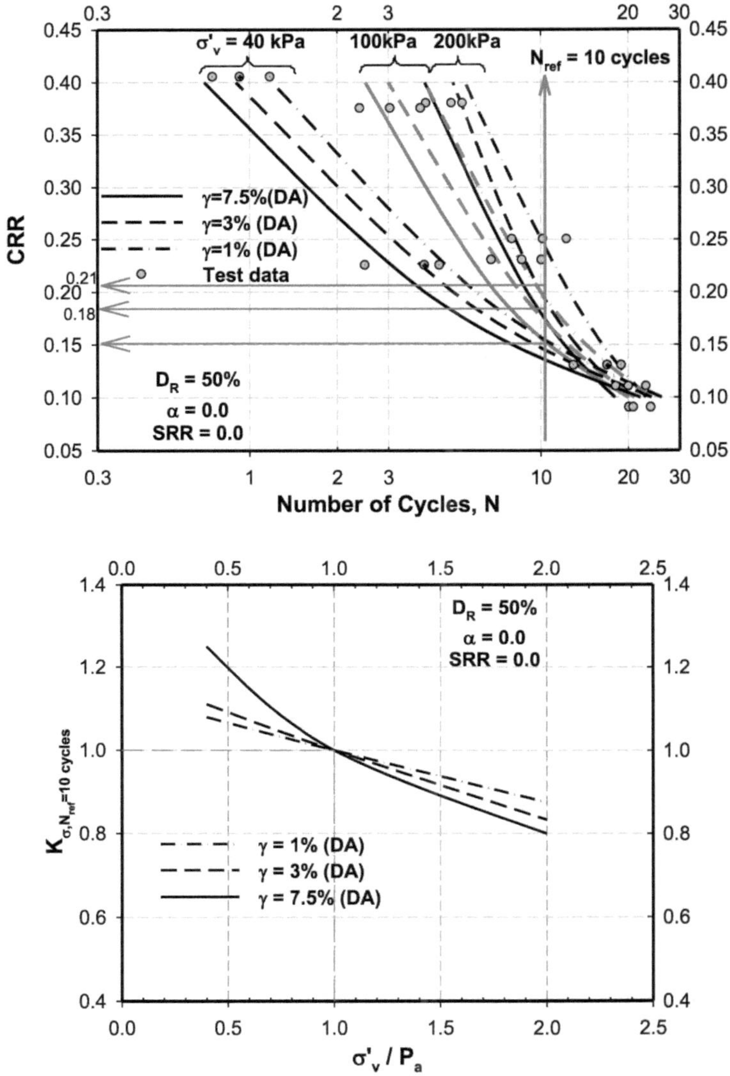

Fig. 14 Strain dependency in K_σ response

to assess the potential shear strain dependent K_α response. For the purpose, cyclic laboratory tests performed on $D_R = 60–64\,\%$ samples, subjected to a confining stress range of 100–105, were studied. As shown in Fig. 15, CRR vs. N curves were developed for $\alpha = 0.0$, 0.10 and 0.265. After normalization with reference CRR value (i.e.: CRR at zero static shear stress state), $K_{\sigma=100kPa,\alpha,SRR}$ values, shown on in Fig. 15 were estimated. Again $K_{\alpha,SRR}$ response is observed to be strain dependent, and specific for $D_R = 60–64\,\%$ range, it decreases with increasing shear strain performance levels.

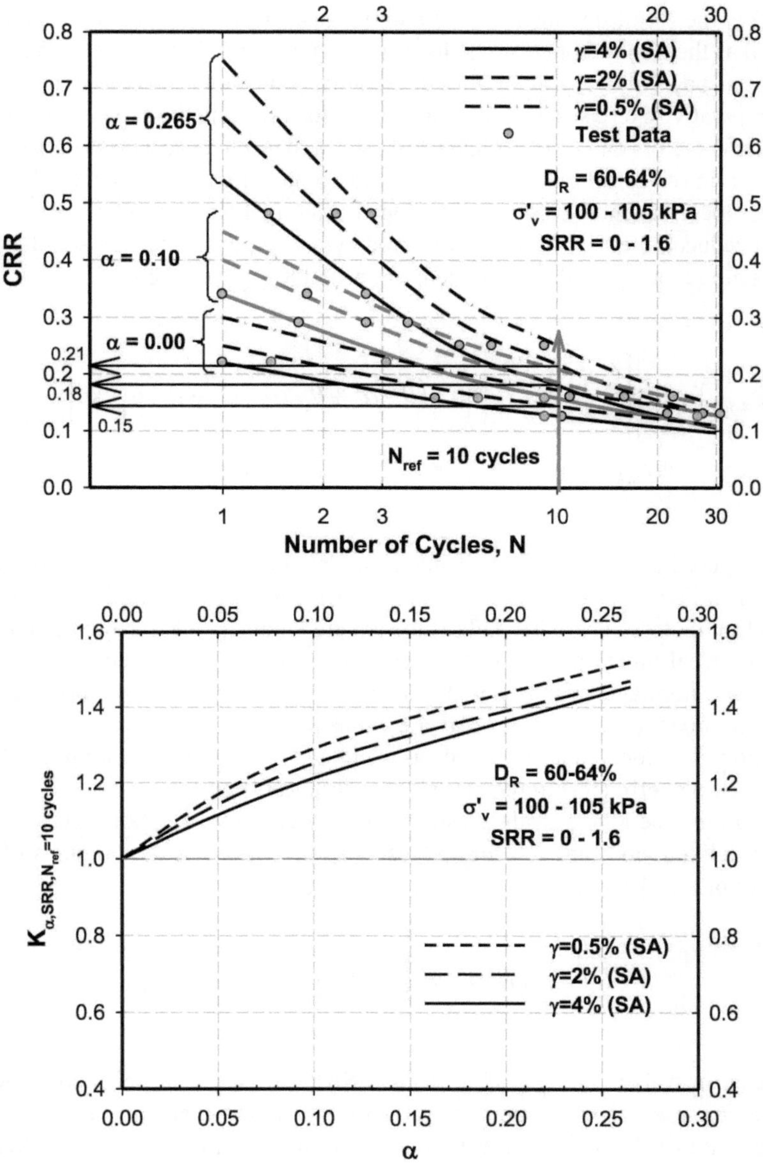

Fig. 15 Strain dependency in $K_{\alpha,SRR}$ response

A Unified Probabilistic $K_{\sigma,\alpha,SRR}$ Stress Correction Framework

The first step in developing a probabilistic model is to select a limit state expression that captures the essential parameters of the problem. The model for the limit state function has the general form $g = g(\mathbf{x}, \Theta)$ where x is a set of descriptive parameters and Θ is the set of unknown model coefficients. Inspired by previous studies and as well as the trends from test results, various functional forms have been tested.

On the basis of cyclic laboratory test results, and conclusions reached as a result of comparative exercises illustrated in the previous section, the dependency of stress correction factors on D_R, confining stress, static shear stress, cyclic shear stress, and shear strain (or corollary r_u) level was assessed and alternative functional forms to express these dependencies were developed. Among them, the following functional form produced the best fit to the observed cyclic response and was adopted as the limit state function for the assessment of unified $K_{\sigma,\alpha,SRR}$ response.

$$\ln\left(K_{\sigma,\alpha,SRR}\right) = \ln\left\{\left[\left(\frac{\sigma'_{v,0}}{P_a}\right)^{-\theta_1 \cdot D_R}\right] \cdot \left[1+\left(\theta_2 \cdot \alpha^2 + \theta_3 \cdot \alpha\right)\cdot\left(D_R - \theta_4\right)\right] \cdot \left[\exp\left(\frac{\theta_5 \cdot SRR}{D_R}\right)\right]\right\} \pm \varepsilon_{\ln K_{\sigma,\alpha,SRR}} \tag{13}$$

The proposed model includes a random model correction term (ε) to account for the possibilities that (1) missing descriptive parameters with influence on stress correction factors may exist; and (2) the adopted mathematical expression may not have the ideal functional form. It is reasonable and also convenient to assume that ε follows a normal distribution with a mean of zero for the purpose of producing an unbiased model (i.e., one that on average makes correct predictions). The standard deviation of ε, denoted as σ_ε, however, is unknown and must be estimated. The set of unknown coefficients of the model, therefore, is $\Theta = (\theta, \sigma_\varepsilon)$.

Assuming the $K_{\sigma,\alpha,SRR}$ values estimated for each test to be statistically independent, the likelihood function for "n" tests can be written as the products of probabilities of the observations.

$$L_{K_{\sigma,\alpha,SRR}}(\theta, \sigma_\varepsilon) = \prod_{i=1}^{n} P\left[g_{K_{\sigma,\alpha,SRR},i}(\cdot) = 0\right] \tag{14}$$

Suppose the values of $D_{R,i}$, $\sigma'_{v,0,i}$, α, and SRR_i at each data point are exact, i.e. no measurement or estimation error is present, noting that $g(...) = \hat{g}(...) + \varepsilon_i$ has the normal distribution with mean $\hat{g}$ and standard deviation σ, then the likelihood function can be written as a function of unknown coefficients as follows. In this equation, is the standard normal probability density function.

$$L_{K_{\sigma,\alpha,SRR}}(\boldsymbol{\theta}, \sigma_\varepsilon) = \prod_{i=1}^{n} \phi\left[\frac{\hat{g}_{K_{\sigma,\alpha,SRR}}(\sigma'_{v,0,i}, D_{R,i}, \alpha_i, SRR_i, \theta)}{\sigma_{\varepsilon_{\ln(K_{\sigma,\alpha,SRR})}}}\right] \tag{15}$$

Consistent with the maximum likelihood methodology, model coefficients are estimated by maximizing the likelihood function. The final forms of the resulting proposed models for single amplitude $\gamma=0.5$, 1 and 3 % performance levels, are presented in Eqs. 16, 17 and 18, respectively. Note that these equations are recommended to be used within the range of $0.3\leq\frac{\sigma'_v}{P_a}\leq 4.0$; $0.0\leq\alpha\leq 0.35$; $0.0\leq SRR\leq 2.0$.

$$\ln\left(K_{\sigma,\alpha,SRR}\right)_{\gamma=0.5\%}=\ln\left[\left(\frac{\sigma'_v}{P_a}\right)^{-0.0026D_R}\left[1+(0.078\alpha^2-0.116\alpha)(D_R-60.1)\right]\exp\left(\frac{-12.42\cdot SRR}{D_R}\right)\right]\pm 0.25 \tag{16}$$

$$\ln\left(K_{\sigma,\alpha,SRR}\right)_{\gamma=1.0\%}=\ln\left[\left(\frac{\sigma'_v}{P_a}\right)^{-0.0032D_R}\left[1+(0.396\alpha^2-0.118\alpha)(D_R-38)\right]\exp\left(\frac{-8.24\cdot SRR}{D_R}\right)\right]\pm 0.196 \tag{17}$$

$$\ln\left(K_{\sigma,\alpha,SRR}\right)_{\gamma=3.0\%}=\ln\left[\left(\frac{\sigma'_v}{P_a}\right)^{-0.0042D_R}\left[1+(0.060\alpha^2-0.015\alpha)(D_R+54.4)\right]\exp\left(\frac{-5.42\cdot SRR}{D_R}\right)\right]\pm 0.158 \tag{18}$$

Along with these closed form solutions, graphical representations of the proposed models are also presented in Fig. 16 for single amplitude $\gamma=0.5$, 1 and 3 % performance levels. This figure clearly addresses the strain- and collectively-dependent nature of $K_{\sigma,\alpha,SRR}$ stress correction factor. It should be noted that even though K_σ, K_α, and K_{SRR} components of $K_{\sigma,\alpha,SRR}$ can be expressed independently, they were assessed, and hence should be used, in a unified manner. Also revealed by these figures, unified $K_{\sigma,\alpha,SRR}$ values are shear strain, relative density, confining stress, static and cyclic shear stress dependent. Thus, use of independent K_σ and K_α values, or a single set of correction factors independent of the expected performance level (i.e.: if a soil layer liquefies or not) is observed to produce biased estimates of the true CRR values. More specifically, with increasing shear strain levels, K_σ and K_{SRR} components increase by 5 to 15 % and 10 to 35 %, respectively, for effective stress range less than 1 atm. The effects are more pronounced for K_α component. Depending on the shear strain level, K_α values may be different by a factor of 0.9 to 1.8.

For the purpose of testing whether any trend (bias) as a function of model input variables (descriptors) is present in residuals (defined as "***ln(measurement/model prediction)***"), residuals vs. problem descriptor (D_R, $\sigma'_{v,0}$, α and SRR) plots were prepared for $\gamma=3$ % model (Fig. 17). Lack of a general trend as a function of any of

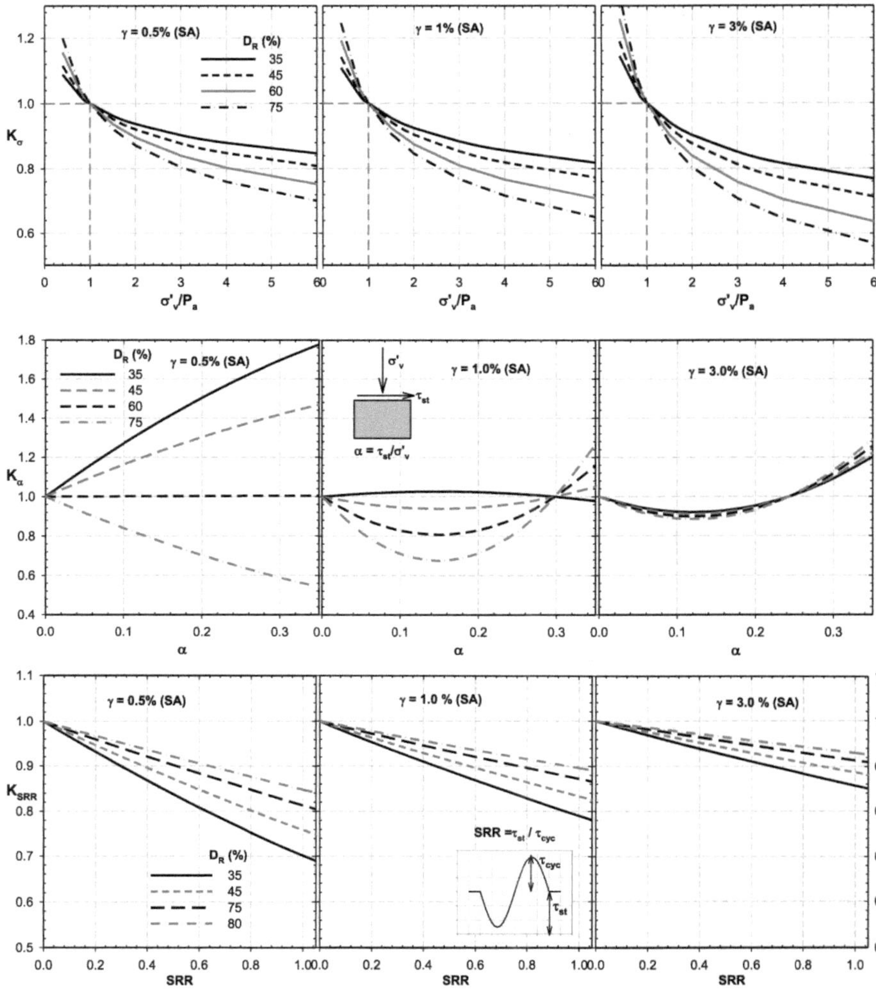

Fig. 16 Proposed correction factors for different shear strain levels

these input variables confirmed the validity of the selected functional forms. Pearson product moment correlation coefficient (R^2) is another descriptor often used to evaluate the predictive success of models. As presented in Fig. 18 in the domain of test vs. model predicted CRR (and $K_{\sigma,\alpha,SRR}$) values for γ=3 % model, an R^2 value of 79.3 % is obtained. This value is considered to be satisfactorily high considering the quite complex nature of the problem.

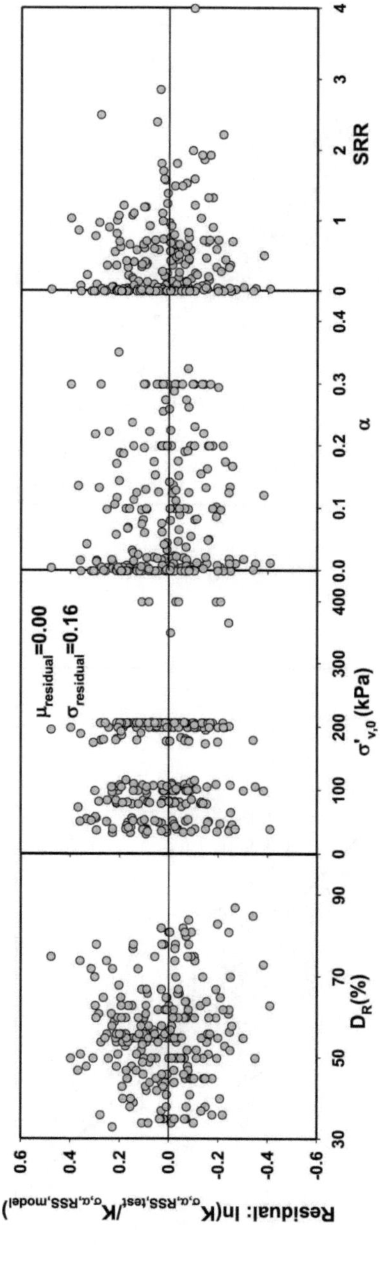

Fig. 17 Variations of residuals as functions of major input parameters for $\gamma = 3\,\%$

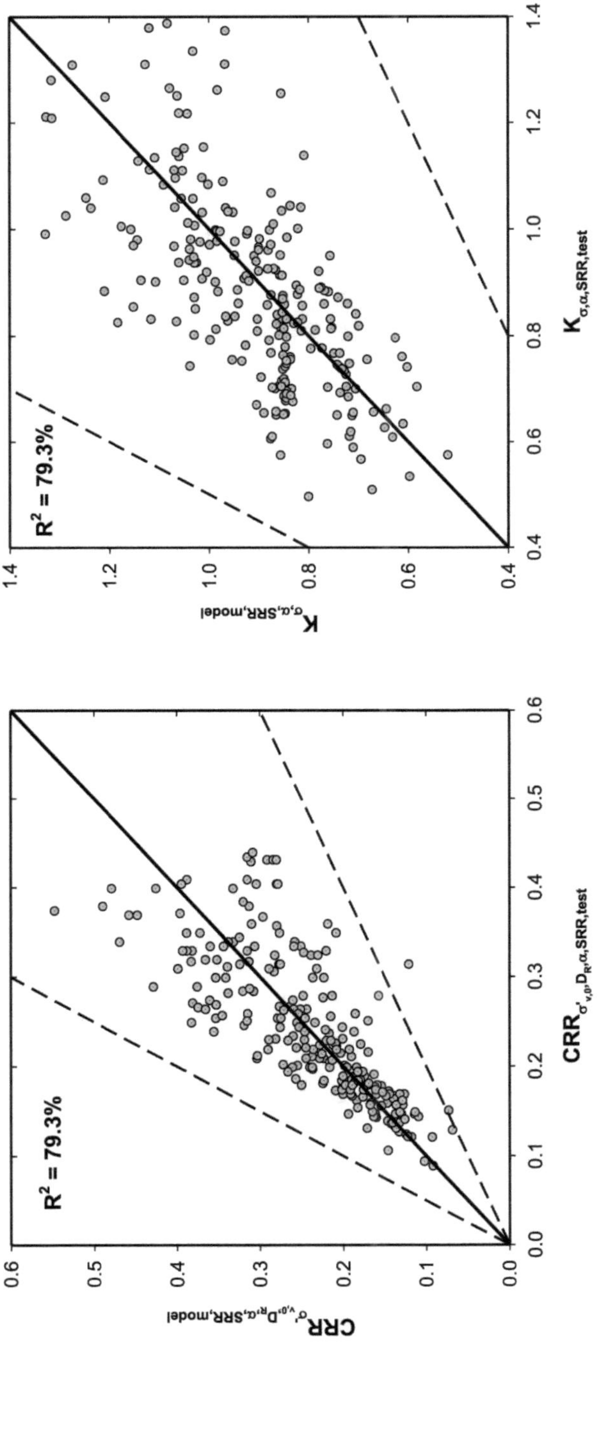

Fig. 18 Measured vs. predicted responses

Comparisons with Available Methods

For the purpose of establishing a basis of comparison with existing K_σ and K_α models, the proposed unified closed form $K_{\sigma,\alpha,SRR}$ solutions given in Eqs. 16 and 18, corresponding to single amplitude shear strain levels of 0.5 % and 3.0 % were solved for $\alpha = SRR = 0$ and $D_R = 40$ and 80 % to develop K_σ correlations. Figure 19 presents these curves along with the recommendations of Youd et al. (2001), Boulanger and Idriss (2004), Çetin et al. (2004). As clearly addressed by this figure, the variations in the current state of K_σ recommendations are bounded by newly proposed K_σ curves corresponding to two relative density and shear strain level combinations: i) $D_R = 40\,\%$ and $= 0.5\,\%$ and ii) $D_R = 80\,\%$ and $\gamma = 3.0\,\%$. Hence, it can be concluded that the wide scatter in currently available K_σ recommendations can be attributed to strain and D_R independent assessment of K_σ models (Fig. 19).

Similarly, a solution for $K_{\alpha,SRR}$ was also prepared for a confining stress of 100 kPa and CSR value of 0.2. Resulting curves are plotted comparatively in Fig. 20. As clearly illustrated by this figure, recommended K_α values are observed to be in good agreement with the average of the recommendations by NCEER i.e. Harder and Boulanger (1997), Vaid et al. (2001), and Boulanger (2003) for single amplitude shear strain level of 0.5 %. For shear strain levels exceeding 0.5 %, there exists a significant disagreement. However, it should be noted that for the purpose of making direct comparisons with existing methods, K_{SRR} component of $K_{\alpha,SRR}$ is neglected. This choice unfortunately makes these comparisons biased and their interpretation very difficult. However, still following general conclusions can be drawn rather fairly:

i. the use of strain independent K_σ values, independent of the expected performance level of a soil site, may produce biased conclusions regarding liquefaction triggering resistance or post liquefaction deformations. In simpler terms, for a soil site subjected to cyclic stress levels smaller than CRR, the use of currently recommended K_σ values may overestimate the expected deformations or shear

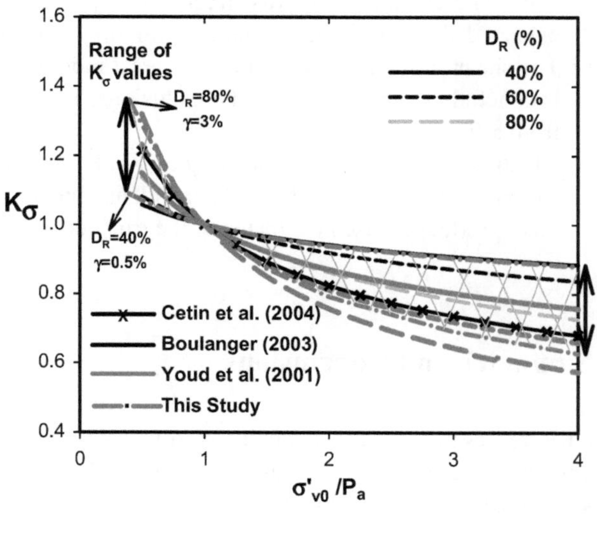

Fig. 19 Comparison of K_σ correction factors

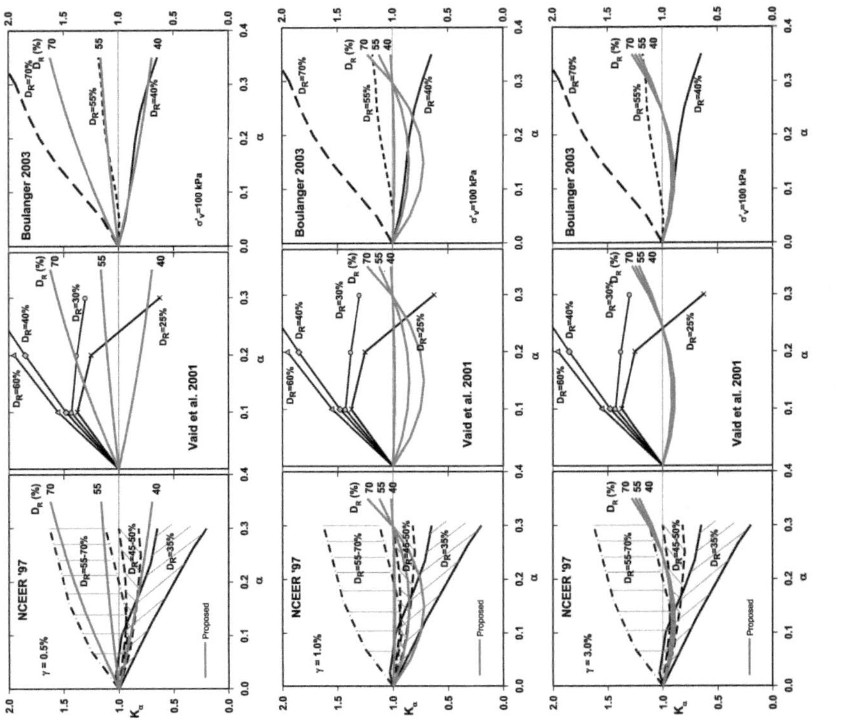

Fig. 20 Comparison of K_{α} correction factors

strain levels. Consistently, for a liquefied soil layer (i.e: CSR>CRR), the use of currently recommended K_{σ} values may underestimate the expected deformations or shear strain levels.

ii. Unlike K_{σ} corrections, there exists lack of agreement regarding trends on how initial static shear stresses may affect cyclic response of saturated cohesionless soils. The extend of this disagreement, depending on the relative density and expected shear strain performance level, may reach up to a factor of 2.5.

iii. The shear strain and cyclic stress reversal independent assessment of K_{α} values is concluded to produce biased conclusions. The extend of this bias could range from a factor of 1.5 to 2.0.

iv. All stress correction components (i.e.: K_{σ}, K_{σ} and K_{SRR}) are observed to be collectively- and relative density-dependent, and of independent assessment of them produces biased estimates of the true performance level.

Summary and Conclusions

CRR values proposed by available seismic soil liquefaction triggering methods are valid at the reference stress state of one atmosphere normal effective stress and zero static shear stress. To convert these CRR values to the ones other than the reference conditions, a series of correction factors is applied to eliminate the variability due to normal (or confining) effective and static shear stresses (i.e. K_σ and K_α corrections) acting on the horizontal plane. In the literature there exists a consensus (i) regarding the need to correct for K_σ and K_α effects, (ii) for stresses less than and greater than 1.0 atm., K_σ corrections are greater than and less than 1.0, respectively, (iii) with increase in α, K_α increases for "dense" soils. However, quantification of these corrections factors still remains as a controversial issue and for critical projects, site specific solutions are recommended (NCEER, 1997 and Youd et al. 2001).

Considering their significance in liquefaction engineering assessments, and inspired by the lack of consensus on even general trends, the main motivation is defined as to develop a unified, probabilistic, strain (performance) compatible approach to evaluate the effects of variations in effective overburden stress and static shear stress on cyclic response of saturated clean sands. For the purpose of compiling a high quality database, a detailed laboratory testing program has been executed, in addition to the assessment of available literature.

On the basis of assessment of compiled database, a probabilistic unified stress correction, $K_{\sigma,\alpha,SRR,}$ framework is proposed. Followings are the major conclusions of this study:

i. K_σ and for K_α (and K_{SRR}) responses should be assessed within a collectively- and strain-dependent unified framework (i.e.: $K_{\sigma,\alpha\text{'}SRR}$). Independent assessment of these correction factors may produce biased estimations of the true performance. Depending on the expected shear strain levels, the extend of this bias may range from conservatism to unconservatism by a factor of as high as 2.5.
ii. In addition to effective confining stress, D_R, static and cyclic shear stresses, and target shear strain levels are concluded to be the major input parameters of the stress correction factors.
iii. Unlike K_σ, corrections, there exists lack of agreement regarding trends on how initial static shear stresses may affect cyclic response of saturated cohesionless soils. The extend of this disagreement, depending on the relative density and expected shear strain performance level, may reach up to a factor of 2.5.
iv. The shear strain and cyclic shear stress reversal (SRR) independent assessment of K_α values is concluded to produce biased conclusions. The extend of this bias could range from a factor of 1.5 to 2.0. Hence, the effects of the presence of static shear stresses (i.e.: K_α effects) should be assessed along with the magnitude of the cyclic shear stresses (i.e.: K_{SRR} effects).
v. The strain dependent nature of stress correction factors require an iterative assessment of CRR or CSR-based liquefaction engineering evaluations.
vi. Even though a strain compatible iterative solution scheme has been recommended, for preliminary performance assessments of non-liquefied cases, stress correction factor, $K_{\sigma,\alpha,SRR}$, can be estimated by Eq. 16, for 0.5 % single amplitude shear strain level. Consistently, for liquefied cases, $K_{\alpha,SRR}$ can be estimated by Eq. 18, for 3.0 % single amplitude shear strain level.

As the concluding remark, within the confines of this chapter, it was shown that compared to the one at the reference stress state, cyclic resistance of cohesionless soils was governed by relative density, static confining and shear stress states, as well as the shear stress reversal and shear strain levels during cyclic loading. Contrary to general state of practice, a collectively-dependent sets of correction schemes need to be used compatible with the expected cyclic shear straining performance. This requires a performance-based assessment of liquefaction engineering problems, which suggests an iterative convergence scheme.

References

Bilge HT, Çetin KÖ (2011) "Effect of vertical effective stress on liquefaction triggering analyses" 7th National Conf. on Earthquake Eng., Istanbul, Turkey

Bolton MD (1986) The strength and dilatancy of sands. Geotechnique 36(1):65–78

Boulanger RW (2003a) "High overburden stress effects in liquefaction analyses". J Geotech Geoenviron Eng ASCE 129(12):1071–1082

Boulanger RW (2003b) "Relating Kα to relative state parameter index". J Geotech Geoenviron Eng ASCE 129(8):770–773

Boulanger RW, Idriss IM (2004) "State normalization of penetration resistances and the effect of overburden stress on liquefaction resistance". In: Doolin D. et al (eds) Proceedings, 11th international conference on soil dynamics and earthquake engineering, and 3rd international conference on earthquake geotechnical engineering, Stallion Press, Singapore, Vol. 2, 484–491

Boulanger RW, Seed RB, Chan CK, Seed HB, Sousa J (1991) "Liquefaction behavior of saturated sands under uni-directional and bi-directional monotonic and cyclic simple shear loading." Geotechnical Report No. UCB/GT/91—08. University of California, Berkeley, California

Castro G, Poulos SJ (1977) "Factors affecting liquefaction and cyclic mobility." J Geotech Eng Div ASCE 103(6):501–516

Çetin KÖ (2000) "Reliability—based assessment of seismic soil liquefaction initiation hazard" PhD dissertation, University of California, Berkeley, Calif

Çetin KÖ, Bilge HT (2012) Performance-based assessment of magnitude (duration) scaling factors. J Geotech Geoenviron Eng 138(3):324–334

Çetin KÖ, Seed RB, Der Kiureghian A, Tokimatsu K, Harder LF Jr, Kayen RE, Moss RES (2004) SPT-based probabilistic and deterministic assessment of seismic soil liquefaction potential. J Geotech Geoenviron Eng. 130(12):1314–1340

Harder LF Jr (1988) "Use of penetration tests to determine the cyclic load resistance of gravelly soils" Ph. D. Thesis, University of California, Berkeley

Harder LF Jr, Boulanger RW (1997) "Application of K_s and K_a Correction Factors" Proc., NCEER Workshop on Evaluation of Liquefaction Resistance of Soils, Report No. NCEER-97-0022, National Center for Earthquake Engineering Research, SUNY Buffalo, N.Y., 167–190.

Hynes ME, Olsen RS (1999) Influence of confining stress on liquefaction resistance. Proc. Int. Workshop on Physics and Mechanics of Soil Liquefaction, Balkema, Rotterdam, 145–151

Konrad J-M (1988) Interpretation of flat plate dilatometer tests in sands in terms of the state parameter. Geotechnique 38(2):263–277

Lee KL, Seed HB (1967a) "Cyclic stress conditions causing liquefaction of sand". J Soil Mech Found Div ASCE 93(1):47–70

Lee KL, Seed HB (1967b) "Dynamic strength of anisotropically consolidated specimens ". J Soil Mech Found Div ASCE 93(5):169–190

Moss RES, Seed RB, Kayen RE, Stewart JP, Der Kiureghian A, Çetin KÖ (2006) "CPT-based probabilistic and deterministic assessment of in-situ seismic soil liquefaction triggering". J Geotech Geoenviron Eng 132(8):1032–1051

Olsen RS (1984) "Liquefaction analysis using the cone penetrometer test (CPT)" Proc. 8th world conference on earthquake engineering, San Francisco, California

Seed HB (1983) "Earthquake-resistant design of earth dams" Proc. of symposium on seismic design of embankments and caverns, Philadelphia, ASCE, NY, 41–64

Seed HB, Idriss IM (1982) Ground motions and soil liquefaction during earthquakes. EERI Monograph, Berkeley, Calif

Seed HB, Peacock WH (1971) "Test procedures for measuring soil liquefaction characteristics". J Soil Mech Fnd Div ASCE 97(8):1099–1119

Seed HB, Tokimatsu K, Harder LF, Chung RM (1984) "The influence of SPT procedures in soil liquefaction resistance evaluations" Earthquake Engineering Research Center Report No. UCB/EERC-84/1, Univ. of California, Berkeley, California

Seed RB, Harder LF Jr (1990) "SPT-based analysis of cyclic pore pressure generation and undrained residual shear strength" Proc. H. Bolton Seed Memorial Symposium

Vaid YP, Chern JC (1983) Effect of static shear on resistance to liquefaction. Soils Found 123(1):47–60

Vaid YP, Chern JC, Tumi H (1985) Confining pressure, grain angularity and liquefaction. J Geotech Eng 111(10):1229–1235

Vaid YP, Finn WDL (1979) Static shear and liquefaction potential. J Geotech Eng Div 105(10):1233–1246

Vaid YP, Stedman JD, Sivathayalan S (2001) "Confining stress and static shear effects in cyclic liquefaction". Can Geotech J 38:580–591

Vaid YP, Thomas J (1995) "Liquefaction and post-liquefaction behaviour of sand". J Geotech Eng ASCE 121(2):163–173

Wu J, Seed RB, Pestana JM (2003) "Liquefaction triggering and post liquefaction deformations of Monterey 0/30 sand under uni-directional cyclic simple shear loading." Geotechnical engineering research report No. UCB/GE-2003/01. University of California, Berkeley, California

Youd TL, Idriss IM, Andrus, RD, Arango I, Castro G, Christian JT, Dobry R, Finn WDL, Harder LF Jr, Hynes ME, Ishihara K, Koester JP, Liao SC, Marcuson WF, Martin GR, Mitchell JK, Moriwaki Y, Power MS, Robertson PK, Seed RB, Stokoe KH (2001) "Liquefaction resistance of soils: summary report from the 1996 NCEER and 1998 NCEER/NSF workshops on evaluation of liquefaction resistance of soils.". J Geotech Geoenviron Eng ASCE 127(10):817–833

Site Amplification Formula Using Average V_s in Equivalent Surface Layer Based on Vertical Array Strong Motion Records

Takaji Kokusho

Abstract Seismic site amplification formula was developed for virtual surface arrays to be consistent with vertical arrays using a number of KiK-net records during recent 8 destructive earthquakes. A correlation between peak spectrum amplification and S-wave velocity ratio (base V_s/surface V_s) improved much better if the surface V_s was evaluated from fundamental mode frequency combined with a thickness of equivalent surface layer in which peak amplification is exerted, rather than using the conventional V_{s30}. Also found was that soil nonlinearity effect during strong earthquakes has only a marginal effect on the surface array amplification observed and evaluated by the formula using small-strain V_s-values. The theoretical basis for the minor nonlinear effect has been discussed by conducting a simple study on a two-layered system with strain-dependent soil nonlinearity. The proposed amplification formula seems also applicable with a satisfactory accuracy to sites where soil liquefaction is involved.

Introduction

Site amplification defined here as Fourier spectrum peak value between ground surface and bedrock depends on several factors; the composition of soil layers, S-wave velocities, soil densities, internal damping of the individual layers. Furthermore, strain-dependent nonlinear properties may affect the amplification in soft soils during strong earthquakes. Among the influencing factors, the ratio of S-wave velocities between a base layer and a surface layer is of utmost importance. Shima (1978) showed by conducting a simple one dimensional wave propagation analysis that the site amplification based on multi-reflection theory of SH wave is almost linearly related to the S-wave velocity ratio (base layer to surface top layer) despite the difference in intermediate soil layers.

In order to evaluate site amplification from bedrock to surface, NIED (National Research Institute for Earth Science and Disaster Prevention, Japan) deployed

T. Kokusho (✉)
Dept. Civil & Environment Eng., Chuo University, Tokyo, Japan
e-mail: Kokusho@civil.chuo-u.ac.jp

A. Ansal, M. Sakr (eds.), *Perspectives on Earthquake Geotechnical Engineering*, Geotechnical, Geological and Earthquake Engineering 37,
DOI 10.1007/978-3-319-10786-8_6

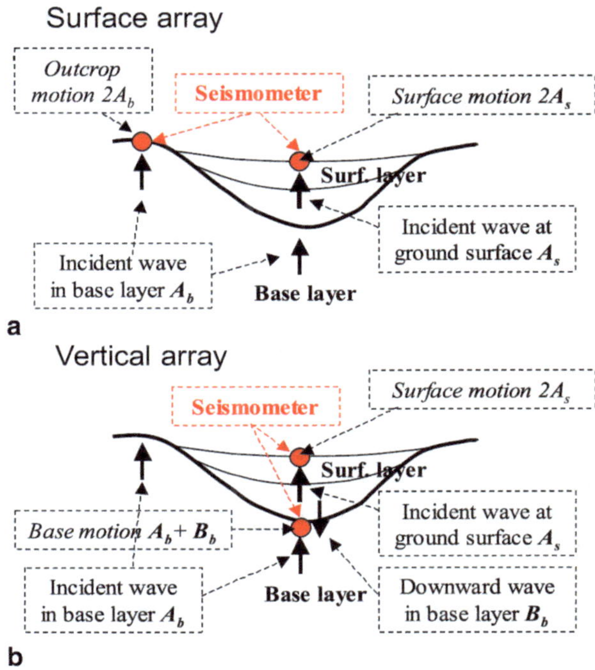

Fig. 1 Surface array (**a**) and vertical array (**b**) for earthquake observations

seven hundreds of vertical array strong motion recording systems called KiK-net all over Japan after the 1995 Kobe earthquake. Each system comprises a pair of 3D accelerometers at a ground surface and a base layer about 100 m deep in most case. The observed records together with associated geotechnical data are easily accessible for international researchers at the web site; http://www.kik.bosai.go.jp/kik/.

In general, there can be two different seismic observation array systems to measure the site amplification between ground surface and base layer as illustrated in Fig. 1a and b. One is a surface array (a) consisting of a set of surface seismometers on different surface geologies in a relatively small area with a common base layer. Seismic records $2A_s$ on a surface soil and $2A_b$ on outcropping base layer allow to directly evaluate site amplification between the two different geologies $2A_s / 2A_b$ if incident seismic wave A_b is assumed basically the same in that area. The amplification between the two records $2A_s / 2A_b$ is what we need for micro zonation to develop a seismic hazard map. The other is a vertical array (b) consisting of surface and down-hole seismometers at the same place. This can evaluate site amplification exactly at the same location, but does not directly provide the amplification $2A_s / 2A_b$ required for seismic zonation. Some modification is necessary to extract the outcrop motion $2A_b$ from the observed base motion $(A_b + B_b)$ which is more or less contaminated by downward wave B_b from overlying layers.

In this context, Kokusho and Sato (2008) evaluated spectrum amplifications from strong motion records obtained by the KiK-net vertical arrays during 3 strong earthquakes occurred in recent years in Japan. The Fourier spectrum ratio $2A_s / (A_b + B_b)$

between surface and base was used to calculate the theoretical transfer function $2A_s/2A_b$. The multiple reflection theory of 1-dimensionally propagating SH wave was employed to remove the effect of downward wave B_b on the measured motion at the base layer. The peak amplifications of the transfer functions $2A_s/2A_b$ were then correlated with S-wave velocity ratios between base and surface which was newly proposed in this research.

In the present paper, the number of earthquakes is increased from 3 to 8 to investigate if their spectrum peak amplifications are correlated in the same manner with the S-wave velocity ratios. In addition, strain-dependent soil nonlinearity effect on the amplification is investigated by comparing main shock records and those of small shocks obtained at the same KiK-net sites and also at the Port Island where extensive liquefaction occurred during the 1995 Kobe earthquake.

Earthquake Records and Average V_S

KiK-net records corresponding to 8 earthquakes (EQ1 to EQ8) listed in Table 1 are investigated in this research. As shown in Fig. 2, the PGA-values in the vertical axis are from 100 to 2500 cm/s^2, while maximum horizontal accelerations at the base are less than 2000 cm/s^2, showing approximately 2–10 times amplification of the peak horizontal accelerations. The depths of the down-hole seismometers in the vertical arrays used here vary from 100 to 330 m except for 1 site (800 m). S-wave velocities in the top and base layers of all the vertical arrays are plotted in the vertical and horizontal axes, respectively, in Fig. 3. S-wave velocity at the base V_{sb}, which is normally stable in each site and does not drastically change from upper or lower layers, diverge from V_{sb}=400 m/s to 3000 m/s among different seismic observation sites of the eight earthquakes as plotted in Fig. 3. In a good contrast, the surface velocities V_s are within a limited extent; between 100 and 400 m/s, around 200 m/s on average in most of the observation sites used here.

Table 1 Earthquakes and KiK-net records used here

EQ No.	Name of EQ	Year	JMA Mag. M_J	NO. of MS records	PGA (gal)
1	Tottoriken-Seibu	2000	7.3	20	>100
2	Giyo	2001	6.4	13	>100
3	Tokachi-Oki	2003	8.0	19	>200
4	Niigataken-Chuetsu	2004	6.8	15	>100
5	Fukuokaken Seiho-Oki	2005	7.0	10	>100
6	Noto-Hanto	2007	6.9	7	>100
7	Niigataken-Chuetsu-Oki	2007	6.8	9	>100
8	Iwate-Miyagi Nairiku	2008	7.2	11	>150

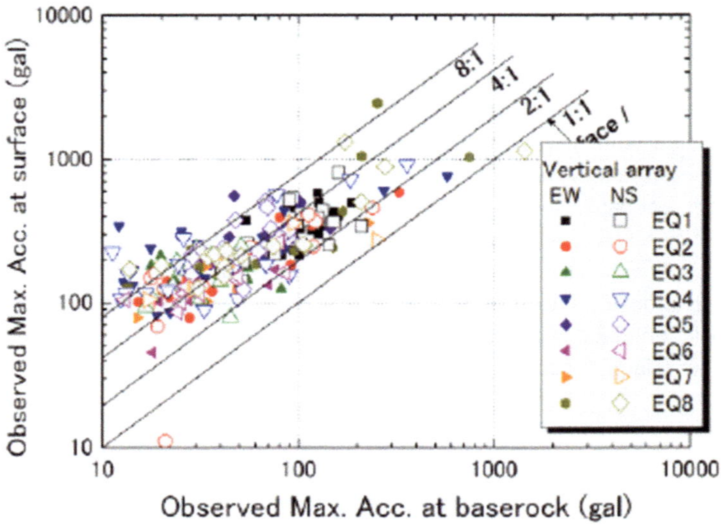

Fig. 2 Max. Acc. at bedrock versus Max. Acc. at ground surface

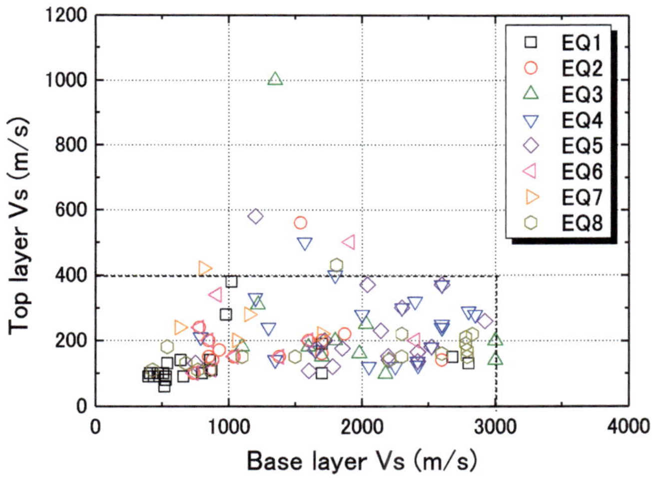

Fig. 3 S-wave velocity Vs at the top and at the base layers

Fourier spectrum ratio $2A_s/(A_b+B_b)$ between surface and base is computed between a ground surface and a base layer for each vertical array site. A typical result for main shock and several aftershocks at a site during EQ3 is shown for the EW direction in Fig. 4a and b. There is a clear difference in the spectrum ratio between the main shock and the aftershocks reflecting strain-dependent soil properties, which result in nonlinear site response to strong shaking. In order to specify soil layers generating peak frequencies in the spectrum ratio, fundamental mode

Fig. 4 Typical spectrum ratios for a main shock and aftershocks during EQ3 (**a**) and peak frequencies computed by 1/4 wave length formula (**b**)

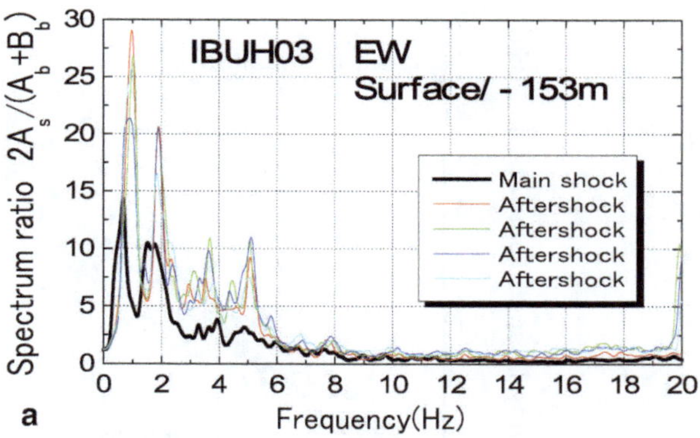

a

Layer NO. i	Thickness H_i(m)	Depth (m)	Vs in each layer Vs_i(m/s)	AverageVs from surface (m/s)	Fund. mode freq. by Eq.(1) f (Hz)	Identification of peaks by observed spectrum peaks
1	4	4	110	110	6.88	3rd peak
2	6	10	320	181	4.54	2nd peak
3	30	40	500	347	2.17	
4	8	48	430	359	1.87	1st peak
5	34	82	510	409	1.25	
6	12	94	650	430	1.14	
7	7	101	870	449	1.10	

b

frequencies f of the layered soil system were calculated by the following 1/4-wave length formula, Eq. (1), based on soil logging data together with S-wave velocities of individual layers.

$$f = 1 \Big/ \left[4 \sum_{i=1}^{n} (H_i / V_{si}) \right] \tag{1}$$

Here, H_i and V_{si} are the thickness and the S-wave velocity of the i-th layer numbered from the top, and H_i / V_{si} is summed up layer by layer down to the base. This frequency f corresponds to an equivalent layer thickness which is equal to 1/4 wave length. The calculated frequency is compared one by one with the peak frequency in the spectrum ratio of observed motions such as in Fig. 4a to identify the equivalent surface layer of thickness $H_s = \sum H_i$ consisting of one or more layers generating the fundamental mode frequency calculated by Eq. (1) as tabulated in Fig. 4b. Note that there can be multiple equivalent surface layers in the same site corresponding to individual peak frequencies.

In Fig. 5, the peak frequencies f calculated by Eq. (1) based on given soil data are taken in the horizontal axis to compare with the frequencies f^* identified in the observed spectrum ratios for the main shock records of the 8 earthquakes. There exists a satisfactory correspondence between the fundamental mode frequency of

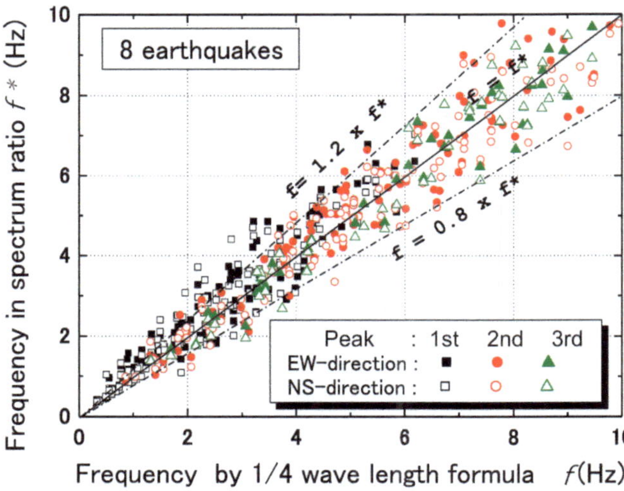

Fig. 5 Peak frequencies calculated by Eq. (1) compared with identified frequencies in observed spectrum ratios for main shock records of 8 earthquakes

soil models f based on Eq. (1) and the peak frequency f^* observed in spectrum ratios irrespective of the order of peaks, 1st to 3rd. The average S-wave velocity $\overline{V_s}$ for each equivalent surface layer can be calculated by Eq. (2) from the fundamental mode frequency f and its thickness $H_s = \sum H_i$.

$$\overline{V_s} = 4H_s f \tag{2}$$

The procedures how to quantify the frequency f and average S-wave velocity is included in the flow chart shown in Fig. 9.

Spectrum Ratios $2A_s/(A_b+B_b)$ Versus $2A_s/2A_b$

In seismic zonation, a transfer function between ground surface records at two different sites $2A_s/2A_b$ is needed, while in vertical array records another transfer function $2A_s/(A_b+B_b)$ is obtained as explained in Fig. 1. Hence, the problem is how to evaluate $2A_s/2A_b$ based on measured motions in the vertical arrays. Figure 6a, b, c, d, e and f show typical examples of the theoretical transfer function $2A_s/(A_b+B_b)$ (thick dashed curves) compared with the other transfer function $2A_s/2A_b$ (thick solid curves) and associated soil profiles at 6 vertical array sites. The installation depths of seismometers are indicated with arrows. At all sites, peak frequencies of $2A_s/(A_b+B_b)$ are mostly compatible with those of observed spectra (EW & NS directions, thin curves), indicating an applicability of one-dimensional soil models in these sites to a certain extent. If $2A_s/(A_b+B_b)$ is compared with $2A_s/2A_b$, the coincidence in peak frequencies is perfect in (a) and good in (b) but gets poorer in

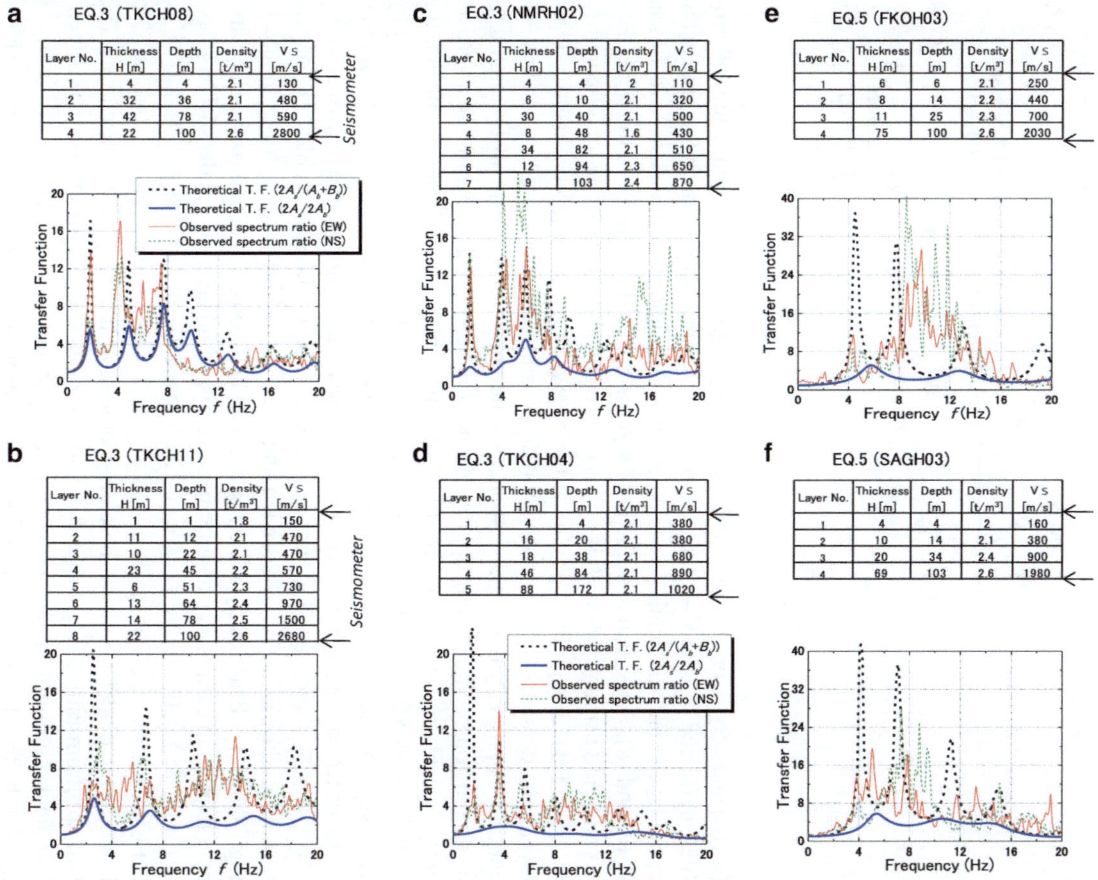

a EQ.3 (TKCH08)

Layer No.	Thickness H [m]	Depth [m]	Density [t/m³]	Vs [m/s]
1	4	4	2.1	130
2	32	36	2.1	480
3	42	78	2.1	590
4	22	100	2.6	2800

b EQ.3 (TKCH11)

Layer No.	Thickness H [m]	Depth [m]	Density [t/m³]	Vs [m/s]
1	1	1	1.8	150
2	11	12	21	470
3	10	22	2.1	470
4	23	45	2.2	570
5	6	51	2.3	730
6	13	64	2.4	970
7	14	78	2.5	1500
8	22	100	2.6	2680

c EQ.3 (NMRH02)

Layer No.	Thickness H [m]	Depth [m]	Density [t/m³]	Vs [m/s]
1	4	4	2	110
2	6	10	2.1	320
3	30	40	2.1	500
4	8	48	1.6	430
5	34	82	2.1	510
6	12	94	2.3	650
7	9	103	2.4	870

d EQ.3 (TKCH04)

Layer No.	Thickness H [m]	Depth [m]	Density [t/m³]	Vs [m/s]
1	4	4	2.1	380
2	16	20	2.1	380
3	18	38	2.1	680
4	46	84	2.1	890
5	88	172	2.1	1020

e EQ.5 (FKOH03)

Layer No.	Thickness H [m]	Depth [m]	Density [t/m³]	Vs [m/s]
1	6	6	2.1	250
2	8	14	2.2	440
3	11	25	2.3	700
4	75	100	2.6	2030

f EQ.5 (SAGH03)

Layer No.	Thickness H [m]	Depth [m]	Density [t/m³]	Vs [m/s]
1	4	4	2	160
2	10	14	2.1	380
3	20	34	2.4	900
4	69	103	2.6	1980

Fig. 6 Typical examples of the theoretical transfer functions $2A_s/(A_b+B_b)$, $2A_s/2A_b$ and associated soil profiles at 6 vertical array sites

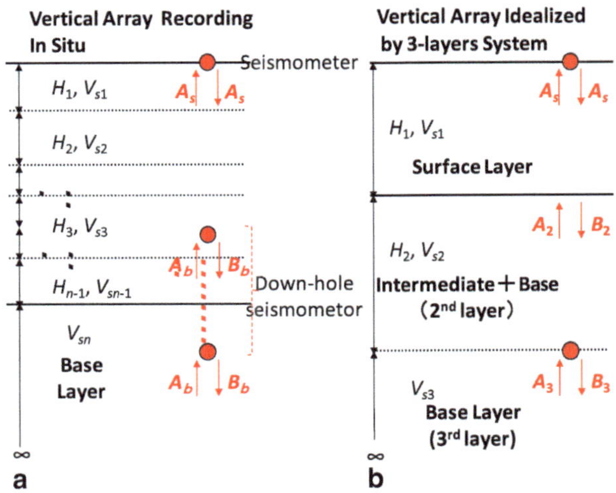

Fig. 7 Vertical array recording in situ (**a**) versus an idealized 3-layerd model (**b**) for a simple analysis

(c), (d) and very poor in (e), (f). The reason may be given by examining the soil profiles. In (a) and (b), the V_s-value at the depth of seismometer is much larger than the upper layers, and the depth is not so deep from a boundary of the clear V_s-contrast. In (c) and (d), the V_s-value at the base layer is not so different from the upper layers and the seismometer depth is not so deep from a boundary of major V_s-contrast. In (e) and (f), though the V_s-value at the base layer is much larger than the upper layers, the depth of seismometer is too deep from the upper boundary of clear V_s-contrast to properly detect the response of the upper layers. This observation tells us a significance of choosing appropriate seismometer depth in deploying a vertical array system considering soil profiles. In deploying down-hole seismometers, the installation depth may sometimes be decided not by technical considerations but by economy. In case of KiK-net, it is around 100 m from ground surface in most case.

In order to find out a basic rule how peak frequencies in the two spectra, $2A_s/(A_b+B_b)$ and $2A_s/2A_b$ are governed by soil profiles, a simple 1D 3-layers system as shown in Fig. 7 has been analyzed. An arbitrary soil profile in situ in Fig. 7a may be idealized by a 3-layered model in (b) consisting of a soft surface layer (1st layer: thickness H_1), an intermediate layer (2nd layer: H_2) and a stiff base layer (3rd layer) with infinite thickness. The impedance ratios are $\alpha_{12}=(\rho V_s)_1/(\rho V_s)_2$ between the 1st and 2nd layer and $\alpha_{23}=(\rho V_s)_2/(\rho V_s)_3$ between the 2nd and 3rd layer. In the model, the down-hole seismometer is at the top of the base layer. The two spectrum ratios, $2A_s/(A_b+B_b)$ and $2A_s/2A_b$, are calculated for typical geotechnical conditions in the KiK-net web-site and shown in the vertical axis in Fig. 8 against the frequency normalized by the first resonant frequency in the surface layer f_1 in the horizontal axis.

In Fig. 8a, where the thickness of the intermediate layer H_2 is parametrically varied with respect to H_1, the spectrum ratio $2A_s/2A_b$ is very stable with its peak frequency $f/f_1 \approx 1.0$ irrespective of the H_2/H_1-ratio because of the low impedance ratios, $\alpha_{12}=0.2$, between the surface and intermediate layer. In contrast, the spectrum

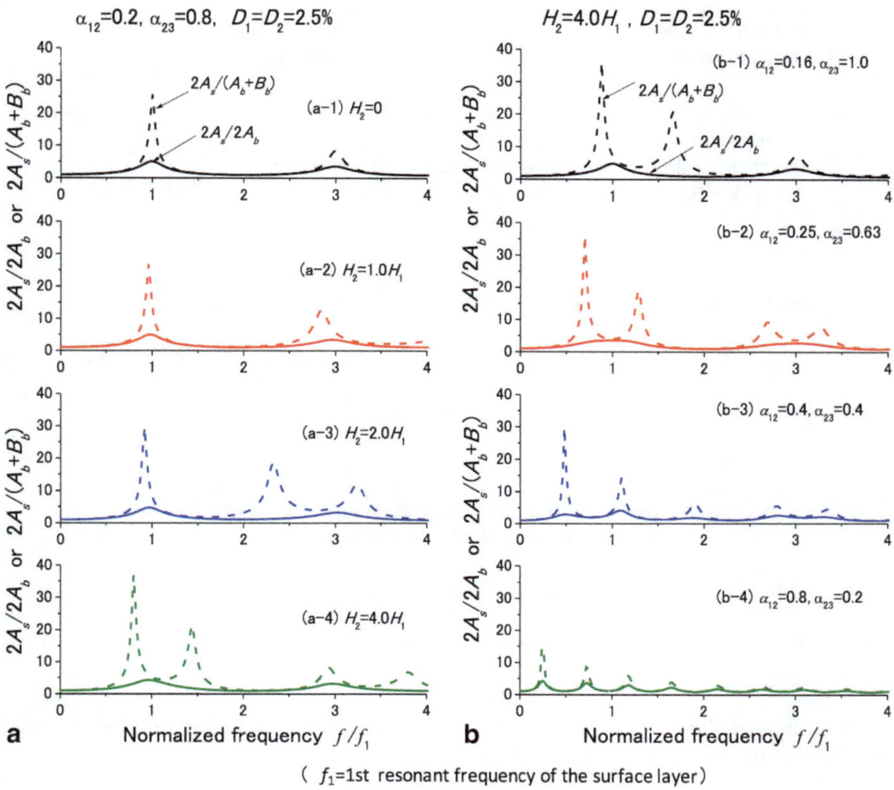

Fig. 8 Two spectrum rations, $2A_s/(A_b+B_b)$ and $2A_s/2A_b$, calculated for a 3-layers system: (**a**) $H_2=0\sim4.0\,H_1$ for $\alpha_{12}=0.2$, $\alpha_{23}=0.8$, (**b**) $\alpha_{12}=0.16\sim0.8$, $\alpha_{23}=1.0\sim0.2$ for $H_2=4.0\,H_1$

ratio $2A_s/(A_b+B_b)$ changes its shape with variable peak frequencies if the H_2/H_1-ratio exceeds 2.0. This is because, in the spectrum ratio $2A_s/(A_b+B_b)$, the depth of seismometer installation serves as a rigid boundary generating a virtual spectral amplification even if it is in the midst of an uniform layer. In Fig. 8b, where the impedance ratios α_{12}, α_{23} are parametrically changing (keeping $\alpha_{12}\times\alpha_{23}=0.16$ constant) with a constant thickness ratio $H_2/H_1=4.0$, the two spectrum ratios together with their peak frequencies are quite different for $\alpha_{23}=1.0$ and $\alpha_{23}=0.63$, though they coincide to each other if α_{23} is nearly equal to 0.4 or smaller. This indicates that a clearer impedance contrast is preferred at the boundary near the down-hole seismometer.

Consequently, in utilizing vertical array data for a micro-zonation, a down-hole seismometer should be in a base layer not too deep from the layer boundary. A clearer impedance contrast at the base boundary is preferable. If the depth is twice or more deeper than the surface layer thickness from a boundary with the clear impedance contrast, virtual peak frequencies tends to appear in $2A_s/(A_b+B_b)$ different from those in $2A_s/2A_b$.

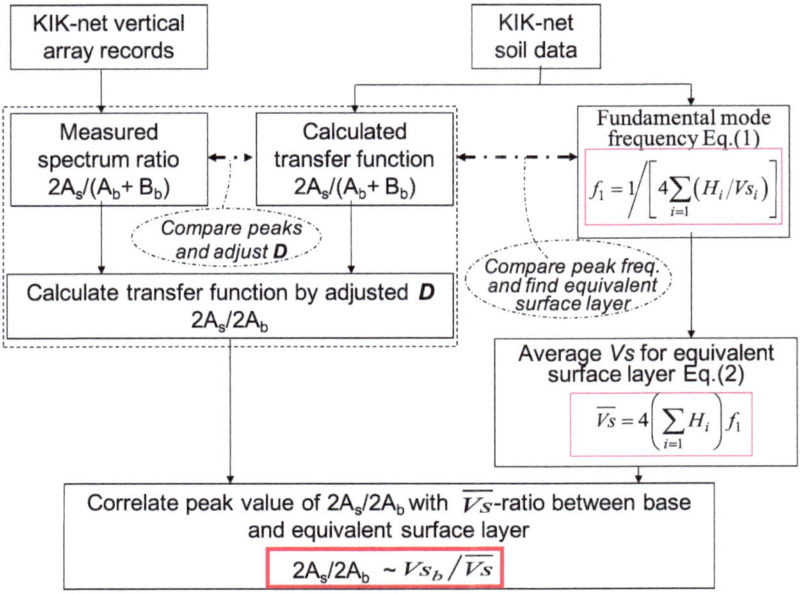

Fig. 9 Flow chart for correlating peak value of $2A_s/2A_b$ with $\overline{V_s}$ -ratio

This indicates that there is an appropriate installation depth for down-hole seismometers; never the deeper, the better.

Peak Amplification Versus V_s–Ratio

A procedure chosen here to evaluate $2A_s/2A_b$ from $2A_s/(A_b+B_b)$ is shown in a flow chart Fig. 9 and described as follows (Kokusho and Sato 2008).

1. A theoretical transfer function, $2A_s/(A_b+B_b)$ is calculated for each site using given soil properties with frequency-independent damping ratio, D, tentatively assumed as 2.5 % in all layers.
2. The function $2A_s/(A_b+B_b)$ is compared with measured spectrum ratio as exemplified in Fig. 10. If a peak in the transfer function can be found at about the same frequency in the spectrum ratio of observed motions, it is identified as the corresponding peak, and the damping ratio common in all layers, assumed as $D=2.5$ % previously, is modified by $D=Q_1/Q_2\times 2.5\%$ to have the identical peak value, where Q_1 is the peak amplitude of the theoretical transfer function, and Q_2 is that of spectrum ratio of the actual records. The values of D calculated in EW and NS directions are averaged.
3. Another theoretical transfer function $2A_s/2A_b$ is computed using the modified damping ratio D based on the same soil layers model. When the peak frequencies

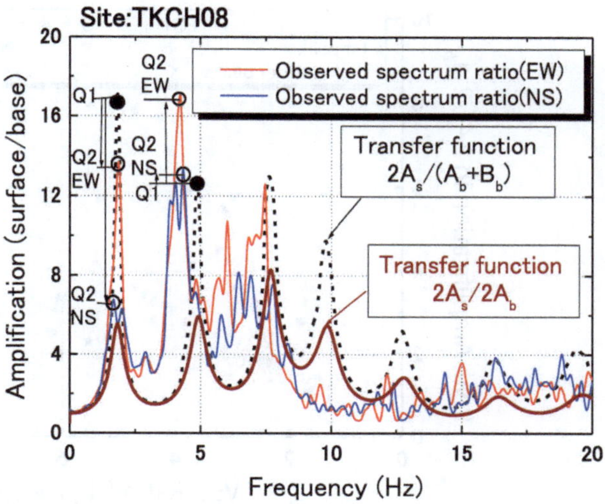

Fig. 10 Comparison of transfer functions, $2A_s/(A_b+B_b)$ and spectrum ratio of observed motions at the same site

of the two functions are different, peak frequencies of $2A_s/2A_b$ are compared with fundamental mode frequencies calculated by Eq. (1) in newly defined equivalent surface layers, and associated average S-wave velocities by Eq. (2) are determined again. The peak values of $2A_s/2A_b$ are adjusted using D-values determined from the nearest peaks of $2A_s/(A_b+B_b)$. If the two transfer functions were very different with completely different peak frequencies such as in Fig. 7d, e, f, that particular site was not used for the data plots. In Fig. 11, damping ratios thus

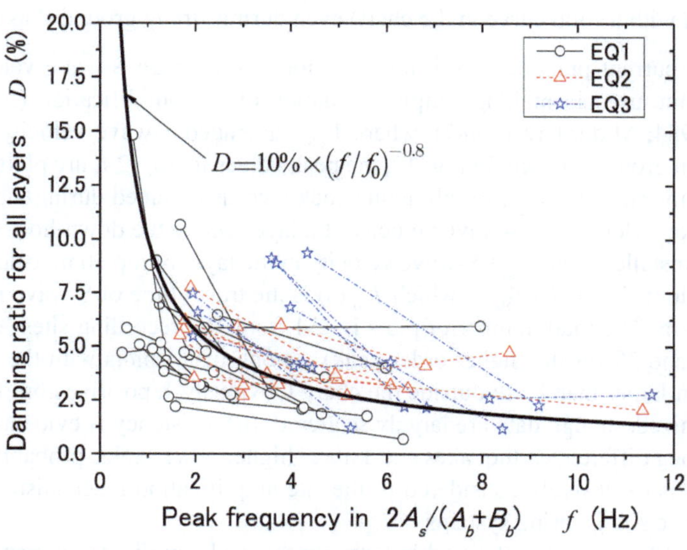

Fig. 11 Modified damping ratio for good matching in peak spectrum values and corresponding frequencies for 3 earthquakes

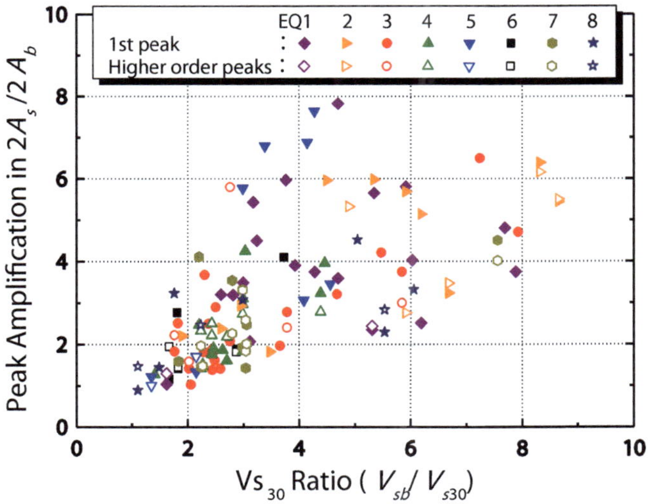

Fig. 12 Peak amplifications of $2A_s/2A_b$ for main shocks of 8 earthquakes versus average S-wave velocity ratios Vs_b/Vs_{30} using Vs_{30} of top 30 m surface soils

modified to have good agreement between observation and theory in spectrum amplifications $2A_s/(A_b + B_b)$ are plotted versus corresponding peak frequencies for three earthquakes, EQ1, EQ2 and EQ3. The plots connected with dashed lines are for the 1st and higher order peaks from the same records. Though very variable, the damping ratio *D* mostly tends to decrease with increasing frequency in higher order peaks. This seems to indicate that the damping ratio is not purely non-viscous but somewhat frequency-dependent (approximated by a function shown with a solid curve in the chart) even during strong ground shaking.

In the current practice of seismic zonation, the average S-wave velocity V_{s30} is sometime used in making simple evaluation of site amplification (Joyner and Fumal 1984; Midorikawa 1987), where V_{s30} = averaged S-wave velocity over top 30 m from ground surface. In Fig. 12, the peak values in $2A_s/2A_b$ are plotted versus the velocity ratio, V_{sb}/V_{s30}, for all main shock records obtained during the 8 strong earthquakes. Here, V_{sb} = S-wave velocity at a layer where the down-hole seismometer was installed, and V_{s30} = S-wave velocity in the layer of top 30 m, calculated by the equation; $V_{s30} = 30/T_{30}$ in which T_{30} (s) is the travel time of S-wave in the soil of top 30 m. The total number of plots is 134 from 102 recording sites (95 for the 1st peak and 39 for the higher order peak), among that 4 plots with the peak amplification larger than 8 are outside the chart. Obviously, a positive correlation can be recognized, though data are largely scattered. Inconsistency is evidently seen in plots among different earthquakes and 1st and higher order peaks, probably because top 30 m is exclusively considered in the site amplification mechanism, ignoring site-specific soil layering systems.

Figure 13 shows the relationship between the peak amplitude corresponding to $2A_s/2A_b$ and the $\overline{V_s}$ - ratio plotted for main shocks of the eight strong earthquakes.

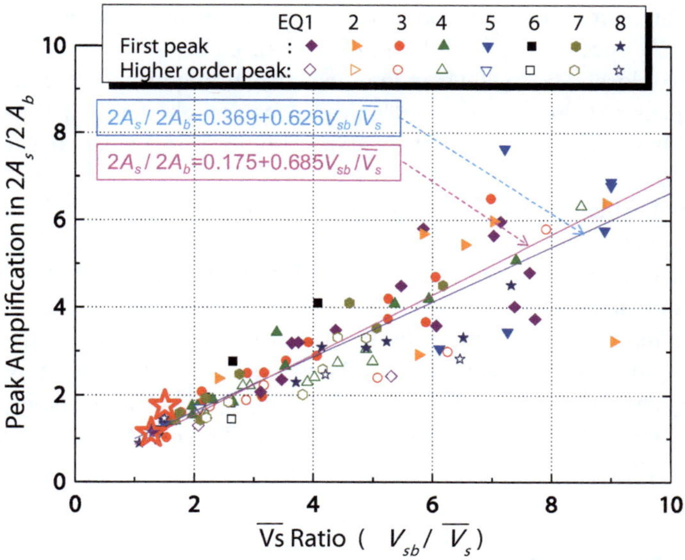

Fig. 13 Peak amplifications of $2A_s/2A_b$ for main shock of 8 earthquakes versus average S-wave velocity ratios $V_{sb}/\overline{V_s}$ proposed here and approximation by empirical equation

The $\overline{V_s}$-ratio, $V_{sb}/\overline{V_s}$, is defined here as a division of the S-wave velocity at a base layer V_{sb} by the velocity $\overline{V_s}$ evaluated by Eq. (2) from the fundamental mode frequencies of Eq. (1) based on S-wave velocities at each site. Totally 131 data points from 95 recording sites are included in this chart (93 for the 1st peak and 38 for the higher order peak, and 16 data points in which $V_{sb}/\overline{V_s} \geq 10$ are outside the chart). Note that a large number of plots are overlapping in the zone of $V_{sb}/\overline{V_s} \leqq 4.0$. In a good contrast to Fig. 12 where Vs_{30} is used, the plots in Fig. 13 using $\overline{V_s}$ show a fairly good correlation for both 1st and higher order peaks despite differences in various influencing factors associated with the eight earthquakes; namely, dominant frequency, shaking duration, regional geology, etc. This indicates the importance to define the average S-wave velocity properly by identifying site by site the equivalent surface layer in which individual peak amplifications are exerted.

A simple linear equation may be derived in Fig. 13 for the data-points of $\overline{V_s} = 4Hf$ (a condition normally encountered), with the regression coefficient $RC = 0.89$, as

$$2A_s/2A_b = 0.369 + 0.626\left(V_{sb}/\overline{V_s}\right) \tag{3}$$

The equation is not so different from what was proposed based on 3 earthquakes (EQ3, EQ4 and EQ5) in the previous research Kokusho and Sato (2008): $2A_s/2A_b = 0.175 + 0.685\left(V_{sb}/\overline{V_s}\right)$. Equation (3) may be conveniently used because of its applicability to a wide variety of base layers with $V_{sb} = 400\,\text{m/s}$ to $3000\,\text{m/s}$. The relative amplification for the same seismic motion in an area underlain by a common base layer is evaluated as follows (Kokusho and Sato 2008);

1. Based on microtremor measurements, decide fundamental frequency f of a site using H/V spectrum ratios (Nakamura 1989). Estimate the thickness of a soft soil or Holocene layer H where the fundamental frequency is exerted, which is sometimes possible based on geological maps or high-density soil logging data available in city areas. Then, the average S-wave velocity can be calculated by $\overline{V_s} = 4Hf$.
2. If, a Vs-profile is obtained by surface wave methods or down-hole methods, $\overline{Vs}$ is readily calculated by Eqs. (1) and (2)
3. Calculate the S-wave velocity ratio $V_{sb}/\overline{V_s}$ from V_{sb} of the common base layer and $\overline{Vs}$ obtained above.
4. Comparing the amplifications by Eq. (3) at two different sites gives the relative amplification between them. To be precise, the amplification by Eq. (3) is slightly changeable depending on the value of V_{sb} to be chosen among different base layers in the two sites, though its effect is ignorable for design purposes.

Soil Nonlinearity Effect

In Fig. 14, the peak amplifications of the first peak for the two kinds of spectrum ratio, $2A_s/2A_b$ and $2A_s/(A_b + B_b)$, are plotted versus $\overline{V_s}$ -ratio, $V_{sb}/\overline{V_s}$, not only for main shocks but also for aftershocks of the 8 earthquakes. Totally 42 sites were selected where the peak frequencies f of observed Fourier spectrum ratio are consistent with the theoretical values, and the f-values in two theoretical transfer functions $2A_s/2A_b$ and $2A_s/(A_b + B_b)$ are comparable to each other at least in the 1st peak. In

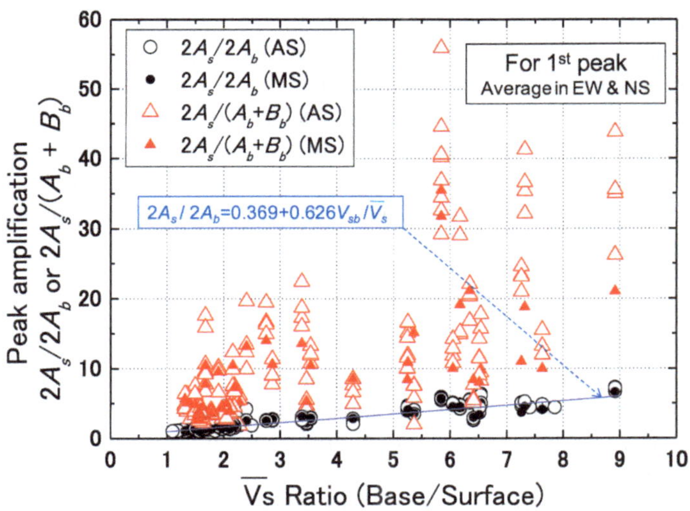

Fig. 14 Peak amplifications of $2A_s/2A_b$ and $2A_s/(A_b+B_b)$ for the first peak of main shocks and aftershocks of 8 earthquakes versus average S-wave velocity ratios $V_{sb}/\overline{V_s}$

most sites, 4 aftershocks were chosen individually to compare with the corresponding main shock. The peak values of $2A_s/2A_b$ are adjusted in the same way as explained before using D-values determined from the nearest peaks of $2A_s/(A_b+B_b)$, and the values in EW and NS directions are averaged. Solid and open circles are peak amplifications of $2A_s/2A_b$ for main shock and aftershocks, and solid and open triangles are peak amplifications of $2A_s/(A_b+B_b)$ for main shock and aftershocks, respectively. For $2A_s/2A_b$, a clear correlation can be recognized between the peak amplification and the $\overline{V_s}$-ratio, which can be approximated by Eq. (3), again. Note that the difference in peak amplifications between a main shock and corresponding aftershocks are really marginal, indicating insignificant effect of soil-nonlinearity. In contrast, the plots for $2A_s/(A_b+B_b)$ are not so well correlated to $\overline{V_s}$-ratio. Furthermore, the peak amplifications of $2A_s/(A_b+B_b)$ for the main shocks are evidently smaller than the corresponding aftershocks in most sites. This indicates that the soil nonlinearity effect is quite dominant in $2A_s/2A_b$ in $2A_s/(A_b+B_b)$.

In order to account for this difference, a simple 2-layers system of a surface layer (Impedance$=\rho_1 V_{s1}$, Damping ratio$=D_1$) underlain by an infinitely thick base layer (Impedance$=\rho_2 V_{s2}$, Damping ratio$=D_2$) shown in Fig. 15a was studied Kokusho and Sato (2008), assuming the impedance ratio for small strain properties as $\alpha=\rho_1 Vs_1/\rho_2 Vs_2=0.3$ The transfer functions $2A_s/(A_b+B_b)$ and $2A_s/2A_b$ can be expressed by the following equations, respectively Kokusho and Sato (2008);

$$2A_s/(A_b+B_b)=2\Big/\left(e^{ik_1^*H}+e^{-ik_1^*H}\right) \tag{4}$$

$$2A_s/2A_b=2\Big/\left[\left(1+\alpha^*\right)e^{ik_1^*H}+\left(1-\alpha^*\right)e^{-ik_1^*H}\right] \tag{5}$$

where $k_1^*=\omega\Big/\left(Vs_1\sqrt{1+2iD_1}\right)$ and $\alpha^*=\alpha\sqrt{(1+2iD_1)/(1+2iD_2)}$.

In taking account the effect of strain-dependent soil properties on the amplification, the shear modulus ratio G/G_0 and the damping ratio D_1 in the surface layer are parametrically changed; $G/G_0=1.0$, 0.65, 0.25 and $D_1=2.5$, 5 and 15 % correspondingly, for strain level of 5×10^{-6}, 1×10^{-4}, 1×10^{-3}; respectively, assuming a typical degradation curve for sand (Seed and Idriss 1970) as indicated in Fig. 15b, while in the base layer $D_2=0$. The calculated results of $2A_s/(A_b+B_b)$ and $2A_s/2A_b$ are shown in Fig. 15c and d, respectively. Here, the amplification between surface and base is shown versus the normalized frequency, f/f_1, where $f_1=$fundamental mode frequency of the surface layer for small strain properties ($G/G_0=1.0$).

Obviously, nonlinear soil properties have a great effect on the peak frequency and the amplification. In particular, the peak frequency becomes evidently lower with increasing shear modulus degradation both in $2A_s/(A_b+B_b)$ (Fig. 15c) and $2A_s/2A_b$ (Fig. 15d). However, in terms of peak amplification, it is noted that the soil nonlinearity has much smaller effect in $2A_s/2A_b$ than in $2A_s/(A_b+B_b)$ for the 1st peak in particular. This is because radiation damping represented by the impedance ratio α^* affects $2A_s/2A_b$. In Eq. (5), the amplification is expressed as $|2A_s/2A_b|\approx|1/\alpha^*|$ at the 1st peak. In contrast, no effect of impedance ratio α^* is involved in $2A_s/(A_b+B_b)$ as indicated in Eq. (4). Under the paramount effect of

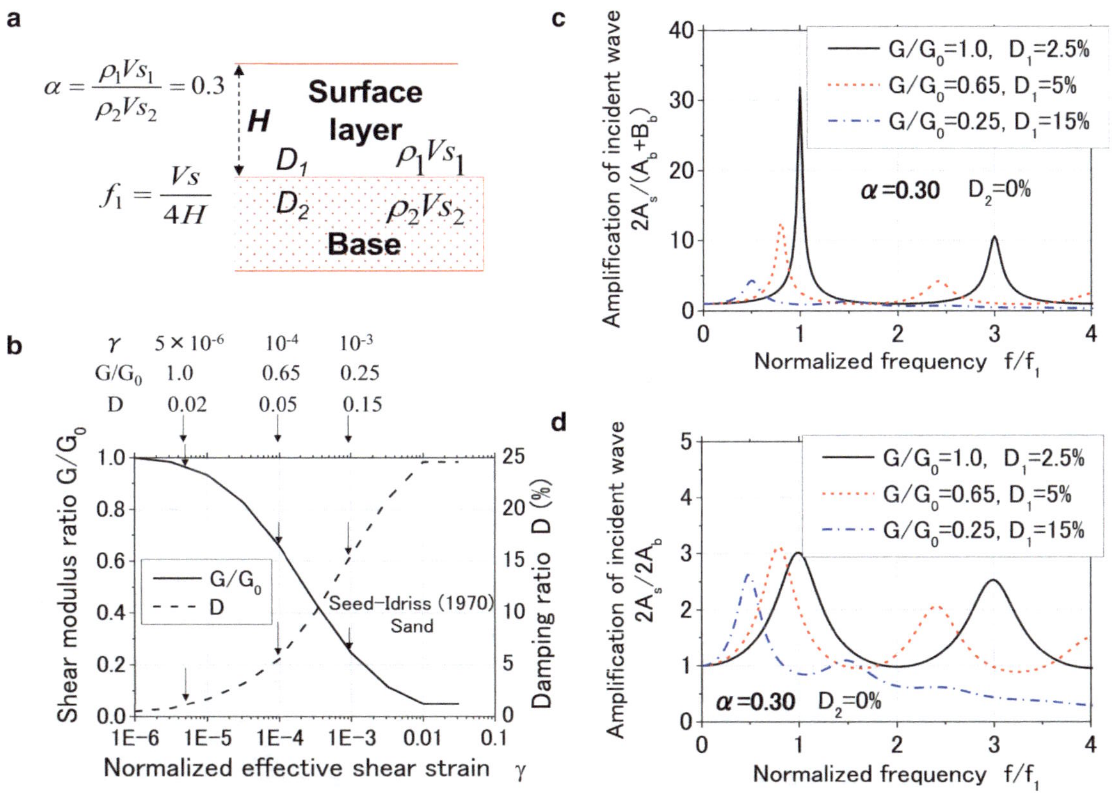

Fig. 15 Two-layers system (**a**), Strain-dependent properties of surface layer (**b**), $2A_s/(A_b + B_b)$ (**c**) and $2A_s\ /\ 2A_b$ (**d**); for different strain levels in surface layer

radiation damping, the difference in the amplification due to strain-dependent properties becomes less dominant. Furthermore, the impedance ratio $\alpha = \rho_1 V_{s1} / \rho_2 V_{s2}$, which becomes smaller with degraded modulus or degraded S-wave velocity in the surface layer, tends to give larger amplification, compensating the effect of increased soil damping during strong earthquakes. Thus, the difference in soil properties between the main shock and aftershocks tends to have smaller influence on the amplification in $2A_s / 2A_b$ than in $2A_s / (A_b + B_b)$ as demonstrated in Fig. 14 by actual seismic data.

Soil Liquefaction Effect

In the above, seismic response of ground was evaluated based on linear transfer function, assuming that strain-dependent soil nonlinearity exerted during earthquakes is not so considerable. Actually, no KiK-net records studied here appear to have experienced strong soil nonlinearity accompanying extensive liquefaction. Hence, Port Island (PI) vertical array records during the 1995 Kobe earthquake, in-

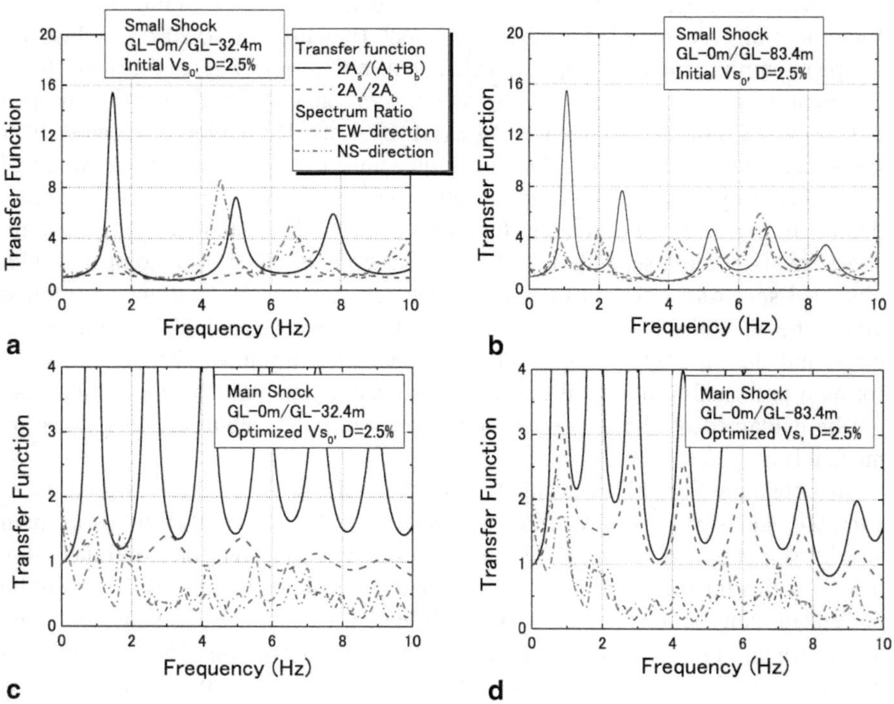

Fig. 16 Transfer functions of Port Island (PI) vertical array; (**a**, **b**) based on logging Vs compared with observed spectrum ratios of small shocks, (**c**, **d**) based on back-calculated Vs for main shock compared with observed spectrum ratios of main shock

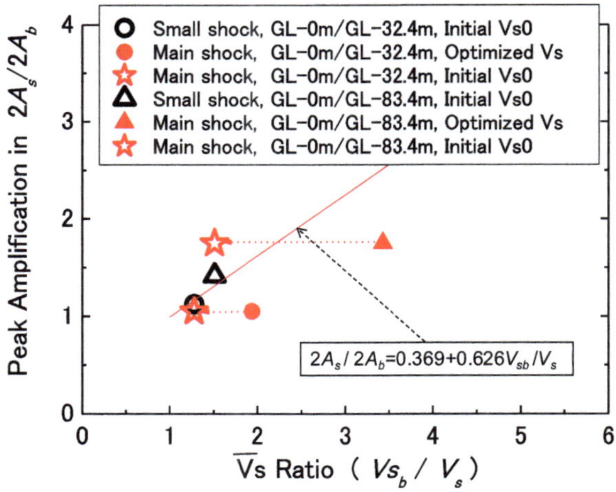

Fig. 17 Peak amplifications of $2A_s/2A_b$ versus $\overline{\text{V}}$s-ratios for small shock and main shock records at Port Island during the 1995 Kobe earthquake

flicted by severe liquefaction in surface fill layer of about 15 m thick, are examined in the same methodology as before in the following.

In Fig. 16a and b, transfer function $2A_s/(A_b + B_b)$ at PI-site calculated between surface and two down-hole levels, GL-32.4 m and GL-83.4 m, using small-strain V_s by wave logging, are compared with corresponding spectrum ratios from small shock motions recorded before the main shock (Kokusho et al. 2005). In Fig. 17, the peak amplifications of $2A_s/2A_b$ calculated in the same way as for many KiK-net records are plotted versus the $\overline{\text{V}}_\text{s}$-ratios based on small strain V_s with an open circle (GL-32.4 m) and an open triangle (GL-83.4 m). These plots are located near the straight line of Eq. (3), indicating satisfactory evaluation of amplification of small shocks by the proposed formula between ground surface and the two different depths. In Fig. 16c and d, transfer functions using back-calculated V_s for the main shock, available in a separate literature (Kokusho et al. 2005), are compared with observed spectrum ratios of the main shock. The peak amplifications are plotted versus the $\overline{\text{V}}_\text{s}$ -ratios based on back-calculated V_s (Kokusho et al. 2005) with a close circle and close triangle in Fig. 17. The plots are evidently lower than the proposed formula presumably due to the effect of extensive soil liquefaction, which may not be explained by the previously mentioned simplified nonlinear amplification mechanism.

In obtaining Fig. 13 by using KiK-net records, small-strain V_s was used to compute values of $\overline{V}_s$ and $\overline{\text{V}}_\text{s}$ -ratios to correlate with peak amplifications of main shocks as mentioned before. If the same procedure is followed for the PI data, too, the $\overline{\text{V}}_\text{s}$ -ratios for small-strain V_s, in lieu of back-calculated large-strain V_s, are to be correlated with the main shock peak amplification values. The plots thus obtained are shown on Fig. 17 and also on Fig. 12 with open star symbols. They indicate that even the amplification in PI reflecting extensive liquefaction may be roughly evaluated by the proposed formula Eq. (3) using small-strain V_s.

Conclusion

Site amplification formula for seismic zoning was developed to be consistent with the vertical array records using a number of KiK-net records during recent 8 destructive earthquakes. Among a number of KiK-net sites, those which show good correspondence between observed spectrum ratios and theoretical transfer functions between surface and base were used in the analysis. It was also pointed out that the two transfer functions for the vertical array $2A_s/(A_b+B_b)$ and for the surface array $2A_s/2A_b$ together with their peak frequencies may significantly differ from each other depending on the soil profile and the installation depth of the down-hole seismometer. Then, average S-wave velocity $\overline{V_s}$ for an equivalent surface layer, generating a peak in observed spectrum ratio between ground surface and a base layer, was introduced to be correlated to the calculated peak amplification of outcrop motions, yielding the following major findings;

1. The spectrum peak amplifications of $2A_s/2A_b$ for outcrop motions plotted versus the $\overline{V_s}$-ratios ($V_{sb}/\overline{V_s}$: $V_{sb}=V_s$ at a base layer) show a good correlation with a small data dispersions between the peak amplification and the velocity ratio both for 1st and higher order peaks, despite differences in influencing factors of individual earthquakes.
2. The correlation mentioned above obtained by utilizing 8 strong earthquakes coincides well with Eq. (3), which was previously proposed. Equation (3) or its modified version in the present paper may be conveniently used for evaluating relative amplification in seismic zonation study covering an area sharing a common base layer with S-wave velocity, Vsb.
3. If the same peak amplifications are plotted versus velocity ratios defined by V_{sb}/V_{s30} (Vs_{30}=average velocity for top 30 m sometimes used in the current seismic zonation practice), the correlation becomes poorer, indicating the importance to define the average S-wave velocity adequately by identifying a site-specific equivalent surface layer in which peak amplifications are exerted.
4. The velocity $\overline{V_s}$ of the equivalent surface layer can be evaluated from Vs-logging data, or if it is unavailable, may be determined from fundamental mode frequency f of a site using *H*/*V* spectrum ratios in micro-tremor measurements together with the thickness of soft soil or Holocene layer H by $\overline{V_s}=4Hf$.

Furthermore, the effect of nonlinear soil properties was studied by comparing peak amplifications for main shock and small shocks recorded at the same sites, revealing the following;

5. Strain-dependent soil nonlinearity has a minor effect on the peak amplifications of $2A_s/2A_b$ for surface arrays (outcrop motions) compared to those of $2A_s/(A_b+B_b)$ for vertical arrays.
6. The logical bases of the minor nonlinear effect on $2A_s/2A_b$ can be explained by a simple 2-layers equivalent linear system considering strain-dependent soil nonlinearity.
7. Even the peak amplification of $2A_s/2A_b$ at Port Island during the 1995 Kobe earthquake where extensive liquefaction occurred may be roughly approximated by the proposed formula.

Acknowledgments NIED (National Research Institute for Earth Science and Disaster Prevention) in Tsukuba, JAPAN, who disseminated numerous KiK-net data, and Kobe city, who provided Port Island vertical array data during the 1995 Kobe earthquake, are gratefully acknowledged. The great efforts in data reduction of voluminous vertical array records by graduate students (K. Sato and K. Ejiri in particular) and undergraduate students (T. Kanamaru and K. Suzuki in particular) of Civil Engineering Department in Chuo University are also very much appreciated.

References

Joyner WB, Fumal TE (1984) Use of measured shear-wave velocity for predicting geologic site effects on strong ground motion. Proc. of 8th World Conference on Earthquake Engineering, vol 2, pp 777–783

Kokusho T, Sato K (2008) Surface-to-base amplification evaluated from KiK-net vertical array strong motion records. Soil Dynamics and Earthquake Engineering. Elsevier, pp 707–716

Kokusho T, Aoyagi T, Wakunami A (2005) In situ soil-specific nonlinear properties back-calculated from vertical array records during 1995 Kobe Earthquake. J Geotech Geoenviron Eng, ASCE 131(12):1509–1521

Midorikawa S (1987) Prediction of iso-seismal map in the Kanto Plain due to hypothetical earthquake. J Struct Eng 33B:43–48 (in Japanese)

Nakamura Y (1989) A method for dynamic characteristics estimation of subsurface using microtremor on the ground surface. QR Railw Techn Res Insti 30:25–33

Seed HB, Idriss IM (1970) Soil Moduli and damping factors for dynamic response analysis, Report EERC 70–10. University of California, Berkeley

Shima E (1978) Seismic Microzoning map of Tokyo. Proc. Second International Conference on Microzonation (1), pp 433–443

Observations from Geotechnical Arrays in Istanbul

Asli Kurtuluş, Atilla Ansal, Gökçe Tönük and Barbaros Çetiner

Abstract Few small earthquakes with local magnitude slightly larger than $M_L = 4$ were recorded by geotechnical downhole arrays that have been recently deployed in the west side of Istanbul. Same events were also recorded by Istanbul Rapid Response Network (IRRN) which comprises of 55 surface strong motion stations in the European side of Istanbul. The strongest one of these earthquakes took place on 12/3/2008 in Çınarcık with local magnitude of $M_L = 4.8$. Even though the observed PGAs were not exceeding 0.01 g, an effort is made to model the recorded response at the downhole array sites as well as the at the IRRN stations using the acceleration records obtained by the deepest sensors, i.e. on the engineering bedrock, at the downhole array sites as input bedrock motions. 1D equivalent linear site response analysis that is generally adopted for site-specific response analysis is used for modelling. Observations from the recorded response and results from 1D modelling of ground response have yielded in general good agreement between the observed and recorded soil response at the station sites.

Introduction

Estimation of site-specific ground motion characteristics has been a crucial issue in the assessment of the vulnerability of existing structures, for retrofit and rehabilitation alternatives as well as in the design of new structures. The scientific aspects contain significant degree of uncertainties and reliable solution of this problem requires development of realistically comprehensive numerical and analytical models that rely on large amount of field evidence. Geotechnical arrays, which are deployed in seismically active areas to record ground motion during earthquakes,

A. Ansal (✉) · A. Kurtuluş
School of Engineering, Özyeğin University, Istanbul, Turkey
e-mail: atilla.ansal@ozyegin.edu.tr

G. Tönük
School of Engineering, MEF University, Ayazağa, Istanbul, Turkey

B. Çetiner
Kandilli Observatory & Earthquake Research Institute, Boğazici University, Istanbul, Turkey

A. Ansal, M. Sakr (eds.), *Perspectives on Earthquake Geotechnical Engineering*,
Geotechnical, Geological and Earthquake Engineering 37,
DOI 10.1007/978-3-319-10786-8_7

are one of the ways to collect field evidence for soil response. The information provided by these arrays is considered valuable for verification, calibration and development of predictive tools that are used in earthquake engineering. Acceleration records obtained during recent earthquakes have demonstrated that geological and geotechnical site conditions such as soil stratification, depth of ground water table, and properties of soil layers could have significant influence on strong motion characteristics on the ground surface (Beresnev et al. 1998; Trifunac and Todorovska 1998; Fukushima et al. 2000). The observations in previous earthquakes indicate that the differences in soil profiles even within relatively short distances can have major effect on the response.

Recently, three geotechnical downhole arrays are deployed in the European side of Istanbul, Turkey (Kurtulus 2011). Each array is composed of one accelerometer on the ground surface and three or four borehole accelerometers at various depths along the soil profile with the deepest sensor located at the engineering bedrock level ($V_S > 750$ m/s). These arrays also provide reference bedrock motion for Istanbul Rapid Response Network (IRRN) comprising of 55 surface instruments operated by Kandilli Observatory and Earthquake Research Institute (KOERI) in the European side of the city (Erdik et al. 2003). Local soil conditions at the downhole array and IRRN station sites, which are characterized by conventional geotechnical methods and in-hole seismic wave velocity measurements, indicate that the majority of sites are classified as NEHRP (2003) Site Class D with the rest being represented with Site Class C (Ansal et al. 2010). A number of small magnitude events have been recorded by these arrays in recent years. Observations from these recently recorded weak motions provide an opportunity to investigate soil response at linear strain range.

Recorded Events

The downhole arrays are relatively new, therefore only a number of small magnitude earthquakes have been recorded so far. Locations of the three downhole arrays, referred as ATK, ZYT and FTH hereafter, as well as IRRN stations are shown in Fig. 1. Maximum ground acceleration recorded at the sites so far does not exceed 0.01 g. Among the recorded earthquakes, most significant are 12.03.2008 $M_L = 4.8$ Çınarcık, 03.10.2010 $M_L = 4.4$ Marmara and 19.05.2011 $M_L = 5.9$ Kütahya events. On the European side of city 10 of the IRRN stations have simultaneously recorded all three events. Locations of the epicenters of these events with respect to the accelerometer arrays are shown in Fig. 2. All are shallow events (depth of ~11 km) generated by different segments of EW strike-slip North Anatolian Fault (NAF). As can be observed from Fig. 2, epicentral distances of the two earlier events are quiet similar (~70 km) with almost perpendicular azimuths while the Kütahya event occurred at a much larger distance (~200 km) yet in similar direction with the Çınarcık event.

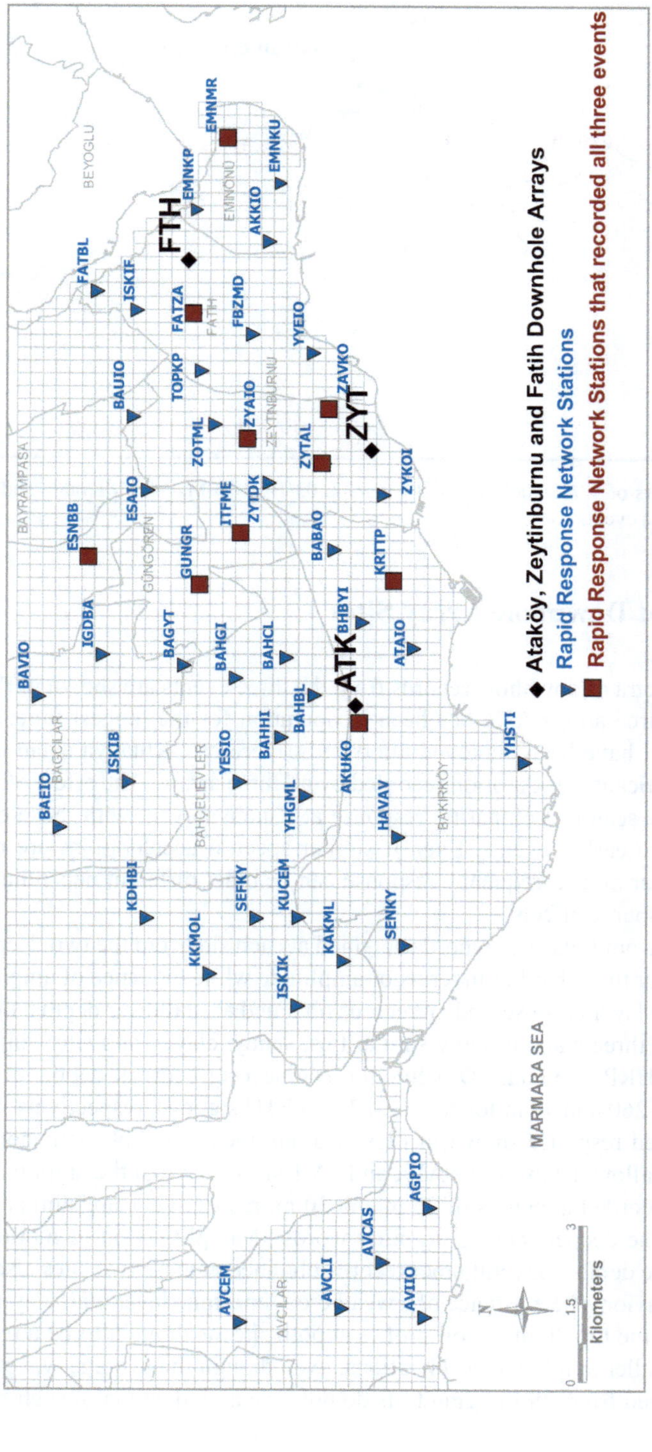

Fig. 1 Locations of geotechnical downhole arrays and Istanbul Rapid Response Network stations

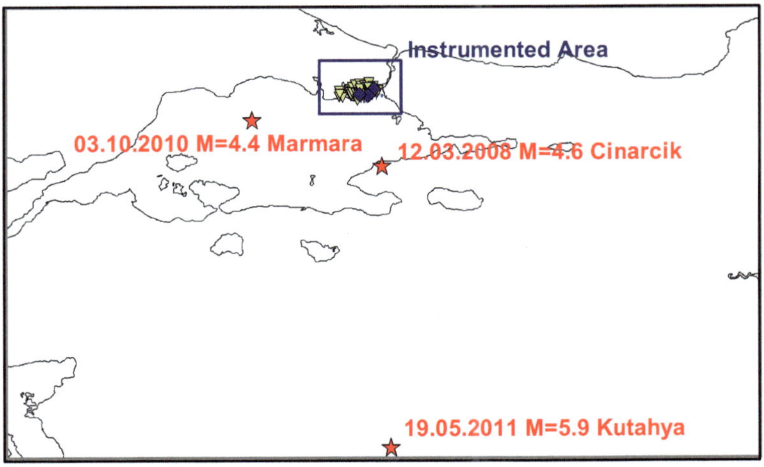

Fig. 2 Epicenters of 12/03/2008 $M_L = 4.8$ Çınarcık, 03/10/2010 $M_L = 4.4$ Marmara and 19/05/11 $M_L = 5.9$ Kütahya events

Response at Downhole Array Sites

Response spectra of downhole records from the three events are shown in Figs. 3-5. Among the three arrays ATK has been in operation for the longest time. Stations ZYT and FTH have been deployed afterwards, therefore, Kütahya event has been the only significant weak motion recorded by all three arrays. The effect of distance can be readily seen by comparing response at Kütahya event with those observed during the two earlier events. Even if similar spectral accelerations are recorded and all at linear range, frequency response can be quiet different depending on the distance and source effects.

Also seen from Figs. 3-5, records obtained at each site from the two deepest sensors are similar to each other in terms of amplitude while evidence of amplification by the surface layers is observed on records obtained at shallower depths. Local soil profiles at all three stations show similarities; sandy, clayey and silty soils classified with NEHRP Site Class D, while the engineering bedrock depths of stations varies as 110, 260 and 40 m for ATK, ZYT and FTH stations, respectively. Looking at the observed response shown in Figs. 3-5, an unexpected phenomenon can be detected at shallow depths at the ATK and ZYT sites such that the amplitude of accelerations recorded at depths of 25 m and 30 m, respectively, are similar to those recorded by the deeper sensors. In other words, deamplification of motion is observed at these depths. Several other records obtained at these two sites also exhibit the same behavior. The existence of a weathered limestone layer with clay-silt-sand interlayers at the top 30 m of soil profile at both sites is thought to be a reason for observing smaller amplitude accelerations, even though the average wave velocity profiles inferred from PS Logging tests do not give an indication of such behavior.

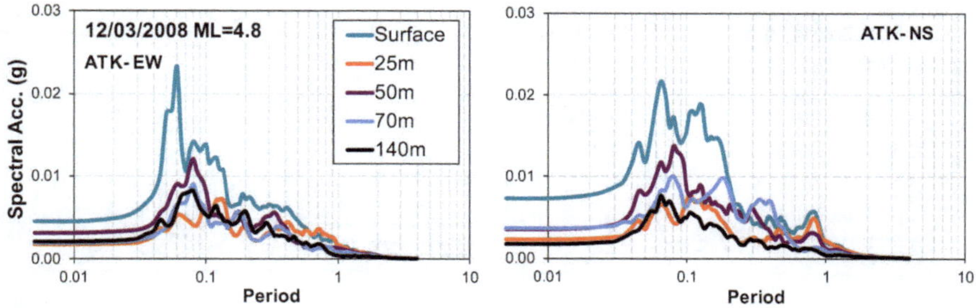

Fig. 3 Acceleration spectra of downhole records during the 12/03/2008 M_L=4.8 Çınarcık event.

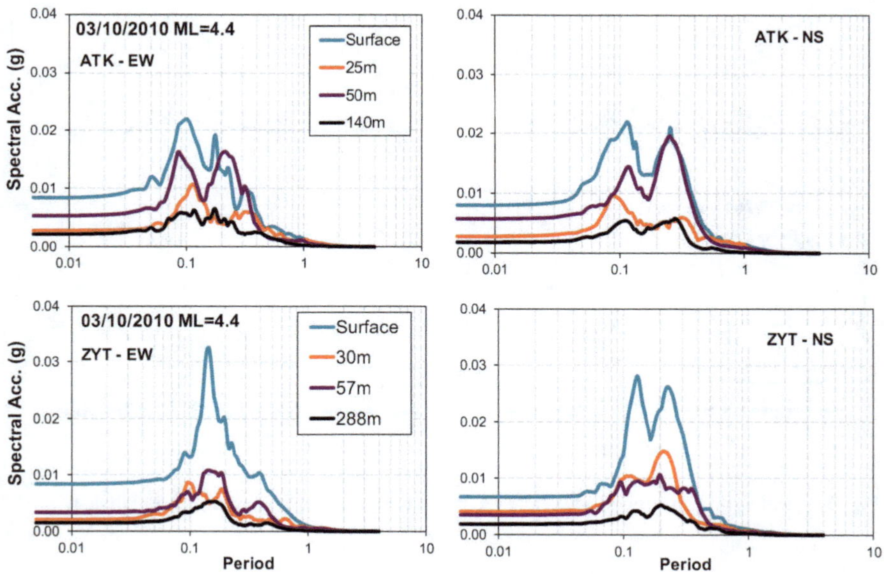

Fig. 4 Acceleration spectra of downhole records during 03/10/2010 M_L=4.4 Marmara event.

Similar trend is not observed in FTH site where the soil profile does not include the 'limestone' layer. It is decided that this shallow layer, which exists at many locations within the old city, requires more investigation.

Average of surface/bedrock spectral ratio at each array site is shown in Fig. 6. Observed spectral ratios show that the average amplification from bedrock to the surface of the soil reaches a factor of 4, 7 and 12 at periods of 0.9, 1.5 and 0.5 s at ATK, ZYT and FTH sites, respectively. At ATK and FTH sites, the values of predominant periods observed from amplification spectra seem to be in agreement with theoretical 1D fundamental periods of 1.1 and 0.6, respectively, that are calculated from the measured wave velocity profiles using the well-known V_S=4 H/T_0

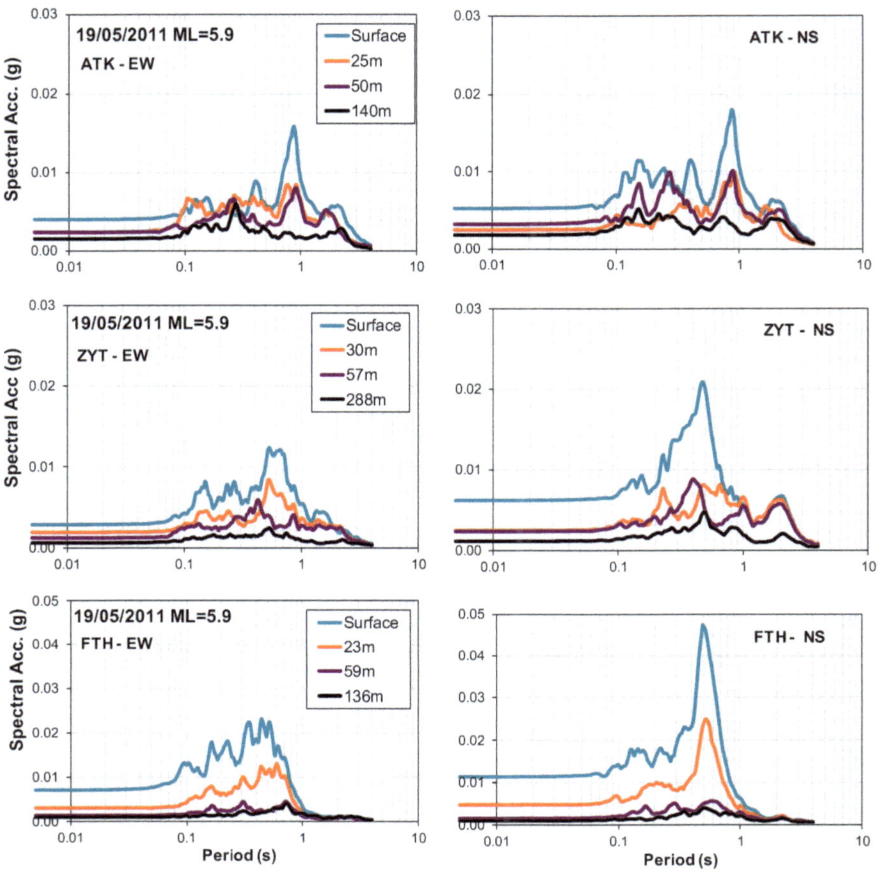

Fig. 5 Acceleration spectra of downhole records during the 19/05/11 $M_L = 5.9$ Kütahya event

relationship. However, calculated fundamental period suggests a longer period (2.7 s) for ZYT than that observed from the records obtained at this site. On the other hand, the observed higher amplification potential of FTH site is related to the similarity of the fundamental period of the site to that of the recorded event (~0.6 s).

Ground responses at ATK, ZYT and FTH downhole array sites during the recorded events are modeled through 1D site response analyses (Shake91, Idriss and Sun 1992) using the acceleration-time histories recorded with the deepest accelerometers as input motions. Comparisons of the modeled and recorded response spectra at ATK, ZYT and FTH sites show that 1D equivalent-linear analysis can successfully model the linear response of soil. Examples of modeled and observed site responses are shown for ATK and ZYT sites during Cinarcik and Marmara events, in Figs. 7 and 8, respectively.

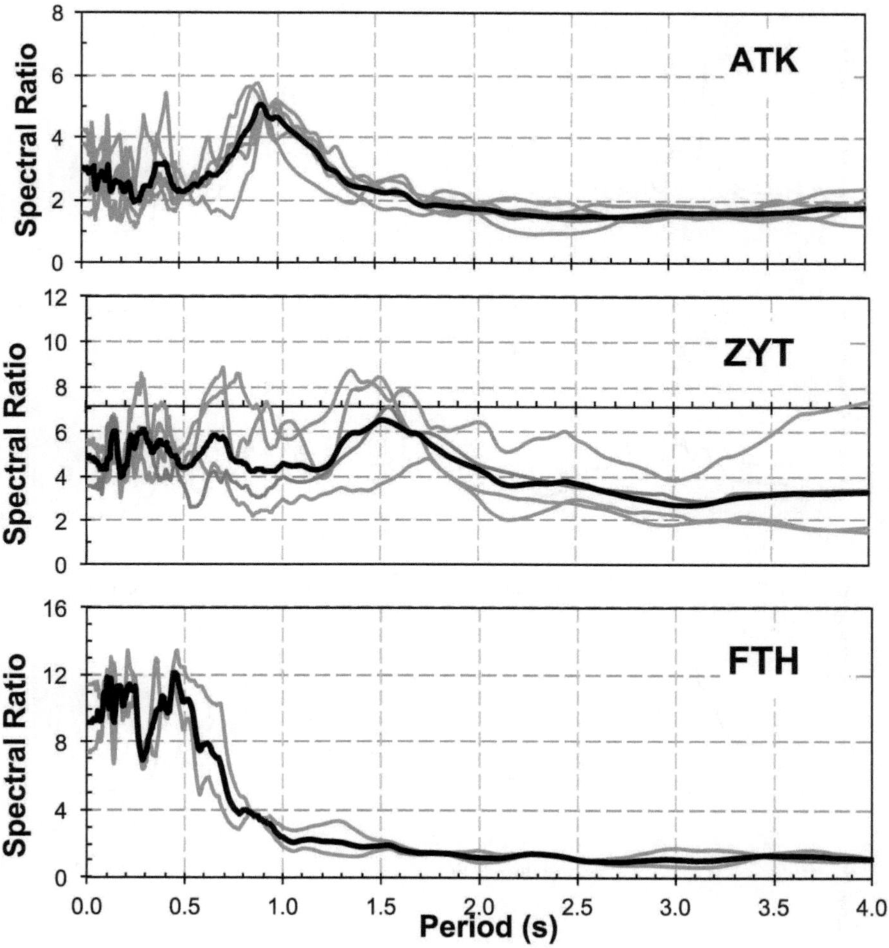

Fig. 6 Average site amplification (surface/bedrock) at ATK, ZYT, and FTH sites

Response at IRRN Stations

Recorded responses at IRRN stations during the same events are shown in Fig. 9, NEHRP soil classes of station sites are also indicated. A significant variation is observed for all three events with average coefficient of variation in the range of 0.6 to 0.8. The variation of average shear wave velocity and engineering bedrock depth in IRRN stations are shown in Fig. 10a. The average shear wave velocity, V_{s30}, indicates relatively uniform distribution over the whole region, while there is a significant variation in depth of engineering bedrock (Fig. 10b).

Variations of PGA (peak ground acceleration) and PSA (peak spectral acceleration) are further compared in Fig. 11 with respect to epicentral distances and NE-

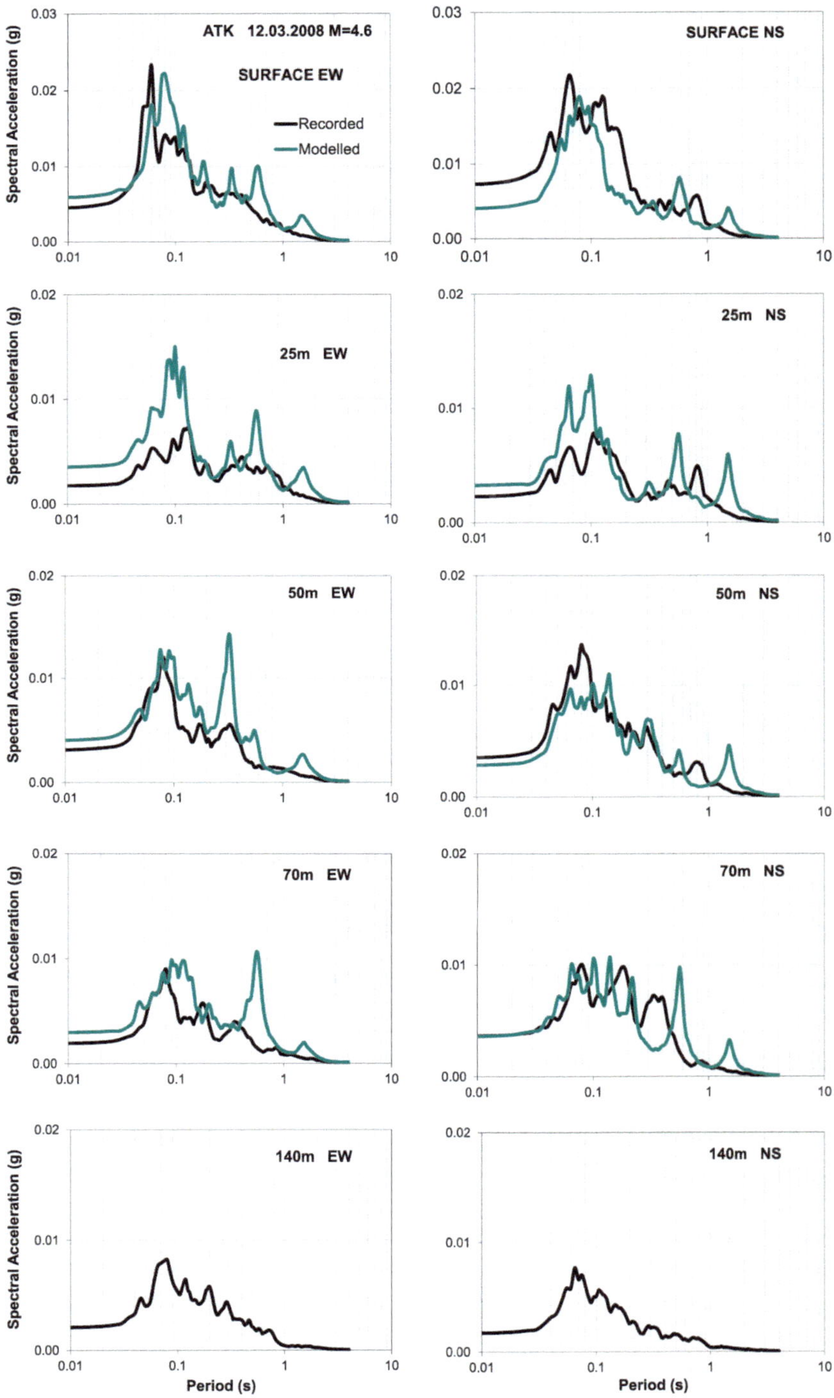

Fig. 7 Comparison of the recorded and modeled response spectra at ATK during the 12/03/2008 $M_L = 4.8$ Çınarcık event

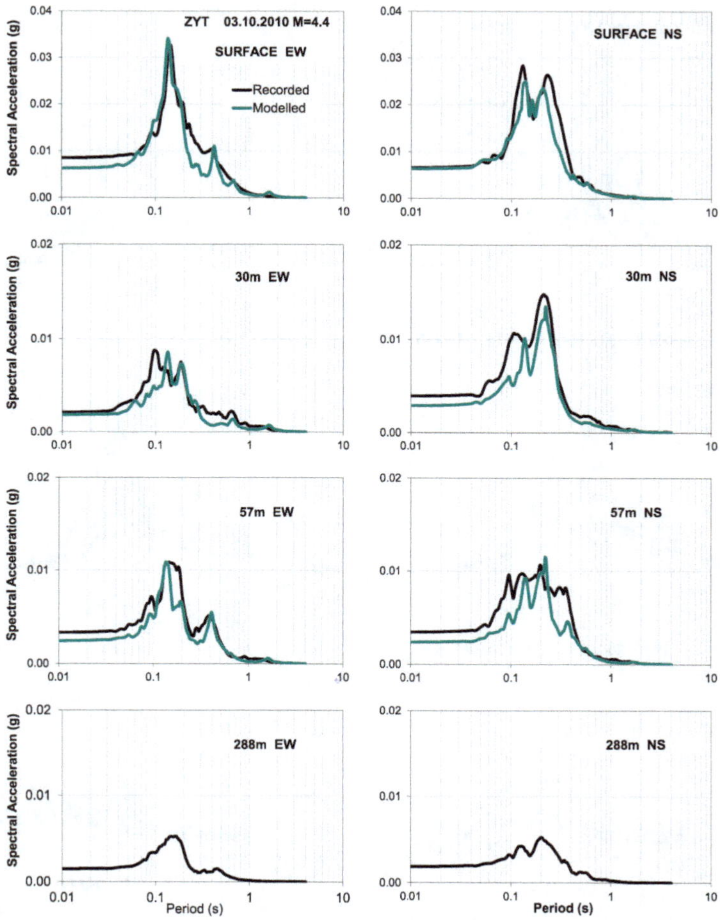

Fig. 8 Comparison of the recorded and modeled response spectra at ZYT during the 03/10/2010 M_L=4.4 Marmara event

HRP site classes. According to the information presented in Fig. 11, it is possible to say that for both ground motion parameters, all three events generate similar range of values with Marmara event producing slightly higher ground shaking. The epicentral distances for the Çınarcık and Marmara events can be considered in the same range (<75 km) compared to Kütahya event occurring at distance slightly greater than 200 km. The effect of distance has introduced differences not in the recorded peak ground accelerations but rather in the frequency content of the response. Response spectra observed during the Kütahya event clearly shows a shift towards larger periods. Another observation is that spectra of motions recorded during the NS directional Çınarcık and Kütahya events exhibit a broader frequency band when compared to those observed during the EW directional Marmara event. Considering the EW strike-slip mechanism of NAF, the difference may be attributed to directivity/path effects.

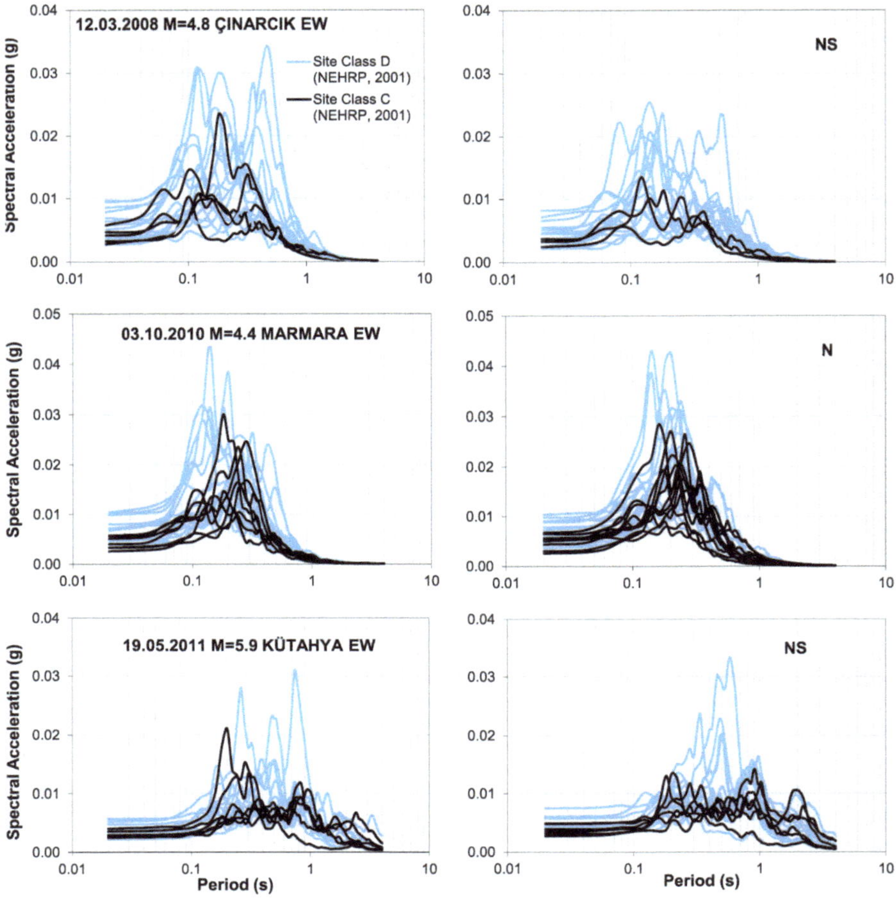

Fig. 9 Acceleration spectra recorded by IRRN stations (in relation to NEHRP Site Class)

On the other hand, the observed high scatter of the ground motions recorded during each event suggests that some part of the observed variation can be related to site effects. The variations of ground motion parameters with respect to NEHRP site classes show some indication of site effects (i.e. stations sites identified as Site C tend to have lower spectral response) but also demonstrate that V_{S30} alone is not a sufficient indicator for amplification potential, as observed from the significant scatter in the spectral response of stations located on Site D soils. As shown in Fig. 10a, almost all strong motion stations have comparable V_{S30} values.

An attempt is made to estimate site-specific ground motion parameters at the IRRN stations by performing 1D response analysis (Shake91) with the assumption that time histories recorded by the deepest accelerometers at ATK, ZYT and FTH downhole arrays can represent bedrock motions for these sites. A comparison of deepest accelerometer records obtained at the sites is shown in Fig. 12. As observed

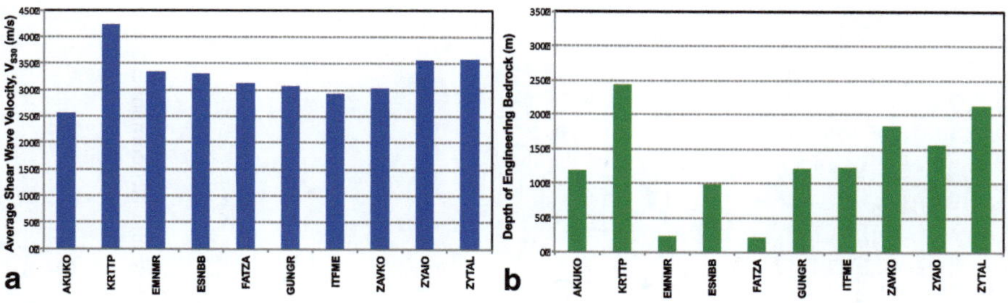

Fig. 10 **a** Variation of average shear wave velocities and **b** depth of engineering bedrock at IRRN stations that recorded all three events

from Fig. 12, the recorded bedrock motions have differences in both frequency and amplitude response. The differences are less pronounced between the ZYT and FTH sites. One possible reason could be that the properties of the engineering bedrock with measured V_S in the range of 1000 m/s do not sufficiently represent the engineering bedrock. Another reason could be the differences in the geological formations at these locations. Bedrock model for the region (OYO 2007) suggests that FTH and ZYT sites are located on Trakya formation, geologically an older age formation than Ceylan formation that underlies ATK site.

Site response modelling is carried out for 10 of the IRRN stations that recorded all three events. In the analysis, the bedrock acceleration time histories recorded at ATK, ZYT and FTH are used as input motions. A comparison of the recorded and modelled PGA and PSA values are shown in Fig. 13. In general, there is a certain agreement between the recorded and the calculated parameters (mean squared error <0.2). However, as seen in Fig. 13, calculated ground motion parameters are different from each other depending on the input acceleration time history. ZYT seems to be providing the best-fitting bedrock motion. One reason could be the associated bedrock geology, another possible reason is the selection of ground motion parameter used in the comparisons.

Conclusions

Few small earthquakes with local magnitude slightly larger than $M_L=4$ were recorded at geotechnical downhole arrays recently deployed in the west side of Istanbul. Istanbul Rapid Response Network (IRRN) comprising of 55 surface strong motion stations in the European side of Istanbul also recorded these events. Observations from the recorded response at the array sites provided some insight to the effect of local soil conditions at the station sites. A highly weathered limestone layer with several sand-silt-clay intrusions that exists at the top 30 m of soil profiles at ATK and ZYT sites somehow seems to be deamplifying the ground motion. Mean-

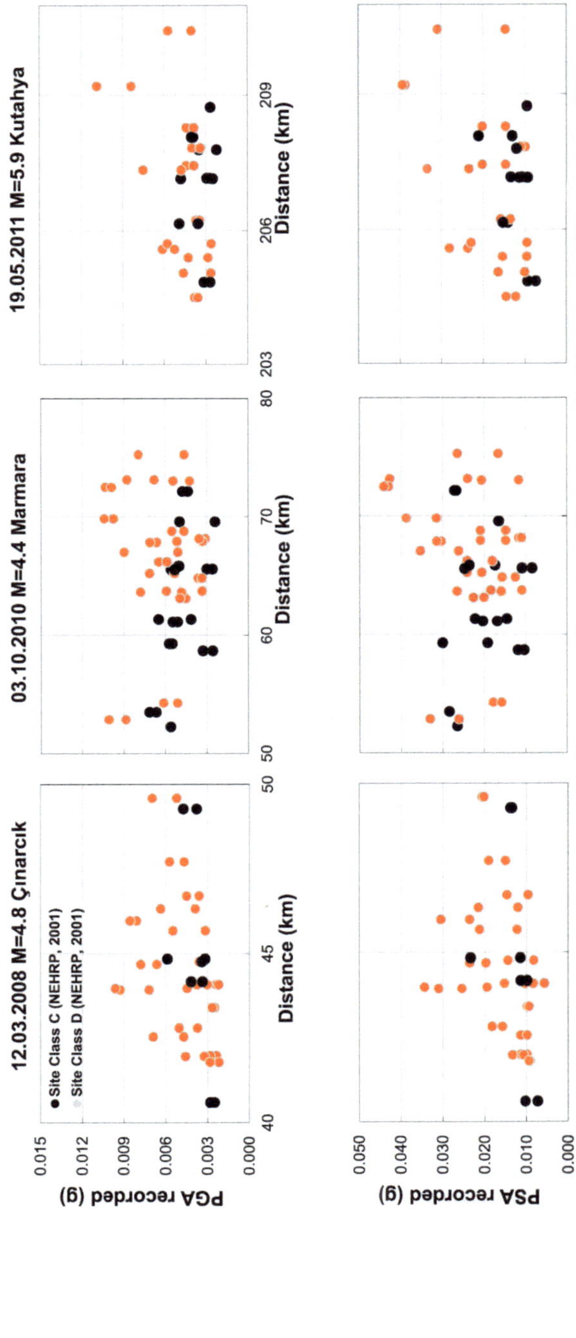

Fig. 11 Variations of PGA and PSA recorded by IRRN stations with epicentral distance and in relation to NEHRP Site Class (both EW and NS components are shown)

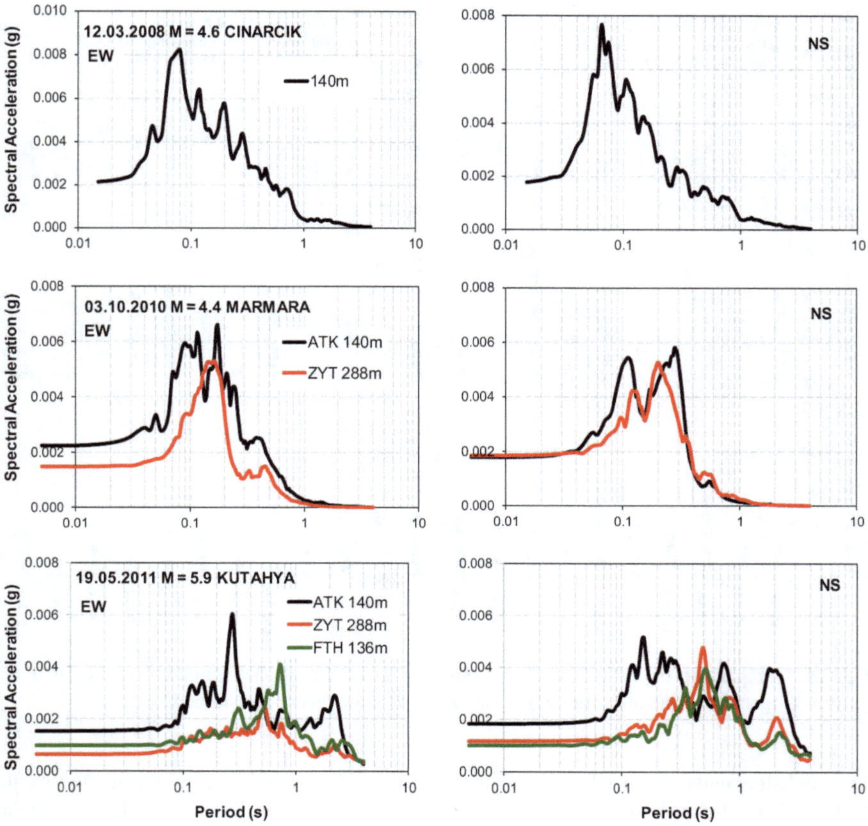

Fig. 12 Acceleration spectra of records obtained from the deepest downhole accelerometers at ATK, ZYT, and FTH sites

while, there is a significant scatter in the recorded response at IRRN stations, which have similar measured values of V_{S30}.

Even though the maximum accelerations on the ground surface do not exceed 0.01 g during the recorded events, an attempt is made to model the recorded response at the downhole array sites and at the IRRN stations. Response at the downhole array sites is successfully modelled with 1D equivalent linear site response analysis (Shake 91). Modelled response at IRRN sites show sensitivity to the selection of reference bedrock motion, with acceleration time history recorded by the deepest sensor at ZYT site providing the best estimation of reference bedrock motion for IRRN stations.

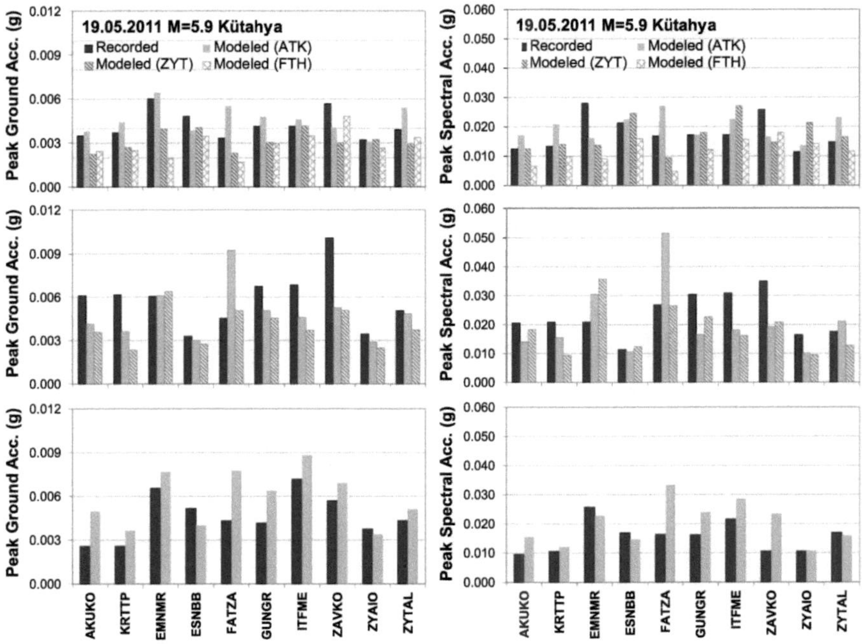

Fig. 13 Comparison of the recorded and modeled ground motion parameters at the 10 IRRN stations (logarithmic mean of the two horizontal components are considered)

Acknowledgements Financial support for installation of downhole arrays are granted by the Turkish Scientific and Technological Research Council (TUBITAK), Istanbul Metropolitan Municipality, GFZ Potsdam and Boğaziçi University. The authors wish to acknowledge Prof. Mustafa Erdik, the coordinator of Istanbul Rapid Response Network, and Prof. Stefano Parolai, co-investigator at Megacity Istanbul Project. Barbaros Çetiner and Ahmet Korkmaz, members of project team for installing downhole arrays, are acknowledged for their tedious work in the field. The support of local municipalities of Bakırköy and Zeytinburnu as well as the personnel at Yunus Emre Cultural Centre and Fatih Mosque is greatly appreciated.

References

Ansal A, Tönük G, Kurtuluş A, Erdik M, Parolai S (2010) Modeling the observed site response from Istanbul strong motion network, Fifth International Conference on recent advances in geotechnical earthquake engineering and soil dynamics, San Diego—May 24–29

Beresnev IA, Field EH, Johnson PA, den KE-A Abeele (1998) Magnitude of nonlinear sediment response in Los Angeles Basin during the 1994 Northridge, California Earthquake. Bull Seism Soc Am 88:1079–1084

Erdik M, Fahjan Y, Özel O, Alcik H, Mert A, Gül M (2003) Istanbul earthquake rapid response and the early warning system. Bull Earthq Eng 1(1):157–163

Fukushima Y, Irikura K, Uateke T, Matsumoto H (2000) Characteristics of observed peak amplitude for strong ground motion from the 1995 Kobe Earthquake. Bull Seism Soc Am 90:545–565

Idriss IM, Sun JI (1992) Shake91, a computer program for conducting equivalent linear seismic response analysis of horizontally layered soil deposits, modified based on the original SHAKE program by Schnabel, Lysmer and Seed (1972), University of California Davis, California

Kurtulus A (2011) Istanbul geotechnical downhole arrays. Bull Earthq Eng 9(5):1443–1461

NEHRP Recommended Provisions for New Buildings and other Structures (2003) FEMA-450, prepared by the Building Seismic Safety Council for the Federal Emergency Management Agency, Washington, DC

OYO, Inc (2007) Production of microzonation report and maps—European Side (South), Report for Istanbul Metropolitan Municipality Microzonation Project

Trifunac MD, Todorovska MI (1998) Nonlinear soil response as a natural passive isolation mechanism- the 1994 Northridge, California Earthquake. Soil Dyn Earthq Eng 17:41–51

Combined Failure Mechanism of a Breakwater Subject to Tsunami during 2011 East Japan Earthquake

Iai Susumu

Abstract In this study, a centrifuge model tests and effective stress analyses are performed on a breakwater subject to Tsunami such as those seriously damaged during 2011 East Japan Earthquake (Magnitude 9.0). The centrifuge model tests at a scale of 1/200 are performed to simulate the failure of a breakwater subject to Tsunami. With the effective stress analyses, this study demonstrates the importance of the mechanism of failure in the rubble mound due to seepage flow of pore water in addition to the force of Tsunami action.

Keywords Breakwater · Failure mechanism · Seepage flow · Tsunami

Introduction

An earthquake with a Japan Meteorological Agency (JMA) magnitude 9.0 hit north east Japan at 14:46, March 11, 2011. JMA named this earthquake '2011 Tohoku-Pacific Ocean earthquake'. This earthquake is the greatest in its magnitude since the modern earthquake monitoring system was established in Japan.

Recorded heights of the tsunamis were higher than 7.3 m at Soma, higher than 4.2 m at Oarai, and higher than 4.1 m at Kamaishi. The impact of the tsunamis is also the greatest since the existing design methodology was adopted for designing breakwaters. The existing design methodology of a breakwater is based on limit equilibrium by considering wave force acting from the lateral side and additional pressure acting underneath the caisson as shown in Fig. 1.

The most typical example of the damage to breakwater was the one at the Kamaishi Harbor. In this example, the breakwater was specifically designed for protection against the impact of tsunami and constructed at the mouth of the Kamaishi harbor at a depth of 63 m or less over the length of 990 m in the northern part and 670 m in the southern part with an opening of 300 m in between. However, the breakwater was devastated by the tsunami. This may be partly due to the failure of

I. Susumu (✉)
Disaster Prevention Research Institute, Kyoto University, Kyoto, Japan

A. Ansal, M. Sakr (eds.), *Perspectives on Earthquake Geotechnical Engineering*, Geotechnical, Geological and Earthquake Engineering 37,
DOI 10.1007/978-3-319-10786-8_8

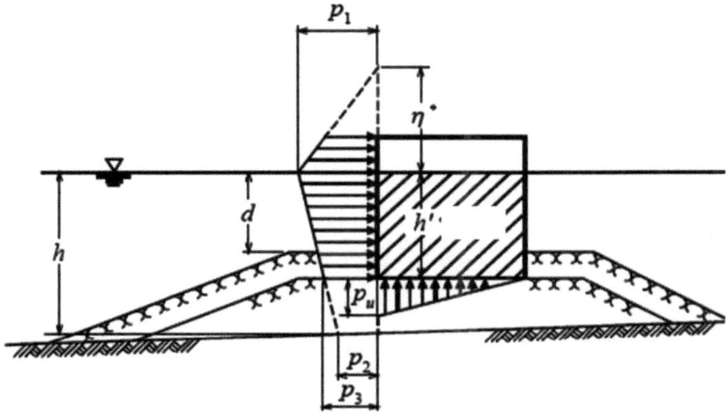

Fig. 1 Existing design procedure of a breakwater

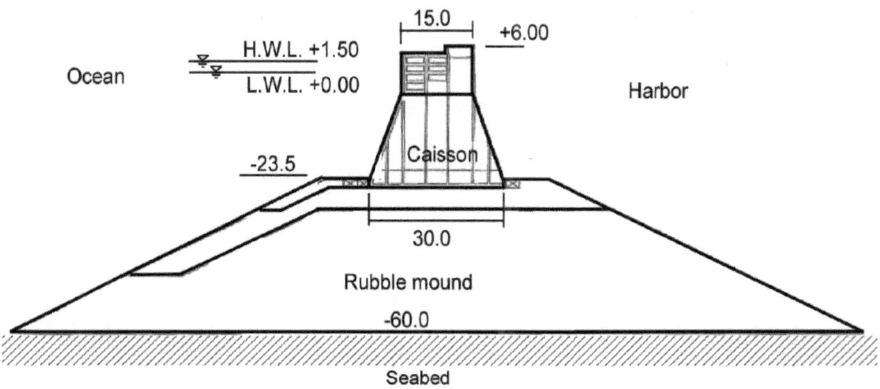

Fig. 2 Cross section of a composite breakwater at Kamaishi Harbour (unit in m)

design procedure then adopted but also a symbolic event for indicating the serious impact by the tsunami.

Figure 2 shows a typical cross section of the break water. The breakwater is a composite of caisson and rubble mound. Based on the visual data (video) monitored from the coastal line, the first arrival of the tsunami was at about 15:24 (about 40 min after the earthquake). The break water was not damaged by this first arrival of the tsunami.

However, there was a continual overtopping of water at the top of the breakwater. At about 15:32, when the height of the tsunami was reduced, serious devastation of the breakwater was visually confirmed by the monitored visual data. At 15:59, the devastation was progressed further and many of the caissons fell into the harbor side from the rubble mound. The rubble mound was also seriously devastated even to the extent that about half of its original body of the rubble mound was lost.

In order to investigate the primary mechanism of failure of this type of a breakwater due to Tsunami, a series of centrifuge model tests and effective analyses

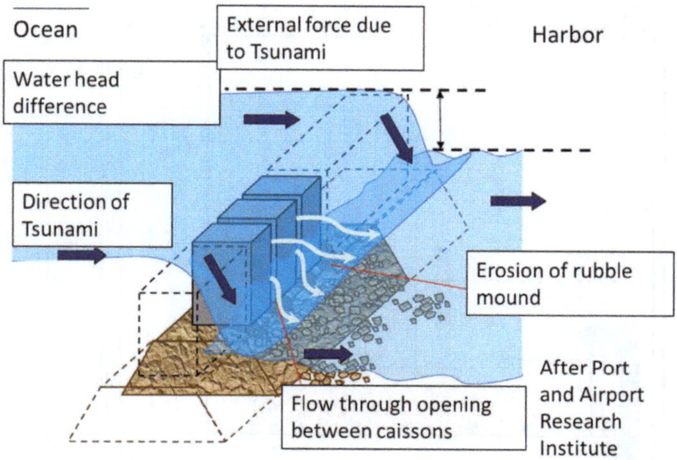

Fig. 3 Primary mechanism of failure hypothesized by a hydrodynamics expert group

are performed in this study. As a straight forward extension of the existing design procedure shown in Fig. 1, the primary mechanism of failure was considered by a hydrodynamics expert group due to erosion of rubble mound by the water flow through opening between caissons as shown in Fig. 3. As parallel efforts to that group, this study was performed with the objective of evaluating the effect of seepage flow through the rubble mound due to water head difference due to Tsunami in combination with the wave force action.

Centrifuge Model Tests

Tsunami Generator in Centrifuge and Model Test Conditions

The centrifuge at Disaster Prevention Research Institute, Kyoto University, (effective radius 2.5 m) was used for the centrifuge model tests in this study. The equipment for simulating Tsunami in the centrifuge is shown in Fig. 4. In this equipment, the remote-control valve at the bottom of the water tank is opened for generating Tsunami-like water flow toward caisson and overflowing water at the other end of the model (left in this figure) is absorbed in a tentative storage pit for reducing the effect of reflecting wave from the (left side) wall of the container.

The centrifuge model tests were performed for a caisson 21 m high in prototype with a scaling factor of 1/200 by adopting the generalized scaling relation (Iai et al. 2005) in a 25 g centrifugal acceleration field. Four cases were performed by varying initial sea water level and water level difference due to Tsunami as shown in Table 1. The material of rubble mound was also varied by using Silica No.1 sand that consists of particles scaled in 1/200 of the prototype rubble and Silica No.4 sand that artificially reduces the effect of seepage flow in the rubble

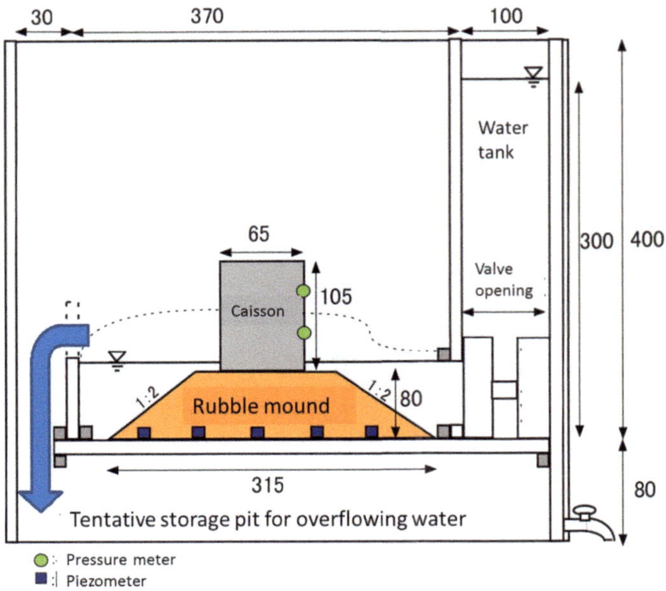

Fig. 4 Equipment for simulating Tsunami in centrifuge model tests (unit in mm in model scale)

Table 1 Model test cases

Case	Mound	Relative density	Sea water level	Water level difference
Case 1	Silica #1 sand	55.5%	6.23 m	14.4 m
Case 2	Silica #1 sand	60.3%	12.6 m	10.3 m
Case 3	Silica #4 sand	58.0%	8.15 m	15.2 m
Case 4	Silica #4 sand	65.8%	13.0	10.2 m

mound. Pure water was used for centrifuge tests. The rubble mound model was made by water pluviation.

Centrifuge Test Results

Figure 5 shows the results of the centrifuge test for Cases-2 and 4 with respect to the deformation. As described earlier, Case-2 was performed for simulating the prototype condition. As shown in (a-1) in this figure, when the Tsunami at the water level difference of 10.3 m acts on the caisson, the caisson remains standing without deformation. At the same time, bubbles appear on the bay side (left side), indicating that a seepage flow of water through the rubble mound begins at this instance. After a certain period of time ((a-2) in the same figure), rubble mound gradually deforms in a bearing capacity failure mode with inclined load together with the caisson toward the direction of Tsunami wave force. The caisson eventually falls

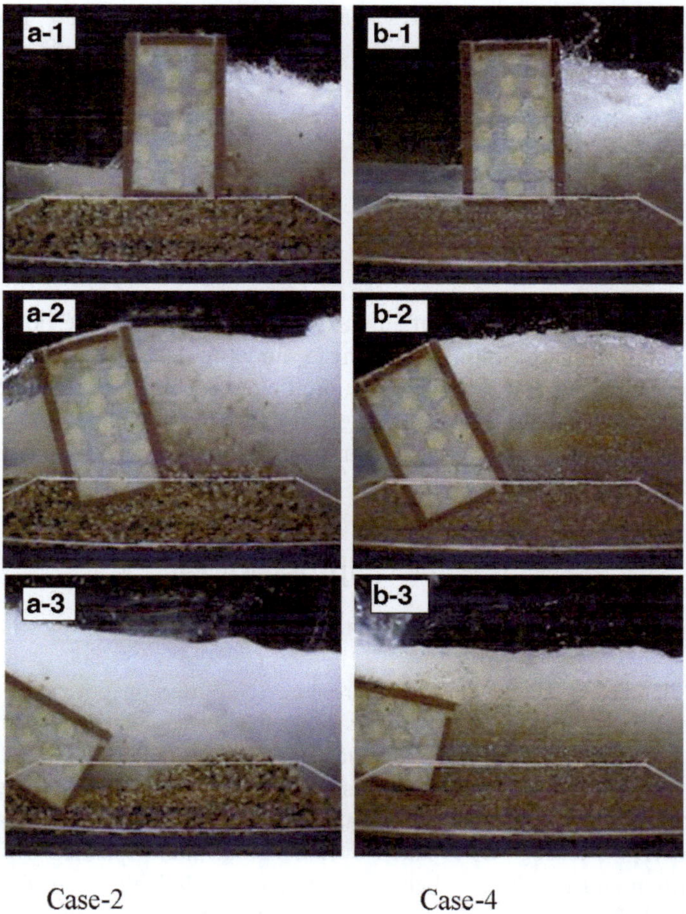

Fig. 5 Deformation of caisson and rubble mound model

down from the rubble mound and the rubble mound exhibits a residual failure mode that resembles the classic circular failure mode.

These failure modes of the caisson and the rubble mound are consistent with those investigated and identified at Kamaishi Harbour after 2011 East Japan earthquake. This fact is indicative of the fact that the mechanism of failure due to tsunami is considered due to the combined effect of seepage flow and wave action. However, the mechanism of erosion progressing from the harbor side also produces the similar failure mode. Thus, the mechanism of failure has to be studied further by performing effective stress analysis as described later.

Case-4 was performed as comparison to Case-2 by artificially reducing the effect of seepage flow in the rubble mound. Even if artificially reduced, the effect of seepage flow still exists so that a combined failure mode is also observed as shown in (b-1) through (b-3) in Fig. 5. However, the failure mode in Case-4 is more or less dominated by sliding of caisson rather than deep failure mode of rabble mound.

This fact indicate that if there is no seepage flow, then sliding failure mode will be dominant as often considered in the existing design procedure described earlier.

Effective Stress Analysis

Effective Stress Model

For the effective stress analysis of a breakwater subject to seepage flow and wave action, the strain space multiple mechanism model with a cocktail glass model is used (Iai et al. 2011). In the strain space multiple mechanism model, the effective stress, defined as extension positive, is given based on a dyad defined by the unit vector **n** along the direction of the branch between the particles in contact with each other and the unit vector **t** normal to **n** as follows:

$$\boldsymbol{\sigma}' = -p\mathbf{I} + \frac{1}{4\pi}\iint q\langle \mathbf{t}\otimes\mathbf{n}\rangle \mathrm{d}\omega \mathrm{d}\Omega \qquad (1)$$

$$\langle \mathbf{t}\otimes\mathbf{n}\rangle = \mathbf{t}\otimes\mathbf{n} + \mathbf{n}\otimes\mathbf{t} \qquad (2)$$

where p denotes effective confining pressure (compression positive), **I** denotes second order identity tensor, q denotes micromechanical stress contributions to macroscopic deviator stress due to virtual simple shear mechanism (called virtual simple shear stress), and $\langle \mathbf{t}\otimes\mathbf{n}\rangle$ denotes second order tensor representing the virtual simple shear mechanism. Out of the double integration, the integration with respect to $\omega (= 0 \text{ through } \pi)$ is taken over a virtual plane spanned by the direction vectors **n** and **t** with $\omega/2$ being the angle of **n** relative to the reference local coordinate defined in the virtual plane, while the integration with respect to the solid angle Ω is taken over a surface of a unit sphere to give a three dimensional average of two dimensional mechanisms.

The integrated form of the constitutive equation, i.e. direct stress strain relationship, is derived by relating the macroscopic strain tensor $\boldsymbol{\varepsilon}$ to the macroscopic effective stress $\boldsymbol{\sigma}'$ through the structure defined by Eq.(1). The first step to derive this relationship is to define the volumetric strain ε (extension positive) and the virtual simple shear strains γ as the projections of the macroscopic strain field to the second order tensors representing volumetric and virtual simple shear mechanisms as follows:

$$\varepsilon = \mathbf{I} : \boldsymbol{\varepsilon} \qquad (3)$$

$$\gamma = \langle \mathbf{t}\otimes\mathbf{n}\rangle : \boldsymbol{\varepsilon} \qquad (4)$$

where the double dot symbol denotes double contraction. In order to take into account the effect of volumetric strain due to dilatancy ε_{d}, effective volumetric strain ε' is introduced by

$$\varepsilon' = \varepsilon - \varepsilon_{\mathrm{d}} \tag{5}$$

where the rate of volumetric strain due to dilatancy is given by the projection of strain rate field to a second order tensor $\mathbf{I}_{\mathrm{d}}$ as

$$\dot{\varepsilon}_{\mathrm{d}} = \mathbf{I}_{\mathrm{d}} : \dot{\boldsymbol{\varepsilon}} \tag{6}$$

The scalar variables defined in Eqs. (4) and (5) as the projection of macroscopic strain field are used to define the isotropic stress p and virtual simple shear stress q in Eq. (1) through path dependent functions as

$$p = p(\varepsilon') \tag{7}$$

$$q = q(\gamma) \tag{8}$$

In the strain space multiple mechanism model, the virtual simple shear mechanism is formulated as a non-linear hysteretic function, where a back-bone curve is given by the following hyperbolic function;

$$q(\gamma) = \frac{\gamma / \gamma_{\mathrm{v}}}{1 + |\gamma / \gamma_{\mathrm{v}}|} q_{\mathrm{v}} \tag{9}$$

The parameters q_{v} and γ_{v} defining the hyperbolic function are the shear strength and the reference strain of the virtual simple shear mechanism, respectively.

The isotropic component in Eq. (7) is defined by a hysteretic tangential bulk modulus depending on the loading/unloading (L/U) condition as

$$K_{\mathrm{L/U}} = -\frac{\mathrm{d}p}{\mathrm{d}\varepsilon'} = r_K K_{\mathrm{U0}} \left(\frac{p}{p_0} \right)^{l_K} \tag{10}$$

where p_0: initial confining pressure, K_{U0}: tangential bulk modulus at initial confining pressure.

Dilatancy in Eq. (5) in the Cocktail glass model is decomposed into contractive component ε_d^c and dilative component ε_d^d as

$$\varepsilon_{\mathrm{d}} = \varepsilon_{\mathrm{d}}^{\mathrm{c}} + \varepsilon_{\mathrm{d}}^{\mathrm{d}} \tag{11}$$

However, in this study, effect of dilatancy of the rubble mound was assumed negligibly small and dilatancy was ignored.

Load Conditions for Simulating Tsunami

The effective stress analysis of a breakwater was performed for prototype scaled from the centrifuge model, including a caisson 21 m high. As shown in Fig. 6, joint element is specified at the bottom of the caisson to allow sliding and separation

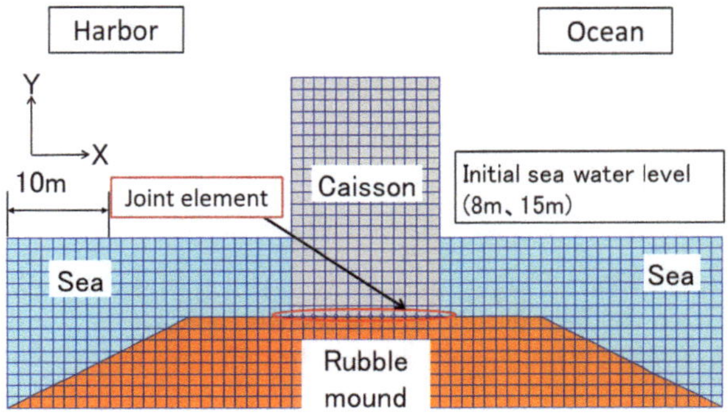

Fig. 6 Finite element mesh of a breakwater for analysis (prototype scale)

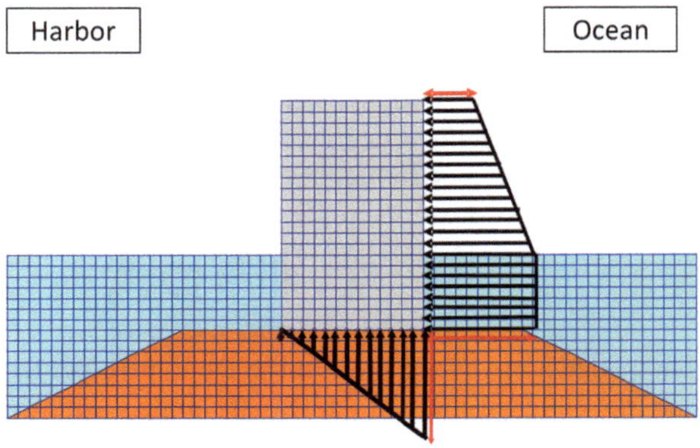

Fig. 7 Load conditions for simulating wave force due to Tsunami

between the caisson and the rubble mound. After static gravity analysis to set the initial conditions for dynamic Tsunami analysis, Tsunami wave force was applied on the caisson as equivalent static distributed force as shown in Fig. 7. In order to analyze the effect of seepage flow due to sea level difference, excess pore water pressure was applied at the rubble mound. External force due to sea level difference is also simulated by equivalent lateral static force on the caisson.

Distribution and magnitude of the wave force due to Tsunami adopted for the analysis (Fig. 7) was determined based on the proposal by Tanimoto et al (1984). This proposal has been also adopted in the existing design procedure of a breakwater (Fig. 1).

The effective stress analyses were performed for Cases-1 through 4 with the material parameters used for the rubble mound shown in Table 2. These parameters were determined by undrained cyclic shear tests using hollow cylinder specimen and by permeability test with constant water level difference. The initial sea levels for Cases-1 through 4 were simplified for the analysis as shown in Fig. 6. The sea

Table 2 Material parameters of rubble mound

Mound	Density	Permeability	Shear modulus	ϕf	Cohesion
Silica #1	1.91(t/m^3)	7.06×10^{-2}(m/s)	6.15×10^4(m/s)	40.8°	0(kPa)
Silica #4	1.90(t/m^3)	1.56×10^{-3}(m/s)	7.67×10^4(m/s)	38.7°	0(kPa)

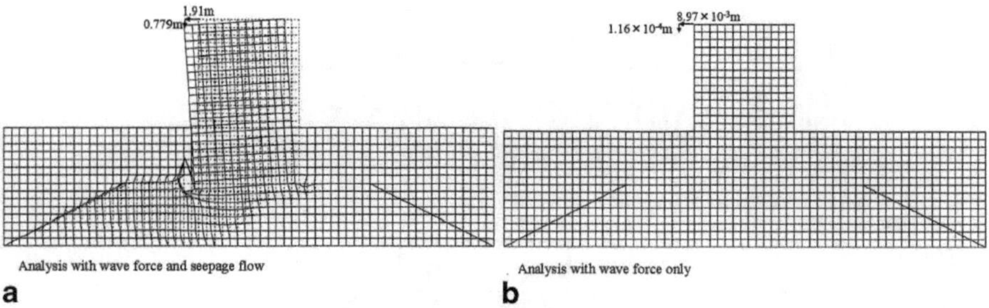

Fig. 8 Residual deformation of a breakwater (a) Analysis with wave force and seepage flow (b) Analysis with wave force only

level differences were also simplified as 15 m for Cases-1 and 3, and 11 m for Cases-2 and 4.

Results of the Effective Stress Analyses

The results of analysis of Case-2 are presented below. Figure 8 (a) shows residual deformation of a breakwater with both Tsunami wave force on the caisson and seepage flow in the rubble mound. The tilting of the caisson is associated with a significant deformation of rubble mound. This mode of failure is consistent with that observed at the centrifuge model test. In comparison to this result, deformation of a breakwater due to Tsunami wave force only is negligibly small as shown in Fig. 8 (b). These results of the analyses indicate that primary mechanism of failure of a breakwater due to Tsunami is combined failure mechanism due to Tsunami wave force and seepage flow in the rubble mound due to water level difference associated with Tsunami.

In addition, distribution of an inverse of safety factor with respect to Mohr-Coulomb failure criterion in the rubble mound is shown in Fig. 9. As shown in Fig. 9(a), the rubble mound beneath the caisson undergoes a compression shear due to the overburden from the caisson before Tsunami. When the Tsunami wave force and seepage flow act on the rubble mound, significant area of cross section of rubble mound are brought close to a shear failure condition as shown in Fig. 9(b). However, if only the Tsunami wave for acts, there is not a significant change in the rubble mound with respect to its stress status. These results also supports the previous notion that the primary mechanism of failure of a breakwater due to Tsunami is combined failure mechanism involving seepage flow in the rubble mound.

Conclusions

In this study, a centrifuge model tests and effective stress analyses are performed on a breakwater subject to Tsunami such as those serious damaged during 2011 East Japan Earthquake (Magnitude 9.0). Both the centrifuge model tests at a scale of 1/200

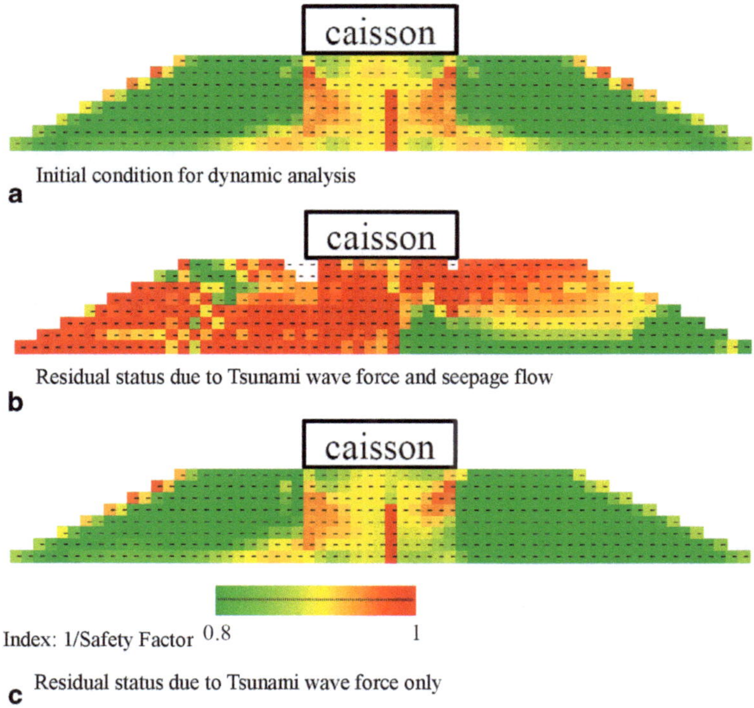

Fig. 9 Distribution of 1/Safety Factor in the rubble mound (a) Initial condition for dynamic analysis (b) Residual status due to Tsunami wave force and seepage flow (c) Residual status due to Tsunami wave force only

and the effective stress analyses demonstrate the importance of the mechanism of failure in the rubble mound due to seepage flow of pore water in addition to the wave force of Tsunami action.

Rereferences

Iai, S., Tobita, T. and Nakahara, T. (2005): Generalized scaling relations for dynamic centrifuge tests, Geotechnique, 55(5): 355–362.

Iai, S., Tobita, T., Ozutsumi, O. and Ueda, K. (2011): Dilatancy of granular materials in a strain space multiple mechanism model. Int. JNumAnalyt Methods Geomechanics 35(3): 360–392.

Tanimoto, K., Tsuruya, K., and Nakano, S. (1984): Tsunami wave force and failure mechanism of bulkhead due to Tsunami during 1983 Nihonkai-chubu earthquake, Proc. 31th Research Conference on Coastal Engineering, 257–261 (in Japanese)

Lessons Learned From Dams Behavior Under Earthquakes

Pedro Simão Sêco e Pinto

Abstract It is important to analyze embankment dam behavior from the past lessons learned. It is noticed that modern embankment dams withstand the design earthquake without significant damages. In spite of this scenario it is important to prevent the occurrence of incidents and accidents of embankment dams under earthquakes and so a deep understanding of the triggering factors is needed.

Well documented case histories from different regions of the world related with embankment dam behaviour were carefully selected and are discussed.

The design and the analysis of dam stability under seismic conditions are addressed. The new trend for performance basis design is to consider 2 levels of seismic actions and to analyse the situation when the limit of force balance is exceeded for high intensity ground motions associated with a very rare seismic event.

The reservoir triggered seismicity (RTS) is linked to dams higher than about 100 m or to large reservoirs (capacity greater than 500×10^6 m^3), rate of reservoir filling and to new dams of smaller size located in tectonically sensitive areas.

Dam monitoring and inspections of dams are presented. Experience has shown that the rational and systematic control of dam safety should consist of several tasks: (i) regular instrumentation measurements; (ii) data validation; (iii) data storage; (iv) visual inspections; (v) safety evaluation; (vi) corrective actions.

The risks associated with dam projects are discussed. The potential risk associated with dams consists of structural components and socio-economic components. The structural components of potential risk depend mostly on storage capacity and on the height of the dam, as the potential downstream consequences are proportional to the mentioned values. Socio-economic risks can be expressed by a number of persons who need to be evacuated in case of danger and by potential downstream damage.

It is important to develop new ways of thinking and strategies to address the future challenges.

Pedro Simão Sêco e Pinto (✉)
National Laboratory of Civil Engineering (LNEC), Geotechnical Engineering,
University of Coimbra, Coimbra, Portugal
e-mail: pinto.pss@gmail.com

A. Ansal, M. Sakr (eds.), *Perspectives on Earthquake Geotechnical Engineering*,
Geotechnical, Geological and Earthquake Engineering 37,
DOI 10.1007/978-3-319-10786-8_9

Keywords Case Histories · Modelling · Analysis · Induced Seismicity · Monitoring · Dam Safety · Risk Analysis

Introduction

From a careful study of dam behaviour for static and seismic scenarios the failure mechanisms are presented. Well documents case histories from many parts of the world related embankment dam behaviour were carefully selected and are discussed. The background of earthquake embankment dam engineering history is presented.

The seismic design and the analysis of dam stability during earthquakes are addressed. The reservoir triggered earthquakes and the causative factors are discussed. Dam monitoring and inspections of dams after earthquakes are presented.

The risks associated with dam projects are discussed. Some future challenges that deserved more consideration are introduced.

Background of Embankment Dam Engineering History

History does not record exactly when irrigation systems and dams were first constructed. Study of ancient Egypt, China, India and Iran dams does reveal that such work in these lands was begun thousands of years ago and provided lifelines on which their civilizations depended. Menes, the first Pharaoh of Egypt, ordered irrigation works to draw from the River Nile. In China, construction of impressive dams was accomplished on the Min River for flood control and diversion of water to nearby farm lands. The sacred books of India cite the very early operation of dams, channels, and wells, evidence that this land may have been the birthplace of the art.

Many of the outstanding waterworks of antiquity eventually declined into disuse because the knowledge of their designers and builders was not preserved by the generations who inherited them.

The ruins of the Sadd-el-Kafara embankment dam were discovered over 100 years ago in the Garawi ravine in Egypt. The dam was built around 2600 BC and was 14 m high and 113 m along the crest (Fig. 1). It is the oldest dam of such size known in the world. The purpose of the dam was to retain the water from rare but violent floods.

The grossly overdesigned cross section of the dam was due to inexperience. Unfortunately, there was no channel or tunnel to divert the river around the dam site while it was being built. As a result, the dam was destroyed while still under construction during one of the rare floods it was intended to control. The consequences of the dam's failure must have been so grave that the Egyptian engineers refrained from further dam construction for about eight centuries.

Fig. 1 The Sadd-el-Kafara Damphoto. (from Schnitter courtesy of A.A. Balkema)

Fig. 2 Anfengtang reservoir dyke. (courtesy ICOLD 2007)

About 5000 years ago, Chinese people started to build earth dikes and embankments for flood prevention. The dykes to form large plain reservoirs were always long, with irregular shape, but small in width and height. One example is the Shaobei, now called Anfengtang reservoir, with a dyke of 24.3 km (Fig. 2).

In southern of India the embankments used gravelly materials and so the outer slopes were considerably steeper. An extreme example was Motitalav dam located north of the city of Mysore with 24 m of high and upstream face protected by ashlar blocks with 1:1.5 slope and downstream face sloped 1:1 (Fig. 3).

Modern cross sections were adopted around 1213 for Anantha dam.

It is interesting to mention that at near Anantha dam an inscription dated 1369 states twelve essential requisites for the construction of a good reservoir (ICOLD 2007):

1. A king (i.e. owner or client) endowed with righteousness, rich, happy and desirous of acquiring fame;
2. A person well versed in hydrology;
3. A reservoir bed or hard soil;
4. A river conveying sweet water from a distance of about 40 Km;
5. Two projecting portions of hills in contact with the river;
6. Between these protecting portions of hills a dam built of compact stone, not too long but firm;

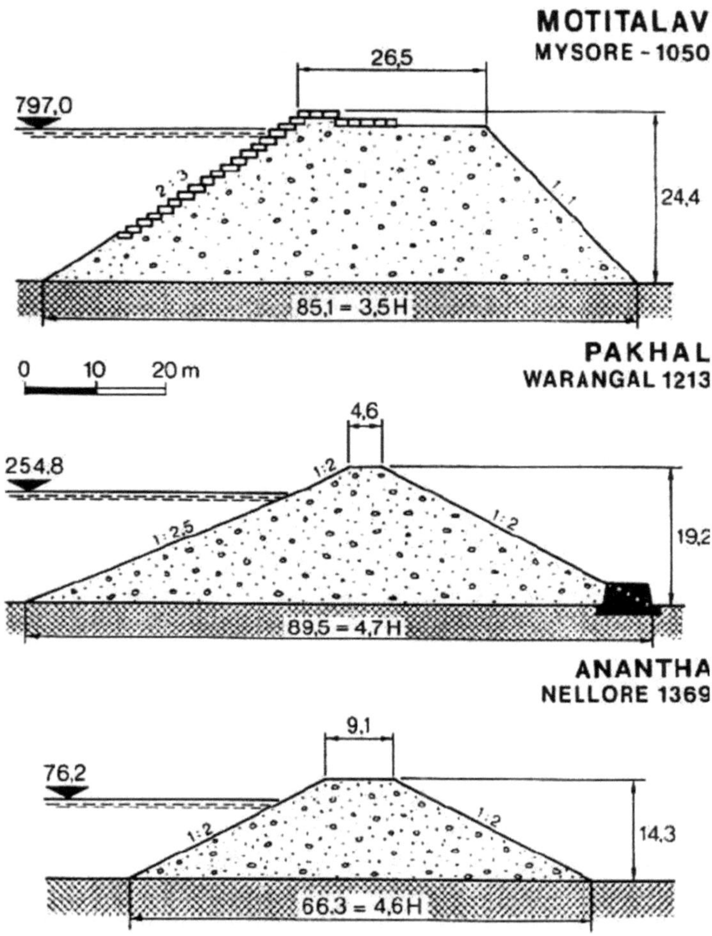

Fig. 3 Cross sections of three South-Indian embankment dams. (after ICOLD 2007)

7. The two extremities of the hills to be devoid of fruit-bearing land (i.e. humus);
8. The bed of the reservoir to be extensive and deep;
9. A quarry containing straight and long stones;
10. Fertile low and level (i.e. irrigable) area in the neighbourhood;
11. A watercourse having strong eddies in the mountain region; and
12. A group of men skilled in the art of dam construction.

The inscription also mentioned that the following six faults should be avoided:

1. Oozing water from the dam;
2. Saline soil;
3. Site at the boundary of two kingdoms;
4. High ground in the middle of the reservoir;

5. Scantly water supply and an extensive area to be irrigated;
6. Too little land to be irrigated and excessive supply of water.

The Shadorvan Dam-Bridge is an outstanding historical remain of Shoshtar, built nearly 1700 years ago and one of the important historical dams in Iran and of the world. It is about 500 m long, 10 m high, with 40 original gates, but only 8 gates remain in northern section and 15 gates in the southern section (Fig. 4).

The dam is 6 m wide and is made of rubble stone and lime mortar.

The Amir dam is a multi purpose dam located on the Kur river, 37 Km northeast of Shiraz, 104 m long, 15 m high and 5 m wide and its construction dates back to 1000 years ago (Fig. 5).

Dam toe is built in the shape of a pillow forming a stilling basin to mitigate river-bed erosion.

The dam is made of rubble stone and lime mortar, the three major purposes of the project were:

- Spanning the river as a bridge;

Fig. 4 Shadorvan dam–bridge in Iran. (Courtesy, ICOLD 2007)

Fig. 5 Amir dam in Iran. (Courtesy ICOLD 2007)

- Diverting water into irrigation canals;
- Running the water mills with the energy of river water.

These dams built in Egypt, China, India and Iran can be considered jewellery fashioned by the hand from the pure gold of the intellect.

Embankment Dam Performance During Earthquakes

The statistics presented by ICOLD (1995), indicates that from 50, 000 large dams (defined as a dam with height more than 15 m or water in the reservoir more than 1 million m^3) approximately 70 % are embankment dams.

The deteriorations exhibited by 1105 embankment dams show the following scenario (Budweg 1997; Fanelli 1994):

1. Deteriorations due to hydraulic structures and operation about 35 %;
2. Deteriorations due to cracking and hydraulic fracturing about 30 %;
3. Slope failure of dams and reservoir about 10 %;
4. Excessive seepage through foundation about 12 %;
5. Inadequate slope protection 7 %;
6. Other situations 6 %.

The distribution of the deteriorations during the dam phases is the following (ICOLD 1995):

1. 20 % during the construction phase;
2. 20 % during the filling of the reservoir;
3. 22 % after the construction phase;
4. 16 % after the period of 5 years following the construction phase;
5. 22 % no identified cases.

From a careful study of dam behaviour during earthquakes occurrences the following failure mechanisms can be selected (Sêco e Pinto 2001):

Sliding or shear distortion of embankment or foundation;
Transverse cracks;
Longitudinal cracks;
Unacceptable seepage;
Liquefaction of dam body or foundation;
Loss of freeboard due to compaction of embankment or foundation;
Rupture of underground conduits;
Overtopping due to seiches in reservoir;
Overtopping due to slides or rockfalls into reservoir;
Damages to waterproofing systems in upstream face;
Settlements and differential settlements;
Slab displacements;
Change of water level due fracture of grout curtain;

Movements on faults under or adjacent to the dam.

A survey of dams behaviour during earthquakes carefully selected is presented in Annex 1.

Lessons Learned

Some interesting case histories are discussed subsequently, in order to absorb the lessons learned from the incidents and accidents that have occurred in earth-rockfill dams, CFRD dams and tailing dams.

Earth-rockfill Dams

High Aswan Dam

High Aswan Dam (HAD) is a rockfill dam with clay core, rockfills shells and a wide grout curtain from the bottom of the clay to rock formation (Figs. 6 and 7). It is 111 m high and 3600 m long, the storage capacity is 162 km^3, the reservoir is the world's third largest reservoir, and is built on main river Nile, the world's longest river (Shenouda 1982).

The Aswan dam benefits Egypt by controlling the annual floods, improving navigation along the river and providing half of Egypt's power supply.

For the design purposes it was considered that Aswan area was not seismic.

On 14 November 1981 a moderate earthquake of magnitude 5.3 occurred about 5.3 km Southwest of the dam.

The sudden occurrence of this earthquake caused a significant concern due to the concentration of population in the valley and along downstream the dam. The evaluation of fault capability of releasing earthquakes in the Aswan area became a high priority problem (Shalaby 1995).

High-gain seismographs and also a network of six portable seismographs surrounding the after shock zone were installed on July 1982 with a telemeter network.

The seismic monitoring and the telemetered network have shown a close association between the Kalabsha fault and the main shock of November 1981 and much of the subsequent local seismicity. It was also concluded that the risk of reservoir triggered seismicity was insignificant.

Seismic stability and potential deformations were assessed by a non linear finite element analysis. The results of the studies show that the occurrence of the largest potential earthquake would not jeopardize the safety and integrity of the dam and its appurtenant structures.

Lesson 1 The identification of tectonic mechanisms, location and description of faults and estimation of fault activity play an important role to assess the involved dam risk.

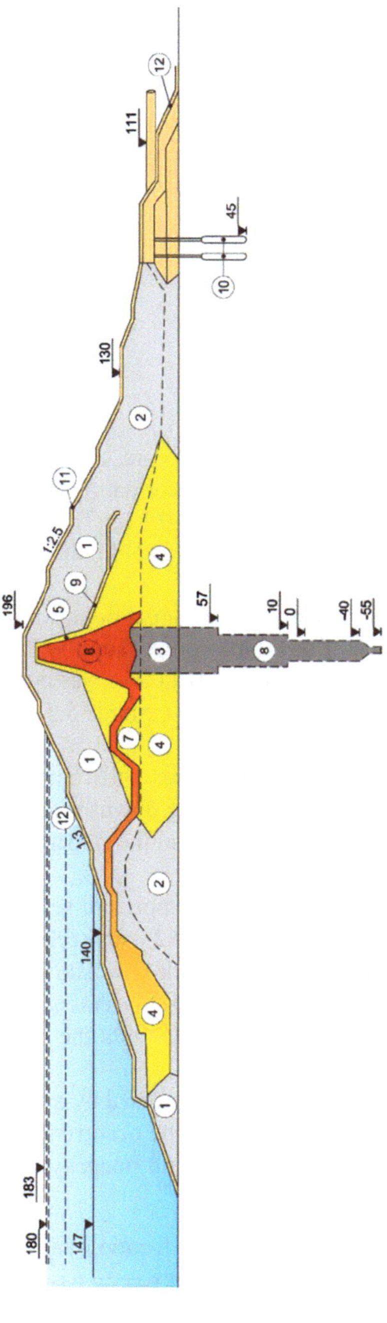

(1) Rockfill of muck.
(2) Sized stones sluiced with sand.
(3) Sized stones sluiced with silt and clay
(4, 4') Dune sand.
(5) Coarse sand.
(6, 7) Clay core and blanket.
(8) Grout curtain.
(9) Three layer filter.
(10) Drainage wells.
(11) D.S. prisms of fines.
(12) Protective layer of big blocks.

Fig. 6 Cross section of dam. (after Shenouda 1982)

Fig. 7 General view of Aswan dam

Aramos Dam

Aramos dam with 42 m high and 220 m long is located in Chile. The dam profile shown in Fig. 8 has a core with fine soils and shells with gravelly sand. Due to the existence of a foundation of sandy material a plastic concrete wall of 80 cm thick and a maximum of 22.5 m depth was built and the sectors of plastic wall that did not reach the foundation was injected (Verdugo and Peters 2009).

During the construction of the embankment dam a bulldozer operating in the river bed sank showing that the ground was susceptible to liquefaction. A external evaluation of the project was performed and the recommended countermeasures have included a berm with 13 m high confining the upstream shell and a battery of drainage columns between the toe of the dam and the spillway was installed.

During March 3, 1985 earthquake the dam exhibited an extremely good behaviour with a maximum settlement of 10 cm. A possible explanation is that the low SPT zones are surrounded by stiffer zones that have reduced the disturbance of loose zones.

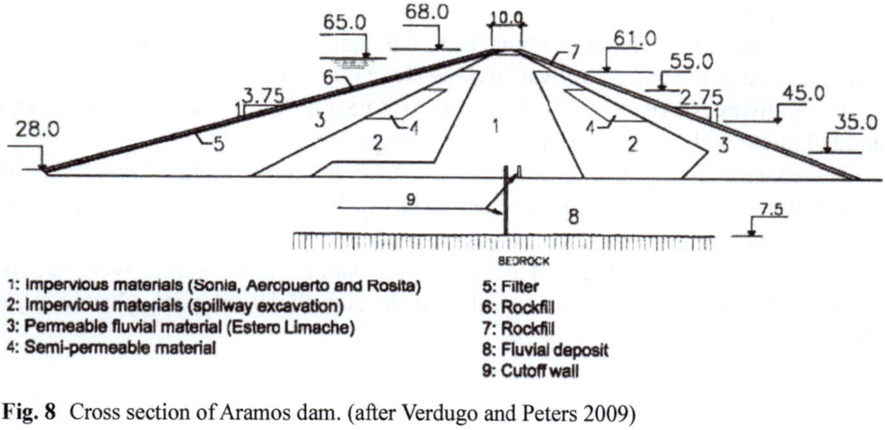

Fig. 8 Cross section of Aramos dam. (after Verdugo and Peters 2009)

Lesson 2 Occurrence of liquefaction in a ground with heterogeneous conditions requires a deeper analysis and a better understanding of the interaction phenomena.

Tapar Dam

Tapar earthen dam is located in Kachch district of Gujarat state, India, severely affected by the massive earthquake of intensity 6.7 on Richter scale on 26th January 2001. Tapar dam has a total length of 4575 m and the maximum height of 17.75 m with a design capacity of 48.81 million cmt was constructed in 1975. Tapar dam is located in the heart of epicenter region and was badly damaged due to earthquake.

The dam is founded directly upon the alluvium that fills the Sakara river basin. The maximum depth of this alluvium is greater than 30 m and a relatively shallow central key trench cut-off beneath the core, coupled with an impervious blanket, is used to reduce under seepage losses. Under seepage is significant nonetheless, and 11 relief wells have been installed at the downstream toe to provide filtered drainage for this under seepage passing through the foundation. During the earthquake, liquefaction occurred in the alluvium beneath several section of the main embankment that caused lateral spreading and translation movement of several sections of the upstream face and berms. Large cracks and fissures occurred on the upstream face. One of the slippages occurred principally through the two lower berms at the upstream toe, and sand and silky sand boil eject was noted in fissures over a considerable area at this location.

Longitudinal cracks (Fig. 9) with apertures as wide as 0.3 to 0.5 m and depth of up to 3 m occurred along several hundred meters of the crest. At the crest near the center of the embankment longitudinal cracking appeared adjacent to the concrete intake tower. Maximum cracking apertures were up to 0.5 m, and cracks were traced to a depth in excess of 5 m.

This scrap was associated with a zone of upstream slippage that extends for approximately 150 m along the central portion of the dam, beginning just north of the concrete intake tower.

In addition to these types of damages to the main earthen embankment, damage also occurred at the gated spillway and the Head regulator.

The spillway is a concrete lined weir channel near the north end of the main earth embankment, with multiple rotating steel weir gates.

The concrete lining was separated in extension by approximately 2 to 10 cm of lateral displacement of the adjacent embankment in both the upstream and downstream direction.

Cracking at the crest (Fig. 10) and faces, and damage to the berms, were largely repaired by excavation and reoccupation of the superficial materials (Yadav et al. 2008).

Fig. 9 Tapar dam Failed Maximum Height channel section. (after Yadav et al. 2008)

Fig. 10 Tapar dam Transverse crack on the crest. (after Yadav et al. 2008)

Lesson 3 Foundations singularities that occurred in dams play an important role in the appearance of both longitudinal and transverse cracks.

Tarbela Dam

Tarbela dam with 143 m high built in Pakistan is an example of a dam that was designed without consideration of Darband fault which was revealed during the dam construction. So the review design with the dam cross sections showed in Fig. 11 has estimated a fault movement of 1.0 to 1.5 m and the core dam was constructed of self healing material with a transition zone, a generous freeboard and a wide chimney drain on the downstream. A general view of the dam is shown in Fig. 12.

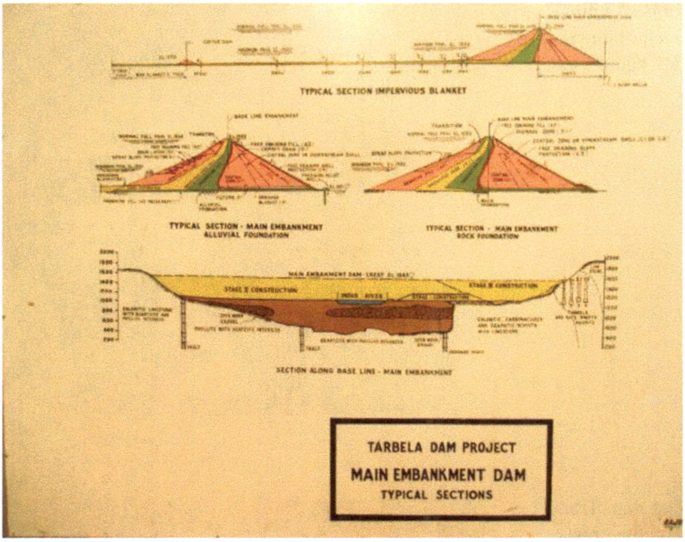

Fig. 11 Tarbela dam-cross sections

Fig. 12 Tarbela dam -general view

Tarbela dam was shaked by October 8, 2005 earthquake with a 7.5 magnitude. The pore pressures and seepage rise were observed in the dam right abutment. Relevant piezometers and seepage points were continuously monitored to know the trend which became normal after a few days.

Due to this event planning and installation of earthquake monitoring and strong motion recording instrumentation for dams and hydropower projects in northern areas of Pakistan was implemented.

Preparation of seismic provisions for Building Code of Pakistan has included:

- Revised seismic zoning map of Pakistan;
- Catalogue of Historical earthquakes of Pakistan;
- A catalogue of available fault plane solutions of earthquakes in Pakistan.

Lesson 4 Dam behaviour during earthquakes contributes to update and implement national codes.

Cariblanco and Toto Projects

Figure 13 shows the location of dams including the main faults where strong motion records where obtained during Costa Rica earthquake January 8, 2009, 6.2 magnitude. It is clear that Cariblanco and Toro projects are located very near of the epicentral area sites (ICE 2009).

Cipreses dam of Cariblanco project has no accelerograph installed, but is located very close (1.7 km) of the surge tank instrument and has shown cracks.

Toro II project has exhibited longitudinal cracks in the crest.

Lesson 5 Dams located near faults can exhibit a good behaviour for strong motions with only minor cracks.

Coihueco and Colbun Dams

Maule earthquake has occurred on 27 February 2010 in Chile, with a Ritcher magnitude 8.8, Coihueco earth dam with 975 m long and 31 m high after the earthquake has exibitted two longitudinal cracks along the dam crest, one in the centre (Fig. 14) and other at the upstream shoulder (Fig. 15). The width of the crest is 5.2 m and the depth of the crack reached at maximum 1.9 m (Yasuda et al. 2010).

Colbun dam has suffered significant displacements. This dam with 116 m high and 550 m long is 100 Km away from the fault zone.

The dam crest has suffered a horizontal displacement of 2 m and a vertical displacement of 1 m (Fig. 16).

Lesson 6 Well designed dam in spite of severe damages have not collapsed.

El Infiernillo Dam

El Infiernillo dam built in Mexico on 1963 is a zoned dam with a thin central clay core and shells with rockfill materials.

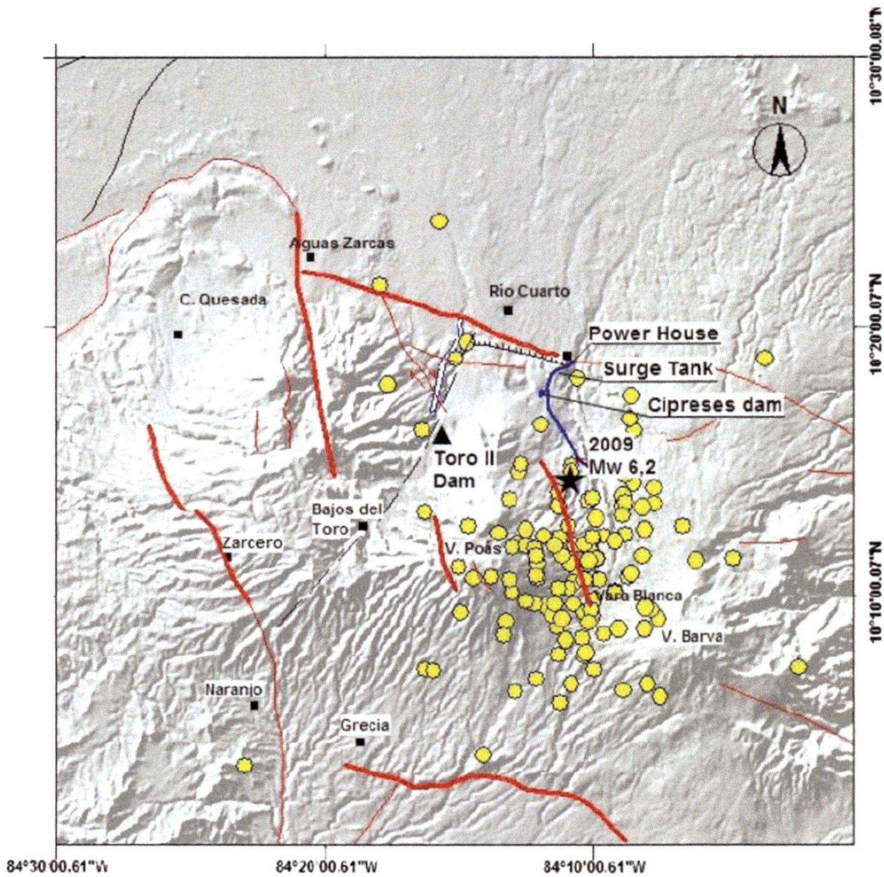

Fig. 13 Epicentral location (*black star*), aftershocks and other small events until January 16 (*yellow circles*). Main faults in the region (in *red*) and location of Cariblanco and Toro projects. (after ICE 2009)

Fig. 14 Coihueco Dam -Longitudinal cracks at the crest

Fig. 15 Coihueco Dam-Downward slope movement at the upstream shoulder

Fig. 16 Colbun earth Dam with 2 m horizontal displacements and 1 m vertical displacements

The difference of deformability between clay core and rockfill shells has favoured stress transference.

A dam cross section is shown in Fig. 17.

Dam body view and general layout are shown in Figs. 18 and 19, respectively.

During earthquake 1981 longitudinal cracks have occurred.

Lesson 7 Due to the stress transfer between clay core and rockfill shells the earthquakes favoured the occurrence of longitudinal cracking.

Oroville Dam

Oroville dam with 1707 m long and 235 m high was built in 1968 (Banerjee et al. 1979). The dam cross section is shown in Fig. 20. The dam has suffered August, 1, 1975 Oroville earthquake with 5.7 magnitude and has exhibited crest settlements of 9 mm and horizontal displacements of 15 mm.

Fig. 17 El Infiernillo dam profile

El Infiernillo Dam

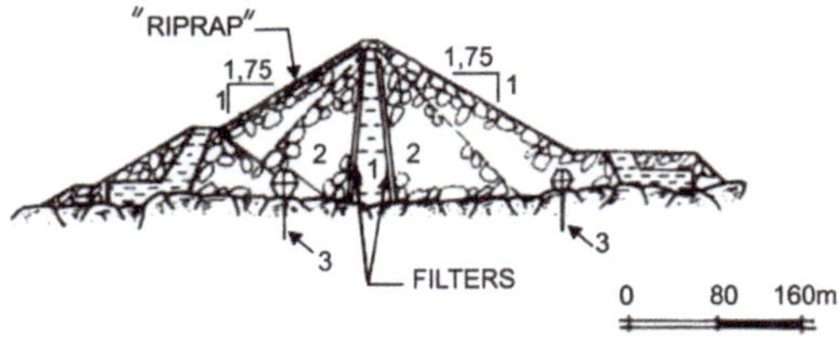

1-Clay core
2- Rockfill shells
3- Grout curtain

Fig. 18 El Infiernillo dam body view

Fig. 19 El Infiernillo general dam layout

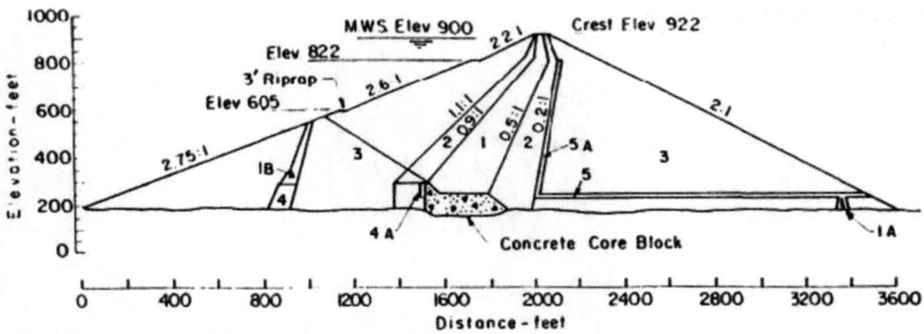

Zone	MATERIAL
1, 1A, 1B	Impervious Core
2	Transition
3	Pervious Shell
4, 4A	Impervious (Abutment Stripping)
5, 5A	Drain (Selected Shell)

Fig. 20 Oroville dam cross section. (after Banerjee et al. 1979).

The conventional pseudo-static analysis with a seismic coefficient of 0.1 g was performed during the design stage.

The Oroville earthquake discloses the existence of a previously unidentified fault in the vicinity of the dam. Due to the concerns related the occurrence of a 6.5 magnitude earthquake, with a hypocentral less than 8.5 km from the dam, 2D and 3D finite element analyses were performed.

To correct simulate the dynamic response of dams in steep triangular canyons a three-dimensional analysis is needed.

Lesson 8 Compacted rockfill dams exhibit a good behaviour during the occurrence of earthquakes.

Keddara Dam

On May 2003 an earthquake of magnitude 6.8, with a depth of 10 km, occurred 40 km East Alger, provoking 2270 deaths. In Keddara rockfill dam, with 106 m high (Fig. 21), located 30 km from the fault, only 1 longitudinal crack and 3 transverse cracks were observed (Fig. 22).

A value of 0.34 g was recorded in rock and the dam was designed for a acceleration value of 0.25 g. No damages were observed in the gallery (Benlala 2003).

Lesson 9 Well designed and constructed rockfill dams exhibit a good behaviour for strong ground motions.

Fig. 21 Keddara dam-downstream view

Fig. 22 Keddara dam-cracks observed at the crest

CFRD Dams

Zipingpu Dam

Zipingpu dam 156 m high and 663.7 m long is one of the largest CFRDs in China was built in 2006 and designed for a peak ground acceleration of 0.26 g (Chen 2008). A dam profile is shown in Fig. 23.

The dam was shaked during May 12, 2008 Wenchian earthquake (magnitude 8.0) and is located 17 km of the epicenter.

During the earthquake the reservoir was low with a volume of 300 Mm^3, when under the normal conditions the reservoir volume is 1100 Mm^3. Due to this situation is difficult to estimate the dam behaviour when the reservoir is full.

The crest of the dam and the concrete face were damaged during the earthquake (Figs. 24 and 25).

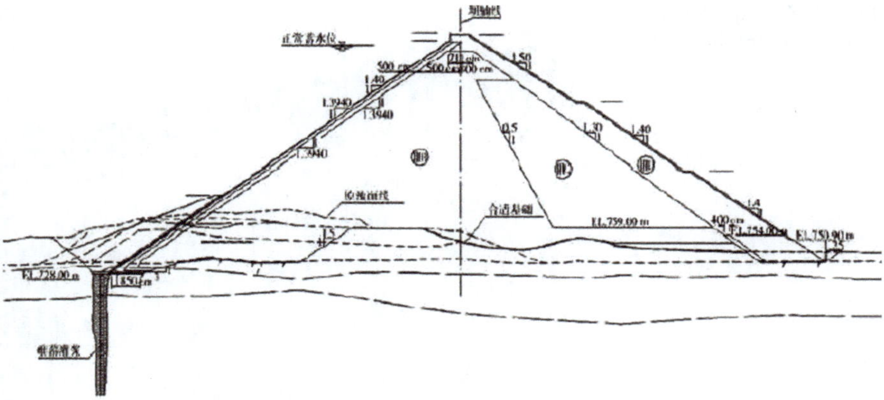

Fig. 23 Zipingpu dam profile. (after Chen 2008)

Fig. 24 Zipingpu dam crest damages. (after Chen 2008)

The dam crest settled 715 mm and had a horizontal deflection of 180 mm.

From the six strong motions instruments only 3 located at the crest were in good conditions at on the crest a peak acceleration of 2 g was recorded.

Lesson 10 In spite the apparent good behaviour of CFRDs (Concrete Face Rockfill Dams) with no failures, during earthquakes, there is still a lack of case histories of CFRDs operating with full reservoir.

Tailing Dams

The main concern of tailing dams located in seismic prone zones is the evaluation of their stability during the occurrence of earthquakes.

Fig. 25 Zipingpu dam joint damages. (after Chen 2008)

For example in Chile there are in operation tailing dams with more than 150 m high, e.g. Las Tortolas dam (Los Broncos mine) with a maximum height of 190 m, Quillayes dam (Los Pelambres mine) with a final height of 198 m and Mauro dam (Los Pelambres mine, Fig. 26) with a maximum final height of 248 m (Barrera et al. 2011).

Fig. 26 General view of El Mauro tailing dam. (after www.ipnews.net)

The design has incorporated the modern soil mechanics and soil dynamics concepts and extensive investigation campaign of high stress triaxial stress were performed with the objective of evaluating the strength, deformation and permeability characteristics of tailings sands under maximum vertical stress close to 4.5 MPa.

Lesson 11 Seismic failures of tailing dams have been associated with the occurrence of liquefaction and it is important to ensure the seismic stability for a large period of time.

Accidents on Dams

During Tohoku earthquake, on March 11, 2011, magnitude Mw = 9.0, almost all of the dams withstood minor to severe ground shaking and retained their reservoirs with generally minor to moderate damage. The exception to this was Fujinuma Dam, an embankment dam located in southern Fukushima Prefecture that failed shortly after the earthquake. The failure of the dam resulted in the uncontrolled release of the entire reservoir, which flowed downstream into a small village and killed 8 people (Matsumoto 2011; Towhata et al. 2011).

Fujinuma Dam had a maximum height of about 18.5 m and had a maximum reservoir volume of approximately 1.5 million m^3 (~ 1200 acre-feet) (Fig. 27). It has sometimes been referred to as Fujinuma-ike, which means it was considered to be a pond-retaining structure because the dam was not on a regulated river (Matsumoto 2011).

Construction of Fujinuma Dam began in 1937 and was halted during World War II. Following the war, the dam was completed in 1949 (Matsumoto 2011; Towhata et al. 2011).

There were actually two embankment dams retaining the reservoir: a 18.5-meter-high main dam and a ~ 6-meter-high auxiliary dam. Figure 28 presents a general

Fig. 27 Aerial view of Fujinuma Main Dam and Auxiliary Dam in 2009

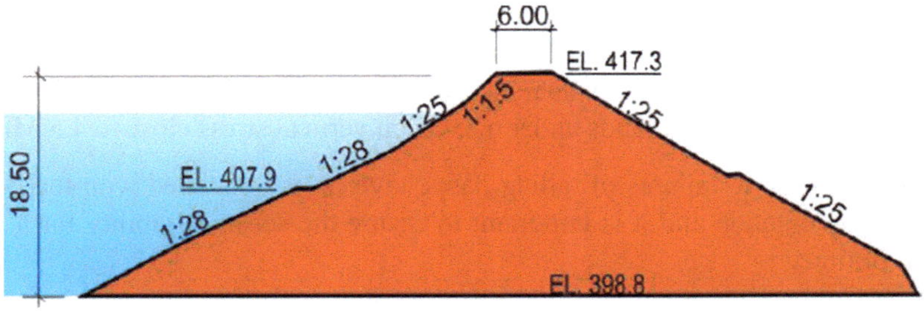

Fig. 28 Cross section of Fujinuma Dam. (after Matsumoto 2011)

cross section of the main dam obtained from Matsumoto (2011). The main dam had a crest width of 6 m and an upstream slope which ranged principally between 2.5:1 to 2.8:1, together with small benches and a relatively steep 1.5:1 upper slope.

The downstream slope had a general slope of 2.5:1 with a small bench at mid-height (Fig. 29). It also had a steepened downstream toe, perhaps indicative of a possible rock or gravel toe. It is believed that the auxiliary dam had a similar geometry. The main dam had a crest length of 133 m.

It is not known what the zoning of the dam was designed to be, although some reports indicate that it was a homogeneous embankment design (Matsumoto 2011). Nor is there any information currently available regarding what type of an amount of foundation treatment was provided at either the main dam or auxiliary dam.

However, a diagram posted on the left abutment of the main dam indicates that the main dam has had a seepage problem in either the upper foundation or lower embankment materials.

The dam was located approximately 80 km from the fault rupture and directly opposite the approximate middle of the 600-km fault rupture zone. The nearest strong motion seismograph, a surface instrument located 2.8 km away in the community of Naganuma, registered a peak ground acceleration of 0.315 g. According

Fig. 29 Pre-earthquake photograph of crest and downstream slope of Fujinuma Main Dam

Fig. 30 View of Fujinuma Dam Embankment fill layers on top of black organic foundation residual soil exposed within downstream slope at left abutment

to Matsumoto (2011), the reservoir was nearly full when the earthquake occurred. The dam crest reportedly was overtopped. The brief reconnaissance by the GEER team resulted in the following observations of the main dam and the breach within it:

1. At least some of the fill was placed in relatively thick lifts (Fig. 30), so the degree of compaction may have been relatively small compared to the level of compaction required for modern dams. Even though most of the fill material appeared to be cohesive in nature, thick lifts of this material may still be relatively loose, weak, and/or subject to strength losses during a long duration of earthquake shaking.
2. There did not appear to be any filters constructed within the dam embankment or foundation. If the dam deformed significantly during the earthquake shaking, transverse cracks could have developed and promoted internal erosion.

Lesson 12 Embankment dams with low degree of compaction are vulnerable to earthquakes.

Conclusion The current State of the Art and State of Practice allow that the design and construction of embankment dams exhibit a good performance record in regards of earthquake shakes with less than 1 % dam failures.

Seismic Design

Introduction

According to Aristotle (384 − 322 B.C.) in his book Meteorologica earthquakes were produced by the dried exhalations (spirits or winds) in caves inside the earth which trying to escape make the earth shake.

Martin Lister in England and Nicolas Lemery in France in seventeenth century were the first to propose that earthquakes were produced by large explosions of inflammable material formed by a combination of sulphur, coal, nitre and other products accumulated in the interior of earth (Udias and Arroyo 2005). The explosive theory was also proposed by Newton's Optics (1718) and the modern scientific ideas consider the earthquake a natural phenomenon.

In France the world was considered a good place in which everything that happened was viewed to be "for the best" and earthquake was considered with optimism. Voltaire in his novel Candide presented a hard attack to this optimistic view point. Also Kant and Rosseau defended the optimist position.

Selection of Design Earthquakes

The selection of seismic design parameters for dam projects depends on the geologic and tectonic conditions at and in the vicinity of the dam site (Sêco e Pinto 2004).

The regional geologic study area should cover, as a minimum, a 100 km radius around the site, but should be extended to 300 km to include any major fault or specific attenuation laws.

When available emphasis should be given to recorded earthquake data. For each event the following data shall be considered: (i) epicentre coordinates; (ii) magnitude (or epicentral intensity); (iii) date and time of occurrence; (iv) focal depth; (v) focal mechanism; (vi) felt area; (vii) accompanying surface effects; (viii) intensity of ground motion induced at the dam site, (ix) source of data; (x) reliability.

The probabilistic approach quantifies numerically the contributions to seismic motion, at the dam site, of all sources and magnitudes larger than 4 or 5 Richter scale and includes the maximum magnitude on each source.

The dam should be designed for Design Earthquake (DE) and Maximum Design Earthquake (MDE). Both depend on the level of seismic activity, which is displayed at each fault or tectonic province (Wieland 2003).

For the OBE only minor damage is acceptable and is determined by using probabilistic procedures (SRB 1990). For the MDE only deterministic approach was used (ICOLD 1983) but presently it is possible to use a deterministic and probabilistic approach. If the deterministic procedure is used, the return period of the event is ignored, if the probabilistic approach is used a very long period is taken (ICOLD 1989).

Two levels for seismic activity, namely MCE (Maximum Credible Earthquake) considering a return period of 500–1000 years and DBE (Design Basis Earthquake) for a return period of 145 years, with a probability of exceeding in 100 years less than 50 %, were proposed by ICOLD (1989).

ICOLD (2002) has considered 3 levels of seismic action, namely: MDE (Maximum Design Earthquake), MCE (Maximum Credible Earthquake) and OBE (Operating Basis Earthquake). Four hazard classes were defined, namely: Low with PGA 0,10 g, Moderate with 0.10 PGA0.25 g, High with PGA 0.25 g (no active faults within 10 Km) and Extreme with PGA 0.25 g (active faults within 10 Km).

In Eurocode 8 (1998a), in general, the hazard is described in terms of a single parameter, i.e. the value a_g of the effective peak ground acceleration in rock or firm soil called "design ground acceleration" expressed in terms of: a) the reference seismic action associated with a probability of exceeding (P_{NCR}) of 10% in 50 years; or b) a reference return period (T_{NCR})=475 years. The seismic action to be taken into account for the "damage limitation requirement" has a probability of exceedance, of 10% in 50 years and a return period of 95 years (Seco e Pinto 2009a).

Tectonic Conditions

Within this framework the tectonic conditions should include tectonic mechanisms, location and description of faults (normal, stryke and reverse) and estimation of fault activity (average slip rate, slip per event, time interval between large earthquake, length, directivity effects, etc), these factors are important to assess the involved risk.

The foundation properties for soil materials are estimated by geophysical tests (crosshole tests, seismic downhole tests and refraction tests), SPT tests, CPT tests, seismic cone and pressurometer tests (ICOLD 2005a).

An active fault is a fault, identified and located, known to have produced historical fault movements or showing geologic evidence of Holocene (11,000 years) displacements and which, because of its present tectonic sitting, can undergo movements during the anticipated life of man-made structures (ICOLD 1998).

The current practice is the deterministic approach in which the seismic evaluation parameters were ascertained by identifying the critical active faults which show evidence of movements in Quaternary time (ICOLD 1998).

Dense recording GPS arrays with sampling rate allow determining deformation rates in seismic active regions. Intrinsic properties of rock at depth have to be obtained in situ by deep drilling into active faults. Computational with high resolution model for stress and deformations in communicating fault systems should be developed. A better exploration of microtremors technique, directivity effects and attenuation laws is needed.

Surface fault breaking i.e. surface slip along an identified fault zone under the dam is considered as the most dangerous tectonic scenario for dam safety.

With the tendency of less favourable dam sites these tectonic conditions are getting increase attention.

The active tectonic movements result on fault breaks and in creep movements. Also block movements have to be considered in the near field of major faults.

Following Sherard et al. (1974) a concrete dam on active faults or near major active faults is not advisable and if a site with fault movements can not be avoided it is recommended to build an embankment dam.

Evaluation of the displacement that could occur along the fault during the lifetime of the dam and the selection of the design details to ensure safety against fault displacement are still difficult problems to be solved (Wieland et al. 2008).

Potentially Liquefiable Soils

To assess liquefaction potential empirical liquefaction charts are given with seismic shear wave velocities and SPT values.

The new proposals integrate: (i) data of recent earthquakes; (ii) corrections due the existence of fines; (iii) experience related with a better interpretation of SPT test; (iv) local effects; (v) cases histories related to more than 200 earthquakes; and (vi) Bayesian theory.

For liquefaction assessment by shear wave velocities two methodologies are used: (i) methods combining the shear wave velocities by laboratory tests on undisturbed samples obtained by tube samplers or by frozen samples; (ii) methods measuring shear wave velocities and its correlation with liquefaction assessment by field observations.

It is important to refer that Eurocode 8 (1998b) considers no risk of liquefaction when the ground acceleration is less than 0.15 in addition with one of the following conditions: (i) sands with a clay content higher than 20% and a plasticity index 10; (ii) sands with silt content higher than 10% and N1 (60) 20; and (iii) clean sands with N1 (60) 25.

For post liquefaction strength relationships between SPT and CPT tests and residual strength were proposed by several authors.

Also to assess the settlement of the ground due to the liquefaction of sand deposits there are some proposals based on the knowledge of the safety factor against liquefaction and the relative density converted to the value of N1.

The remedial measures against liquefaction can be classified in two categories (TC4 ISSMGE 2001; INA 2001): (i) the prevention of liquefaction; and (ii) the reduction of damage to facilities due to liquefaction.

For the selection of the remedial measure it is important to consider: (i) Potential efficiency; (ii) Technical feasibility; (iii) Impact on structure and environmental; (iv) Cost-effectiveness; and (v) Innovation (Sêco e Pinto 2006).

More recently it is recognized that gravelly material can liquefy.

The behavior of Keenleyside dam with foundation composed of sands and gravel was investigated. Due many uncertainties in the assessment multiple methods both field tests (SPT tests, Becher Penetration tests, shear wave velocities) and laboratory tests (triaxial and permeability tests) were used (Yan and Lun 2003).

Performance Basis Design

The new trend for performance basis design is to consider 2 levels of seismic actions and to analyse the situation when the limit of force balance is exceeded for high intensity ground motions associated with a very rare seismic event (Sêco e Pinto 2009b).

Table 1 Acceptable level of damage in performance based design

Acceptable level	Structural	Operational
Degree I: serviceable	Minor or no damage	Little or no loss of serviceability
Degree II. repairable	Controlled damage	Short term loss of serviceability
Degree III: near collapse	Extensive damage in near collapse	Long term or complete loss of serviceability
Degree IV collapse	Complete loss of structure	Complete loss of serviceability

For the design two basic requirements are defined: (i) Non collapse requirement (ultimate limit states) i.e. after the occurrence of the seismic event the structure shall retain its structural integrity, with respect to both vertical and horizontal loads, and adequate residual resistance, although in some parts considerable damage may occur, (ii) Minimization of damage (serviceability limit state) after seismic actions with high probability of occurrence during the design life of the structure some parts can undergo minor damage without the need of immediate repair. The structure shall be designed and constructed without the occurrence of damage and the associated limitations of use, the costs of which would be disproportionately high in comparison with the costs of the structure itself (Towhata 2008).

Acceptable level of damages in performance based design shown in Table 1 are specified by a combination of structural and operational damage (Iai 2009).

Analysis of Dam Stability During Earthquakes

Introduction

Principles are the fundamental of dam safety. The dam shall safely retain the reservoir and any stored solids and pass environmentally acceptable flows as required for all loading conditions ranging from normal to extreme loads, commensurate with the consequences of failure.

Practices and Procedures suggest methodologies that may be used to meet the Principles.

Table 2 presents the seismic criteria probabilistic approach recommended by Canada.

New Zealand and UK have suggested a return period of 10,000 years for SEE for Extreme and High risk dams, 3000 years for Moderate risk and 1000 years for Low risk.

Table 2 Seismic criteria-probabilistic approach recommended by Canada

Consequence class of dam	EQ Design Ground Motion (EDGM-Mean Annual Exceedance Probability)
Low	1/500
Significant	1/1000
High	1/2500
Very High	1/5000
Extreme	1/10,000

Experimental Models

Experimental methods are used to test predictive theories and to verify mathematical models. Nevertheless some limitations they are useful for physical modelling in geotechnique.

The most popular techniques for embankment dams are shaking table and centrifuge models.

Japan E-Defense 3D Shaking table with 15 m long 20 m wide with payload capacity of 1200 t and with maximum accelerations of 9 m/s^2 is one of the largest facilities in the world (Tokimatsu et al. 2007).

2D shakers were installed in UC Davis and RPI geotechnical centrifuges with robots.

A review of the existing testing facilities for earthquake research in order to address new scientific topics in earthquake engineering was performed by Taucer (2005).

In order to clarify dynamic progressive failure of dams subjected to strong motions, shaking table tests considering simultaneously upstream-downstream and vertical direction were performed by Masukawa et al. (2004). The deformations were measured with laser displacement devices, video cameras and special designed ground strain instruments installed within the embankment. In Fig. 31 an overview of shaking table tests is presented.

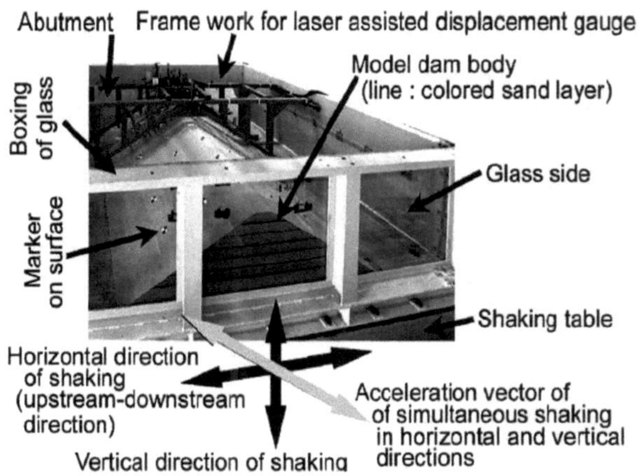

Fig. 31 Overview of shaking table tests. (after Masukawa et al. 2004)

A special model tests was carried out at LNEC (National Laboratory of Civil Engineer) for 185 m high Steno arch dam, to be built in Greece on a fault, where displacements up to some decimetres could occur. Earthquakes up to 6.5 magnitude could happen within 50 Km from the dam site during the lifetime of the dam.

The model shown in Fig. 32 reproduces at a 1:250 scale the exact shape of the arch, its pulvino and special joints as well as the surrounding rock and abutments with the fault.

The design criteria require that the arch should withstand with no serious damages small fault movements up to some centimeters and not collapse under larger movements of some decimetres (Gilg at al. 1987).

The tests have verified the good behaviour of the arch and the fundamental role played by the joints in so severe foundations conditions.

Mathematical Models

The following dynamic analysis of embankment dams is used (Sêco e Pinto 1993):

1. pseudo-static analyses;
2. simplified procedures to assess deformations;
3. dynamic analysis.

The pseudo-static analyses assume a rigid or elastic behavior for the material (Ambraseys 1960) and have the limitation that the seismic coefficient acts in one direction for an infinite time.

Simplified procedures to assess deformations were proposed by Newmark (1965), Sarma (1975), Makdisi and Seed (1977) and Bray (2007) and have given reasonable answers in areas of low to medium seismicity.

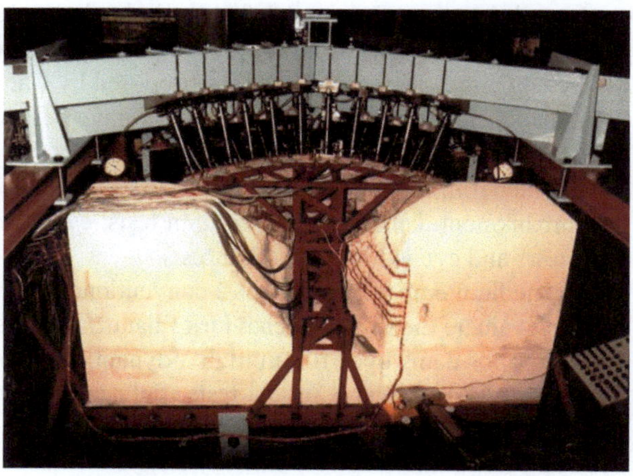

Fig. 32 Steno arch dam model. (Courtesy LNEC)

Newmark's original sliding block model considering only the longitudinal component was extended to include the lateral and vertical components of earthquake motion by Elms (2000).

The use of dynamic pore pressure coefficients along with limit equilibrium and sliding block approaches for assessment of stability of earth structures during earthquakes was demonstrated by Sarma and Chowdhury (1996).

For large dams where strong earthquakes have occurred more sophisticated methods were used (Seed 1980; ICOLD 2001).

Several finite element computer programs assuming an equivalent linear model in total stress have been developed for 1D-SHAKE code (Schanabel et al. 1972), 2D-LUSH code (Idriss et al. 1973; Lysmer et al. 1974) and pseudo 3D- TLUSH code (Lysmer et al. 1975).

Since these models are essentially elastic the permanent deformations cannot be computed by this type of analysis and are estimated from static and seismic stresses with the aid of strain data from laboratory tests (cyclic triaxial tests or cyclic simple shear tests) (Sêco e Pinto 1993).

For embankment dams a value of 5 % axial strain is used as allowable deformation.

To overcome these limitations, nonlinear hysteretic models with pore water pressure generation and dissipation have been developed using incremental elastic or plasticity theory.

The incremental elastic models have assumed a nonlinear and hysteretic behavior for soil and the unloading-reloading has been modeled using the Masing criterion and incorporate the effect of both transient and residual pore-water pressures generated by seismic loading implemented in TARA 3 and DESRA codes (Lee and Finn 1978; Finn 1987).

For the models based on the theory of plasticity two particular formulations appear to have a great potential for multidimensional analysis: the multi-yield surface model implemented in DYNAFLOW code (Prevost 1993) and the two-surface model (Mröz et al. 1979).

A modified cam-clay model for cyclic loading taking into account that when saturated clay is unloaded and then reloaded the permanent strains occur earlier than predicted by the cam-clay model was proposed by Carter et al. (1982). The predictions exhibit many of the same trends that have been observed in laboratory tests involving the repeated loading of saturated clays.

For the definition of the constitutive laws the following laboratory tests are used for embankment dams: resonant column tests, cyclic simple shear tests, cyclic triaxial tests and cyclic torsional shear tests.

For medium embankment dams a conventional pseudo-static analysis method is used to evaluate the seismic behavior of dams (Ambraseys 1960; Seed and Martin 1966), but for dams over 100 m high a dynamic analysis including computational analysis (modal analysis), model tests, field measurements and prototype tests is recommended (ICOLD 1975).

A flowchart that integrates stability analysis of dams, monitoring and case histories is shown in Fig. 33.

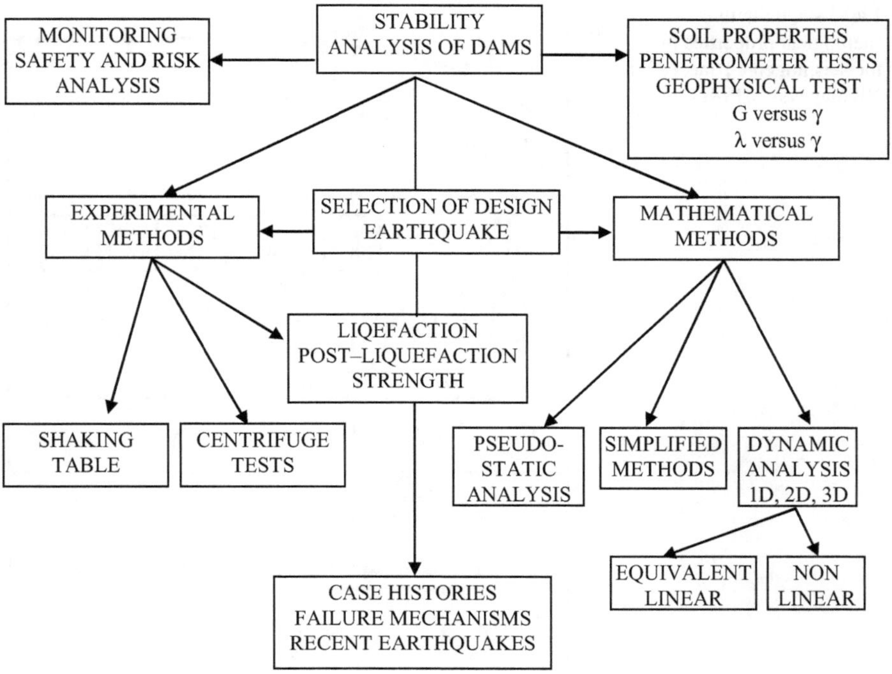

Fig. 33 Flowchart for embankment dams

Reservoir Triggered Seismicity

Man—made earthquakes caused by the filling of reservoirs have drawn the attention of designers concerned with dam safety (ICOLD 2010).

The reservoir triggered seismicity (RTS) is linked to dams higher than about 100 m or to large reservoirs (capacity greater than 500×10^6 m^3), rate of reservoir filling and to new dams of smaller size located in tectonically sensitive areas. This means that the causative fault is already near to failure conditions and so the added weight stresses and pore pressures propagation due to reservoir impounding, can trigger the seismic energy release.

The earthquakes that have occurred around the few dams by mere accident cannot definitely be attributed to dam or water load, which is insignificant, compared to the earth mass.

The detection of reservoir induced seismicity may be performed in two phases (ICOLD 1999): (i) phase 1 includes on historical seismicity and surveys of reservoir and surrounding geological structures, aiming at identification of possible active faults; and (ii) the second phase is carried out starting at least 1 or 2 years prior to the impounding with the installation of a permanent network of seismometers and other measures such as precise levelling, use of instrumentation to detect active fault movements, and reservoir slope stability studies.

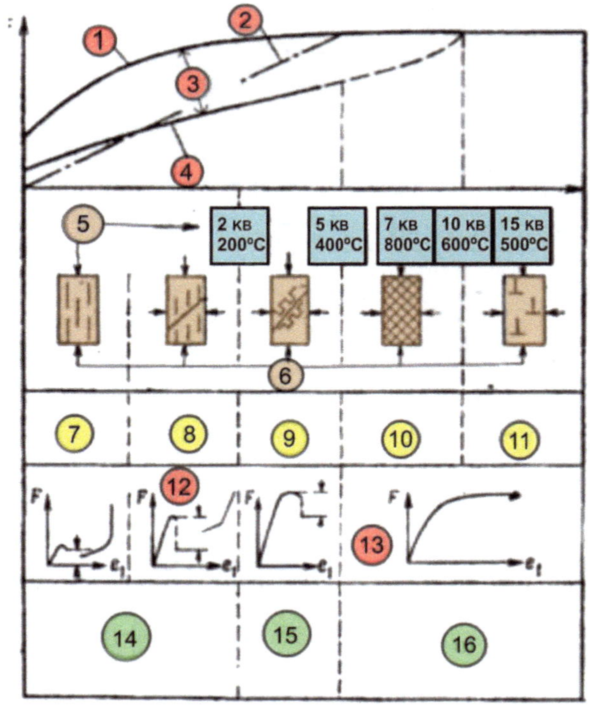

Fig. 34 Qualitative rheological properties in the seismogenic zone. (after Nikolayevskiv 1982)

Seismological observations established at Bhakra, Pong and Ramanga dams in the Hymalayan terrain have not registed any increase in seismicity due to impounding of reservoir.

In the hydro-mechanical model it is assumed that the large scale fault consists of many smaller fault patches which may slip independently but are mechanically coupled to their neighbours (so called "block-spring model"). Whenever shear slip occurs, the shear stress resolved on the patch is reduced by the amount of stress drop and shear stress on the neighbouring patches is increased due to mechanical coupling. Due to stress redistribution, slip on an individual patch may cause overcritical stress conditions on neighbouring patches, thus triggering an "avalanche" of simultaneous slip events. The strength of a seismic shear event can be described by its seismic moment which is proportional to the shearing area, or in other words to the number of patches slipping simultaneously.

In Fig. 34 is shown a qualitative presentation of rheological properties in the seismogenic zone of the rock crust (Nikolayevskiy 1982) where: (1) Failure envelop; (2) Dry friction; (3) Dilatancy zone; (4) Elastic limit; (5) Granite; (6) Mechanisms of failure; (7) Vertical Fractures; (8) Inclined cracks; (9) Localized deformation; (10) Pseudo-plasticity; (11) Plasticity; (12) Stress drops; (13) No stress drop; (14) Brittle failure; (15) Stick-slip region and (16) Plastic failure.

Table 3 presents some examples of dam sites where induced earthquakes with magnitude higher than 5 in the Ritcher scale have occurred (Sêco e Pinto 2006).

Table 3 Examples of dams with induced seismicity

DAM	Country	Type	Height (m)	Reservoir volume (x 10^6 m^3)	Year of impounding	Induced seismicity		Prior seismicity
						M	year	
Marathon	Greece	gravity	63	41	1930	5	1938	Moderate
Hoover	U.S.A.	arch-gravity	221	36,703	1936	5	1939	–
Kariba	Zimbabwe/ Zambia	arch	128	160,368	1959	5,8	1963	low
Haifengkiang	China	buttress	105	10,500	1959	6,1	1962	aseismic
Koyna	Índia	gravity	103	2708	1964	6,5	1967	low
Kremasta	Greece	embankment	165	4750	1965	6,3	1966	moderate
Roi Constantine	Greece	embankment	96	1000	1969	6,3		moderate
Oroville	U.S.A.	embankment	236	4298	1967	5,7	1975	moderate
Nurek	Tajikistan	embankment	330	11,000	1972	5	1977	Moderate
Tarbella	Pakistan	embankment	143	14,300	1974	5,8	1996	low
Aswan	Egypt	embankment	111	163,000	1974	5.3	1981	aseismic
Polyphyton	Greece	embankment	112	2244	1974	6.7	1995	aseismic
Mornos	Greece	embankment	126	640	1961			aseismic

The question of maximum magnitude to be ascribed to reservoir triggered seismicity is difficult to clarify but it seems in the range of 6–6.3.

An interesting overall picture is shown in Fig. 35 taken from USCOLD Report (1997). Monitoring the RTS activity is recommended for large dams and reservoirs. In order to distinguish between background seismicity and RTS monitoring before impounding is recommended. For obtained reliable epicentral locations and hypocentral depths, a local array of stations is required (ICOLD 2005a).

The International Symposium on Reservoir Induced Seismicity 95 held in Beijing, China, referred to 120 RIS cases from 29 countries with 22 in China, 18 in USA and 12 in India.

Most of the existing reservoirs are aseismic, i.e. no correlation with triggering.

Out of the existing 50,000 dams and reservoirs for only 120 a correlation has been reported with a relevant seismic event.

The seismic phenomena are taking place in the brittle and fractured part of the earth crust in which water is circulating and that such phenomena are spent in the underlying plastic mass.

The difference between a reservoir triggered earthquake and a natural earthquake is that the reservoir triggered earthquake has a relatively high likelihood of occurring within the first few years after the filling of the reservoir or when the reservoir reaches the maximum level.

These earthquakes have a shallow focus and their epicentres are closed to the dam sites or reservoirs.

One interesting case study is Kariba dam that has a storage capacity of 180 km^3 of water when at a level of 484.6 m above sea level. A general view of Kariba dam is shown in Fig. 36. This volume of water in Kariba translates into a mass of water of about 180 billion t. The sudden imposition of this mass on the underlying geology of the basin would be expected to lead to some tectonic readjustment and some subterranean movement. The mass of the stored water notwithstanding, the Kariba basin is located in the southern extremity of the African Rift Valley system, a longitudinal zone of faulting and recent earth movement.

Thus it can be stated that the Kariba basin was an area prone to tectonic movement well before the establishment of the Kariba Dam.

The risks to the dam wall associated with earth movement and the analysis of the risk of dam failure as considered in the original dam design can be derived from seismic activity since the dam was constructed.

In the studies presented only the seismic data was analysed. The data was studied for spatial and temporal distribution. Shallow earthquakes have large surface waves that propagate along the earth's surface for long distances.

Surface wave amplitudes get smaller as the earthquake source depth increases. Thus in the case of earthquakes with deep-focus, surface waves are usually too small to give a reliable value of the surface wave magnitude. In such a case the body wave magnitude that is based on the amplitude of the P-waves is used. P-waves propagate inside the earth and are referred to as body waves (mb) giving rise to the name body wave magnitude. The P waves are not seriously affected by the focal depth, as is the case with the surface waves.

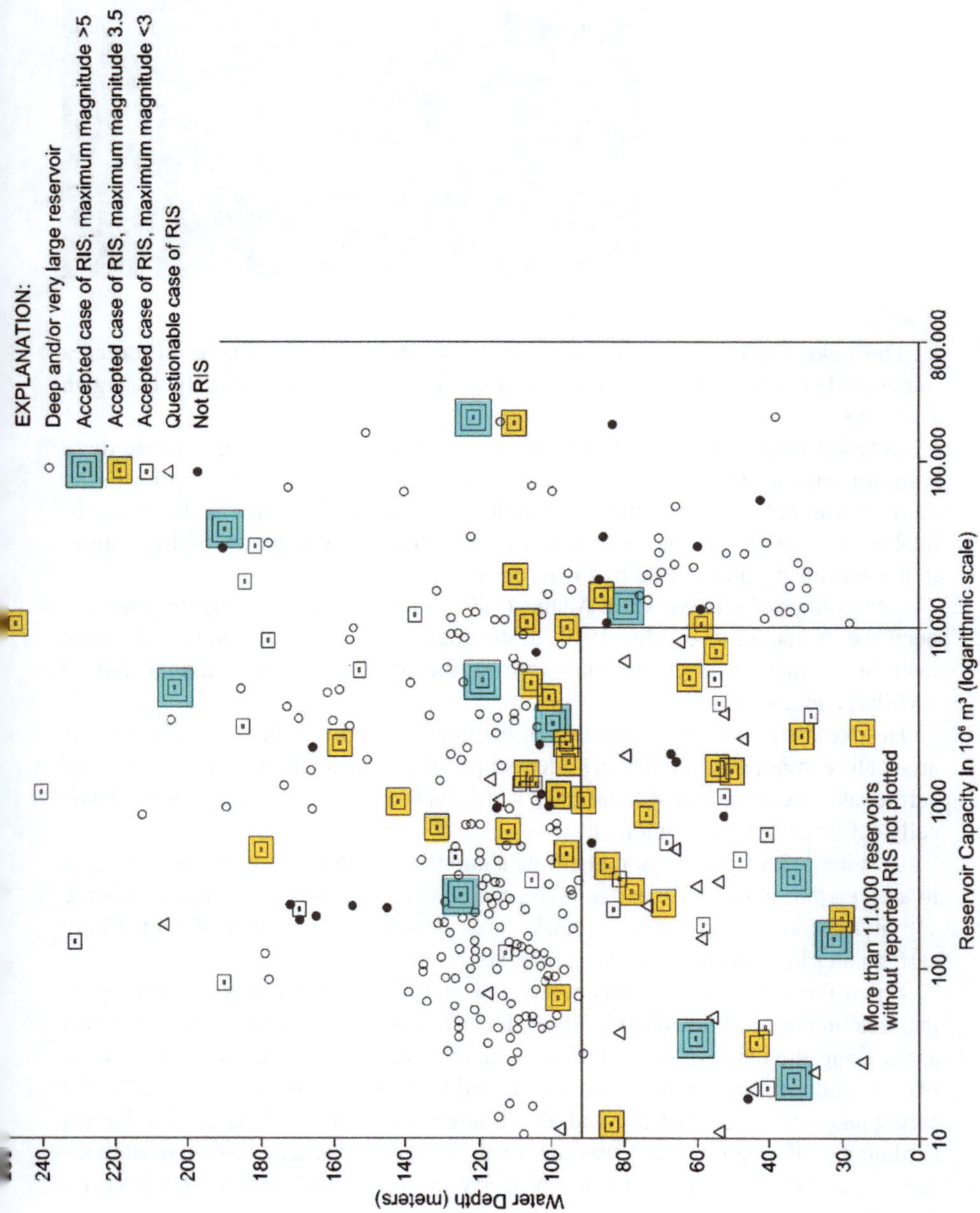

Fig. 35 Scattergraph of RTS cases (after USCOLD 1997)

Fig. 36 General view of Kariba dam

The Lake Kariba area is highly faulted with normal faults. Faults in the area under study continue in a north-easterly direction into Zambia to join the Luangwa Rift zone.

Detailed information on pre-impoundment seismicity is lacking because there were no seismic stations around or close enough to the reservoir area prior to the dam. It was not until 1959 that a seismic network was installed to detect earthquakes. As a result it is not possible to fully establish the exact nature of the changes in the seismic regime caused by the reservoir.

According to Ambraseys and Adams (1991) an earthquake of magnitude 6.0 Ms occurred on the 28 th of May 1910, in the southern part of Lake Kariba. Seismic instruments that are at great distances from the source can detect earthquakes of such large magnitudes.

Despite there being no seismograph equipment within or close to the reservoir area before impoundment, the tectonic set up of the area and the anecdotal reports of earthquake occurrence prove that Lake Kariba is located in an area that was seismically active prior to impoundment.

Despite lack of pre impoundment data and the fact that the reservoir is located in an area that is naturally earthquake prone, these studies concluded that seismic activity at and around the Lake Kariba was mainly due to impoundment. Many earthquakes have occurred in the area of study since 1959.

When the impoundment was completed in 1959, recorded seismic activity in the area increased and peaked in 1963. The 1963 activity was concentrated in two areas along the dam wall: the neck area, and in the Matusadona sub-basin. These two areas constitute a region that was found to have high incremental stress. All earthquakes of magnitude 5.0 and above have occurred in this area. The largest earthquake, of magnitude 6.1 was recorded in 1963. This earthquake caused damage to the dam structure and some property in nearby settlements. However, no casualties were reported.

The following causes of the earthquakes occurring at Kariba were suggested that near the western end of the lake, failure of pre-stressed, previously active faults were triggered by the addition of stress or due to fluid pressure or both; and that near and under the Sanyati basin failure of the pre-stressed faults were triggered by shear stresses of 1–2 bars resulting from the lake load. It was concluded that the increase of fluid pore-pressure due to the flow of water from the reservoir to the surrounding rocks played an insignificant role in the triggering of the earthquakes.

The observed initial seismic activity at Kariba has been correlated to the water level fluctuations. Thus they concluded that the triggering mechanism for the earthquakes was the increase of pressure in pores and rock fractures.

Earthquakes with magnitudes greater than 4.9 have occurred in the Lake Kariba area; earthquakes of such magnitudes can pose hazards to life and property of nearby populations.

In the case of large reservoirs such as Kariba, which have a tendency to significantly modify the seismicity in their vicinity, such anticipated changes have to be taken into account when doing the hazard analysis.

This is especially so in areas of relatively low seismicity. In areas of relatively high seismicity the designs usually takes seismicity into considerations anyway.

Though it has been documented that the 6.1 magnitude of 23 September 1963 caused some damage to the dam structure and some property in nearby settlements, no elaboration on the damages could be found. The occurrence of surface rupturing accompanying a triggered earthquake on a fault passing through the dam itself is the greatest risk. The accuracy of the epicentres as determined by the Zimbabwean seismological network is not high enough to enable association of earthquakes with a particular fault. Hence it is not possible to know which fault is active or to know the position of the fault relative to the dam wall.

The available pre- and post-impoundment seismic data do not show any trends that can be used to ascertain or predict the magnitude of future earthquakes and tremors.

In future it is advisable to have a network of seismic stations close to and in the reservoir area prior to building a dam the size of Kariba in any seismically active area.

Monitoring and Inspections

The detailed definition of the monitoring scheme cannot be made on the solid basis of the features of the dam, because many external factors are to be taken into account when safety problems are considered.

The risk potential is associated with the structural components and with social-economic components. The structural components depend of the reservoir capacity and the heigth of the dam.

The social economic risks are related with the number of persons that need to be evacuated and with the potential downstream damages.

The risk factors are classified in three classes, which are referred respectively to actions, to the structure or to values affected by hazards. The arithmetic average of all indices falling in a given class forms an overall risk factor for the class; in this way we define, respectively, an environmental factor E, a reliability factor F, a potential human/economic hazard factor R. Lastly a global risk index a_g, is developed by taking the product of the three partially factors E, F, R, (ICOLD 1981).

Experience has shown that the rational and systematic control of dam safety should consist of several tasks:

- frequent visual inspection by staff in charge of the observation system;
- periodic visual inspection by specialist;
- regular instrumentation measurements;
- data validation;
- data storage;
- visual inspections;
- safety evaluation;
- corrective actions.

Visual inspections are compulsory after exceptional occurrences, such as important earthquakes, big floods and total or nearly total drawdowns of the reservoir (ICOLD 1988; Sêco e Pinto 1993).

There are two steps in performing dam inspection: (i) an immediate inspection by the dam operator; (ii) follow-up inspection by dam engineering professional (ICOLD 2008b).

During inspections the following aspects deserve attention:

Dam Body—(i) upstream face (slope protection, vegetative growth, settlement, debris, burrows and unusual conditions); (ii) downstream face (signs of movement, seepage or wet areas, vegetative growth, condition of slope protection, burrows or unusual conditions); (iii) crest (surface cracking, settlement, lateral movement, camber).

Spillway—(i) approach channel (vegetation, debris, slides, slope protection); (ii) control structures (apron, crest, walls, gates, bridge, chute, stilling basin, outlet channel).

Outlet Works—(i) inlet works, emergency control facility, outlet conduit, service control facility, stilling basin.

Reservoir—Log Boom, landslides, other.

Access Road—Condition of pavement, ditches, bridge.

If an earthquake of moderate or high Richter magnitude occurs an immediate inspection of the dam shall be done following these procedures (ICOLD 1988):

1. If the dam is damaged to the extent that there is increased or new flow passing downstream immediately implement failure or impending failure procedures as previously planned.
2. If abnormally reduced flow is present at the upstream end of the storage, immediately inspect the river course for possibility of upstream damages due to landslide. If such is the case, implement failure or impending failure procedures.
3. Make an estimate of the characteristics of the earthquake.
4. Immediately conduct a general overall visual inspection of the dam.

5. If visible damage has occurred but has not been serious enough to cause failure of the dam, quickly observe the nature, location and extent of damage and report all the information to the supervisory office for a decision on further actions.
6. Make additional inspections at any time because of possible aftershocks.
7. During inspection the following aspects deserve attention: (i) cracks, settlements and seepage located on abutments or faces of the dam; (ii) drains and seeps for increased flow or stoppage of flow; (iii) outlet works or gate misalignment; (iv) visible reservoir and downstream areas for landslides, new springs and sandboils and rockfalls around the reservoir and in downstream areas; (v) for tunnels and conduits, observe whether silt, sand, gravel, rock or concrete fragments are being carried in the discharge stream.
8. Continue to inspect and monitor the facilities for at least 48 h after the earthquake because delayed damage may occur.
9. A secondary inspection 2 weeks to a month after the initial inspection should be made.
10. A schedule of very frequent readings should be followed for at least 48 h after the earthquake.
11. If failure is imminent, warning to downstream residents is essential. All measures should be used to reduce storage in the reservoir.

Sketches, photographs, videos, may help to describe the nature and extent of any damage.

If an earthquake is observed at or near a dam with a Ritcher magnitude greater than and within a radial distance shown in Table 4 immediately conduct a general overall inspection of the dam and major appurtenant structures.

If the dam is damaged with increased new flow passing downstream or there are signs of imminent failure immediately implement Emergency Action Plan.

Following Australia dam guidelines for all dams that have experienced a MMI (Maximum Magnitude Intensity) of 4 or greater the response required has adopted the guidelines of Table 5.

Related earthquakes alarm systems, shake maps are generated automatically.

Re-Evaluation of Dam Safety

Due to the periodic evaluation of dam safety there is a need for a revision and update of hydrologic studies, geologic studies and seismic studies taking into account the current state of the art.

An interesting case history is Matahina dam where after a careful evaluation of the dam safety there was a need to strength the dam in order to accommodate the settlements and leakage effects.

Matahina dam built in New Zealand on 1968 (Galloway 1967) is a rockfill embankment 80 m high, with a central clay core and rockfills shells.

Table 4 Recommended inspections based in magnitude versus distance

Magnitude	Distance (Km)
>4.0	25
>5.0	50
>6.0	80
>7.0	125
>8.0	200

Table 5 Australia dam guidelines (ICOLD 2008b)

Response Level	Likely Impact	Response Required
A	<MMI 4	Inspect dam at next routine inspection
B	MMI 4	Inspect dam within 18 h
C	MMI 5	Inspect dam within 6 h
D	MMI 6	Inspect dam immediately
E	MMI 7 or greater	Inspect dam immediately

After the first reservoir filling a great concentration of dirty water has appeared in the right bank. From the investigations was concluded that the seepage was due to the differential settlements and the high susceptibility of core cracking and was consequently repaired.

In 1987 the dam was exposed to strong seismic shaking (peak horizontal crest acceleration 0.42 g) due to M=6.3 earthquake, located at the Edgecumbe fault.

The dam is sited across the Waiohau active fault, 80 km long, with proven surface breaking during the Holocene.

For the dam safety evaluation, taking into account the new seismic studies, an earthquake with M=7.2 was selected considering the surface rupture of Waiohau fault, crossing the dam site.

The value of 3.0 m in oblique was thus selected for the fault surface displacement as 2:1 (i.e. 2.7 m horizontal and 1.3 m vertical displacement). Such displacements would result in major cracking of the dam body inducing piping and leakage

The dam cross-section and the proposed strengthening (shadowed zone) is shown in Fig. 37 and consists in excavating significant part of the downstream shoulder and on keeping the existing core. The post seismic leakage control is to be ensured by placing a wide zone of filter, transition and drainage materials with 5.0 m minimum of thickness. The new crest will be approximately 40 m with the crest being heightened to accommodate any settlement due to shaking and maintain sufficient freeboard (ICOLD 1999).

In Fig. 38 is shown a general view of the dam after the rehabilitation.

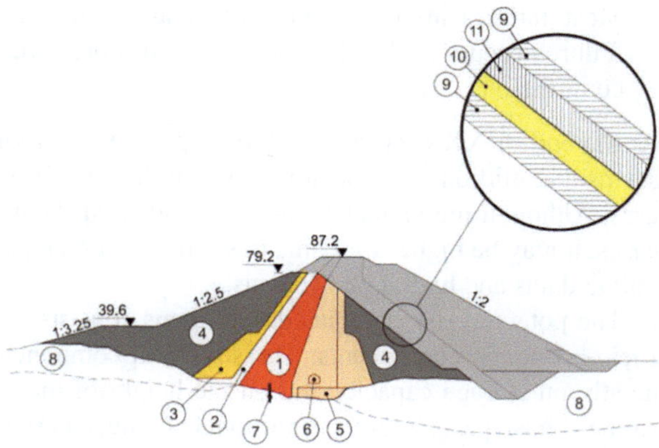

Fig. 37 Mathanina Dam cross section (New Zealand). *1* Core. *2* Inner transition. *3* Outer transition. *4* Rock-fill shoulders. *5* Drainage blanket. *6* Inspection gallery. *7* Cut-of wall. *8* Alluvium. *9* Transition. *10* Filter. *11* Drain. (ICOLD 1999)

(1) Core
(2) Inner transition
(3) Outer transition
(4) Rockfill shoulders
(5) Drainage blanket
6) Inspection gallery
(7) Cut-of wall
(8) Alluvium
(9) Transition
(10) Filter
(11) Drain

Fig. 38 General view of Mathanina dam

Risk Analyses

Introduction

The findings of dam failures statistical analysis of data show that (ICOLD 1995):

i. The percentage of failure of large dams has been falling over the last four decades, 2.20% of dams built before 1950 failed, failures of dams built since 1951 are less than 0.5%;

ii. Most failures involve newly built dams. The greatest proportion of 70% of failures occurs in the first 10 years and more especially in the first year after commission.

At the end of XX century, one billion people was living downstream of dams. It seems that millions may be at risk within the next 50 years as a result of dam failures. Although the annual failure probability of dams is lower than 10^{-6} in most cases, it may be higher for dams in seismic areas subject to sudden failures such as tailing dams and hydraulic fill dams.

The potential risk associated with dams consists of structural components and socio-economic components. The structural components of potential risk depend mostly on storage capacity and on the height of the dam, as the potential downstream consequences are proportional to the mentioned values (ICOLD 1989). Socio-economic risks can be expressed by a number of persons who need to be evacuated in case of danger and by potential downstream damage.

Risk management comprises the estimation of the level of risk and exercising adequate control measures to reduce the risk when the level is not tolerable (Caldeira et al. 2005).

The essence of risk management and the role of quantitative risk assessment (QRA) within the context of risk management are shown in Fig. 39 (Ho et al. 2000).

ICOLD has introduced the potential risk of dam associated with capacity, height, evacuation requirements and potential downstream damage considering these 4 hazard classes (ICOLD 2009).

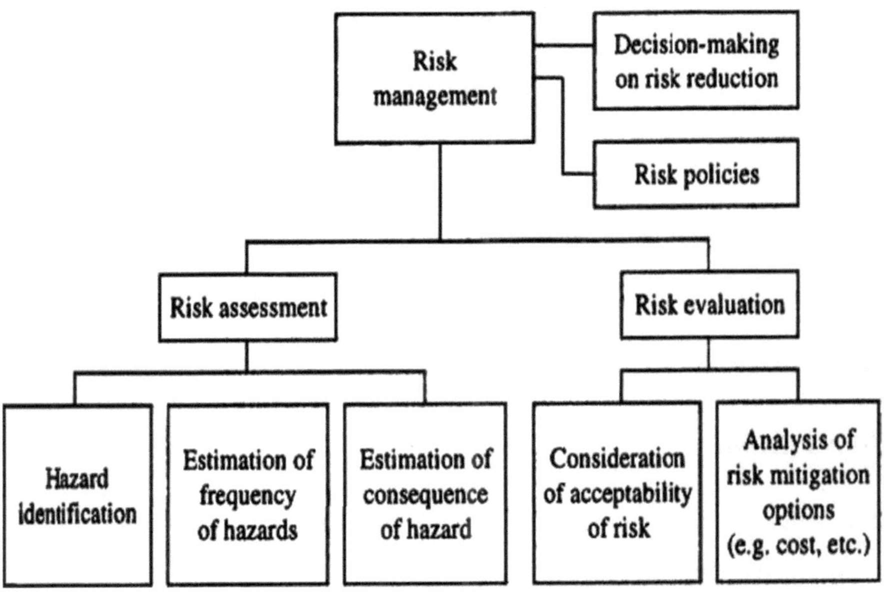

Fig. 39 Framework for risk management (after Ho et al. 2000)

Table 6 Risk factor (ICOLD 2009)

Risk factor	Extreme	High	Modera te	Low
Capacity (hm3)	120 (6)	1201 (4)	1-0-1 (2)	0.1 (0)
Height (m)	45 (6)	45 – 30 (4)	30 – 15 (2)	15 (0)
Evacuation require ments	1000 (12)	1000 – 100 (8)		None (0)
Potential downstream damage	High (12)	Moderate (8)	Low (4)	None (0)

Table 7 Risk class (ICOLD 2009)

Total risk factor	Risk class
0–6	I (Low)
7–18	II (Moderate)
19–30	III (High)
31–36	IV (Extreme)

Tables 6 and 7 are convenient to define risk associated with dams. Four risk factors are separately weighted as low, moderate, high or extreme.

The weighting points of each of the four risk factors, shown in brackets in Table 6 are summed to provide the Total Risk Factor as:

Total Risk Factor = Risk Factor (capacity)
+Risk Factor (height)
+Risk Factor (evacuation requirements)
+Risk Factor (potential downstream damaged).

The link between the Total Risk Factor and Risk Class is giving in Table 7.

There is a rich discussion related Failure Modes and Effect Analysis (FMEA), Failure Mode, Effects and Critically Analysis (FMECA), Event Tree Analysis (ETA), Fault Tree analysis (FTA) (ICOLD 2005b).

Structural Reliability Methods permit the calculation of failure probabilities of the mechanisms. Probabilities are calculated using the methods of the modern reliability theory such as Level III Monte Carlo, Bayesian theory, Level II advanced first order second moment calculations.

Dam owners, regulatory authorities and consultants have been carrying out risk analyses for many years. Its purpose is to identify the main real risks associated with each type and height of dam for all circumstances and can be conducted: (i) in extensive risk analysis of very large dams, to substantiate reliably the probabilities chosen in fault trees using Monte Carlo simulation technique; (ii) in simplified risk analysis of smaller dams, to focus low-cost risk analysis on a few main risks; (iii) and in identifying possibilities for reducing these risks through low-cost structural or non-structural measures (Lempérière,1999).

In Fig. 40 is illustrated the risk analysis proposed by AGS (2000) where annual probability of failure versus consequence of failure is shown not only for dams, but

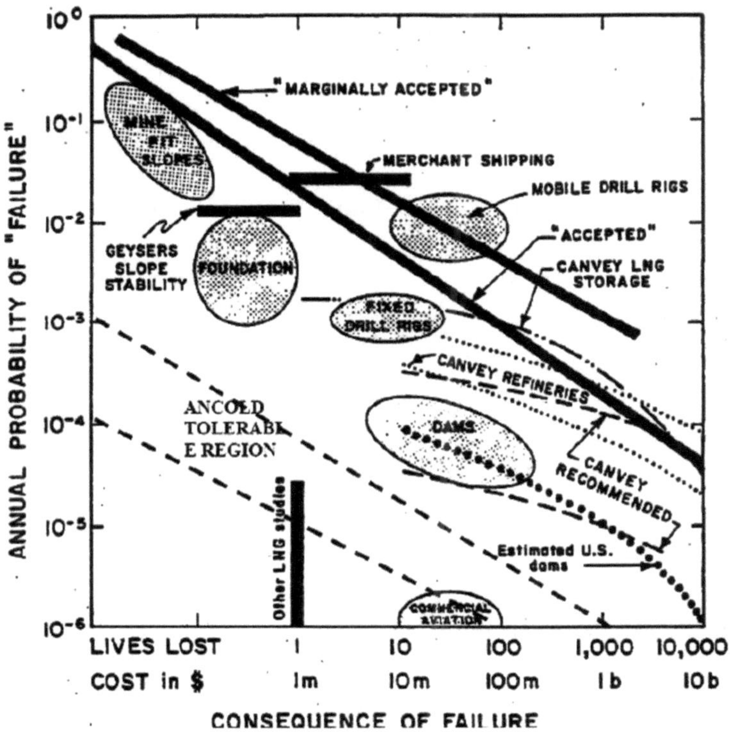

Fig. 40 Risk analysis proposed by AGS (2000)

also for other activities i.e. mines, slope stability, foundations, merchant shipping, drilling, storage, commercial aviation, etc.

The main components of risk management are risk assessment (risk analysis and risk evaluation), risk mitigation (Fig. 41) and risk control (risk reduction, emergency actions) and decision (Salmon and Hartford 1995; Seco e Pinto 2002).

Consideration of human behavior is essential when assessing the consequence of failures: well organized emergency planning and early warning systems could decrease the number of victims and so the study of human behavior plays an important role in assessment of risk analysis (Sêco e Pinto 1993).

We should never forget the contribution of Voltaire and the book Candide published in 1759, after the Lisbon earthquake (1755), for the change from the intellectual optimism and potential fatalism that is a necessary condition for the construction of future scenarios in a risk analysis context.

The results of a risk analysis can be used to guide future investigations and studies, and to supplement conventional analyses in making decisions on dam safety improvements.

Increasing confidence in the results of risk analyses can lead to a better cost-effective design and construction, satisfy our personal needs providing a better insight of the different factors of the design and give more confidence to our decisions (Sêco e Pinto 2002).

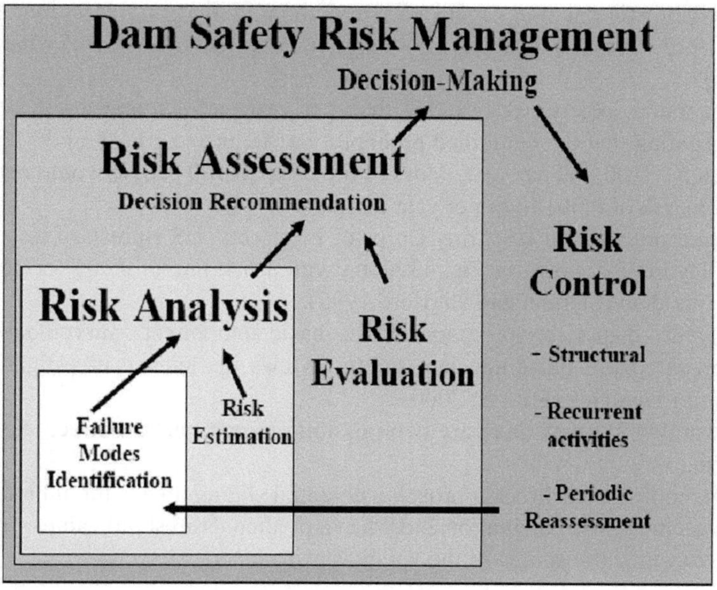

Fig. 41 Incremental hazard criteria (after Salmon and Hartford 1995)

There are several uncertainties in seismic hazard and seismic input in material properties, in structural modelling, in dynamic analysis and in performance criteria.

Due to large uncertainties in predicting the seismic behaviour of dams it is recommended to increase the resilience to earthquake loading instead reducing the uncertainties in seismic hazard, or material properties or using more sophisticated methods of seismic analysis.

First order methods such as the First Order Second Moment (FOMS) and the First Order Reliability Method (FORM) have received significant exposure (e.g. Nadim 2002; Duncan 2000) in recent years as relatively simple methods for estimating the probability of events occurring in geotechnical analysis.

The basic objective is as follow: given statistical data (mean and standard deviation) for key geotechnical input parameters (e.g. strength parameters c' and tan, seepage parameters k, settlement parameters E) what are the statistics (mean and standard deviation) of the key output quantities (e.g. factor of safety FS, flow rate Q, settlement d).

In the case of the output parameter, if these statistics are combined with an assumed probability density function, the probability of events such as slope failure, excessive flow rates, excessive settlements, etc, can be estimated.

While these methods are relatively easy to implement and give useful qualitative and sensitivity information about the input and output parameters, they are based on an underlying assumption of a Taylor Series truncated after the linear terms-hence first order.

In Fig. 42 is illustrated the framework for dam safety assessment in USA decomposed in 4 steps namely: risk identification, risk estimation, risk evaluation and risk treatment.

A probabilistic risk assessment addresses three fundamental questions (Salmon and Hartford 1995): (i) what can go wrong? (ii) how likely is it?; (iii) what damage will it do?

In general, a society risk of 0.001 lives per year per dam appears to be acceptable. Assuming that the combined probability of failure to a PMF and MCE is approximately 1/100,000 per year, a loss of life of up to 100 people would result in an acceptable risk of 0.001 lives per year per dam.

The past practices of US Army Corps of Engineers, US Bureau of Reclamation and BC Hydro are shown in Fig. 43 along with a risk line of 0.001 (or 10^{-3}) lives per year per dam (Salmon and Hartford 1995).

B.C Hydro dam safety program has five basic components: surveillance, emergency preparedness planning, dam safety reviews, deficiency investigations and capital improvements (Stewart 2000).

For warning systems there are two possibilities approaches: direct and indirect monitoring.

For example in the direct approach a potential sliding area is monitored by simple displacement instrumentation and when a predicted threshold value of displacement is exceeded the people of the valley is evacuated.

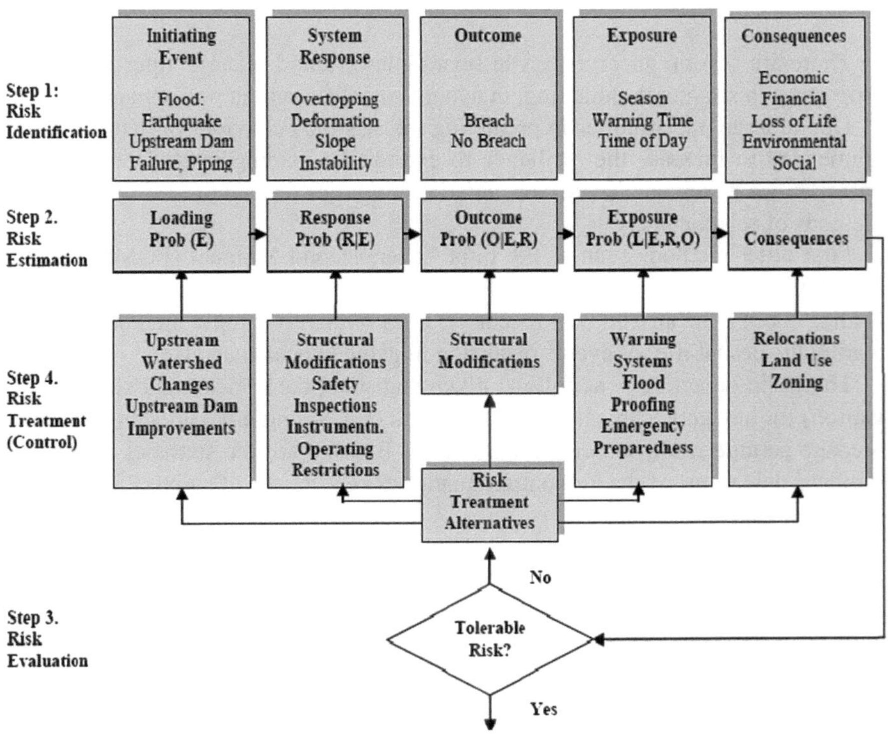

Fig. 42 Framework for dam safety assessment in USA

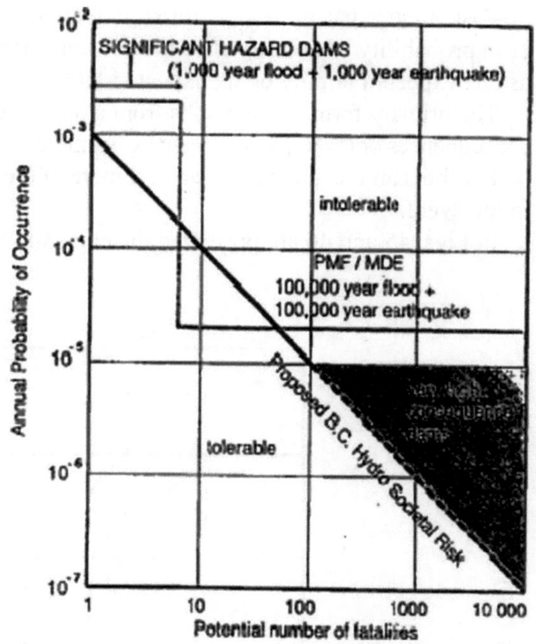

Fig. 43 Incremental hazard criteria (after Salmon and Hartford 1995)

An example of indirect warning system is the city of Hong Kong where an early warning system has been used for over 15 years and people were educated to recognize report landslide symptoms (cracking, reactivation of spring lines, surface runoff, etc).

The possible avenues for warming systems are shown in Fig. 44.

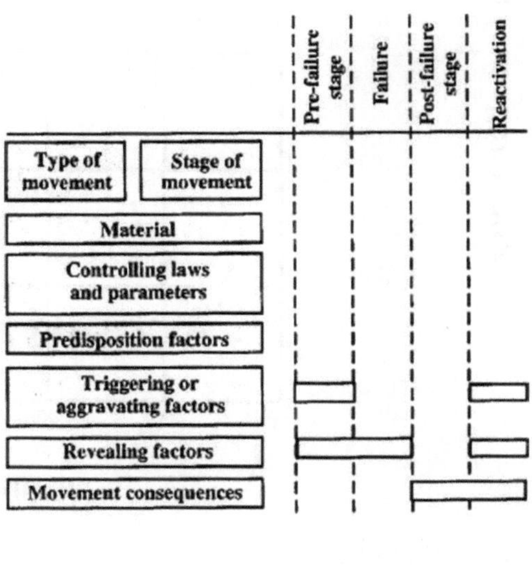

Fig. 44 Possible avenues for warning systems

Past practice has been to require relatively large increases in reliability (decreases in probability of failure) when the consequences exceed some fixed criteria such as one expected fatality or six expected fatalities.

The primary form of the results from a dam safety analysis is a set of probability-consequences or (f, N) pairs, which are estimated in the end branches of an event tree.

For the dam risk analysis f is an estimate of the probability that N fatalities occur in the event.

In Figs. 45 and 46 are presented the curves F-N.

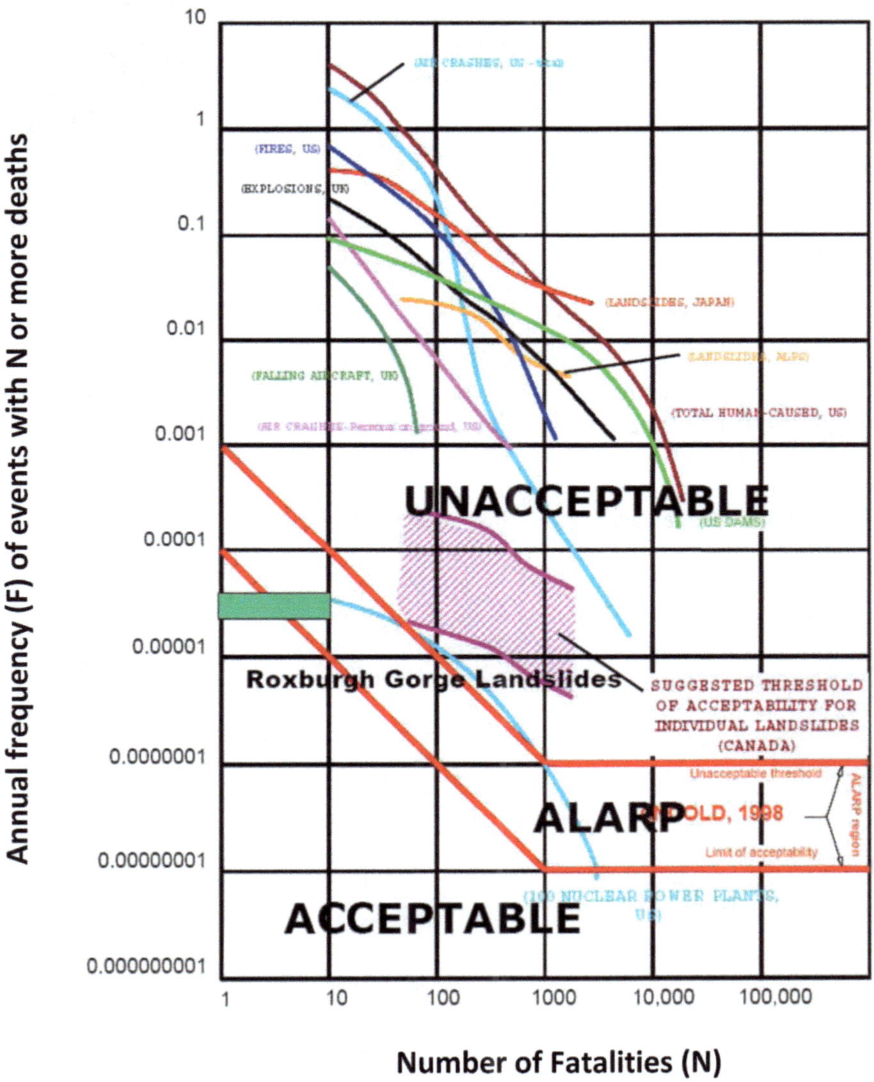

Fig. 45 Curves F-N

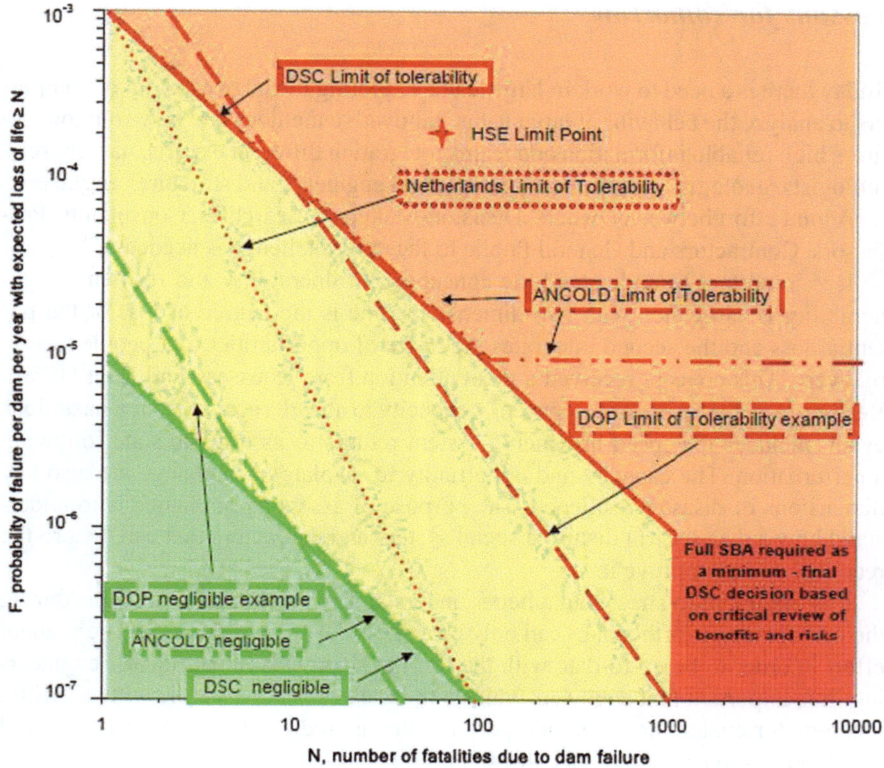

Fig. 46 Curves F-N

Future Challenges. Lessons for Tomorrow

The following topics deserve more consideration and can be considered future challenges for a better understanding of seismic dams behaviour:

1. The main concerns are linked with existing dams, dams that have not designed to sustain earthquakaes or dams built using design criteria and methods of analyses, that are considered as out dated today;.
2. Coupled models with non linear analyses and pore water pressure generation and dissipation models;
3. Incorporation of fault movements effects in dam foundations behaviour;
4. Failure of tailing dams that currently reach more than 200 m high and reservoirs with more than one billion tons of slimes due to the occurrence of liquefaction and the increase of the resistance due to ageing effects of the deposits;
5. Cost benefit analysis of seismic strengthening of existing dams.

Lessons for Tomorrow

Today there is a need to work in large teams exploring the huge capacity of computers to analyze the behavior of large dams. Innovative methods and new solutions require high reliable information and teams integrating different experts, namely seismologists, geologist, geophysics, geotechnical engineers and structures engineers.

A joint effort between Owners, Decision-Makers, Researchers, Consultants, Professors, Contractors and General Public to face this challenge is needed.

It is important to understand the concepts of vulnerability and resilience. Vulnerability is associated with two dimensions, one is the degree of loss or the potential loss and the second integrates the range of opportunities that people face in recovery. This concept received a great attention from Rousseau and Kant (1756). Resilience is a measure of the system`s capacity to absorb recover from a hazardous event. Includes the speed in which a system returns to its original state following a perturbation. The capacity and opportunity to recolate or to change are also key dimensions of disaster resilience. The purpose of assessing resilience is to understand how a disaster can disturb a social system and the factors that can disturb the recovery and to improve it.

It is important to stress that a better understanding of embankment dams during the occurrence of earthquakes can only be achieved by a continuous and permanent effort in order to be up-to-date with the last developments in earthquake engineering. It is important that engineers educate themselves and the Public with scientific methods for evaluating risks incorporating the unpredictable human behavior and human errors in order to reduce disasters.

From the analysis of past dam incidents and accidents occurred during the earthquakes it can be noticed that all the lessons have not deserved total consideration, in order to avoid repeating the same mistakes. We need to enhance a global conscience and to develop a sustainable strategy of global compensation how to better serve our Society. The recognition of a better planning, early warning, quality of evacuation that we should take for extreme events which will hit our civilization in the future. Plato (428 – 348 BC) in the Timaeus stressed that destructive events that happened in the past can happen again, sometimes with large time intervals between and for prevention and protection we should followed Egyptians example and preserve the knowledge through the writing.

We should never forget the 7 Pillars of Engineering Wisdom: Practice, Precedents, Principles, Prudence, Perspicacity, Professionalism and Prediction. Following Thomas Mann we should enjoy the activities during the day, but only by performing those will allow us to sleep at the night.

Also it is important to narrow the gap between the university education and the professional practice, but we should not forget that Theory without Practice is a Waste, but Practice without Theory is a Trap. Kant has stated that *Nothing better that a good theory,* but following Seneca *Long is the way through the courses, but short through the example.* I will add through a careful analysis of Case Histories.

Within this framework all the essential steps of good dam analyses, whatever the type of material is involved shall be performed with a sufficient degree of accuracy that the overall results can be extremely useful in guiding the engineer in the final assessment of seismic stability. This final assessment is not made by numerical results but shall be made by experienced engineers who are familiar with the difficulties in defining the actions, namely the design earthquake, and the material characteristics, who are familiar with the strengths and limitations of analytical procedures, and who have the necessary experience gained from studies of past performance.

In dealing with these topics we should never forget the memorable lines of Hippocrates:

The art is long
and life is short
experience is fallacious
and decision is difficult.

Annex 1

Table A1 Behavior of embankment dams during earthquakes

Dam	Country	L(m)	H (m)	Construction year	Dam characteristics	Construction technique	Earthquake characteristics	Dam behaviour	Reference
Aramo	Chile	220	42		Zoned dam	Compacted fill	3 March, 1985	Settlements 10 cm	Verdugo and Peters(2009)
Assuan	Egypt	3600	111		Zoned clay core rockfill	Compacted fill	November 1981 M = 5.3	Aseismic design	Shenouda (1982) Shalaby (1995)
Baihé	China	960	60		Zoned with sandy gravel materials and slope clay core		Tangshan 1976 I = 6	Slope of the upstream shell	Wenshao (1987)
Ciprese	Costa Rica						January 8, 2009 M = 6.2	Good behavior, minor cracks	ICE(2009)
Coihueco	Chile	975	31		Zoned dam	Compacted fill	February, 27, 2010 M = 8.8	Downstream slope	
Colbun	Chile	550		116	Zoned dam	Compacted fill	February, 27, 2010 M = 8.8	Horizontal displacement 2 m, vertical displacement 1 m	
Douhe	China	6000	22	1970	Homogeneous		Tangshan 1976 I = 6 horiz. accel. 0.4 g	Longitudinal cracks crest settlement due foundation liquefaction	Liu et al. (1979) Shen (1981)

Table A1 (continued)

Dam	Country	L(m)	H (m)	Construction year	Dam characteristics	Construction technique	Earthquake characteristics	Dam behaviour	Reference
El Caracol	México		126	1985	Zoned with rockfill shells and clay core	Compacted fill	1985 Earthquake September 19 M=8.1	Transverse and longitudinal deformations, crest settlement 160 mm	Ulloa (1987)
El Infiernillo	México	350	148	1963	Zoned with rockfill shells and clay core	Compacted fill	1979 and 1981 Earthquakes M=7.6	Longitudinal cracks 0.60 m depth	Tamura (1986) Resendiz et al. (1982)
Fujinima	Japan		18.5	1937	Homogenous	Low compacted fill	March, 11, 2011 M=9,0	Overtopping of the crest	Matsumoto(2011) Towhata et al.(2011)
Gokçe	Turkey		50				August 17, 1999, M=7.4		
Keddara	Argelie		110		Rockfill zoned dam	Compacted fill	May, 2003 M=6.8	Only small cracks on the crest	Benlala(2003)
Kuzuryu	Japan	355	128	1964	Zoned with rockfill shells and sloping clay core	Compacted fill	1969 Earthquake M=6.6		Nose and Baba (1980)
La Marquesa	Chile	220	10	1943	Zoned with silty shells and impervious core	Compacted fill 85–88 % modified proctor	1985 Earthquake March 3 M=7.8	Upstream and downstream shells slopes. Liquefaction of sandy material	Retamal et al. (1989)

Table A1 (continued)

Dam	Country	L(m)	H (m)	Construction year	Dam characteristics	Construction technique	Earthquake characteristics	Dam behaviour	Reference
La Villita	México	420	60	1968	Zoned with rockfill shells and central clay core		1979 and 1981 Earthquakes $M=7.1–7.6$ $a_{max}=0.31–0.38$ g	Longitudinal cracks with 150 m length and 0.50 m depth	Tamura (1986) Resendiz et al. (1982)
Leyroy Anderson	U.S.A.	370	72	1950	Zoned with rockfill shells and impervious core	Compacted core rockfill shells without compaction	Morgan Hill 1984 amax=0.42 g	Longitudinal cracks with 300 m length and 2 m depth	Gazetas (1987)
Long Valley	U.S.A.	200	60	1941	Homogeneous with silty sand material with gravels	Compacted fill 93 % modified AASHO	1980 Earthquake May 27 M=6	Springs on downstream toe, cracks	Seed (1980) Lai and Seed (1985)
Mahio	Japan		106	1961	Zoned with rockfill shells and central core	Compacted core rockfill shells without compaction	Nagano prefecture September 14 1984 M=6.8	Settlements of upstream shell	Yonezawa et al. (1987)
Malpaso	Peru	152	78	1936	Rockfill with upstream concrete membrane	Dumped rockfill	1938 Earthquake October 10 I=6.6 Mercalli modified scale	Crest settlement (76 mm) and downstream displacements 51 mm	Ambraseys (1960)

Table A1 (continued)

Dam	Country	L(m)	H (m)	Construction year	Dam characteristics	Construction technique	Earthquake characteristics	Dam behaviour	Reference
Matahina	New Zeland	400	86		Zoned with rockfill shells and impervious core	Compacted fill	1987 Earthquake March 2 M=6.3	Upstream shell has settled 800 mm, downstream shell has settled 100 mm with 250 mm tilting for downstream	Matsumoto et al. (1985) Gillon (1988)
Miboro	Japan	405	131	1960	Rockfill	Compacted fill	1961 Earthquake M=7.2 acceleration=0.25 g	Settlement of 30 mm displacement for downstream of 50 mm	Nose and Baba (1980)
Minase	Japan	665		1964	Rockfill with upstream concrete membrane	Dumped rockfill	Niigata 1964 M=7.5	Damages of the membrane joints, cracks on the crest, increase of seepage	Matsumoto et al. (1985)
Oroville	U.S.A.	1707	235	1968	Zoned with gravel shells and slope core	Compacted fill	Oroville 1975 M=5.7	Crest settlements 9 mm	Banerjee et al. (1979)
S. Fernando	U.S.A.	664		1940	Homogeneous with sandy silty and clay sandy materials	Hydraulic fill	S. Fernando Fev. 9 1971 M=6.6	Longitudinal cracks Liquefaction	Seed et al. (1973) ICOLD (1975)

Table A1 (continued)

Dam	Country	L(m)	H (m)	Construction year	Dam characteristics	Construction technique	Earthquake characteristics	Dam behaviour	Reference
Sheffield	U.S.A.	220	7.5	1923	Homogeneous with silty sand and upstream concrete membrane	Compacted fill	Santa Barbara June 19 1925 M=6.3	Dam failure	Seed et al. (1969)
Tarumizu	Japan	256	43	1976	Zoned with rockfill shells and central clay core material	Compacted fill	Miyagi-Ken-Oki 1978 M=7.4	No apparent damages Calculated crest acceleration 0.36 g	Yanagisawa and Fukui (1980)
Tapar	India	4575	17.8	1975	Homoge nous	Compacted fill	Jan. 26, 2001 M=6.7	Longitudinal and transverse cracks	Yadav et al.(2008)
Tarbela	Pakisthan		143		Zoned dam	Compacted fill	Oct 8, 2005 M=7.5	High pore pressure	
Vidra	Romania		123		Zoned with rockfill shells and central clay core material		Vrancea March 4, 1977 M=7.2 a_h=0.2 g	No apparent cracks	Priscu (1979)
Wangwu	China	761	40		Zoned with sandy shells and clay core	Dumped fill	Bohai Wan 1969 I=6	Liquefaction and slope of upstream shell	Wenshao (1987)
Zipinghu	China	663	156	2006	CFRD	Compacted fill	May 12, 2008 M=8.0	Crest and concrete face damages	Chen(2008)

H—dam height L—dam length I—earthquake intensity M—earthquake magnitude

References

Ambraseys NN (1960). The seismic stability of earth dams. Proceeding 2nd World Conference on Earthquake Engineering, Tokyo, Vol II, pp 331–336

Ambraseys NN, Adams PD (1991) Reaapraisal of major African earthquakes using Raleigh-wave inverse and body-wave modelling

Australia Geomechanics Society (2000) Landslide risk management concept and guidelines. Australia geomechanics society. Sub-Committee on Landslide Risk Management. Australia Geomechanics 35(1):49–92

Banerjee NG, Seed HB, Chan CK (1979) Cyclic behavior of dense coarse-grained materials in relation to the seismic stability of dams. Report n UCB/EERC 79/13. University of California, Berkeley

Barrera S, Valenzuela L, Campana J (2011) Sand tailings dams. Design, construction and operation. Proceedings Tailings and Mine Waste 2011, Vancouver

Benlala A (2003) Le barrage de Keddara face au seisme du 21 Mai 2003 et la securite du Grand Alger Suite au seisme de Boumerdes du 21 Mai 2003. Question 83-Seismic Aspects of Dams, ICOLD, Montreal

Bray JB (2007) Simplified seismic slope displacements procedures. In Pitilakis K (ed) Proceeding of 4th International Conference on Earthquake Geotechnical Engineering-Invited Lectures. Thessaloniki, pp 327–353

Budweg FMG (1997) Incidents and failures of dams. 18th International Conference of Large Dams. Firenze, pp 751–819

Caldeira L, Pimenta L, Silva Gomes A (2005) Framework of risk analysis and their application to embankment dams (in portuguese). Seminar on Dams- Technology, Safety and Interaction with the Society, Lisbon, pp 569–585

Carter JP, Booker JR, Wroth CP (1982) A critical state soil model for cyclic loading. Soil mechanics transient and cyclic loads. Constitutive relations and numerical treatment. Wiley, pp 219–252

Chen H (2008) Considerations of dam safety after Wenchuan earthquake in China. Proceeding 14th World Conference on Earthquake Engineering. Special Session S13 Seismic Aspects of Large Embankment and Concrete Dams, Beijing, China

Duncan J (2000) Factors of safety and reliability in geotechnical engineering. Journal of Geotechnical and Geoenvironmental Engineering 126(4):307–316

Elms D (2000) Refinements to the Newmark Sliding Block Model. Paper n 2132. 12th WCEE, Auckland, New Zealand

Eurocode 8 (1998a) Design provisions for earthquake resistance of structures. Part 1

Eurocode 8 (1998b) Design provisions for earthquake resistance of structures- Part 5: Foundations, Retaining Structures and Geotechnical Aspects

Fanelli M (1994) Safety assessment and improvement of existing dams. Question 68, General Report. XVIII International Congress on Large Dams, pp 1523–1583

Finn WDL (1987) Finite Element Handbook Chap. 3—Geomechanics. Mc. Graw-Hill Editors H. Hardestuncer

Gazetas G (1987) Seismic response of earth dams: some recent developments. Soil Dynamics and Earthquake Engineering 6(1):2–47

Galloway JHH (1967) Installed instruments in Matahina earth dam. Proceeding 5th Aust. N.Z. Conference Soil Mech Found Eng, pp 147–152

Gilg B, Indermaur W, Matthey F, Pedro J, Azevedo M, Ferreira E (1987) Special design of steno arch dam in Greece in relation with possible faults movements. LNEC, Memória n 695

Gillon MD (1988) The observed Seismic behavior of the Matahina dam. Proceeding 2nd ICCHGE. Vol I, St. Louis, pp 841–848

Ho K, Leroi E, Roberbs B (2000) Quantitative risk assessment: application, myths and future direction. GEO Eng, 2000, Melbourne, pp 269–312

Iai S (2009) Discussion session on performance criteria for designing geotechnical. In: Kokusho T, Tsukamoto Y, Yoshimine M (ed) Proceeding of performance-based design in earthquake geotechnical engineering, Tokyo, pp 265–267
ICE (Instituto Costrarricense de Electricidad) (2009) The Costa Rica earthquake of January 8, 2009. Strong Motion Recorded in Dams or Site Projects
ICOLD (1975) A review of earthquake resistant design of dams. Bulletin 27
ICOLD (1981) Automated observation for the safety control of dams. Bulletin 41
ICOLD (1983) Seismicity and dam design. Bulletin n 46
ICOLD (1988) Inspection of dams. Following earthquake—guidelines. Bulletin n 62
ICOLD (1989) Selection Seismic parameters for large dams—guidelines. Bulletin 72
ICOLD (1995) Dam failures, statistical analysis. Bulletin n 99
ICOLD (1998) Neotectonics and dams. Guidelines and case histories. Bulletin 112
ICOLD (1999) Seismic observation of dams. Guidelines and case studies. Bulletin 113
ICOLD (2001) Design features of dams to effectively resist seismic ground motions. Bulletin 120
ICOLD (2002) Earthquake design and evaluation of structures appurtenant to dams. Bulletin 123, Committee on Seismic aspects of dam design, ICOLD, Paris
ICOLD (2005a) Dam foundations. Geologic considerations. Investigation methods. Treatment. Monitoring. Bulletin n 129
ICOLD (2005b) Risk assessment in dam safety management. Bulletin 130
ICOLD (2007) Historical review on ancient damss. Bulletin
ICOLD (2008b) Inspection of dams. Following earthquake guidelines. Bulletin n 62. 2008 Revision
ICOLD (2009) Selection Seismic parameters for large dams—guidelines. Bulletin 72. 2009 Revision
ICOLD (2010) Reservoirs and Seismicity. State of knowledge. Bulletin 137
Idriss IM, Lysmer J, Hwang R, e Seed HB (1973) Quad-4. A computer program for evaluating the seismic response of soil structures by variable damping finite elements. Report n UCB/EERC 73—16, University of California, Berkeley
INA (International Navigation Association) (2001) Seismic design guidelines for port structures. A.A. Balkema Publishers
Kant I (1756) Writings on the Lisbon earthquake. Portuguese edition of the three essays about earthquakes published in Konigsberg, Almendina, 135 p
Lai SS, Seed HB (1985) Dynamic response of long Valley Dam in Mammoth Lake. Earthquake series of May 25–27 1980. Report N UCB/EERC 85/12, University of California, Berkeley
Lempériere F (1999) Risk analysis: What sort should be applied and to which dams? The International Journal on Hydropower Dams 6:128–132
Lee MK, Finn WLL (1978) DESRA 2, Dynamic effective stress response analysis of soil deposits with energy transmitting boundary including assessment of liquefaction potential. Soil Mechanics series n 38. Department of Civil Engineering, University of British Columbia, Vancouver, Canada
Liu Z, Wang W, Yang X, Yu P, Gao F (1979) Preliminary investigation of damages in the earth dam of Douhe reservoir caused by Tangshan earthquake of 1976. the Haike River Design Institute, Beijing, P.R.C
Lysmer J, Udaka T, Tsai C, Seed HB (1974) A computer program for approximate 3D analysis of soil-structure interaction problems. Report n UCB/EERC 74—4. University of California, Berkeley
Lysmer J, Udaka T, Seed HB, Hwang R (1975) LUSH 2- A computer program for complex response analysis of soil-structure systems. Report n UCB/EERC 75—30. University of California, Berkeley
Makdisi FI, Seed HB (1977) A simplified procedure for estimating earthquake induced deformations in dams and embankments. Report n EERC 79—19. University of California, Berkeley
Masukawa S, Yasunaka M, Kohgo Y (2004) Dynamic failure and deformations of dam-models in shaking table testes. Proceeding of 13th World Conference on Earthquake Engineering, Vancouver, Canada

Matsumoto N (2011) Fujinuma dam inspection after 2011 Tohoku earthquake. 4th Report
Matsumoto N, Takahashi M, Sato F (1985) Repairing the concrete facing of Minase Rockfill dam. 15th ICOLD, Lausanne, Vol IV, Q.59, R.13, pp 203–225
Mroz Z, Norris VA, Zienkiewicz OC (1979) Application of an anisotropic hardening model in the analysis of elasto-plastic deformation of soils. Advanced Topics and New Developments in Finite Element Analysis, Portoroz, pp 18–22
Nadim F (2002) Probabilistic methods for Geohazard problems. In: R Poettler et al (eds) State-of-the-art in probabilistics in geotechnics: technical and economical risk estimation. pub VGE Essen. Germany, pp 333–350
Newmark NM (1965) Effects of earthquakes on dams and embankments. Geotechnique 15(2):139–160
Nikolayevskiy VN (1982) Earth crust, dilatancy and earthquakes. Section of book on focal mechanics of earthquakes. Kir, Moscow
Nose M, Baba K (1980) Dynamic behaviour of Rockfill dams. Dams and earthquakes. Proceeding of a Conference at I.C.E., pp 69–78
Prevost JH (1993) Nonlinear dynamic response analysis of soil and soil-structures interacting systems. Proceeding of the seminar on soil dynamics and geotechnical earthquake engineering, Lisbon, Sêco e Pinto, Publisher A. Balkema
Priscu R (1979) The behaviour of Romanian dams during the Vrancea earthquake of March 4, 1977. 13th ICOLD, New Delhi, Vol II Q51, pp 995–1012
Resendiz D, Romo PM, Moreno E (1982) El Infiernillo and La Villita dams: Seismic behavior. JGED, ASCE 108(GT1):109–131
Retamal E, Musante H, Ortigosa P (1989) The behaviour of earth dams in Chile during the 1985. Proceeding 12th ICSMFE. Rio de Janeiro,Vol 3, pp 1995–2000
Salmon GM, Hartford DND (1995) Risk analysis for dam safety. International Water Power Dam Construction, March, pp 42–47
Sarma SH (1975) Seismic stability of earth dams and embankments. Geotechnique 25(4):743–776
Sarma SH, Chowdhury R (1996) Simulation of Pore pressure in earth structures during earthquakes. 11th WCEE, Acapulco, Mexico
Schnabel PB, Lysmer J, Seed HB (1972) Shake: a computer program for earthquake response analysis of horizontally layered sites. Report n UCB/EERC 72—12. University of California, Berkeley
Sêco e Pinto PS (1993) Dynamic analysis of embankment dams. In: Sêco e Pinto (ed) Soil dynamics and geotechnical earthquake engineering seminar. Balkema Publisher. pp 159–269
Sêco e Pinto PS (2001) Dam engineering- earthquake aspects. Special Lecture. Proceeding of 4th ICSDEE, San Diego
Sêco e Pinto PS (2002) Some reflections about risk analysis of geotechnical structures. Proceeding of 12th Danube- European Conference Geotechnical Engineering, Passau, pp 41–46
Sêco e Pinto PS (2004) Dams, embankment and slopes. General Report. Proceeding of 5th ICCHGE, New York
Sêco e Pinto PS (2006) Dam engineering-new challenges. Schezy Lecture, Budapest, 33 p
Sêco e Pinto PS (2009a) Interaction between eurocode 7-geotechnical design and eurodoe 8-design for earthquake resistance of geotechnical design concept. Special Lecture. In: Honjo Y, Susuki M, Hara T, Zhang F (eds) Proceedings of the 2nd International Symposium on Geotechnical Safety Risk, Gifu. Gifu Conference, pp 51–66
Sêco e Pinto PS (2009b) Seismic behaviour of geotechnical structures-past, present and future. In: Kokusho T, Tsukamoto Y, Yoshimine M (eds) Proceeding of performance-based design in earthquake geotechnical engineering, Tokyo, pp 341–349
Seed HB (1980) Earthquake-resistant design of earth dams. Symposium and practice of dam engineering, Bangkok, pp 41–60
Seed HB, Martin GR (1966) The Seismic coefficient in earth dam design. JSMFD, ASCE 92(SM3):59–83
Seed HB, Lee KL, Idriss IM (1969) Analysis of sheffield dam failure. JSMFD, ASCE 95(SM6):1453–1490

Seed HB, Lee KL, Idriss IM, Makdisi F (1973) Analysis of the slides in the San Fernando dams during the earthquake of Feb 9 1971. University of California. Report N.EERC 73—2

Shalaby A (1995) High Aswan dam and Seismicity. Proceeding of the International Symposium on Seismic and Environmental Aspects of Dams Design Earth, Concrete and Tailing Dams, Chile, pp 507–516

Shen ZJ (1981) Dynamically coupled percolation and deformation analysis of earth dams. Proceeding ICRAGEESD. Vol I, St. Louis, pp 389–394

Shenouda WK (1982) Quality control and testing methods as exercised in the Aswan high dam. 14th International Conference on Large Dams, Rio de Janeiro, Q.55. R. 58, pp 1047–1065

Sherard JL, Cluff LS, Allen CR (1974) Potentially active faults in dam foundations. Geotechnique 24(3):367–428

SRB—Subcommission of Dam Codes (1990) Dams safety code (in portuguese)

Stewart R (2000) Dam risk management. GEO Eng. Melbourne, pp 721–748

Tamura C (1986) Behaviors of dams in Mexico September 19, 1985. Bulletin of Earthquake Resistant Structure Research Center N 19. Inst. of Industrial Science. University of Tokyo

Taucer F (2005) Recent advances and future needs in experimental earthquake engineering. CASCADE Series Report N 7, Published by LNEC, Lisbon (ISBN 972-49-1971-4)

TC4 ISSMGE (2001) Case histories of post-liquefaction remediation. Committee on Earthquake Geotechnical Engineering

Tokimatsu K, Susuki H, Tabata K, Sato M (2007) Three dimensional shaking table tests on soil pile structures models using e-defense facility. 4th International Conference on Earthquake Engineering, June 25–28, Thissaloniki, Greece

Towhata I (2008) Geotechnical earthquake engineering. Springler

Towhata I et al (2011) On Gigantic Tohoku pacific earthquake in Japan. Earthquake news. Bulletin of the international society for soil mechanics and geotechnical engineering, Vol 5, Issue 2. April

Udias A, Arroyo AL (2005) The Lisbon earthquake of 1755 in Spanish contemporary authors. 250th Anniversary of the 1755 Lisbon Earthquake, Lisbon, pp 19–29

Ulloa CR (1987) Effect of the September 1985 earthquakes on dams built on the Balsas River. El Caracol Dam. Comission Federal de Electricidad. Report n 64

USCOLD (1997) Reservoir Triggered Seismicity

Verdugo R, Peters G (2009) Observed Seismic behavior of three Chilean large dams. In: Kokusho T (ed) Earthquake geotechnical case histories for performance based design. pp 409–431

Voltaire (1759) Candide. Oeuvres completes, 9 vol, edit. Moland, Paris 1877

Yanagisawa E, Fukui T (1980) Performance of the Tarumizu Rockfill dam during strong earthquake. Proceeding 7th WCEE. Instanbul, Vol 8, pp 133–140

Yadav SM, Mishra R, Samtani BK (2008) Rehabilitation of earthquake affected Tapar Dam, Gujarat, India. Proceeding of the 12th International Conference of IACMAG, Goa, India

Yan L, Lun KY (2003) Liquefaction of Gravelly soils. Q 83, 21 st ICOLD, Montreal

Yasuda S et al (2010) Geotechnical damage caused by the 2010 Maule, Chile, earthquake. ISSMGE Bulletin, Vol 4, Issue 2

Yonezawa T, Uemura Y, Okmoto I (1987) An analysis of the dynamic behavior of a Rockfill dam during an earthquake with waves of high frequency. International Symposium of Earthquakes Dams Beijing, Vol 1, pp 324–338

Wenshao W (1987) Lessons from earthquake damages of earth dams in China. International Symposium on Earthquakes Dams Beijing China, Vol 1, pp 243–257

Wieland M (2003) Seismic aspects of dams. General Report Q 83, 21 st ICOLD, Montreal

Wieland M (2008) Review of Seismic design criteria of large concrete and Embankment dams. 73rd Annual Meeting of ICOLD, Tehran, Iran

Nonlinear Deformation Analyses of Liquefaction Effects on Embankment Dams

Ross W. Boulanger, Jack Montgomery and Katerina Ziotopoulou

Abstract Nonlinear deformation analyses (NDAs) are used for evaluating the potential effects of liquefaction on the seismic performance of embankment dams. NDAs provide insights on potential failure modes and serve as a basis for generating fragility curves for risk analyses. Confidence in these applications of NDAs may be significantly improved by establishing calibration and analysis protocols that reduce undesirable sources of variability between analysis results obtained by different analysts or different models. A key protocol for improving the quality of NDAs is the calibration and validation of a constitutive model using single-element simulations with the types of loading paths potentially important to the structure's response. These simulations provide a basis for evaluating the strengths and weaknesses of the constitutive models and for identifying likely causes of differences between different modeling approaches. Examples of single-element simulations are presented for several constitutive models and loading paths to illustrate some of the important features for NDAs of embankment dams affected by liquefaction. The benefits of NDA studies also depend on the numerical procedures and results being sufficiently documented so that knowledgeable users can critique and interpret the results. Insights from parametric analyses should be adequately demonstrated so that members of the project team can exercise judgments regarding their practical implications. Recommendations on NDA documentation requirements are presented, along with an example of how such documentation can be important for the interpretation of NDA results. Critical examinations of NDA models and modeling procedures are necessary for identifying weaknesses, fostering improvements, and increasing confidence in their use for evaluating the seismic performance of embankment dams affected by liquefaction.

R. W. Boulanger (✉) · J. Montgomery · K. Ziotopoulou
Department of Civil and Environmental Engineering,
University of California at Davis, California, USA
e-mail: rwboulanger@ucdavis.edu

A. Ansal, M. Sakr (eds.), *Perspectives on Earthquake Geotechnical Engineering*,
Geotechnical, Geological and Earthquake Engineering 37,
DOI 10.1007/978-3-319-10786-8_10

Introduction

I (lead author) was pleased and honored that Professor Ishihara asked me to speak about embankment dams at this conference. I first met Professor Ishihara during the 1995 Kobe earthquake reconnaissance efforts, and I have ever since been grateful for his kind guidance, discussions, and support. The subject of liquefaction effects on dams fits well with this conference because Professor Ishihara's contributions to all aspects of liquefaction, from fundamentals through design, are deeply engrained in our current understanding and engineering practice. In addition to his technical contributions, Professor Ishihara has influenced us all through his enduring example of dedicated service to the international community.

Nonlinear deformation analyses (NDAs) have become widely used for evaluating the potential effects of liquefaction on the seismic performance of existing embankment dams and the design of remedial measures. NDAs provide insights on deformation patterns, interactions with appurtenant structures, and other aspects of behavior important for assessing the expected performance of embankment dams across a range of seismic loading levels.

A parallel development has been the increased use of risk analyses and risk-informed processes for evaluating the seismic safety of embankment dams. Risk analyses and risk-informed processes have been used in dam safety programs by the US Bureau of Reclamation (USBR) since 1997 (Snorteland and Dinneen 2008) and US Army Corps of Engineers (USACE) since 2005 (Halpin and Harkness 2008). These processes facilitate a more complete understanding of system performance and key sources of uncertainty (through sensitivity analyses), and provide an improved basis for decisions regarding risk mitigation strategies. Risk analyses often utilize the results of NDAs in examining potential seismic failure modes and developing the needed fragility relationships.

The choice of an NDA model, including the numerical platform and the constitutive model, may depend on the nature of the specific problem, the likely physical response of the structure, and the anticipated benefits that the selected model might provide. Constitutive models for liquefiable soils range from relatively simplified, uncoupled, cycle-counting models (e.g., Dawson et al. 2001) to more complex plasticity models (e.g., Wang et al. 1990; Cubrinovski and Ishihara 1998; Yang et al. 2003; Byrne et al. 2004; Dafalias and Manzari 2004). Evaluations of NDA models against physical model tests and case histories involving liquefaction have shown that realistic simulations of the dynamic response of many types of structures can only be obtained using more complex constitutive models that can approximate behaviors such as cyclic mobility. Simple practice-oriented models have nonetheless proven useful in the evaluation of dams because the cyclic resistance ratio (CRR) and post-liquefaction residual shear strength, both of which are input parameters for these types of models, have first-order impacts on the potential deformations. Simple practice-oriented models can, for example, be sufficient to demonstrate that a poorly-constructed dam on a weak foundation is in need of remediation or that a well-constructed dam is unlikely to deform significantly. A complex model that is well validated offers the potential for more realistic simulations and greater insights, which may be of particular value when the outcome of the analysis is less predictable. Thus, the continued development of more realistic constitutive models that are applicable to a broad range of conditions, are easy to calibrate and use, and are well validated is a long-term need for advancing the use and value of NDAs in practice.

Increasing our confidence in the use of NDAs requires reducing variability in results obtained by different analysts using the same or different constitutive models. The fact that NDA results sometimes vary significantly depending on the user and model (e.g., Perlea and Beaty 2010; Finn and Wu 2013) raises questions about the reliability of NDA results, and also illustrates that the validation process (i.e., comparisons to case histories and physical models) must be applied to both the model and application protocols. Tasks toward this goal include identifying the sources of such differences and then establishing practices that eliminate undesirable and unnecessary sources of variability.

Calibration and validation of a constitutive model for liquefiable soils using single-element simulations for the types of loading paths potentially important to a structure are essential for: (1) evaluating what the constitutive models do and do not do well, and (2) identifying possible causes of differences between different modeling approaches. The characterization of liquefiable cohesionless soils relies on common in-situ test measurements, such as Standard Penetration Test (SPT) and Cone Penetration Test (CPT) penetration resistances, shear wave velocity (V_s) measurements, and index test results (e.g., fines content). The calibration and validation of constitutive models for these soils then includes comparing single-element simulations with the extensive body of experimental and case history based correlations for various aspects of soil behavior under dynamic loading. The objective is to know when a constitutive model produces results that are within the range of responses that have been observed experimentally for similar soils, and to know when it does not.

The results of NDAs can be sensitive to details of the numerical procedures, boundary conditions, input motion specifications, and other details of the overall modeling procedure. It is therefore essential that these details be sufficiently documented so that a knowledgeable user can critique and interpret the results, either during the review process or years later during subsequent examinations of the dam.

Decisions based on results of an NDA must be guided, in part, by judgments based on the physics of the phenomena, the site characterization, the basis for all input parameters, an understanding of the constitutive model and modeling approach, and the insights gained from a well-defined set of parametric analyses. The documentation of the insights extracted from the NDA results is important for enabling other team members and reviewers to exercise their judgments regarding the practical implications of the results.

The purpose of this paper is to examine some issues important for the advancement of NDAs for evaluating the seismic performance of embankment dams affected by liquefaction. The topics covered are general in nature although there is an emphasis on practices in the United States. Risk analysis practices for dams are reviewed to provide a context for the role of NDAs in these broader assessments of seismic safety for embankment dams. The calibration of constitutive models for liquefiable soils and the importance of demonstrating elemental behavior are discussed in detail, including illustrations of some features of behavior which require further research and development. Recommendations on NDA documentation requirements are presented, along with an example of how such documentation can be important for the interpretation of NDA results. It is suggested that the critical examination of NDA models and modeling procedures is necessary for identifying weaknesses, fostering improvements, and increasing confidence in their use for evaluating the seismic performance of embankment dams affected by liquefaction. Similar recommendations and conclusions can be drawn for cases involving other problem soils, such as soft and sensitive clays or plastic silts; the details for NDA modeling of such cases are, however, outside the scope of the present paper.

Potential Failure Modes Analyses, Risk Analyses, and NDAs for Liquefaction Effects

The past 15 years have seen significant advances in the use of risk analyses and risk-informed processes for evaluating the seismic safety of embankment dams (e.g., Snorteland and Dinneen 2008; Halpin and Harkness 2008). An overview of risk analyses is presented to provide context for the role of NDAs in the broader assessment of seismic safety for embankment dams affected by liquefaction.

The performance of embankment dams during earthquakes has ranged from excellent for well-compacted modern dams to poor for some dams constructed of, or on, weak liquefiable soils. Recent experiences from the 2010 $M_w = 8.8$ Maule Earthquake in Chile, for example, included the failure of five mine tailings dams due to liquefaction (Verdugo et al. 2012). Reviews of past performances of embankment dams during earthquakes can be found in USCOLD (1992, 2000) and USSD (2013).

The effects of liquefaction on the Lower and Upper San Fernando dams in the 1971 San Fernando earthquake are illustrative of two different levels of performance, from the dramatic slide in the upstream shell of the Lower San Fernando dam (Fig. 1a) to the more moderate deformations that developed in the Upper San Fernando dam (Fig. 1b). These two case histories are frequently used as part of validation studies for NDA modeling procedures; a review of the literature identified 26 papers describing separate NDA studies of one or both of these dams.

A potential failure modes analysis (PFMA) is an important step in dam safety and risk analysis practices developed by USBR (2011) and USACE (2011). A PFMA seeks to identify, describe, and evaluate site-specific failure modes, including the sequence of events required for any mode to progress toward a breach of the dam. This sequence of events then becomes the basis for the event trees in the risk analysis. The PFMA for Isabella Dam, as described by Serafini and Rose (2010), provides an example of USACE practices.

a b

Fig. 1 San Fernando Dams after the 1971 San Fernando earthquake: **a** Lower dam (CA DSOD files), and **b** Upper dam (LADWP files, courtesy of Craig Davis)

Potential failure modes for an embankment dam can be initiated by liquefaction-induced deformations in the dam, its foundation, its abutments, or surrounding areas which may affect either the stability of the embankment or any upstream or downstream appurtenant works (e.g., intakes, control towers, access bridges, outlet works). Some potential failure modes associated with liquefaction-induced deformations include the following.

- Overtopping of the dam due to vertical slumping of the crest.
- Seepage-induced erosion through cracks that form in the dam or at contacts between the dam and its abutments or any embedded structures.
- Seepage-induced erosion into or along outlet works, conduits, spillways, or other embedded structures that are damaged or affected by deformations of the dam or surrounding areas.
- Delayed overtopping of the dam if embankment deformations have reduced the available freeboard and/or disrupted the ability of outlet works to control reservoir levels and repairs of the dam or outlet works cannot be completed before the next significant water inflow event.

The sequence of events required for any individual failure mode to progress toward a breach is examined in detail as part of the PFMA.

Potential failure modes associated with seismic, hydrologic, and other loading conditions are best examined at the same time. The consideration of all potential failure modes together can be important for identifying possible remedial actions that address more than one source of risk. The risk analysis performed for Success Dam (Anderson et al. 2011; Bowles et al. 2011; Ruthford et al. 2011) is an example of USACE practices for evaluating risks associated with both seismic (including liquefaction effects) and hydrologic (e.g., overtopping, seepage, piping) loading.

The risk associated with liquefaction-induced dam deformations depends on features that affect the probabilities for any potential failure mode to progress to a breach and on the features that affect the consequences of a breach. The probabilities

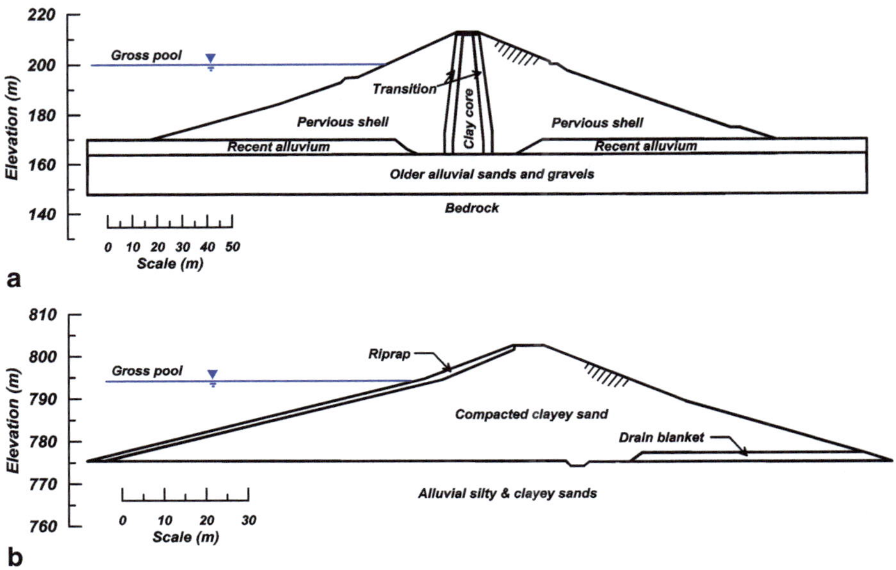

Fig. 2 Cross-sections for two dams having liquefiable soils in their foundations: **a** Success Dam with its central core and transition zones (Anderson et al. 2011), and **b** Isabella Auxiliary Dam with its homogenous section and blanket drain (Serafini and Rose 2010)

of liquefaction-induced deformations resulting in a breach may be affected by a number of site-specific factors, including the following items.

- Magnitude and pattern of deformations in the dam, both transversely and longitudinally
- Available freeboard
- Internal zoning and properties of the dam (e.g., Fig. 2a, b):
  - filter compatibility of adjacent zones
  - vulnerability of transition zones or shell materials to hold cracks open to large depths
  - ability of drains or shell materials to transmit large water flows without erosion
  - erodability of materials in each zone
- Topography of contacts between the dam, its abutments, and any embedded structures where dam deformations may be manifested as tension cracks or separations

Factors which may affect the potential consequences of a breach include:

- The reservoir stage curve
- Rapidity of failure
- Flood inundation maps
- Number and location of the populations at risk
- Monitoring and on-site inspection capabilities

- Intervention capabilities and preparedness
- Emergency action plans and warning systems
- Effects of the earthquake on other infrastructure

There is considerable uncertainty in each of the above factors and sometimes they are estimated through expert elicitations rather than through quantifiable analyses (e.g., USBR 2011; USACE 2011). The range of dam deformations at which risks become significant depends on the above factors and considerations. The assessment of this range can benefit from a full risk appraisal where both the probability of failure and resulting consequences are evaluated. It follows that the specification of allowable deformation criteria (e.g., a maximum allowable crest settlement) for a deterministic evaluation should be developed on a dam-specific basis.

Mitigation of the risks associated with liquefaction-induced deformations can include a combination of actions that reduce the dam deformations (e.g., ground improvement, addition of berms, removal and replacement of problem soils), reduce the probability of a potential breach given the estimated deformations (e.g., a reservoir restriction to increase freeboard, installation of filters and drains), or reduce the potential consequences of a breach (e.g., improved emergency action plans). A risk mitigation plan may include interim risk reduction measures which can be implemented quickly (e.g., reservoir restriction, improved emergency action plans) relative to the time required for major dam modifications. The environmental, economic, and social impacts of each alternative action or measure are important considerations in the final selection and development of risk mitigation plans.

The results of NDAs are often used to generate fragility relationships across a range of seismic hazard levels as inputs to risk analyses (e.g., Ruthford et al. 2011). The fragility relationships should account for the uncertainty in the computed magnitudes and patterns in the dam deformations. This uncertainty is sometimes accounted for by considering a range of site characterization scenarios (e.g., the selection of representative properties for different zones and strata), a sufficient suite of ground motion input time series, and an allowance for epistemic uncertainty (e.g., allowing for actual displacements to range from ½ to 2 times the computed values; Bowles et al. 2011). Thus, it is important that the NDA models and modeling procedures are properly validated to provide reasonable and unbiased deformations across a broad range of loading conditions.

The interpretation of risk analysis results should be tempered by judgments based on a thorough understanding of the factors controlling the calculated risks, the assumptions made at each step of the analysis, and the basis for the tolerable risk guidelines (e.g., Munger et al. 2009). Sensitivity analyses can be important for identifying the primary factors affecting the computed risk at a system level and to understand the major sources of uncertainty in the risk models. The role of the risk analyst must therefore include extracting insights into what is important and summarizing those insights for the project team.

Calibration and Validation of Constitutive Models for Liquefiable Soils

The results of an NDA for a dam affected by liquefaction can be strongly affected by the performance of the constitutive model used for the liquefiable soils. For this reason, it is vital that the calibration and validation of the constitutive model be documented in sufficient clarity that its strengths and limitations are well recognized and quantified.

The characterization of potentially liquefiable soils relies predominantly on in-situ test measurements, such as the Standard Penetration Test (SPT), Cone Penetration Test (CPT), and shear wave velocity (V_s) measurements, along with index test results on disturbed samples (e.g., fines content). More detailed laboratory tests, such as triaxial or direct simple shear (DSS) tests, are almost never available due to the problems with overcoming sample disturbance effects and the challenge of identifying representative samples from highly heterogeneous deposits (e.g., Fig. 3).

The selection of representative SPT or CPT penetration resistances for different zones or strata is often an important step in determining the properties to which the constitutive models will be calibrated. The selection of representative values depends on the size of the potential failure surfaces which could affect the dam and the expected distribution of soil properties across that scale; e.g., are looser soils restricted to local narrow channels or present as continuous horizontal strata. The selection of representative values should therefore be driven by a geologic model for the depositional environment, with exposures and trenches often providing valuable insights (e.g., Fig. 3). In most cases, it is preferable that a detailed site characterization report be prepared and reviewed prior to commencing NDAs because the judgments and evaluations of the site characterization procedures can control the NDA outcomes. Repeating NDAs in response to adjustments in the site characterization can consume considerable engineering effort. Preliminary conclusions based on poorly-defined NDAs can also divert engineering resources into unnecessary tasks.

The calibration of constitutive models for liquefiable soils in practice relies on the extensive body of experimental and case history data embodied in correlations for the different aspects of liquefaction behaviors (e.g., Ishihara 1996; Idriss and Boulanger 2008). The loading behaviors of importance to the performance of dams

Fig. 3 Interlayering and channeling of alluvium exposed in an exploratory trench

include (1) the shear modulus and damping characteristics which affect the dynamic response before and after liquefaction triggering, (2) the rate of pore pressure generation as a function of cyclic loading, (3) the cyclic mobility behavior which controls the progressive accumulation of cyclic shear strains and deformations after liquefaction has been triggered, and (4) the residual shear strength which affects post-earthquake embankment stability and deformations. For the same stratum, each of these behaviors can also vary significantly across the broad range of overburden and horizontal static shear stress conditions encountered beneath a single dam.

The calibration process involves iterating on model parameters until the responses obtained in single-element simulations are in reasonable agreement with appropriate correlations. For example, the cyclic resistance ratio (CRR) against liquefaction for a sand under a vertical effective consolidation stress (σ'_{vc}) of 1 atm (101 kPa) during an $M_w = 7.5$ earthquake is usually determined from a liquefaction triggering correlation (e.g., Fig. 4), based on the stratum's representative penetration resistance. The constitutive model parameters for that stratum are then calibrated so that a single-element simulation of undrained cyclic DSS loading under $\sigma'_{vc} = 1$ atm with an imposed cyclic stress ratio (CSR) equal to the target CRR value (determined from Fig. 4) causes liquefaction in 15 uniform loading cycles. This one step is, however, just the first of many aspects of behavior that the constitutive model needs to be evaluated against (e.g., Cubrinovski 2011, Wang et al. 2012).

A major challenge in the calibration of a constitutive model is that not all aspects of the simulated behavior can be adjusted by the model parameters. Instead, some

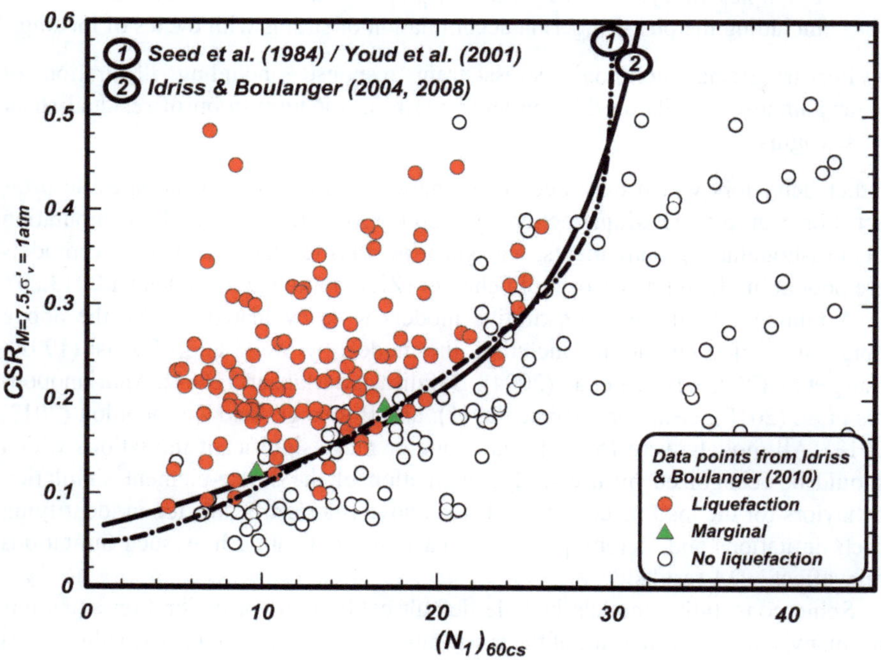

Fig. 4 Two SPT-based liquefaction triggering correlations (after Boulanger et al. 2012)

aspects are controlled by the underlying functional forms of the constitutive equations, such that the model may have certain limitations that cannot be overcome without implementing fundamental changes to the constitutive model. The onus is on the user to know when a constitutive model produces responses that are at least within the expected range of responses, as observed experimentally for similar soils across the broad range of applicable loading conditions, and to know when it does not.

A thorough evaluation (or validation) of a constitutive model for liquefiable soils therefore requires a mixture of calibration and examination of its behavior under the broad range of loading conditions of potential importance to performance of the dam. It is recommended that such an evaluation include, as a minimum, single-element DSS simulations for the following behaviors.

- Dynamic properties:
  - Shear modulus reduction (G/G_{max}) and equivalent damping ratios versus cyclic shear strain amplitude in drained strain-controlled cyclic loading with shear strain amplitudes ranging from 10^{-4} to 3 %
- Liquefaction triggering and post-triggering cyclic mobility:
  - Cyclic resistance ratio (CRR) versus number of uniform loading cycles in undrained cyclic loading, with the number of cycles ranging from less than 5 to more than 50, and including its dependence on:
    - vertical effective consolidation stress, σ'_{vc}
    - coefficient of lateral earth pressure at rest, K_o
    - horizontal static shear stress ratio, $\alpha = \tau_{hv}/\sigma'_{vc}$
  - Examples of cyclic stress-strain responses with $\alpha = 0.0$, 0.1, 0.2, and 0.3, including the post-triggering accumulation of strains with cycles of loading
- Post-triggering monotonic stress-strain response, including illustration of any imposed limits on dilation and/or criteria for imposition of residual shear strengths

Other behaviors which may need examination will depend on the specific project. For example, post-liquefaction reconsolidation strains affect the computation of post-liquefaction settlements, and yet most elastic-plastic constitutive models are poor at modeling this aspect of behavior (Ziotopoulou and Boulanger 2013a,b).

A number of different constitutive models were evaluated against the above range of loading conditions, including the models by Wang and Makdisi (1999), Yang et al. (2003), Byrne et al. (2004), Dafalias and Manzari (2004), Andrianopoulos et al. (2010), Beaty and Byrne (2011), and Boulanger and Ziotopoulou (2012, 2013). All models were found to have one or more significant limitations which should be recognized by users. Documentation of the single-element simulation behaviors for the loading conditions listed above is a significant step in identifying such limitations and thereby provides a basis for evaluating how such limitations may affect NDA results.

Some constitutive models include default calibrations or calibration equations for many, and sometimes all, of the constitutive model parameters. The calibrations for model parameters are usually interdependent such that changing any one parameter (e.g., one controlling G_{max}) can affect the calibration of another (e.g., one controlling the CRRs). For this reason, documentation of single-element simulation behaviors should be provided even if using default calibrations.

The following sections present examples of some common constitutive model limitations. These examples are intended to illustrate the importance of documenting single-element simulation behaviors as part of any NDA study for dams.

CRR Versus Number of Equivalent Uniform Loading Cycles

The relationship between CRR and number of uniform loading cycles (N) describes the relative effects of small versus large loading cycles, which affects how a model responds to irregular loading histories of varying durations (the sequence of irregular loading cycles is another influencing factor). This aspect of behavior is important when evaluating dam responses to earthquakes across a range of hazard levels with corresponding variations in ground motion time series characteristics. The relationship between CRR and N can generally be approximated using a power function with a negative exponent of b. The results of experimental studies suggest that the value of b can reasonably be expected to be in the range of 0.2 to 0.37 for most sands, although smaller b values have been obtained in some cases (e.g., Toyoura sand in Fig. 5).

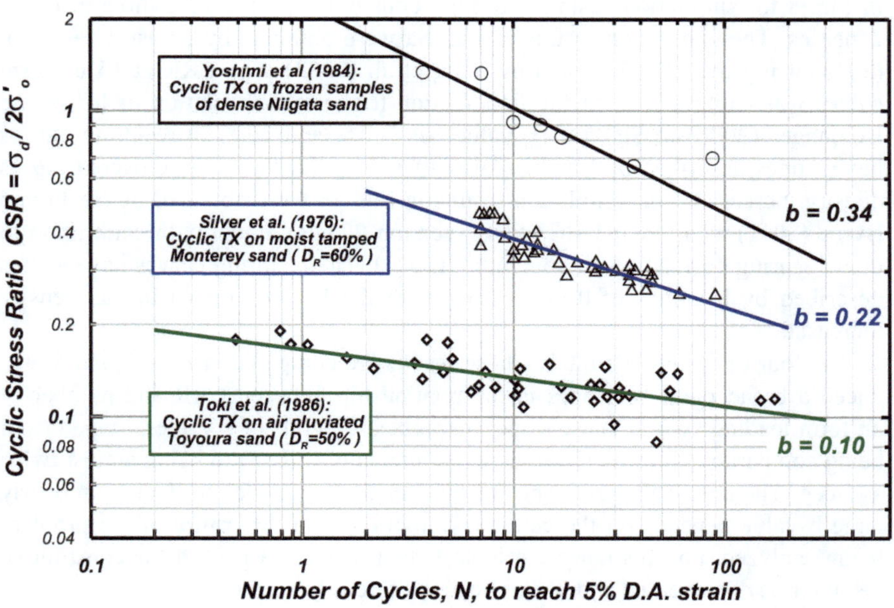

Fig. 5 CRR versus number of loading cycles for three sands covering the range of slopes observed for most sands reported in the literature (Ziotopoulou and Boulanger 2012)

The first step is performing single-element DSS simulations to obtain the stress-strain responses. The definition of "triggering" may depend on the constitutive model being used, because some models lock-up (stop accumulating shear strains) after reaching $r_u = 100\,\%$ (or some limiting value close to 100 %) when the stress cycles are applied symmetrically (i.e., $\alpha = 0$). This is illustrated by the cyclic stress-strain responses obtained for two related constitutive models; the first model (Dafalias and Manzari 2004) locks up at shear strains of about 0.3 % (Fig. 6a) whereas the second model (Boulanger and Ziotopoulou 2012, 2013; Ziotopoulou and Boulanger 2013a) is able to progressively accumulate shear strains after $r_u \approx 100\,\%$ first occurs (Fig. 6b). Thus, a triggering criterion for the simulations may need to be based on r_u for a model that locks up, or on a failure strain amplitude (e.g., a maximum single amplitude shear strain of 3 %) for a model that does not. Note that the second model was based on the framework of the first, with modifications to the functional forms of the constitutive equations being necessary to improve this and other features of behavior.

Additional single-element simulations for a range of imposed cyclic stress ratios can then be used to develop the relationship between CRR and number of uniform loading cycles, as illustrated in Fig. 7 for the same two models. The first model (Fig. 7a) has slopes described by b = 0.55 to 0.61, which are about twice the values typically observed in experiments. The second model (Fig. 7b) has slopes described by b = 0.24 to 0.28, which are in the midrange of experimentally observed values. Again, the second model was based on the framework of the first, with modifications designed to improve these types of behaviors.

The relationship between CRR and number of uniform loading cycles is shown in Fig. 8 for simulations using two other constitutive models for three different densities. The simulation results in Fig. 8a were obtained by Khosravifar et al. (2013) using the PDMY model by Yang et al. (2003). The CRR at 15 uniform cycles was calibrated to fit the liquefaction triggering correlation of Idriss and Boulanger (2008) at SPT $(N_1)_{60}$ values of 5, 15, and 25. The resulting curves have slopes described by b values of 0.40–0.54, with the b value decreasing as density increased. The simulation results in Fig. 8b were obtained by Beaty and Byrne (2011) using their UBCSAND (version 904aR) model with calibration to the triggering correlation in Youd et al. (2001). The resulting curves have slopes described by b values of 0.34 to 0.41, with the b value increasing as density increased.

The four constitutive models whose results are compared in Figs. 7 and 8 produced a broad range of slopes in the relationship between CRR and number of uniform loading cycles. These slopes, expressed in terms of b values, ranged from being comparable to those observed experimentally (Fig. 5) to being almost twice as steep. The practical significance of the constitutive model producing an overly large b value depends on the earthquake motions and the nature of the problem being analyzed. For this reason, it is important that this aspect of the constitutive behavior be documented and its implications carefully considered.

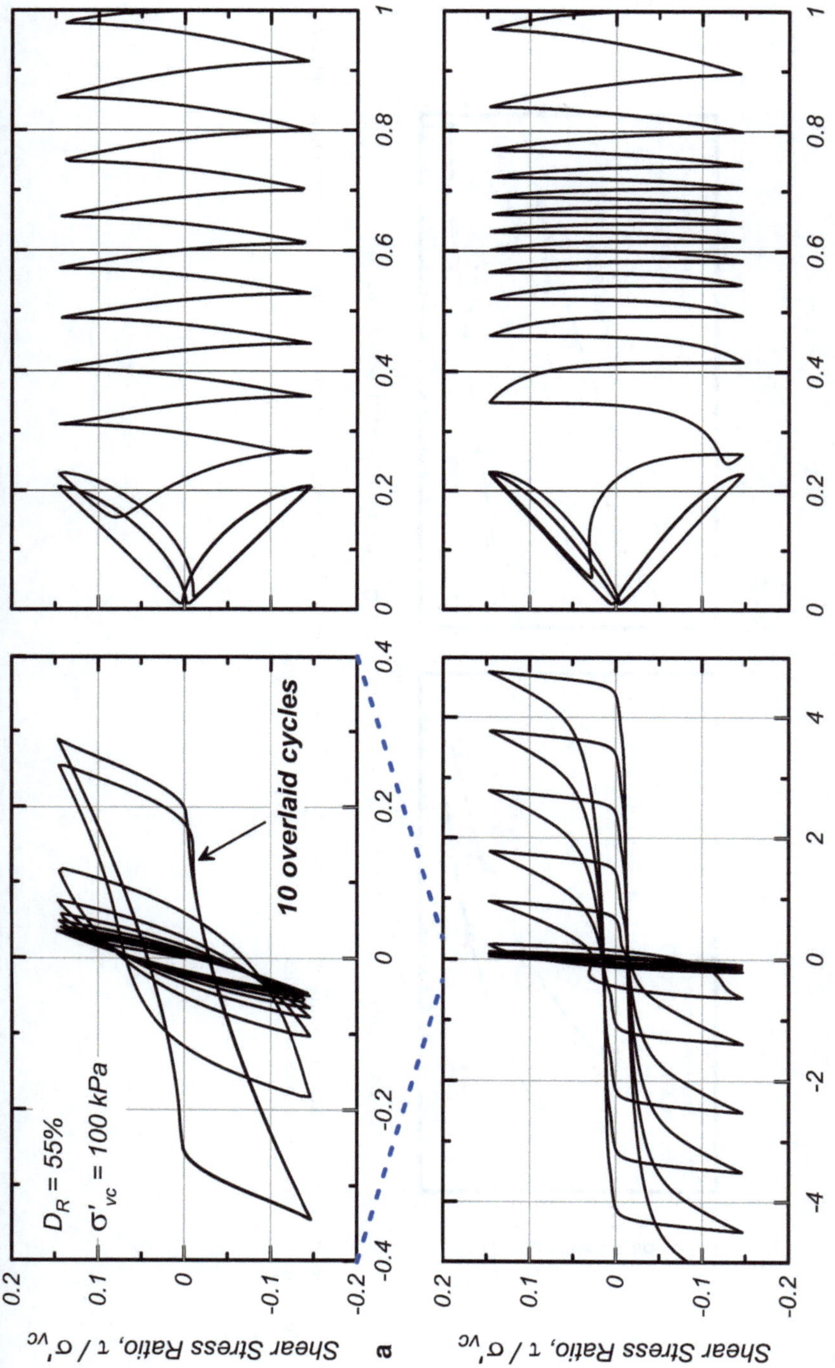

Fig. 6 Stress-strain responses from single-element simulations of undrained cyclic direct simple shear tests **a** using an implementation of the Dafalias and Manzari model (2004), and **b** using the model by Boulanger and Ziotopoulou (2012, 2013)

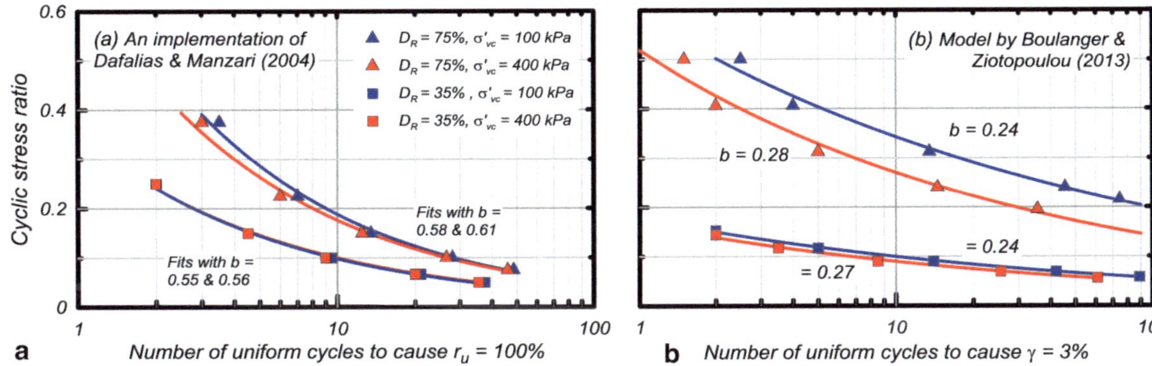

Fig. 7 CRR versus number of uniform loading cycles to cause liquefaction at two relative densities and two overburden stresses for **a** an implementation of the Dafalias and Manzari (2004) model, and **b** the model by Boulanger and Ziotopoulou (2012, 2013)

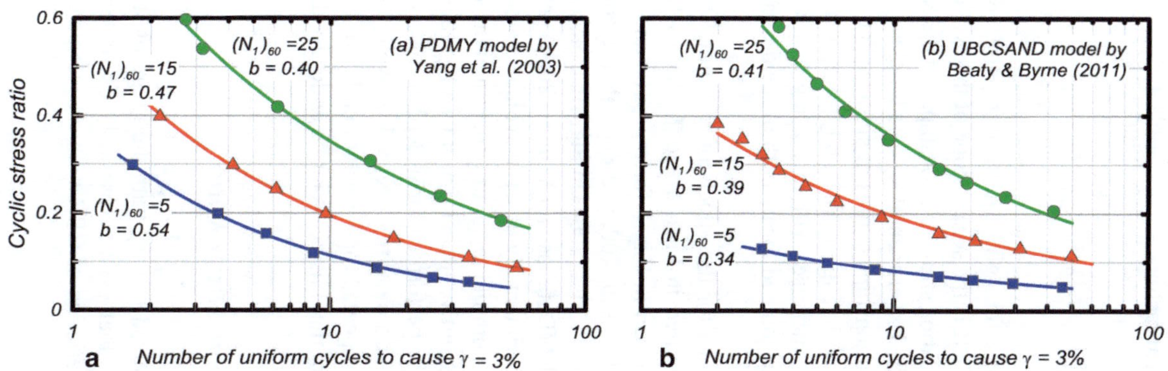

Fig. 8 CRR versus number of uniform loading cycles to cause liquefaction for calibration at three SPT $(N_1)_{60}$ values with $\sigma'_{vc} = 100$ kPa for: **a** PDMY model by Yang et al. (2003), and **b** UBCSAND by Beaty and Byrne (2011)

CRR Versus Overburden Stress

The CRR can also be expected to vary with the range of σ'_{vc} values that exist across a dam's footprint. Seed (1983) described the experimental variation of CRR with σ'_{vc} in terms of an overburden factor K_σ:

$$K_\sigma = \frac{CRR_{\sigma'}}{CRR_{\sigma'=1}}$$

where $CRR_{\sigma'}$ is the CRR value for a given value of σ'_{vc} and $CRR_{\sigma'=1}$ is the CRR value when $\sigma'_{vc}=1$ atm. The CRR is the secant slope of the cyclic strength (as a cyclic shear stress) versus consolidation stress relationship, and the K_σ factor describes the curvature of that relationship. Experimental K_σ data for clean sands are compared to two common K_σ design relationships in bins of similar $CRR_{\sigma'=1}$ values (as a proxy for similar relative densities) in Fig. 9. The body of experimental data show the dependence of CRR on σ'_{vc} increases with increasing relative density of the sand. This aspect of behavior is accounted for by both of the design relationships shown in Fig. 9.

Single-element simulations for the effects of σ'_{vc} on CRR were performed for the same two models compared in Figs. 6 and 7. Curves for CRR versus number of loading cycles are shown in Fig. 7 for $\sigma'_{vc}=100$ and 400 kPa, whereas the equivalent K_σ values for $\sigma'_{vc}=100$, 400, and 800 kPa are compared to the design relationship by Boulanger and Idriss (2004) in Fig. 10. Comparing Figs. 9 and 10, it is evident that the first model (Dafalias and Manzari 2004) produces K_σ values that are outside the range of experimentally observed values, whereas the second model is in reasonable agreement with the range of experimentally observed values. The advantage of having a constitutive model that tracks K_σ values well is that model calibration can reasonably be performed for wider bins of overburden stress conditions in the field. If a model does not track K_σ values well, it means that the model calibration may need to be repeated for a greater number of overburden stress bins. In either case, it is important that this behavior be understood and documented.

CRR Versus Static Shear Stress Ratio

The CRR can also be expected to vary with the range of static shear stress ratio ($\alpha = \tau_{hv} / \sigma'_{vc}$) conditions that exist across a dam's footprint. Seed (1983) described the experimental variation of CRR with α in terms of a static shear stress factor, K_α

$$K_\alpha = \frac{CRR_\alpha}{CRR_{\alpha=0}}$$

where CRR_α is the CRR value for a given value of α and $CRR_{\alpha=0}$ is the CRR value when $\alpha=0$. Experimental results on a range of sands at confining stresses less than

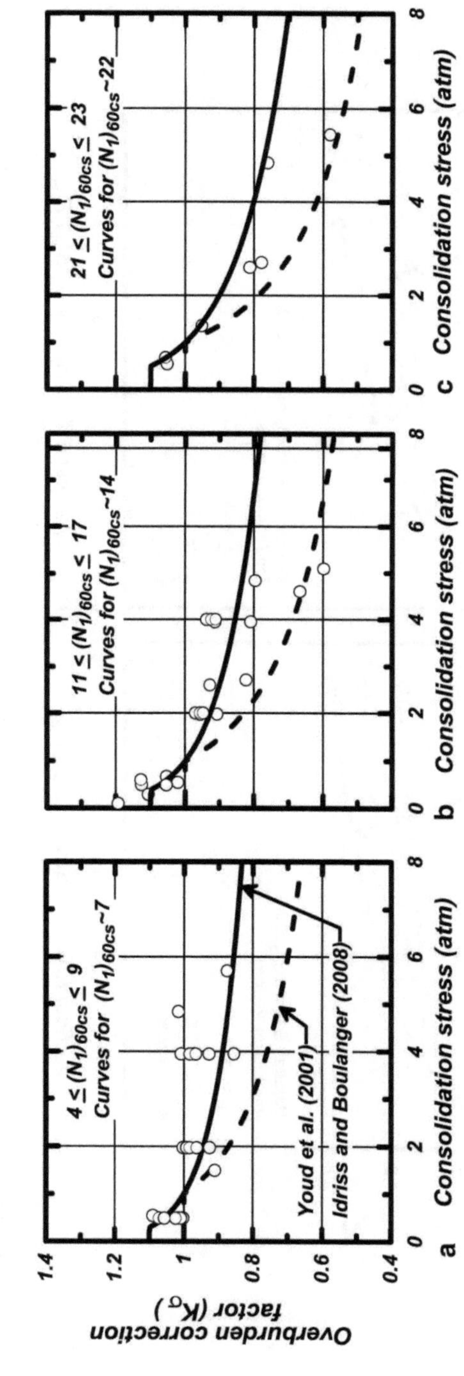

Fig. 9 Experimental K_σ data for clean sands compared to two design relationships in bins of comparable $(N_1)_{60cs}$ and corresponding $CRR_{\sigma'=1}$ values (Montgomery et al. 2012)

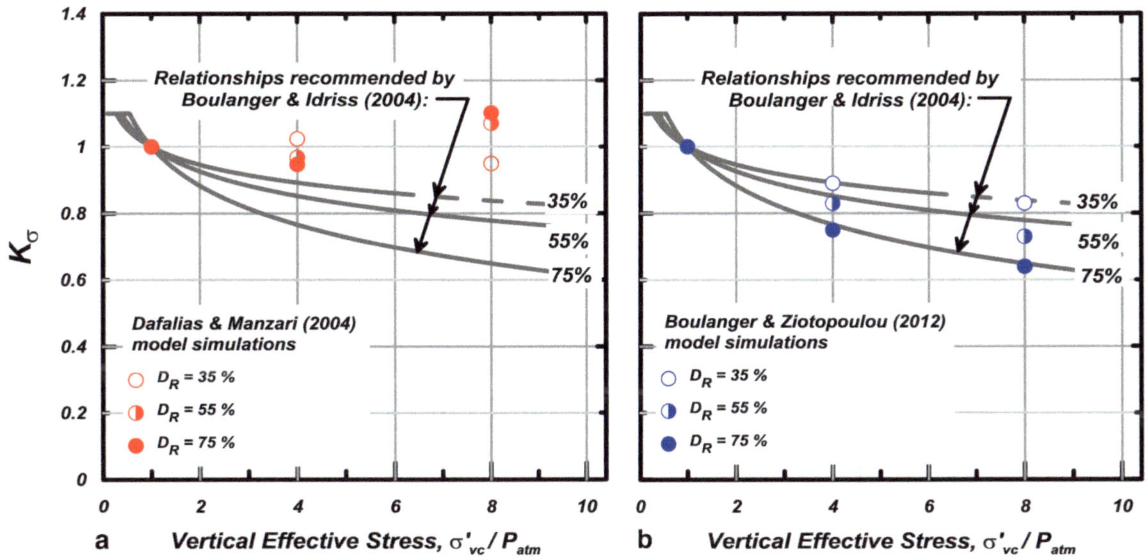

Fig. 10 K_σ values from simulations using two constitutive models compared to the design relationship by Idriss and Boulanger (2008) **a** model by Dafalias and Manzari (2004), and **b** model by Boulanger and Ziotopoulou (2012, 2013)

about 300 kPa showed that CRR (for some strain failure criterion) would decrease with increasing α for loose sands and increase with increasing α for dense sands. The pattern of results for sands tested at confining stresses less than about 300 kPa are summarized in Fig. 11a (Harder and Boulanger 1997). For sands tested across a broader range of confining stresses, Boulanger (2003) showed that the trends were better related to some measure of state, such as the relative state parameter index as shown in Fig. 11b. The results shown in Fig. 11 were determined using a failure criterion of 3 % single-amplitude shear strain to define the CRR values; the excess pore water pressures in sand reach their limiting value at this level of shear strains and thus generally will not increase with further increases in the maximum shear strains.

Simulated stress-strain responses for DSS loading of sand with $D_r = 55\,\%$, $\sigma'_{vc} = 100$ kPa, and $\alpha = 0.2$ are shown for the model by Yang et al. (2003) in Fig. 12a and the model by Boulanger and Ziotopoulou (2012, 2013) in Fig. 12b. The stress paths are both reasonable, showing that $r_u = 100\,\%$ ($p' = 0$) cannot be reached for these dense of critical sands when the cyclic loading is insufficient to cause a full shear stress reversal. The peak r_u values for the two models differ because of their differences in unloading behaviors, with the peak r_u values being about 0.75 for the Yang et al. model and about 0.87 for the Boulanger and Ziotopoulou model. Both models accumulate shear strains in the direction of the static shear stress, consistent with experimental observations, but at different rates per cycle of loading.

Single-element simulations using these two models were repeated for $D_r = 35$, 55, and 75 % with $\sigma'_{vc} = 100$ kPa to define the effect of α on CRR. The CRR values were determined for a shear strain failure criterion of 3 %. The results are presented in terms of equivalent K_α values in Fig. 13a for the Yang et al. (2003) model and Fig. 13b for the Boulanger and Ziotopoulou (2012, 2013) model. The simulated behaviors deviate significantly from the trends observed experimentally (Fig. 11), with neither model able to reasonably reproduce the observed increases in CRR with increasing α for $D_r = 75\,\%$ or decreases in CRR with increasing α for $D_r = 35\,\%$.

Single-element simulations using two versions of the UBCSAND model were presented by Beaty and Byrne (2011) to illustrate the improvements in responses obtained with their newer version. The equivalent K_α values obtained using the older version 904a, as shown Fig. 14a, deviated greatly from the experimentally observed results in Fig. 11. The results obtained with this model included significant kinks in the trends and an almost inverted trend for the sands at $(N_1)_{60}$ values of 10, 15, and 25 (i.e., the curves shift lower with increasing denseness). In contrast, the results obtained using the newer version 904aR, as shown in Fig. 14b, are in reasonable agreement with the experimentally observed trends. It is noteworthy that Beaty and Byrne (2011) identified the limitations in the 904a version during a project involving the seismic analysis of an embankment dam, which provides an early example of this particular aspect of model validation. The α behavior was a critical item at this dam due to the relative magnitudes of the static shear stress and cyclic shear stress ratios. They used the findings of that study to guide model improvements which resulted in their 904aR version. For models such as UBCSAND

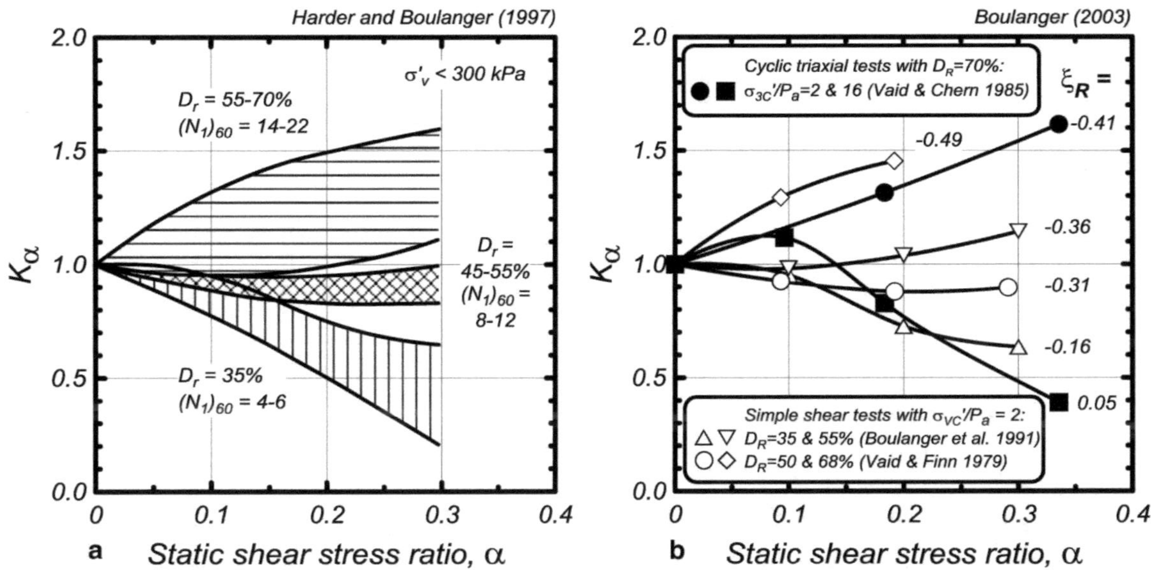

Fig. 11 Experimental trends between CRR and static shear stress ratio (α) depicted in terms of K_α relationships **a** for sands at different D_r and σ'_{vc} < 300 kPa (Harder and Boulanger 1997), and **b** for sands at different relative state parameter index, ξ_R, values. (Boulanger 2003)

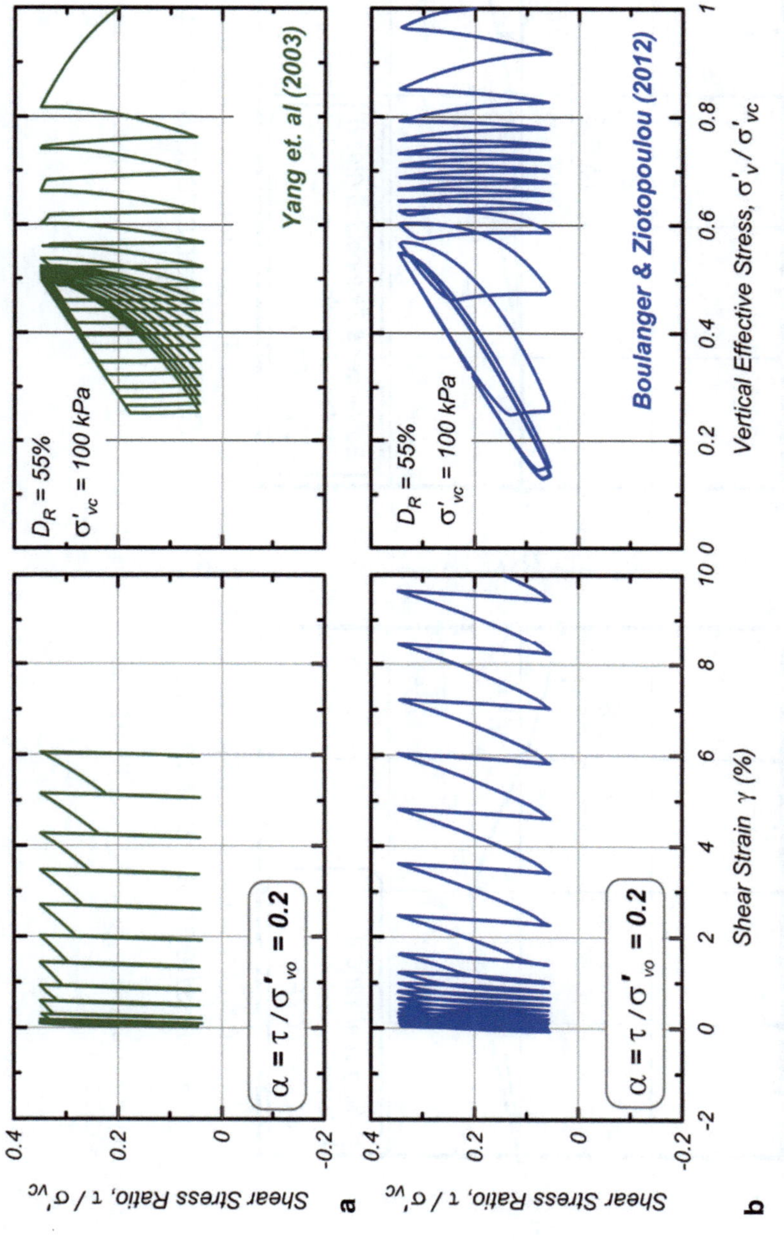

Fig. 12 Single-element responses for undrained cyclic DSS loading on sand at $D_r = 55\,\%$, $\sigma'_{vc} = 100$ kPa, and $\alpha = 0.2$ simulated using: **a** PDMY model by Yang et al (2003), and **b** PM4Sand model by Boulanger and Ziotopoulou (2012, 2013)

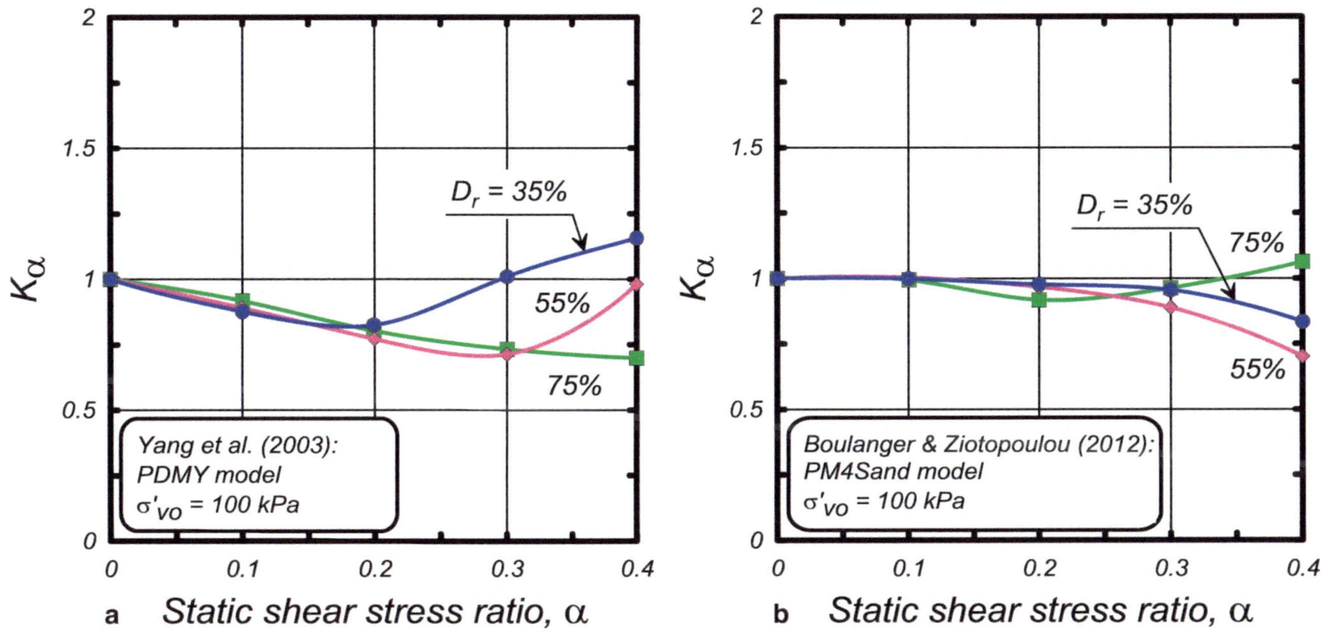

Fig. 13 Single-element simulation results for undrained cyclic DSS loading using: **a** PDMY model by Yang et al. (2003) and **b** PM4Sand model by Boulanger and Ziotopoulou (2012, 2013)

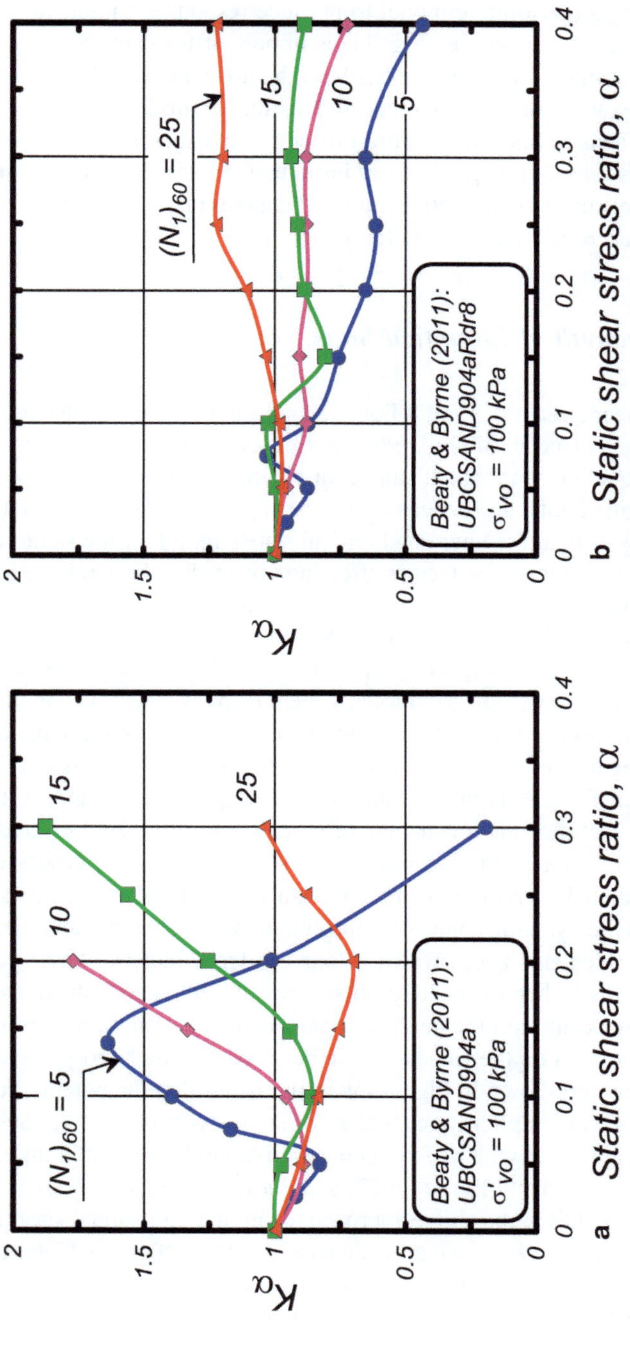

Fig. 14 Single-element simulation results for undrained cyclic DSS loading using two versions of UBCSAND **a** version 904a, and **b** version 904aRdr8 (Beaty and Byrne 2011)

that are periodically updated and modified, it is clearly necessary to document both the model name and version when performing an analysis.

The ability of a constitutive model to produce responses similar to the experimentally observed K_α responses (Fig. 11) is important for dams because α values are often in the range of 0.1 to 0.3 under both the upstream and downstream shells of a dam. The examination of a number of constitutive models suggests that simulation of K_α effects is one of the more difficult challenges for many constitutive models. For this reason, it is particularly important that single-element simulations be used to demonstrate a constitutive model's behavior under these types of loading conditions as part of any NDA study for a dam.

Residual Strength of Liquefied Soils

The residual shear strength (S_r) of a liquefied soil in the field may be significantly affected by the mechanism of void redistribution as illustrated in Fig. 15. The local loosening of liquefied sand which can occur when pore water seepage is impeded by a lower permeability layer means the post-liquefaction shear strength can be significantly lower than the undrained critical state strength of the same sand at its pre-earthquake void ratio. The factors affecting void redistribution have been demonstrated through a range of physical modeling studies (e.g., Kokusho 1999, 2003; Kulasingam et al. 2004; Malvick et al. 2008) and numerical modeling studies (e.g., Yang and Elgamal 2002; Naesgaard et al. 2005; Seid-Karbasi and Byrne 2007). This localization process represents a complex system response that: (1) complicates the interpretation of case histories and development of design correlations, and (2) is currently outside the ability of NDA procedures to adequately simulate in practice.

The approach for simulating residual shear strengths in an NDA is also complicated by the fact that some constitutive models are formulated to model cyclic mobility behaviors alone and by concerns that void redistribution processes cannot be reliably predicted regardless of the constitutive model. For these reasons, it is common to run the dynamic analysis to the end of strong shaking, and then replace any "liquefied" elements/zones with a Mohr Coulomb material having $\phi = 0$ and $c = S_r$. The analysis is then resumed to obtain any additional post-shaking deformations which may occur with the newly assigned residual strength properties. In US practice, it is common for the values of S_r to be based on case history based correlations such as the one shown in Fig. 16, with the belief that the potential effects of void redistribution are implicitly accounted for in the case history data (Seed 1987). The use of this approach with different constitutive models on different projects is described in Naesgaard and Byrne (2007) and Perlea and Beaty (2010).

An important detail in the above approach to simulating residual shear strengths is the criterion used to define what elements/zones have "liquefied" during strong shaking. The shear stresses that exist within a dam and its foundation are often large enough that many elements/zones do not undergo shear stress reversals and thus

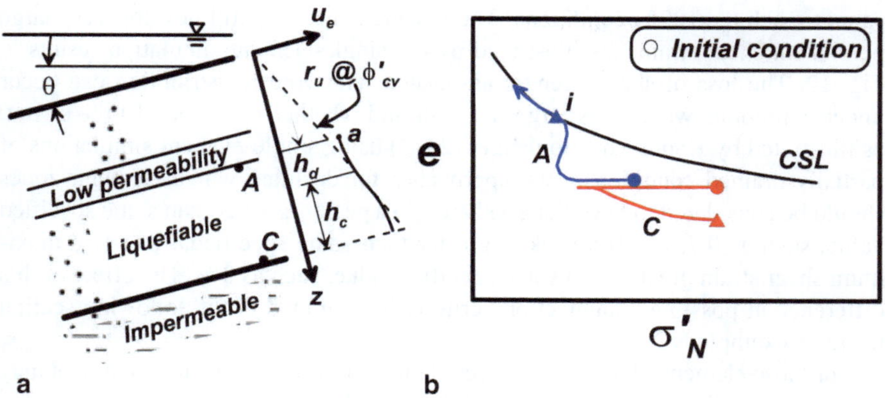

Fig. 15 Void redistribution in layered soil profiles (Kulasingam et al. 2004) **a** infinite slope with low permeability layer over a liquefiable sand layer, and **b** void ratio changes for points at the top and bottom of the confined sand layer

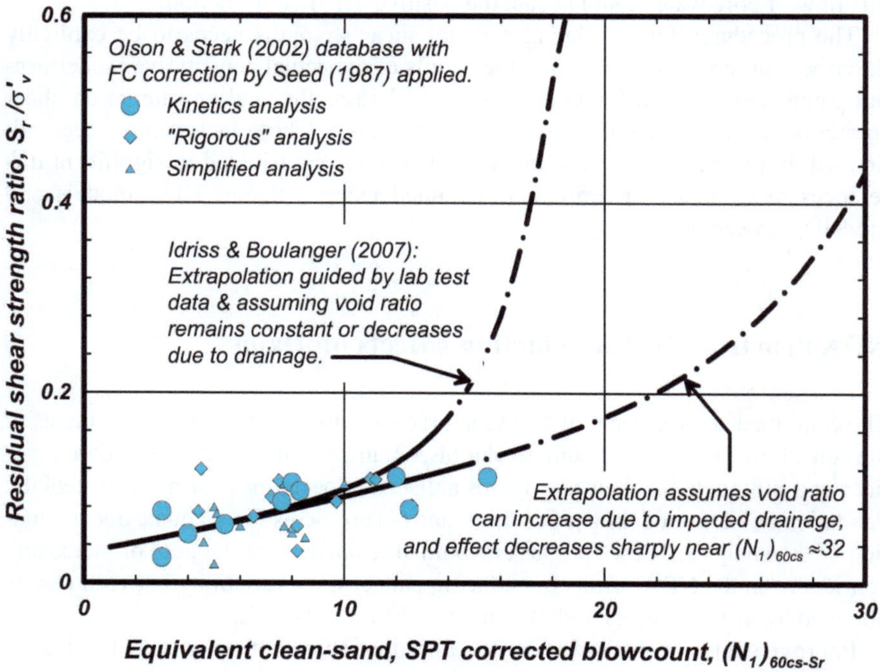

Fig. 16 Residual shear strengths from case history analyses by Olson and Stark (2002) and the design relationships recommended by Idriss and Boulanger (2007) for $\sigma'_v < 400$ kPa [use $(N_{1,}\xi)_{60}$ at higher σ'_v]. (Boulanger and Idriss 2011)

cannot reach $r_u = 100\,\%$ conditions. These same zones can still develop very large permanent shear strains, as illustrated by the single-element simulation results in Fig. 12. The loss of shear strength associated with void redistribution can occur under conditions where α is large or small and whether r_u reaches $100\,\%$ or not, as illustrated by Kamai and Boulanger (2012) using single-element simulations of partially drained conditions. Two approaches for deciding which elements/zones should be considered to have liquefied are: (1) a peak r_u greater than some specified value, such as 0.7, or (2) a peak r_u greater than some specified value or a maximum shear strain greater than some specified value, such as $3\,\%$. The effect of this difference in possible "liquefaction" criteria is examined in the following section using an example NDA.

For those elements that do not trigger "liquefaction," the constitutive model may be capable of dilating indefinitely and thus developing large shear resistances as excess pore pressures become negative. The strength of these dilating zones can compensate for the loss of strength in the "liquefied" zones, and thus have a large effect on the computed post-shaking slope deformations. It is common, however, to limit the shear strengths of such non-triggered zones to their pre-earthquake drained strengths because local drainage patterns are difficult to predict with confidence and this flow of pore water could negate the positive benefits of dilation.

The procedures for simulating residual shear strengths needs to be explicitly described for any NDA, including the details of the actual constitutive models, input parameters, criteria for assigning residual strengths, and any limits on shear strengths in areas of negative excess pore pressures. These factors have been addressed in a variety of ways in practice, which has contributed to significant differences in estimated deformations obtained using different NDA models and modeling procedures.

NDA Practices for Liquefaction Effects on Dams

The continued advancement of NDAs in practice will require: (1) improvements in the constitutive models and numerical tools, (2) improvements in the standard practices and protocols for their use, (3) established expectations for more systematic validation of NDA modeling procedures, and (4) protocols for adequate documentation of the analysis. These four tasks would all contribute to the goal of increasing confidence in the NDA results by reducing unnecessary variability between results obtained by different users with the same or different models.

Improvements in constitutive models for liquefiable or other soils are needed for more realistic modeling of important behaviors and to reduce the undesirable differences in NDA results that arise from limitations in the constitutive models. The previous section illustrated how single-element simulations can be used to validate a constitutive model against the broader body of knowledge about liquefaction behaviors, or to at least identify where a constitutive model has inherent limitations that could be important to the NDA results. The effect of constitutive models on NDA results can be expected to reduce as the limitations are addressed and the models more fully simulate expected behaviors.

Improvements in the standard practices for performing and documenting NDAs are important for quality control purposes, identifying causes of differences in responses obtained using different modeling procedures, and extracting insights that benefit everyone involved with the project. NDA results can be sensitive to minor input errors or differences in procedures, yet finding these errors or procedural differences can be very difficult given the complexity of the analysis procedures. Similarly, the reasons why different modeling procedures can sometimes affect, and other times not affect, deformations can be difficult to understand unless the responses are tracked in great detail. The ability to address these issues is improved whenever the NDA procedures and results are documented with a greater level of completeness and clarity.

Systematic validation of NDA modeling procedures is needed to increase our confidence in their general use and provide a better assessment of potential biases and dispersion for inclusion in risk analyses. Validation of an NDA modeling procedure requires validating the components (e.g., the constitutive models using single-element simulations) as well as the entire NDA procedure. This includes such items as the approaches used to establish material properties, static stresses and initial pore water conditions, static and dynamic boundary conditions, element sizes, and viscous damping requirements. Validation of the entire procedure requires a fairly large number of cases across a range of performance levels and should include physical models (because they provide detailed measurements and well-constrained conditions) and case histories (because they provide geologic and structural complexity). Multiple cases are needed because any single case may provide a challenge for relatively few components of the NDA procedure, while being insensitive to other components. For example, the slide in the upstream shell of the Lower San Fernando Dam is relatively easy to simulate provided that the assigned residual strength is significantly smaller than required for static equilibrium; the ability to predict the triggering of liquefaction in much of the shells is relatively insensitive to the modeling details because the materials are loose and the shaking was strong. As such, evaluation of an NDA procedure against the Lower San Fernando Dam case alone provides a necessary but insufficient degree of validation. Analysts should be familiar with the details of the past validation studies for the NDA modeling procedures they are using, and they should recognize that deviating from the details of those established procedures may necessitate an updating of the validation studies.

Engineering reports summarizing the results of an NDA study also provide an important means of communicating with the engineers, owners, and regulators who may be involved with evaluating the seismic performance and safety of an embankment dam in the future. Again, the utility of such reports for future studies depends on the complete and transparent documentation of the NDA practices that were followed and the clarity with which the computed responses are presented and interpreted.

NDA Example for a Hypothetical Dam

The results of an NDA for a hypothetical dam are used to illustrate how a small, but important, detail can have a significant effect on computed deformations. This example is designed to illustrate the experiences and issues encountered on other dam projects where liquefaction of alluvium under the shells was a concern.

The dam cross-section (Fig. 17) includes a clay core, gravelly sand shells, a layer of alluvium under both shells, and a downstream gravel berm. The shells and alluvium are modeled using the constitutive model by Boulanger and Ziotopoulou (2012, 2013) with properties based on SPT $(N_1)_{60}$ values of 25 and 18, respectively; these materials are initially dense-of-critical state for the range of stresses encountered throughout the dam and its foundation. The clay core and gravel berm are modeled as Mohr-Coulomb materials with the clay core having $s_u/\sigma'_{vc}=0.35$ (with $\phi=0$) and the gravel berm having $\phi'=40$ degrees (with $c'=0$). The input motion is the modified Pacoima record from Seed et al. (1975). The other details of the analysis are omitted for brevity because they are not essential for the purposes of this example.

Contours of initial static shear stress ratio ($\alpha = \tau_{hv}/\sigma'_{vc}$) are shown in Fig. 18 at the end of the static analysis. These results show that α ranges from 0.1 to 0.4 throughout the alluvium under both the upstream and downstream shells. As shown later, these shear stress conditions are important for understanding the element responses during and after dynamic loading and for relating the results to the single-element simulations presented in the previous section.

Contours of maximum shear strain (γ) and maximum r_u at the end of strong shaking are shown in Fig. 19a, b, respectively. These contours show that most of the alluvium under the upstream shell (left side) developed maximum r_u greater than 90 % and maximum γ greater than 14 %. The alluvium under the downstream shell also developed maximum γ greater than 14 %, but the maximum r_u was only greater than 70 % closer to the core and was between 50 and 70 % closer to the toe.

The stress-strain and stress path responses for two points in the alluvium under the downstream shell (i.e., the points labeled zone 2 and zone 3 on Fig. 19), are shown in Fig. 20. The stress paths show that both zones reached limiting pore pressure conditions under the static shear stresses imposed on them, with the stresses for both zones stabilizing near their failure envelopes. The static shear stress ratio in zone 2 is not as large as for zone 3, and as a result zone 2 developed a higher r_u (about 80 %) compared to zone 3 (about 50 %). Both zones accumulated similar maximum γ of about 18 %. These element responses are qualitatively similar to those shown previously in Fig. 12 for single-element simulations.

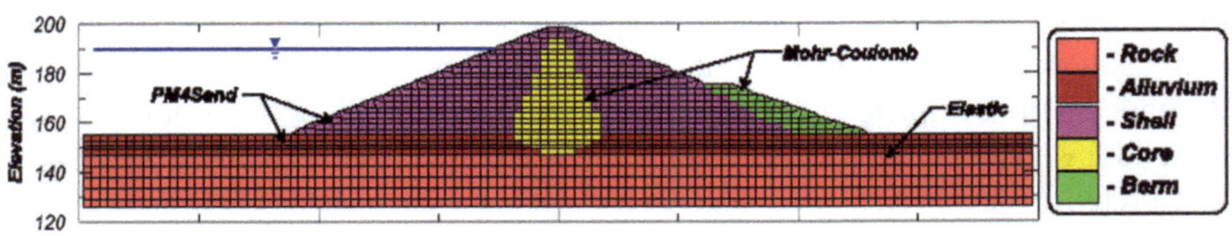

Fig. 17 Cross section of a hypothetical dam

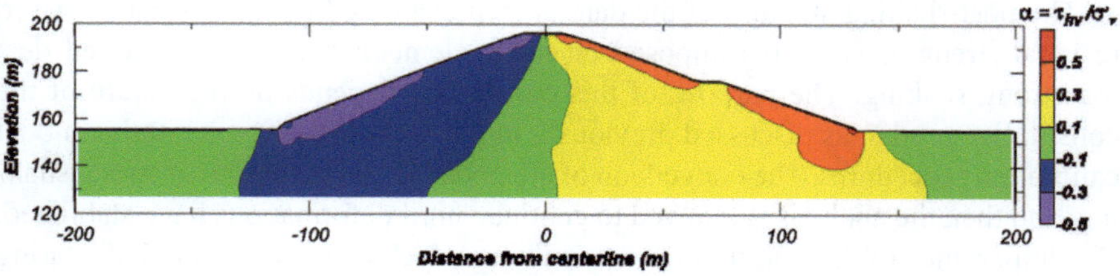

Fig. 18 Horizontal static shear stress ratios under steady seepage conditions

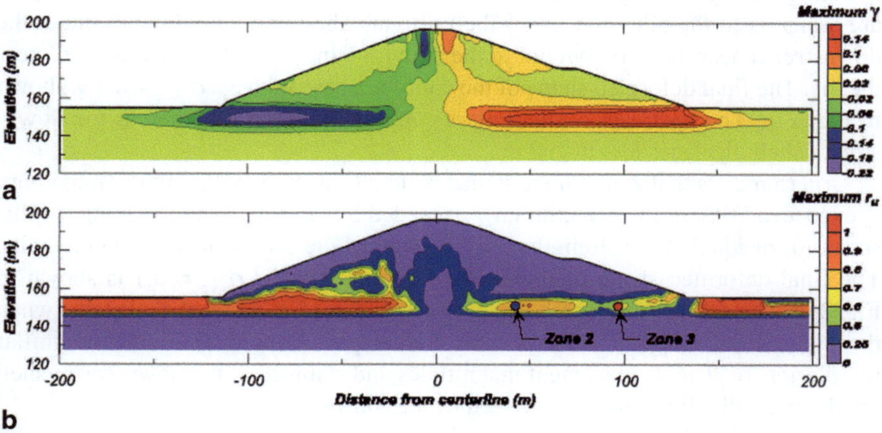

Fig. 19 Deformed shape at the end of strong shaking showing **a** contours of maximum γ, and **b** contours of maximum r_u

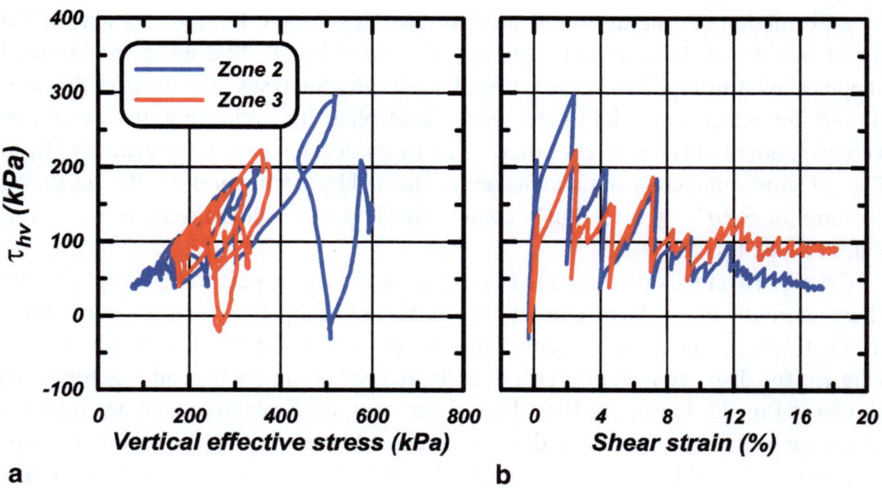

Fig. 20 Responses for two elements (zones) in the alluvium under the downstream shell **a** horizontal shear stress versus vertical effective stress, and **b** horizontal shear stress versus shear strain

The post-shaking stability of the dam is then examined with case-history based residual strengths (Fig. 16) imposed on those elements/zones that liquefied during strong shaking. The purpose of this conversion depends on the nature of the constitutive model, as discussed previously in the section about constitutive model calibration procedures. The conversion of elements occurs at the end of strong shaking, and then the analysis is allowed to continue until deformations have stabilized. The importance of the criterion for determining which elements "liquefied" during strong shaking is illustrated for two scenarios.

Criterion 1a is that zones are assigned residual shear strengths if their maximum r_u during strong shaking exceeded 70%. In this scenario, residual shear strengths are assigned to the alluvium under the upstream shell and the alluvium under the downstream near the core, but not to the alluvium under the downstream shell near the toe. The final deformed shape of the dam for a case with $S_r/\sigma'_{vc} = 0.1$ is shown in Fig. 21a. The upstream shell displaces approximately 5 m, whereas the downstream shell displaces less than 2 m.

Criterion 2a is that zones are assigned residual shear strengths if their maximum r_u exceeded 70% or their maximum γ exceeded 3% during strong shaking. In this scenario, residual shear strengths are assigned to the alluvium under both shells. The final deformed shape of the dam for a case with $S_r/\sigma'_{vc} = 0.1$ is shown in Fig. 21b. The downstream shell displaces more than 15 m and is still unstable when the analysis was stopped. The upstream shell displaces about 5 m, which is similar to criterion 1a. The displacement magnitudes and stability of the downstream shell are dramatically different from those for criterion 1a.

The sensitivity of the dam crest settlements to the residual shear strength ratio for these two criteria is shown in Fig. 22. Crest settlements are greater for criterion 2a than for criterion 1a at the same strength ratio because more of the alluvium is assigned residual shear strength for criterion 2a. Crest settlements are most affected by deformations of the upstream shell for criterion 1a and by deformations of the downstream shell for criterion 2a, as was illustrated by the differences in deformed shapes shown in Fig. 21. These types of sensitivity analyses also identify the conditions for which crest deformations are controlled by cyclic mobility behaviors versus residual shear strength values. For example, the results for criterion 1a in Fig. 22 shows that crest settlements are controlled by cyclic mobility during strong shaking for $S_r/\sigma'_{vc} \geq 0.15$, but become controlled by residual strengths for S_r/σ'_{vc} values much less than 0.15.

Criteria 1b and 2b are the same as criteria 1a and 2a, respectively, except that the shear strengths of the "nonliquefied" zones are now capped by their pre-earthquake drained values. The displacements for criterion 1b are slightly greater than for 1a, whereas the displacements for criterion 2b are approximately the same as for 2a, as shown in Fig. 22. In criteria 1b and 2b, shear strengths associated with negative excess pore pressures are not relied on, as discussed previously in the section on constitutive model calibration. These results illustrate that dilation of the alluvium near the downstream toe had been developing negative pore water pressures for the case based on criterion 1a, and this was contributing to stability of the downstream slope. Limiting the soil shear strength to be no greater than its pre-earthquake drained

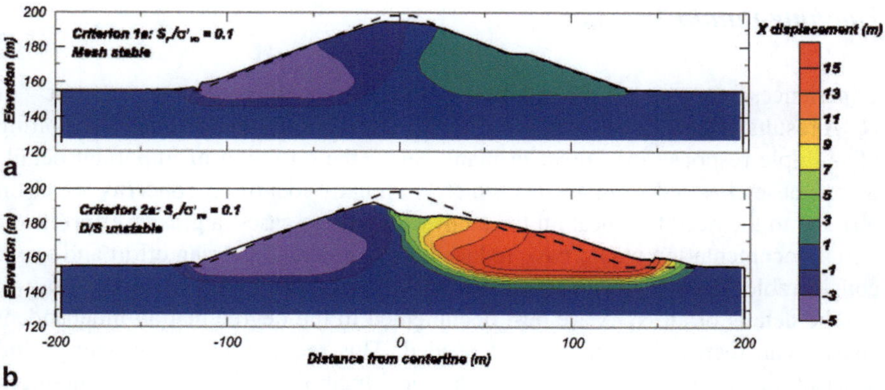

Fig. 21 Post-shaking deformations for residual shear strengths assigned using two different criterion: **a** Criterion 1a is $r_u > 70\,\%$, and **b** Criterion 2a is $r_u > 70\,\%$ or $\gamma > 3\,\%$

strength had relatively small effects on the crest settlement for this example, but it can be significant in other situations or for other response measures.

The results in Fig. 22 further point out the need for: (1) clear documentation on the NDA procedures so that details, such as the difference between criterion 1 and criterion 2, are fully appreciated, and (2) further research on residual shear strengths and the NDA procedures which are used to model these strengths. The uncertainties associated with estimating residual shear strengths and the sensitivity of dam deformations to their values means that this is an area of fundamental and practical research with potentially high impact on dam safety evaluations.

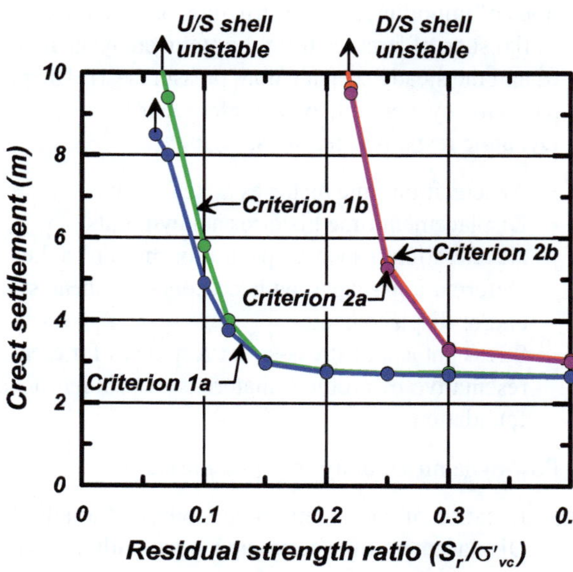

Fig. 22 Dam crest settlement versus residual shear strength ratio for the alluvium based on two criteria for triggering the assignment of residual strengths: Criterion 1 is $r_u > 70\,\%$, and Criterion 2 is $r_u > 70\,\%$ or $\gamma > 3\,\%$. Criteria 1b and 2b include limits that undrained strengths not exceed drained strengths in non-triggered zones

Documentation Practices

Experiences on numerous projects have shown that the interpretation and review of NDA results often requires access to the full details of the analysis and the plotting of multiple response measures. In many cases, the omission of important details or responses has led to questions which eventually identified concerns which in turn led to the need to repeat all the analyses. In these cases, a greater effort in the initial documentation would have reduced the overall engineering effort and saved considerable time.

The details of an NDA are best documented to the degree that an independent analyst can recreate the analyses if needed. This requires complete listings and descriptions of: material properties; numerical methods and parameters; the mesh and any re-meshing details; the method for initializing static stress conditions; the method for initializing pore water pressure conditions; all boundary conditions for static and dynamic phases including how the input ground motions are applied; any special procedures or steps for imposing residual strengths or restricting undrained strengths to drained values; any deviations between the processes used in the analyses versus those used in the supporting validation studies; and any other analysis-specific details of potential importance to the results. Inclusion of the input files for one baseline analysis case can be particularly helpful in enabling reviewers to check these and other details.

The presentation of results should include most all measures of response for at least one baseline analysis case to facilitate the review and interpretation process. It is usually beneficial to include the following items for at least one case.

- Single-element simulations of the constitutive model's responses using the adopted material parameters. These simulations should include the loading conditions of importance to the structure, as discussed previously.
- Initial state of stresses from the static analysis, including contours of σ'_v, K_o, τ_{hv}, α, and the steady seepage pore pressures (u). The pore pressures should be compared to any available piezometer records.
- Dynamic response measures:
  - Acceleration time histories at key points
  - Displacement time histories at key points
  - Acceleration response spectra for motions at key points
  - Deformation shapes with contours of shear strains, excess pore pressures, and/or displacements
  - Stress paths and stress-strain responses for elements or zones which are representative of problem materials or problem locations within the dam or its foundation
- Post-dynamic stability response measures:
  - Location of zones determined to have "liquefied" during earthquake
  - Displacement time series appended with the dynamic responses
  - Deformation shapes with contours of shear strains, excess pore pressures, and/or displacements

Each of the above measures of response can be useful for identifying simple input errors or clarifying certain features of behavior. In some cases, the responses may be unremarkable, but they remain valuable for having documented that nothing has been overlooked. Other measures of response or greater levels of detail may be needed in other situations, depending on the modeling procedure, estimated responses, and details of the project. For embankments that have experienced previous earthquakes, comparisons of computed and observed responses for those events can sometimes provide valuable insights and bounds on the input parameters and modeling procedures.

The insights from NDAs are generally only realized when there are sufficient parametric analyses to identify the controlling behaviors and primary sources of uncertainty. Parametric studies may be used, for example, to evaluate the sensitivity of deformations to the cyclic mobility behavior (CRR and strain accumulation), the residual shear strengths, the shear stiffness distribution, the permeabilities and permeability contrasts, ground motions including the effect of polarity and vertical motions, reservoir levels, properties of nonliquefiable materials, numerical parameters (e.g., time step size, Rayleigh damping), and other parameters which are important for any specific project. The formal presentation of the parametric analyses depends on the specifics of the project, but in general the results do not need to be presented in the same detail as the baseline case described above. Instead, the results can be summarized in the form that best illustrates how specific parameters are affecting the important response measures.

Concluding Remarks

Nonlinear deformation analyses (NDAs) are a valuable tool for evaluating the potential effects of liquefaction on the seismic performance of embankment dams, including the examination of potential failure modes and the development of fragility curves for risk analyses. The required complexity of an NDA model may depend on the nature of the specific problem, but the continued development of more realistic models that are applicable to a broad range of conditions, are easier to calibrate and use, and are well validated is a long-term need for advancing and improving the use of NDAs in practice.

Calibration and validation of constitutive models through single-element simulations for the range of loading paths of potential importance to a structure are essential for: (1) understanding and documenting the strengths and weaknesses of the constitutive models, (2) selecting models appropriate for specific evaluations, and (3) understanding likely causes of differences between different modeling approaches. Recommendations were presented on the types of single-element simulations which should be required for validation of constitutive models used for liquefiable soils on embankment dam projects.

Increasing confidence in the use of NDAs requires reducing unnecessary variability between results obtained by different users with the same or different models. Achieving this goal will require improvements in the NDA models, improvements in the calibration and modeling procedures, and establishment of systematic validation protocols which can better quantify potential biases and dispersion. An example NDA of an embankment dam affected by liquefaction illustrated: (1) how seemingly small details can significantly affect potential deformations, and (2) how research on residual shear strengths of liquefied soils remains an important need for embankment dams.

Quality documentation of an NDA study is an important part of facilitating improvements in practice and increasing confidence in NDA findings. The documentation of the procedures and parameters needs to be sufficiently detailed so that the analyses can be recreated by others. Computed responses for select cases and results of parametric analyses need to be presented in enough detail to facilitate their review and to identify the controlling parameters and primary sources of uncertainty. Quality documentation improves the value of the NDA to the project and ensures that the findings will be useful to owners, regulators, and other consultants, both now and in the future. Recommendations were presented on the types of details which should be included in the documentation of NDAs for dams affected by liquefaction.

Critical examination of NDA models and modeling procedures is necessary for identifying weaknesses, fostering improvements, and increasing confidence in their use for evaluating the seismic performance of embankment dams affected by liquefaction. Similar recommendations and conclusions can be drawn for cases involving other problem soils (e.g., soft and sensitive clays or plastic silts), despite differences in the specific details of the NDA models and modeling procedures. It is hoped that the discussions presented in this paper will contribute to furthering such examinations.

Acknowledgments Portions of the work presented herein were prepared in cooperation with the California Department of Water Resources for research funded under contract 4600009523. The authors appreciate the comments of Drs. Richard Armstrong, Mike Beaty, and Erik Malvick. The simulations for evaluating K_α behaviors for a number of constitutive models were performed by Monika Maharjan. The authors greatly appreciate all of the above support and assistance.

References

Anderson L, Ruthford M, Perlea V, Serafini D, Montgomery J, Bowles D (2011) Risk assessment of success dam, California: flood related potential failure modes. GeoRisk 2011, ASCE, pp 902–909

Andrianopoulos KI, Papadimitriou AG, Bouckovalas GD (2010) Bounding surface plasticity model for the seismic liquefaction analysis of geostructures. Soil Dyn Earthquake Eng 30:895–911

Beaty MH, Byrne PM (2011) UBCSAND constitutive model: version 904aR. Documentation report: UBCSAND constitutive model on Itasca UDM Web Site, February 2011

Boulanger RW (2003) Relating K_α to relative state parameter index. J Geotech Geoenviron Eng ASCE 129(8):770–773

Boulanger RW, Idriss IM (2004) State normalization of penetration resistances and the effect of overburden stress on liquefaction resistance. In: D Doolin et al (ed) Proceeding 11th international conference on soil dynamics and earthquake engineering, and 3rd international conference on earthquake geotechnical engineering. Stallion Press, Vol 2, pp 484–491

Boulanger RW, Idriss IM (2011) Cyclic failure and liquefaction: current issues. 5th international conference on earthquake geotechnical engineering, Santiago, Jan 10–13

Boulanger RW, Ziotopoulou K (2012) PM4Sand (Version 2): a sand plasticity model for earthquake engineering applications. Report No. UCD/CGM-12/01, Center for Geotechnical Modeling, Department of Civil and Environmental Engineering. University of California, Davis, p 100

Boulanger RW, Ziotopoulou K (2013) Formulation of a sand plasticity plane-strain model for earthquake engineering applications. J Soil Dyn Earthq Eng Elsevier 53:254–267

Boulanger RW, Seed RB, Chan CK, Seed HB, Sousa JB (1991) Liquefaction behavior of saturated sands under uni-directional and bi-directional monotonic and cyclic simple shear loading. Geotechnical engineering report No. UCB/GT/91-08, University of California, Berkeley, p 521, August

Boulanger RW, Wilson DW, Idriss IM (2012) Examination and re-evaluation of SPT-based liquefaction triggering case histories. J Geotech Geoenviron Eng ASCE 138(8):898–909

Bowles D, Ruthford M, Anderson L (2011) Risk assessment of success dam, California: evaluation of operating restrictions as an interim measure to mitigate earthquake risk. Geo-Risk 2011 ASCE, pp 586–599

Byrne P, Park S, Beaty M, Sharp M, Gonzalez L, Abdoun T (2004) Numerical modeling of liquefaction and comparison with centrifuge tests. Can Geotech J 41(2):193–211

Cubrinovski M (2011) Seismic effective stress analysis: modelling and application. 5th international conference on earthquake geotechnical engineering, Santiago, Jan 10–13

Cubrinovski M, Ishihara K (1998) State concept and modified elastoplasticity for sand modelling. Soils Found 38(4):213–225

Dafalias YF, Manzari M (2004) Simple plasticity sand model accounting for fabric change effects. J Eng Mech ASCE 130(6):622–634

Dawson EM, Roth WH, Nesarajah S, Bureau G, Davis CA (2001) A practice oriented pore pressure generation model. 2nd FLAC symposium on numerical modeling in geomechanics, Lyon

Finn WDL, Wu G (2013) Dynamic analyses of an earthfill dam on over-consolidated silt with cyclic strain softening. Seventh international conference on case histories in geotechnical earthquake engineering, Chicago, April 29-May 4

Halpin E, Harkness A (2008) USACE dam safety program-transition to risk management. The sustainability of experience-investing in the human factor. 28th Annual USSD Conference, Portland, April 28-May 2, pp 251–267

Harder LF, Boulanger RW (1997) Application of K_α and K_α correction factors. In: Youd TL, Idriss IM (eds) Proceedings of the NCEER workshop on evaluation of liquefaction resistance of soils. Technical Report NCEER-97-0022, National Center for Earthquake Engineering Research, SUNY, pp 167–190

Idriss IM, Boulanger RW (2004) Semi-empirical procedures for evaluating liquefaction potential during earthquakes. In: Doolin D (ed) Proceeding 11th international conference on soil dynamics and earthquake engineering, and 3rd international conference on earthquake geotechnical engineering. Stallion, Vol 1, pp 32–56.

Idriss IM, Boulanger RW (2007) SPT-and CPT-based relationships for the residual shear strength of liquefied soils. In: Pitilakis KD (ed) Earthquake geotechnical engineering, 4th international conference on earthquake geotechnical engineering-invited lectures. Springer, The Netherlands, pp 1–22

Idriss IM, Boulanger RW (2008) Soil liquefaction during earthquakes. Monograph MNO-12. Earthquake Engineering Research Institute, Oakland, pp 261

Ishihara K (1996) Soil behaviour in earthquake geotechnics. Oxford University Inc., New York

Kamai R, Boulanger RW (2012). Single-element simulations of partial-drainage effects under monotonic and cyclic loading. J Soil Dyn Earthq Eng 35:29–40

Khosravifar A, Boulanger RW, Kunnath SK (2013) Effects of liquefaction on inelastic demands on extended pile shafts. Earthq Spectra EERI, on-line ahead of print

Kokusho T (1999) Water film in liquefied sand and its effect on lateral spread. J Geotech Eng ASCE 125(10):817–826
Kokusho T (2003) Current state of research on flow failure considering void redistribution in liquefied deposits. Soil Dyn Earthq Eng 23(7):585–603
Kulasingam R, Malvick EJ, Boulanger RW, Kutter BL (2004) Strength loss and localization at silt interlayers in slopes of liquefied sand. J Geotech Geoenviron Eng ASCE 130(11):1192–1202
Malvick EJ, Kutter BL, Boulanger RW (2008) Postshaking shear strain localization in a centrifuge model of a saturated sand slope. J Geotech Geoenviron Engi ASCE 134(2):164–174
Montgomery J, Boulanger RW, Harder LF Jr (2012) Examination of the Kα overburden correction factor on liquefaction resistance. Report No. UCD/CGM-12-02, Center for Geotechnical Modeling, Department of Civil and Environmental Engineering, University of California, Davis, p 44
Munger DF, Bowles DS, Boyer DD, Davis DW, Margo DA, Moser DA, Regan PJ, Snorteland N (2009) Interim tolerable risk guidelines for US Army corps of engineers dams. Managing our water retention systems. 29th Annual USSD conference, Nashville, April 20–24, pp 1125–1141
Naesgaard E, Byrne PM (2007) Flow liquefaction simulation using a combined effective stress-total stress model. Proceedings Ottawa 2007, Canadian Geotechnical Society, paper 577.
Naesgaard E, Byrne PM, Seid-Karbasi M, Park SS (2005) Modeling flow liquefaction, its mitigation, and comparison with centrifuge tests. Proceedings, performance based design in earthquake geotechnical engineering: concepts and research, geotechnical earthquake engineering satellite conference, Osaka, September 10, pp 95–102
Olson SM, Stark TD (2002) Liquefied strength ratio from liquefaction flow case histories. Can Geotech J 39:629–647
Perlea VG, Beaty MH (2010) Corps of Engineers practice in the evaluation of seismic deformation of embankment dams. Fifth international conference on recent advances in geotechnical earthquake engineering and soil dynamics, San Diego, paper SPL–6
Ruthford M, Perlea V, Serafini D, Beaty M, Anderson L, Bowles D. (2011) Risk assessment of Success Dam, California: earthquake induced potential failure modes. Geo-Risk 2011 ASCE, pp 923–930
Seed HB (1983) Earthquake resistant design of earth dams. Proceeding symposium on seismic design of embankments and caverns, ASCE, pp 41–64
Seed HB (1987) Design problems in soil liquefaction. J Geotech Eng ASCE 113(8):827–845
Seed HB, Lee KL, Idriss IM, Makdisi F (1975) The slides in the San Fernando dams during the earthquake of February 9, 1971. J Geotech Eng ASCE 101(7):651–688
Seed HB, Tokimatsu K, Harder LF Jr, Chung R (1984) The influence of SPT procedures in soil liquefaction resistance evaluations. Earthquake Eng. Research Center Rep. No. UCB/EERC-84/15. University of California, Berkeley
Seid-Karbasi M, Byrne PM (2007) Seismic liquefaction, lateral spreading, and flow slides: a numerical investigation into void redistribution. Can Geotech J 44(7):873–890
Serafini DC, Rose RS (2010) Overview of the Isabella dam potential failure modes workshop. Collaborative management of integrated watersheds. 30th Annual USSD conference, Sacramento. April 12–16, pp 1341–1353
Silver ML, Chan CK, Ladd RS, Lee KL, Tiedemann DA, Townsend FC, Valera JE, Wilson JH (1976) Cyclic triaxial strength of standard test sand. J ASCE 102 GT 5:511–523
Snorteland N, Dinneen E (2008) Using risk to make decisions, prioritize resources, and measure performance for water resources facilities at the U.S. Bureau of Reclamation. The sustainability of experience-investing in the human factor. 28th Annual USSD Conference, Portland, April 28-May 2, pp 229–249
Toki S, Tatsuoka F, Miura S, Yoshimi Y, Yasuda S, Makihara Y (1986) Cyclic undrained triaxial strength of sand by a cooperative test program. Soils Found 26:117–128
USCOLD (1992) Observed performance of dams during earthquakes. US Committee on Large Dams, Denver, July, p 129
USCOLD (2000) Observed performance of dams during earthquakes. US Committee on Large Dams, Vol II. Denver, October, p 162

USACE (2011) Safety of dams-policy and procedures. ER 110-2-1156, US Army Corps of Engineers, Washington, October 28, p 430
USBR (2011) Dam safety risk analysis best practices training manual, U.S. Department of the Interior, Bureau of Reclamation, in cooperation with the US Army Corps of Engineers, Denver, Version 2.2, April
USSD (2013) Observed performance of dams during earthquakes, Volume III. US Society on Dams, Denver, p 130 (in-press)
Vaid YP, Finn WDL (1979) Static shear and liquefaction potential. J Geotech Div ASCE 105(GT10):1233–1246
Vaid YP, Chern JC (1985) Cyclic and monotonic undrained response of saturated sands. Advances in the art of testing soils under cyclic conditions, ASCE, pp 120–147
Verdugo R, Sitar N, Frost DJ, Bray JD, Candia G, Eldridge T, Hashash Y, Olson SM, Urzua A (2012) Seismic performance of earth structures during the February 2010 Maule, Chile, earthquake: dams, levees, tailings dams, and retaining walls. Earthq Spectra EERI 28(S1):S75–S96
Wang ZL, Makdisi FI (1999) Implementing a bounding surface hypoplasticity model for sand into the FLAC program. FLAC and numerical modeling in geomechanics. A.A. Balkema, Netherlands, pp 483–490
Wang ZL, Dafalias YF, Shen CK (1990) Bounding surface hypoplasticity model for sand. J Eng Mech ASCE 116(5):983–1001
Wang Z-L, Makdisi FI, Ma F (2012) Effective stress soil model calibration based on in situ measured soil properties. J Geotech Geoenviron Eng ASCE 138(7):869–875
Yang Z, Elgamal AW (2002) Influence of permeability on liquefaction-induced shear deformation. J Eng Mech 128(7):720–729
Yang Z, Elgamal A, Parra E (2003) Computational model for cyclic mobility and associated shear deformation. J Geotech Geoenviron Eng ASCE 129(12):1119–1127
Yoshimi Y, Tokimatsu K, Kaneko O, Makihara Y (1984) Undrained cyclic shear strength of dense Niigata sand. Soils Found, Jpn Soc Soil Mech Found Eng 24(4):131–145
Youd TL et al (2001) Liquefaction resistance of soils: summary report from the 1996 NCEER and 1998 NCEER/NSF workshops on evaluation of liquefaction resistance of soils. J Geotech Geoenviron Eng 127(10):817–833
Ziotopoulou K, Boulanger RW (2012) Constitutive modeling of duration and overburden effects in liquefaction evaluations. Proceedings 2nd International Conference on Performance-Based Design in Earthquake Geotechnical Engineering, May 28–30, Taormina
Ziotopoulou K, Boulanger RW (2013a) Calibration and implementation of a sand plasticity plane-strain model for earthquake engineering applications. J Soil Dyn Earthq Eng 53:268–280
Ziotopoulou K, Boulanger RW (2013b) Numerical modeling issues in predicting post-liquefaction reconsolidation strains and settlements. Proceeding 10CUEE, March 1–2, Tokyo

Effects of Liquefaction on Ground Surface Motions

Steven L. Kramer, Bita Astaneh Asl, Pelin Ozener and Samuel S. Sideras

Abstract Soil liquefaction has been responsible for extensive damage to buildings, bridges, pipelines and other critical infrastructure in many past earthquakes. The need to predict the potential for initiation of liquefaction and of its various effects has led to greatly improved understanding of the mechanics of the liquefaction process. Most of the efforts made toward evaluating liquefaction hazards have been directed toward various aspects of ground failure, including post-liquefaction settlement, lateral spreading, and the development of flow slides. The occurrence of liquefaction, however, can also affect ground surface motions, and hence the seismic response of structures founded at or near the ground surface. This paper reviews the process of liquefaction and the manner in which its occurrence is typically detected. The hydraulic conditions required for the manifestation of surface effects are reviewed, and limitations in the inference of liquefaction triggering from the presence or absence of such effects are discussed. The effects of liquefaction triggering on soil stiffness, as they pertain to ground response, are described and discussed. Procedures for identification of the timing of liquefaction triggering are reviewed, and illustrated using a procedure based on the Stockwell transform.

The effects of post-triggering response on ground surface motions are interpreted in terms of response spectra. The abilities of four advanced site response programs to compute ground motions in liquefiable soils are examined and evaluated from the standpoints of pre- and post-triggering response. The programs were applied to the Wildlife case history, and to a hypothetical soil profile containing 7 m of liquefiable soil. The programs all predicted consistent response up to the point of triggering,

S. L. Kramer (✉) · B. Astaneh Asl · S. S. Sideras
University of Washington, Seattle, WA, USA
e-mail: kramer@uw.edu

B. Astaneh Asl
e-mail: astaneh@uw.edu

S. S. Sideras
e-mail: sideras@uw.edu

P. Ozener
Yildiz Technical University, Istanbul, Turkey
e-mail: tohomcu@yildiz.edu.tr

A. Ansal, M. Sakr (eds.), *Perspectives on Earthquake Geotechnical Engineering*,
Geotechnical, Geological and Earthquake Engineering 37,
DOI 10.1007/978-3-319-10786-8_11

but their predictions of post-triggering response were highly variable. The implications of post-triggering response for the evaluation of ground surface response spectra are discussed.

Keywords Liquefaction · Response spectrum · Site response · Dilation

Introduction

Soil liquefaction has been known to cause damage to buildings, bridges, dams, highways, pipelines, and other critical elements of infrastructure. Most liquefaction damage is associated with ground failure, i.e., permanent lateral and vertical deformations. However, the occurrence of liquefaction causes substantial changes in the character of a soil deposit, which can have significant effects on the ground motions transmitted to the surface. The nature of ground motions at sites containing potentially liquefiable soils can affect potential damage to structures even if ground failure does not occur. The intent of this paper is to review basic concepts of liquefiable soil behavior, discuss the modeling of that behavior, and illustrate its effects on ground surface motions with an emphasis on response spectral characteristics.

Liquefaction

The behavior of liquefaction-susceptible soils subjected to cyclic loading has been studied for nearly 50 years. Laboratory studies applying constant-amplitude, periodic (generally harmonic) shear stresses have been used by numerous researchers to investigate the rate of pore pressure generation, and its dependence on factors such as soil density, initial effective confining pressure, and shear stress amplitude. Early studies showed that rates of pore pressure generation increased, all other things being equal, with decreasing density and increasing (normalized) shear stress amplitude; the results of laboratory tests are typically reported in terms of curves relating number of cycles to liquefaction to normalized shear stress amplitude, e.g., cyclic stress ratio, CSR (Fig. 1), and curves showing the rate of pore pressure generation up to the point of initial liquefaction (Fig. 2). Later experimental investigations showed that laboratory liquefaction resistance was also influenced by a number of other factors (e.g., soil fabric, prior strain history, etc.) that existed in the field but could not be reliably replicated in the laboratory. As a result, the evaluation of liquefaction resistance moved from a laboratory basis to a basis rooted in the interpretation of field case histories.

Laboratory tests still provide very useful insight into the behavior of liquefiable soils, and are of great value in developing constitutive models that describe their behavior. In the early stages of a typical cyclic loading test, the generation of excess pore pressure causes the stiffness of the soil to decrease (Fig. 3a) and the effective

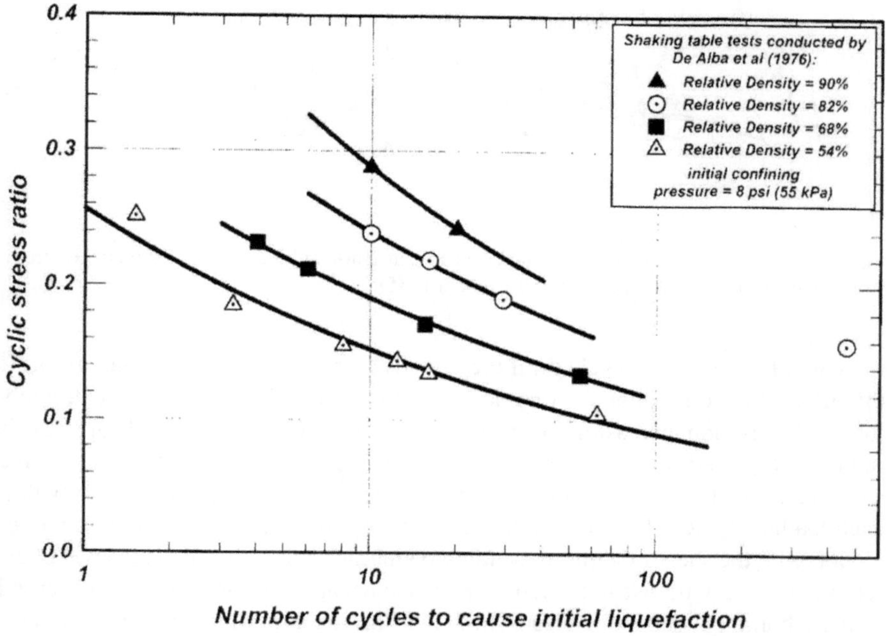

Fig. 1 Liquefaction resistance curves from laboratory tests (after Idriss and Boulanger 2008)

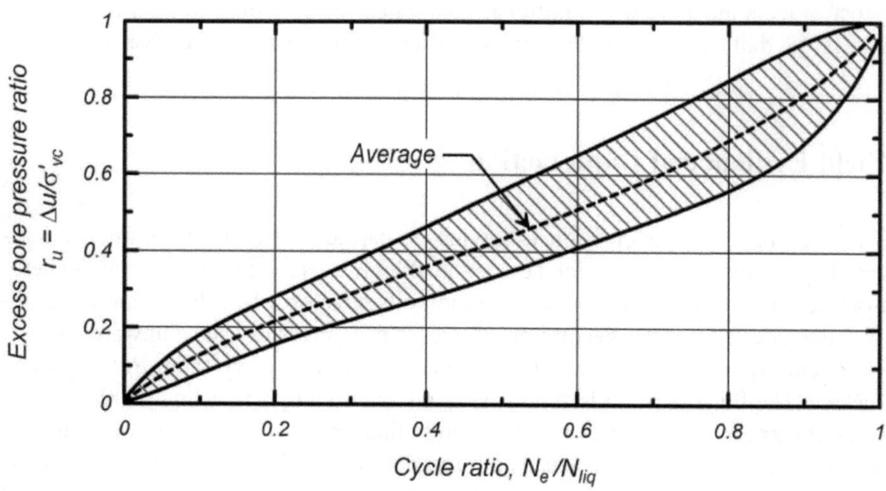

Fig. 2 Rate of pore pressure generation in liquefiable soils (Idriss and Boulanger 2008)

stress path to move to the left (Fig. 3b). The pore pressure initially increases quickly (as in Fig. 2), but its rate of generation then slows until the stress path begins to cross the phase transformation line (PTL) identified by Ishihara (1985). At the PTL, the soil changes from contractive to dilative behavior so that the effective stress increases under increasing loading above the PTL. Upon stress reversal above the PTL, the soil becomes more contractive and the rate of pore pressure generation

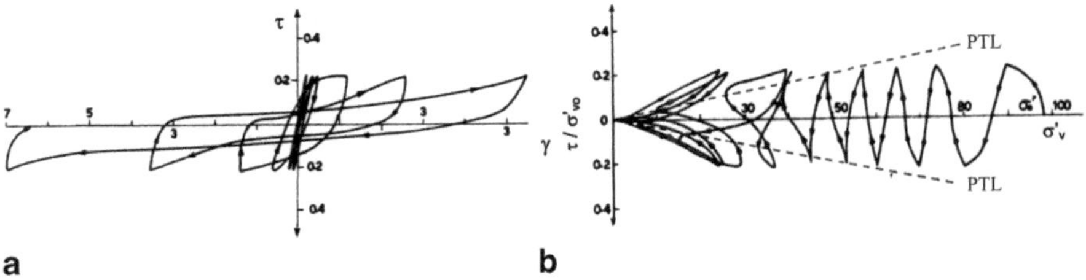

Fig. 3 Behavior of liquefiable soils subjected to constant-amplitude cyclic loading: (**a**) stress-strain, and (**b**) stress path behavior (after Ishihara 1985)

increases from cycle to cycle until the effective stress eventually reaches zero. At this point, the soil exhibits a very low shear modulus but, unless it is extremely loose, dilates upon increasing shear stress and gains stiffness until the shear stress is reduced. Repeated constant amplitude loading cycles results in a stabilized effective stress path in which the soil moves from contractive to dilative states twice within each loading cycle. The stress-strain behavior follows the effective stress path behavior, with the shear modulus depending on the strain level but also the level of effective stress, giving rise to the familiar "banana-shaped" hysteresis loops observed in the laboratory. Laboratory tests show that the shear strain amplitude continues to increase even when the effective stress path has stabilized, indicating that some additional mechanism is affecting the stiffness of the soil under these conditions. Most models attribute this additional component of softening to alteration of the soil fabric that occurs when the soil is at a state of nearly zero effective stress.

Field Evidence of Liquefaction

The occurrence of liquefaction following an earthquake is most commonly indicated by the presence of surficial evidence of high subsurface pore pressure generation, i.e., the presence of sand boils. Sand boils develop when the hydraulic gradients that result from the generation of excess pore pressure cause upward flow of sufficient volume and speed to lift particles of soil and carry them to the ground surface. The liquefiable soils at most sites are overlain by a non-liquefiable crust of some thickness, so the porewater flow must find or create cracks or holes through which it can flow to the surface. The formation of sand boils, then, is largely a hydraulic problem. If the non-liquefied crust is too thick, i.e., if the top of the liquefied layer is too deep, sand boils will not form even if liquefaction does occur. Similarly, if the liquefied layer is too thin, it will not produce enough porewater to cause sand boils to develop. On the other hand, it is possible that a thick, shallow layer of liquefiable soil can develop high pore pressure, say $r_u = 80\,\%$, and produce sand boils even if the high degree of softening and weakening normally associated with initial liquefaction do not occur. Ishihara (1985) illustrated this concept using empirical observations, primarily based on case history data from the 1976 Tangshan and

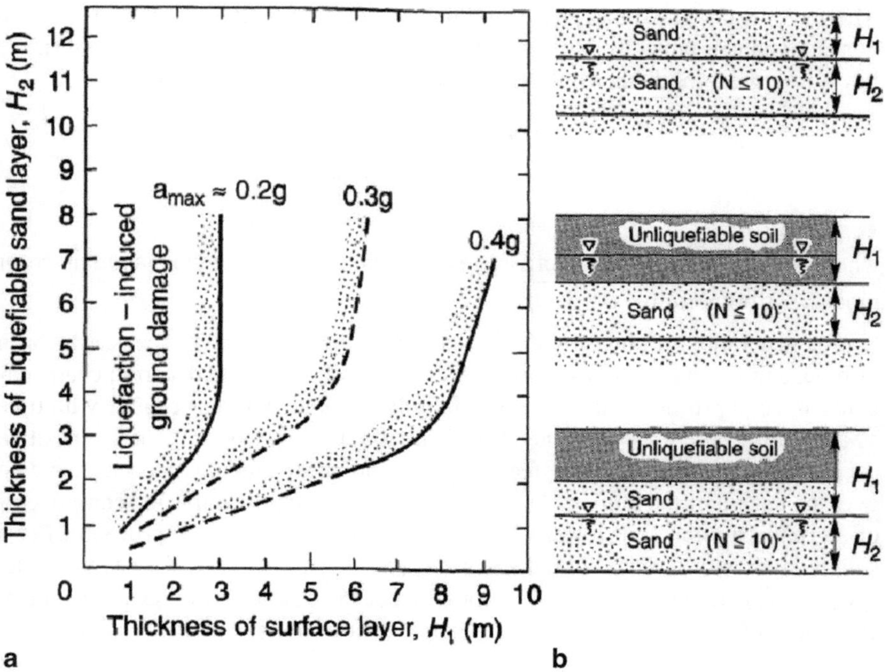

Fig. 4 Relationship between thickness of liquefiable layer and thickness of overlying layer at sites for which surface manifestation of level-ground liquefaction has been observed (after Ishihara 1985)

1983 Nihonkai earthquakes, in terms of the relative thicknesses of liquefiable and overlying non-liquefiable layers, as shown in Fig. 4. Thus, although normally a very good indicator of liquefaction, the presence of sand boils is neither a strictly necessary nor sufficient indicator of the occurrence of liquefaction.

Conceptual Effects of Liquefaction on Ground Motions

The mechanical behavior of coarse-grained soil has long been known to be closely related to effective stresses. Historically, the greatest attention has been paid to the influence of effective stresses on shear strength, but their influence on stiffness is also of critical importance for dynamic response. As pore pressure increases, effective stress decreases. The low-strain shear modulus is typically taken to vary with mean effective stress, σ'_m, as

$$G_{\max} = G_{ref}\left(\frac{\sigma'_m}{p_a}\right)^n \tag{1}$$

where G_{ref} is the maximum shear modulus at a mean effective stress of 1 atm and n is an exponent typically found to be approximately 0.5. Assuming that value, the low-strain shear modulus would vary with pore pressure ratio, r_{u}, as

$$G_{\max} = G_{ref}\sqrt{\frac{\sigma'_{mo}(1-r_u)}{p_a}} \tag{2}$$

where σ'_{mo} = initial mean effective stress, which indicates that the maximum shear modulus is proportional to $\sqrt{1-r_u}$.

The generation of pore pressure leads to a change in effective stress, which, in turn, leads to a change in shear modulus of the soil. Thus, the dynamic characteristics of a soil profile containing potentially liquefiable materials change with time as a result of pore pressure generation. Liquefiable soils tend to generate positive excess pore pressure, so the effective stress and shear modulus tend to decrease over the course of the motion. As a result, the most strongly amplified components of a ground motion shift toward lower frequencies as the motion proceeds. For a simple layer with constant initial shear wave velocity, V_{so}, the initial characteristic site period is given by $T_{\mathrm{so}} = 4\,H/V_{\mathrm{so}}$. From Eq. (2), the characteristic site period under conditions of increasing pore pressure ratio can be expressed as

$$T_s = \frac{4H}{V_s} = \frac{4H}{\sqrt{G/\rho}} = \frac{4H}{\sqrt{G_o(\sigma'_m)^{0.5}/\rho}} = \frac{4H}{\sqrt{G_o(\sigma'_{mo}(1-r_u))^{0.5}/\rho}} \tag{3}$$

Thus, the low-strain characteristic site period will increase in proportion to $(1-r_u)^{-0.25}$ as the pore pressure ratio increases. The variation of both characteristic site period and fundamental frequency with pore pressure ratio is illustrated in Fig. 5. The effects of pore pressure generation on characteristic site period can be seen to be relatively modest at pore pressure ratios less than about 0.9—for this simplified, conceptual case, the characteristic site period has only increased by 50 % at a pore pressure ratio of 0.8, and does not double until the pore pressure ratio reaches a value of about 0.94. These simple calculations indicate that (a) the low-strain stiffness of the soil decreases relatively slowly until pore pressure levels become quite high, and (b) it is possible to generate high pore pressures without seeing the high levels of softening typically associated with the triggering of liquefaction.

The softening of soil elements due to generation of pore pressure, and accompanying period-lengthening of soil profiles, does not continue smoothly until the pore pressure ratio is very close to 1.0. The softening process is interrupted by transient instances of stiffening associated with dilation during the phase transformation process. Depending on the density of the soil, the degree and rate of stiffening may be weak or strong. Because the rate of stiffening increases as the shear stress increases, significant stress pulses at the base of a soil profile become superimposed with time since the stress from the latter part of the pulse propagates at a higher velocity than the stress from the early part of the pulse. The result is a sharp, short-duration pulse that Kutter and Wilson (1999) initially termed "de-liquefaction shock waves," but

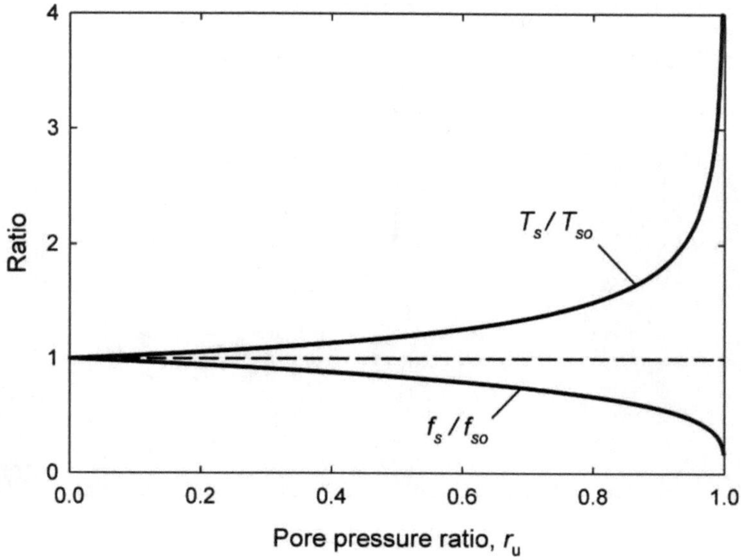

Fig. 5 Variation of ratio of characteristic site period and frequency with changes in low-strain stiffness due to pore pressure generation

which will be described as dilation pulses in this paper. An example of the formation of dilation pulses, taken from the output of a nonlinear, effective stress site response analysis, is shown in Fig. 6. The rate of dilation is a function of the density of the soil and the extent to which the fabric of the soil has been disturbed by large strain amplitudes. If the soil is moderately dense and has not yet reached a large strain amplitude, dilation under a strong stress pulse may lead to large acceleration pulses. In some cases, the peak ground acceleration may be produced by a strong dilation pulse occurring near, or even after, the initiation of liquefaction.

Empirical Effects on Ground Motions

The effects of liquefaction can be manifested in different ways in ground surface motions. The tendency for period lengthening as pore pressure increase can lead to the increasing prevalence of low frequency motions in the latter part of a ground motion, although changes in the incoming wave train from p-waves to s-waves to surface waves also tends to cause some period-lengthening. When liquefaction is triggered, however, there is often an abrupt change in ground motion frequency, as would be suggested by the anticipated rapid drop in natural frequency at very high pore pressure ratios (Fig. 5). In some cases, the change in frequency content is so obvious that it can readily be seen in ground surface accelerograms. Figure 7 shows recorded time histories from the Kawagishi-cho apartment buildings in the 1964 Niigata earthquake (NS component), Kushiro Port in the 1993 Kushiro Oki earthquake

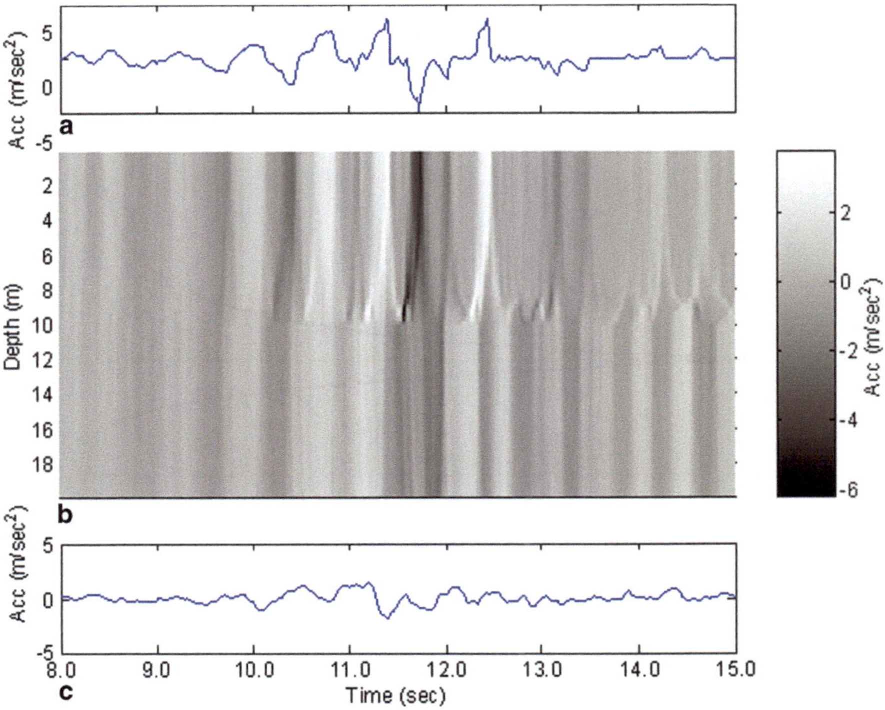

Fig. 6 Computed response of hypothetical, two-layer soil profile to the Yerba Buena input motion: (**a**) ground surface acceleration, (**b**) acceleration vs time and depth, and (**c**) input acceleration time history

(NS component), and the Wildlife liquefaction array in the 1987 Superstition Hills earthquake (NS component). In each case, a relatively sudden change in frequency content can be observed—the changes occur at approximately 7–11, 14–17, and 15–18 s for the Kawagishi-cho, Wildlife, and Kushiro Port records, respectively.

In other cases, changes in frequency content may be difficult to detect at sites where extensive surficial evidence of liquefaction has been observed. This can occur when the liquefiable soil is underlain by a soft non-liquefiable soil so that the ground surface motion is dominated by low frequencies even before liquefaction is initiated. Figure 8 shows recorded time histories from the CCCC station in the February 22, 2011 Christchurch earthquake (EW component), the Harbor Island station in the 2001 Nisqually earthquake (NS component), and IBRH20 station in the 2012 Tohoku earthquake (EW component). In each case, sand boils or other evidence of liquefaction were observed at or very near the location of the recording instrument; the recorded time histories, however, show no apparent evidence of an abrupt change in frequency content. The CCCC and Harbor Island sites show relatively little high-frequency motion virtually from the beginning of the record, which is consistent with the relatively low stiffnesses of their respective soil profiles. The IBRH20 record has a very long duration with high frequencies appearing to extend over the duration of the event.

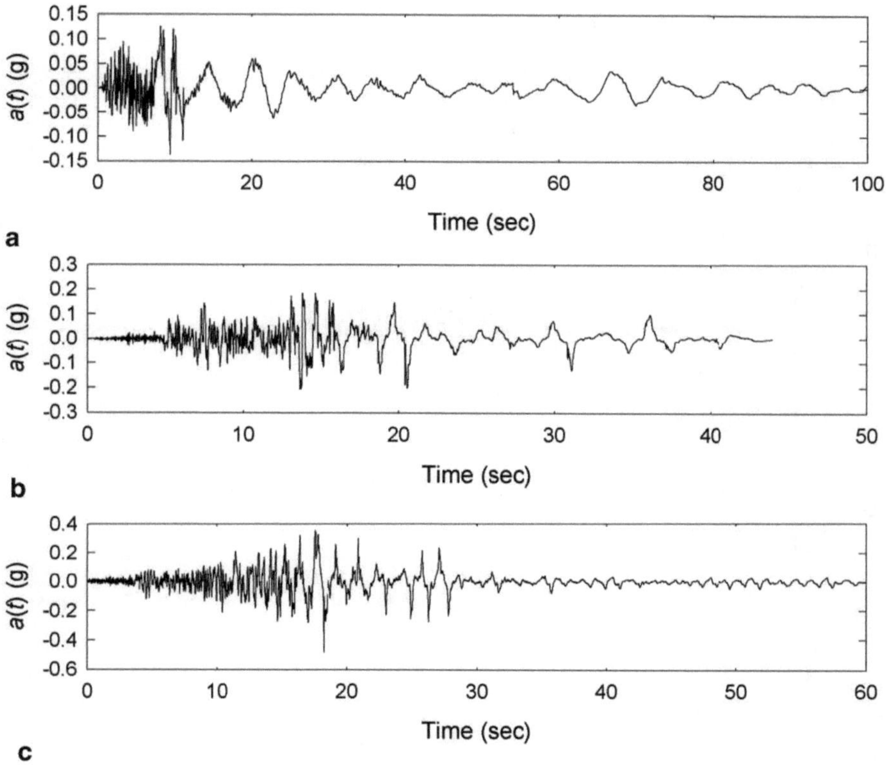

Fig. 7 Acceleration time histories showing clear visual evidence of rapid changes in frequency content of sites exhibiting surficial evidence of liquefaction; (**a**) Kawagishi-cho, (**b**) Wildlife, and (**c**) Kushiro Port

Time of Liquefaction Initiation

The nature of the effects of liquefaction on ground surface motions is influenced by the behavior of a soil deposit before and after the triggering of liquefaction. Because the properties of a liquefiable soil can change so drastically after the triggering of liquefaction, knowledge of the timing of liquefaction can aid in the interpretation of its effects on ground surface motions.

The relatively gradual softening of a liquefiable soil at pore pressure ratios less than about 0.9 should be expected to produce some degree of period-lengthening in a ground surface motion. However, the frequency content of a typical earthquake motion generally changes even without the generation of significant pore pressure. The p-waves that generally arrive first at a site typically have a higher frequency than the s-waves that arrive shortly thereafter. The s-waves are, in turn, often followed by surface waves that may oscillate at lower frequencies than s-waves. Therefore, the effects of gradual softening of a soil deposit on its ground surface motion may be difficult to separate from the effects of changes in frequency content of the base motion that excites the soil deposit.

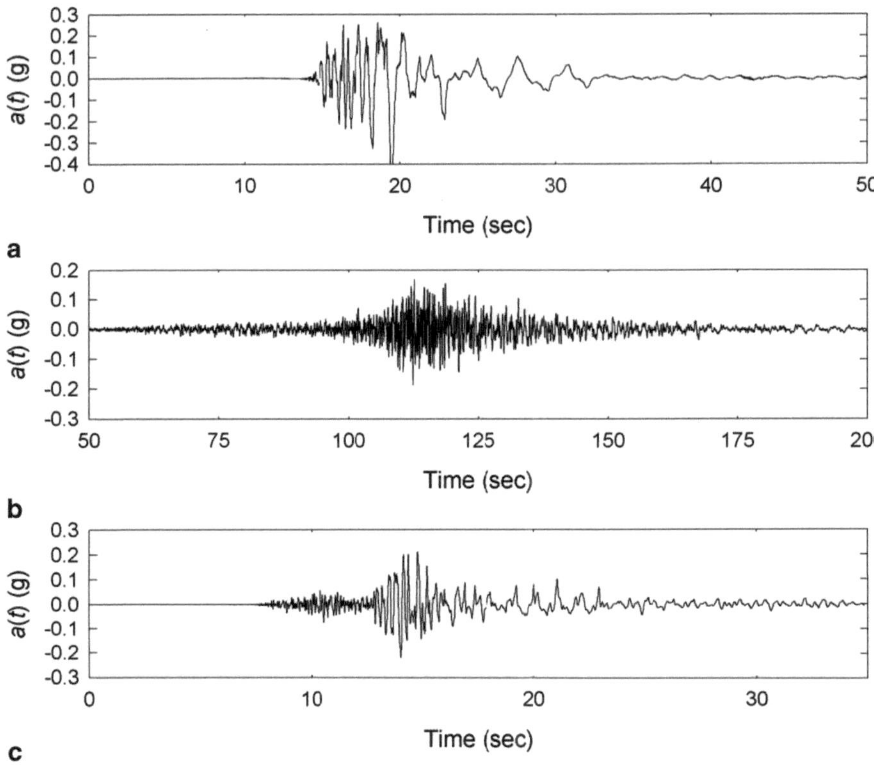

Fig. 8 Acceleration time histories showing little visual evidence of rapid changes in frequency content at sites exhibiting surficial evidence of liquefaction; (**a**) CCCC, (**b**) IBRH20, and (**c**) Harbor Island

A number of formal procedures for detection of liquefaction have been proposed (Kostadinov and Yamazaki 2001; Suzuki et al. 1998; Miyajima et al. 1997; Takada and Ozaki 1997). These procedures, however, do not offer much insight into determination of the time at which liquefaction is initiated. For cases not as obvious as Kawagishi-cho and Kushiro Port, a triggering detection procedure was developed. The triggering procedure is formulated to capture significant changes in the frequency content of an earthquake ground motion considering both horizontal components of the motion. At its current stage of development, it cannot be used as an automated, consistent, stand-alone indicator of initial liquefaction since it cannot distinguish between frequency content changes associated with p-s or s-surface wave conversions. It does, however, provide a useful tool to aid in the interpretation of ground motions. The procedure is based on the hypothesis that higher-frequency components will decrease and lower-frequency components will increase upon the triggering of liquefaction.

Time-frequency representations of the recorded motions are usually created using short-time Fourier transforms (STFTs) and wavelet transforms. The STFT

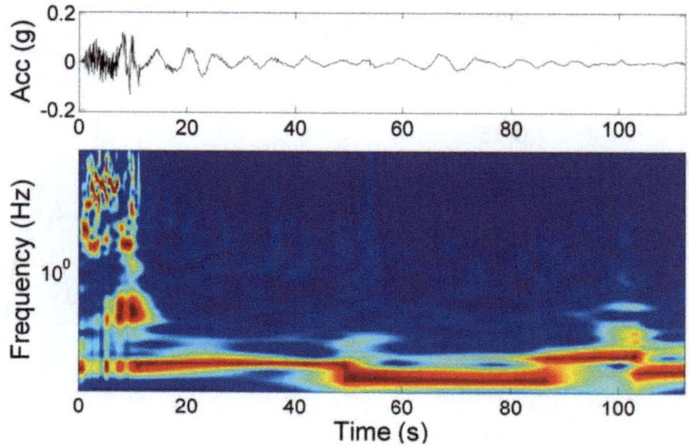

Fig. 9 Normalized Stockwell spectrogram for EW component of Kawagishi-cho record

provides good frequency resolution but poor time resolution, and the wavelet transform provides good time resolution but poor frequency resolution. The Stockwell transform (Stockwell et al. 1996) combines elements of both STFTs and wavelet transforms to provide a balanced resolution of both time and frequency. The triggering detection procedure is based on examination of normalized ground motion amplitudes in frequency bands above and below the fundamental frequency of the site. Figure 9 shows a normalized Stockwell spectrum for the EW component of the Kawagishi-cho record from the 1965 Niigata earthquake. At a time of 10–11 s, which is well beyond the strongest portion of the motion, the normalized amplitudes of the high-frequency components quickly drop and the amplitudes of the low-frequency components increase and persist for the remainder of the motion. Liquefaction is considered to have been triggered at this time. Figure 10 shows values of response spectral ordinates at 10, 11 s, and the end of the complete motion; the response spectra at the time of triggering can be seen to be very nearly equal to the response spectrum obtained from the entire motion, so the effects of post-triggering response on the response spectrum are limited.

Figure 11 shows the normalized Stockwell spectrum for the NS component of the IBR014 motion recorded during the 2011 Tohoku earthquake. This spectrum also shows a rapid reduction in high-frequency amplitudes followed very closely by a persistent increase in low frequency amplitude. In this case, liquefaction is interpreted as being triggered at 105 to 107 s. Note that strong shaking continues for well over a minute after liquefaction has been triggered. Figure 12 shows ground surface response spectra at 105, 107 s, and at the end of the complete motion. The response spectra at the time of triggering can be seen to be significantly lower at both short (<0.1 s) and long (>0.2–0.5 s) periods, but to be identical to the final spectrum at intermediate periods. Figure 13 shows the ratios of final spectral accelerations to spectral accelerations at the time of triggering. The ratio for the case where liquefaction was triggered late in the record (Kawagishi-cho) can be seen to be equal to 1.0 for periods below about 1.2 s and then to be somewhat greater than 1.0 at longer periods. The ratio for the IBR014 record, where liquefaction occurred relatively early in the event, is significantly greater than 1.0 at periods below 0.1 s and above

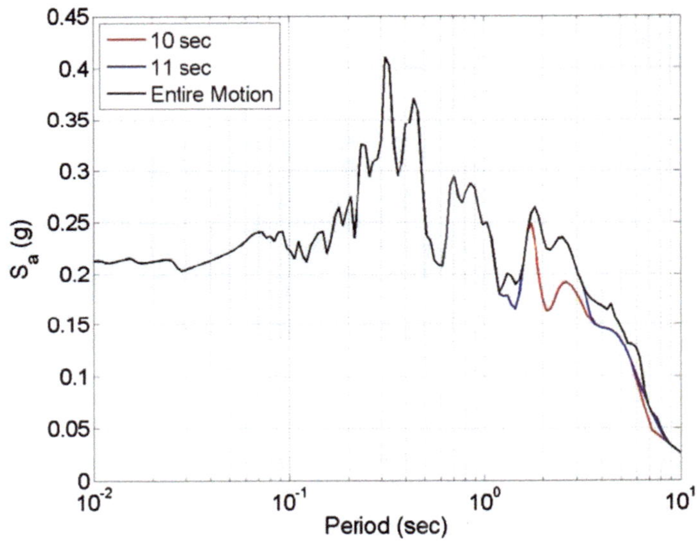

Fig. 10 Computed response spectra at times of 10, 11 s, and at the end of the EW component of the Kawagishi-cho record

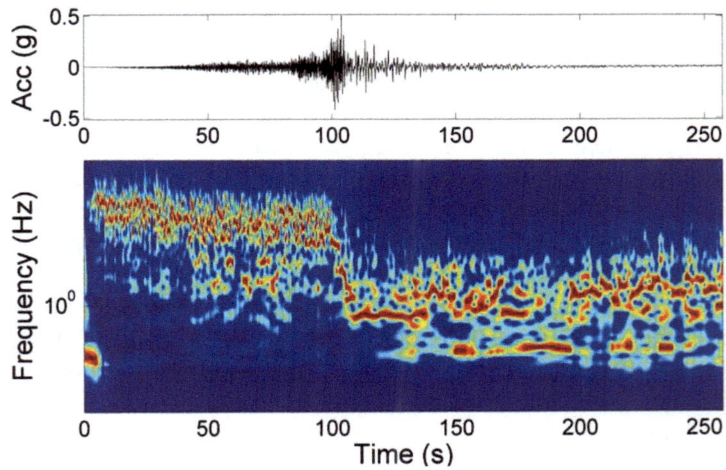

Fig. 11 Normalized Stockwell spectrogram for __ component of IBR014 record

0.5 s. Because significant energy is delivered to the IBR014 site after triggering of liquefaction, the low-period portion of the final IBR014 spectrum is controlled by the dilation pulses that occur after triggering. Similar observations have been made with other recorded motions.

The triggering detection procedure can also be used to identify motions that do not appear to be influenced by liquefaction. Kik-Net Station IBRH20, located in the town

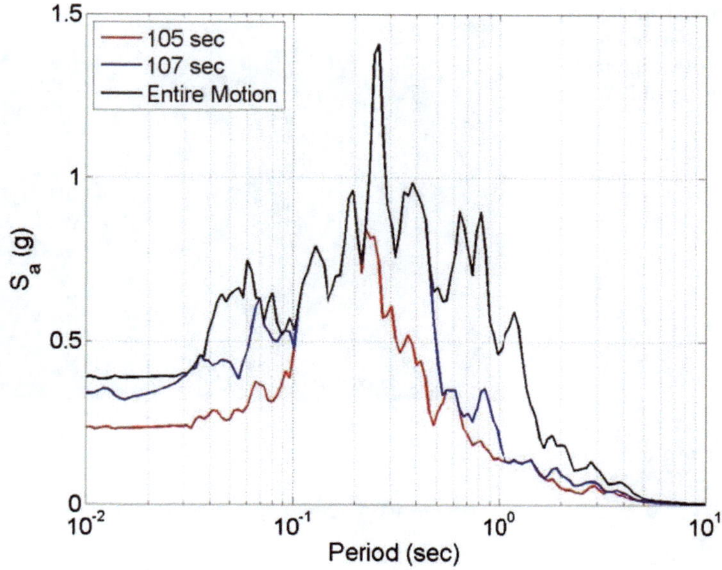

Fig. 12 Computed response spectra at times of 105, 107 s, and at the end of the NS component of the IBR014 record

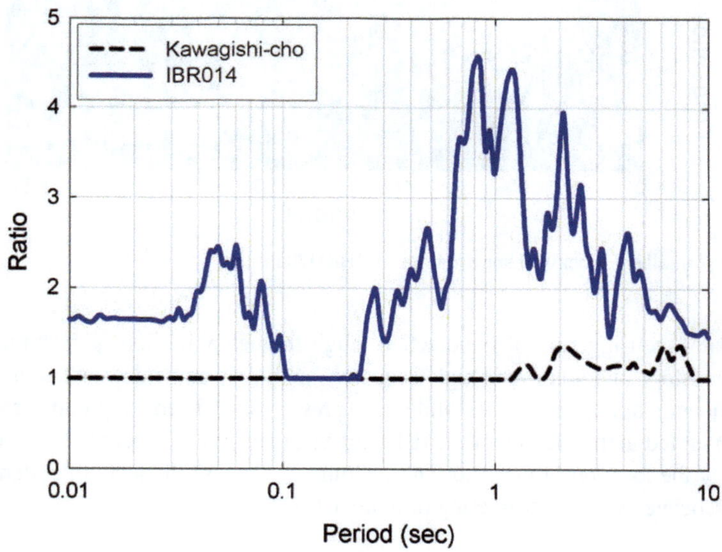

Fig. 13 Variation of response spectral ratio for Kawagishi-cho and IBR014 records

of Choshi near the mouth of the Tone River, is located in a field where numerous sand boils were observed (Fig. 14) following the 2011 Tohoku earthquake. The recorded motion was analyzed with the triggering detection procedure. As illustrated in Fig. 15, the motion does not show the type of frequency content shift that would be expected to

Fig. 14 Sand boils observed near IBRH20 recording instruments (background). Photo courtesy of R.W. Boulanger

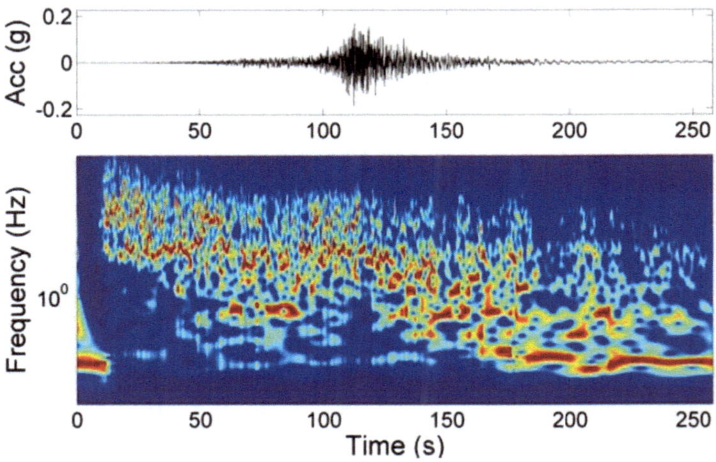

Fig. 15 Normalized Stockwell spectrogram for IBRH20 motion

accompany the triggering of liquefaction. High frequency motions persist throughout the time history, and no drop in high frequency amplitude coincides with an increase in low frequency amplitude. Thus, while pore pressures sufficient to produce sand boils were generated in the vicinity of the IBRH20 site, they do not appear to have been high enough at the location of the recording instrument to cause the level of softening normally associated with the triggering of liquefaction.

Numerical Analysis

A systematic evaluation of the effects of liquefaction on ground surface motions cannot be made solely on the basis of recorded motions because there are not enough recordings of motions affected by liquefaction to do so. However, numerical analyses have developed to the point where they can model the response of potentially liquefiable soil profiles up to and now after the triggering of liquefaction.

Numerical prediction of the response of liquefiable soil deposits requires a nonlinear, effective stress analysis with a relatively sophisticated constitutive model. The constitutive model for liquefiable soils must model the nonlinear, inelastic behavior of the soil skeleton in shear as would be required for a total stress analysis of the soil. However, the behavior of liquefiable soils is dominated

by pore pressure generation so the constitutive model must also model the volumetric behavior under cyclic loading conditions. The former of these requirements is generally relatively straightforward—many total stress nonlinear models are available and generally capable of representing the total stress behavior of sandy soils for nonlinear site response analysis. Perhaps the most difficult issue in modeling total stress behavior is the simultaneous representation of stiffness degradation and energy dissipation. Many total stress models rely on a backbone curve and unloading-reloading “rules” such as the Masing rules (Masing 1926) or Cundall-Pyke hypothesis (Pyke 1979). Such rules relate the shape of an unloading curve to the backbone curve; as a result, the shape of a hysteresis loop, which controls the damping ratio, is uniquely related to the rate of stiffness degradation. In most cases, models calibrated to match a given rate of stiffness degradation (e.g., a particular modulus reduction curve) will overpredict the expected damping (i.e., the corresponding damping curve). Different approaches to the solution of this problem have been proposed (Muravskii 2004; Phillips and Hashash 2009; Arefi et al. 2012), but have not yet been implemented in widely used nonlinear site response codes.

Early nonlinear, effective stress models (e.g., Martin et al. 1975) sought to predict the cycle-by-cycle generation of excess porewater pressure up to the point of liquefaction and paid relatively little attention to the behavior of the soil after the initiation of liquefaction (other than to note that the soil became very soft and weak). The rate of pore pressure generation up to the point of initiation was characterized (Fig. 2) and used as the basis for the cycle-by-cycle methods; these models did not attempt to replicate the pore pressure (hence, normal stress and stiffness) changes that would occur within individual loading cycles. Such models account for the generation of porewater pressure by softening and weakening the soil as a function of pore pressure ratio. Since the pore pressure increases monotonically, the stiffness decreases monotonically and can reach very low values at high pore pressure ratios. When such high levels of pore pressure develop, the soil can become so weak that its ability to transmit shear stress is limited—beyond that point, the liquefied layers act to isolate the overlying soil from significant shaking.

In recent years, more advanced constitutive models (e.g., Iai et al. 1992; Manzari and Dafalias 1997; Yang et al. 2003; Iai et al. 2011; Boulanger and Ziotopoulou 2012) have shown the ability to predict pore pressure fluctuations observed both within and between individual cycles prior to, during, and after the initiation of liquefaction in constant-amplitude, harmonic loading laboratory tests. These models, when implemented into nonlinear site response codes, offer the potential to represent the known behavior of liquefiable soils up to and after the triggering of liquefaction more accurately than the simpler, cycle-by-cycle models. They can also reproduce important features of liquefiable soil profile response that have only been recognized relatively recently.

Nonlinear Codes

The numerical modeling of nonlinear, effective stress-based site response with consideration of phase transformation behavior is relatively new, and not many computer programs capable of performing such analyses are currently available. In order to provide a representative indication of liquefiable soil behavior and to obtain an indication of the variability in predictions of such behavior, four different computer programs were used to investigate site response of liquefiable soil profiles—FLAC, FLIP, OpenSees, and PSNL.

FLAC is a large, commercial, multi-purpose finite difference program with dynamic analysis capabilities. In this study, FLAC is being used with the PM4sand constitutive model of Boulanger and Ziotopoulou (2012). FLIP is a large, multi-purpose finite element program developed by Kyoto University, the Port and Harbour Research Institute, and the Coastal Development Institute of Technology in Japan. FLIP models liquefiable soil using the cocktail glass model of Iai et al. (2011). OpenSees is a large, open-source finite element program developed by the Pacific Earthquake Engineering Research (PEER) Center. Analyses were performed using the Pressure-Dependent Multi-Yield (PDMY) model of Elgamal et al. (2003) to represent potentially liquefiable soils. Finally, PSNL is a one-dimensional, nonlinear, effective stress program written by the first author. All of these programs model the nonlinear, inelastic behavior of soils and can represent phase transformation behavior of potentially liquefiable soils.

Application to Wildlife Array

Vertical arrays have been installed in potentially liquefiable soil profiles at a number of locations around the world. To date, relatively few have been subjected to earthquake shaking sufficiently strong to trigger liquefaction. The most well-known and widely studied of these arrays is the previously described Wildlife Liquefaction Array (WLA). The Wildlife site consisted of 2.5 m of lean clay/silt underlain by 1 m of sandy silt above 3.3 m of loose, silty sand. The silty sand was, in turn, underlain by highly plastic clay within which a downhole instrument was placed at a depth of 7.5 m. A series of piezometers was installed in the silty sand. The Wildlife array was subjected to strong shaking in 1987 from the Elmore Ranch and Superstition Hills earthquakes. The Elmore Ranch earthquake of November 23, 1987 was a M6.2 event epicentered 23 km west of the WLA and did not produce surficial evidence of liquefaction. The Superstition Hills earthquake, a M6.6 event that occurred about 11 h later, produced sand boils, ground fissures, and permanent lateral displacements at the site. The WLA piezometers recorded pore pressure signals that appeared lower than expected, and that have been the subject of some controversy over the years (Earth Technology Corporation 1991; Hushmand et al. 1991; Hushmand 1992; Youd and Holzer 1994; Scott and Hushmand 1995).

The four site response codes were all used to analyze the Wildlife profile. The downhole record of the NS component of the Superstition Hills event was applied as a rigid base motion at a depth of 7.5 m and propagated upward through the overlying materials. Figure 16 shows the recorded and computed ground surface acceleration histories.

The computed acceleration histories are consistent with each other and with the recorded motion up to the time (about 13–15 s) at which point each code predicts very high pore pressure, i.e., the triggering of liquefaction. After triggering, the codes all transition to longer period motions with reduced high frequency content except for instances of dilation-induced acceleration pulses, some of which are followed by decaying high-frequency transients. The timing of the dilation-induced pulses generally matches that of the recorded motions, but the amplitudes of the pulses vary considerably. Some of the codes predict acceleration pulses at extremely high frequencies; these spikes may be of numerical origin rather than representing the physical behavior of the soil profile. These results, however, confirm the

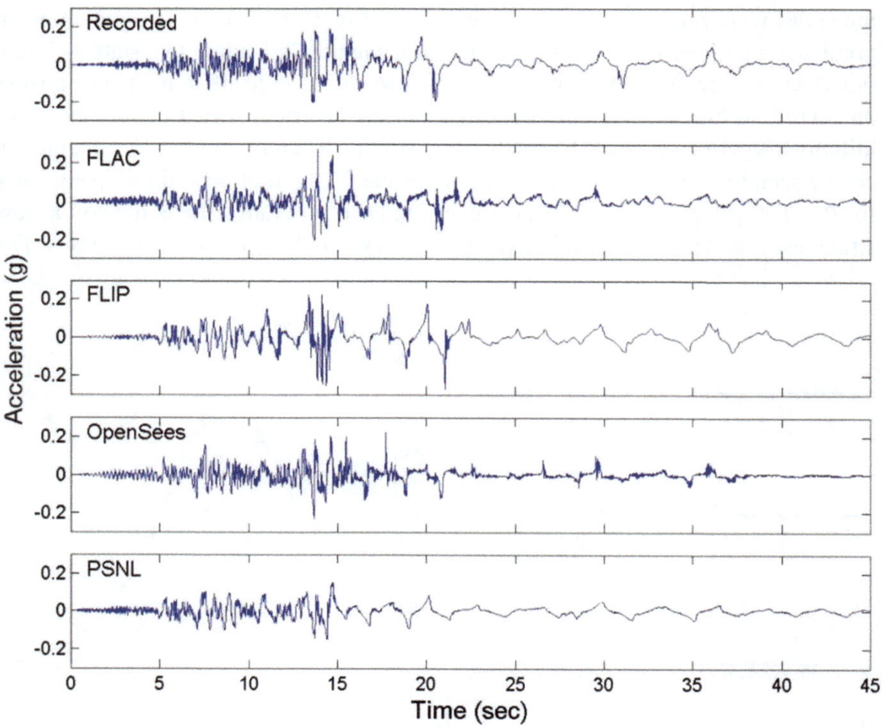

Fig. 16 Recorded and predicted motions for Wildlife NS ground surface motion from 1987 Superstition Hills earthquake

abilities of the four codes to model the primary characteristics of liquefiable soils—pore pressure generation accompanied by general softening with intermittent phase transformation-induced stiffening.

Application to Generic Sites

To investigate the influence of post-triggering response on ground surface response spectra, a generic site was subjected to a series of 30 ground motions scaled to different amplitudes. The ground motions were spectrally matched to a target spectrum defined as the median rock outcrop (V_s=760 m/sec) spectrum for a M_w 7.5 reverse fault event at a distance of 20 km. The soil profile and input motion spectra scaled to a *PGA* of 0.20 g are shown in Fig. 17; the same motions were scaled to *PGA* values of 0.05, 0.10, 0.15, 0.20, 0.25, 0.30, 0.40, and 0.50 g and applied to FLAC, FLIP, OpenSees, and PSNL models of the generic site.

The computed responses for a typical input motion scaled to PGA values of 0.20 and 0.50 g are shown in Figs. 18 and 19. Because these results apply to a single motion and the relative performances of the four codes varied from record to record in a manner that has not yet been studied, the plotted time histories are not associated with their respective codes in the presentation of subsequent results; instead, the codes were randomly assigned identifiers of A, B, C, and D. The programs all predicted high levels of excess pore pressure, eventually leading to a state of initial liquefaction. The times at which liquefaction were triggered are indicated by arrows on each time history; it should be noted that the layers in which liquefaction was initially triggered were not always the same for the different programs. Liquefaction can be seen to be triggered relatively late for the 0.20 g motion and relatively early for the 0.50 g input motion. After triggering, each program softens the soil at low strain levels and models the dilation that occurs when the shear stress increases. The

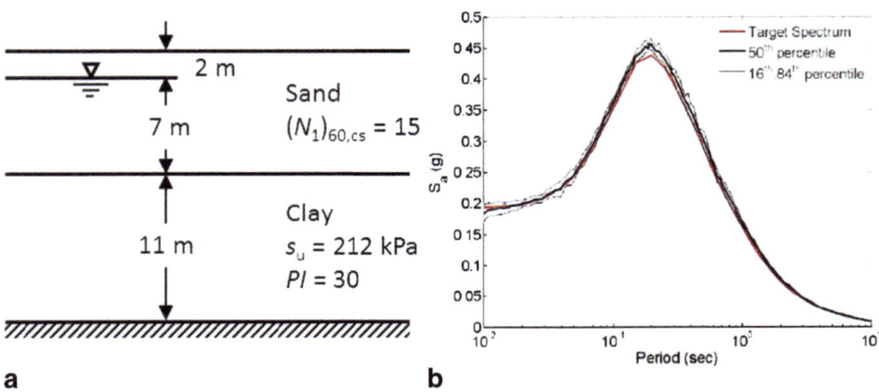

Fig. 17 (**a**) Generic profile used for site response analyses, and (**b**) response spectra for 30 input motions scaled to *PGA*=0.20 g

rates at which the soil stiffens while dilating, however, are different and that difference affects the amplitudes of subsequent dilation-induced acceleration pulses. FLAC and OpenSees tend to stiffen quickly upon dilating, thereby producing relatively strong acceleration pulses that, for many records, exceeded the pre-triggering PGA values.

Response spectra for the 0.20 and 0.50 g input motions were computed at the time of liquefaction triggering and for the entire ground surface motion. Figures 20 and 21 show the response spectra for the two conditions—the red curve represents the response spectral accelerations at the time of liquefaction triggering, and the blue curve represents the response spectrum for the entire motion. At periods where only the red curve is visible, the spectral acceleration at the time of triggering is equal to the spectral acceleration for the entire motion. For the 0.20 g motion, in which case liquefaction is triggered relatively late in the motion, the two spectra are nearly identical for Codes A, B, and D, differing only at periods greater than about 1 s where the post-triggering response produces slightly higher long-period spectral accelerations. The high dilation-induced acceleration spikes predicted by Code C (Fig. 18), however, significantly exceed the pre-triggering peak acceleration and cause the low-period portion of the full Code C spectrum to greatly exceed the spectrum at the time of triggering. The same general behavior is observed

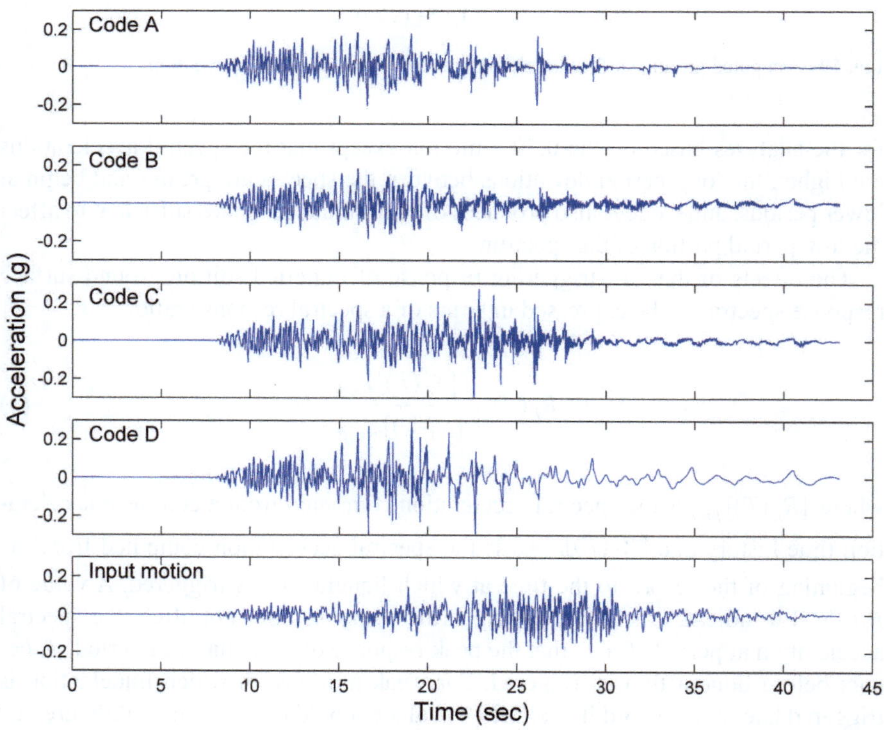

Fig. 18 Computed ground surface accelerograms for 0.20 g rock outcrop motion

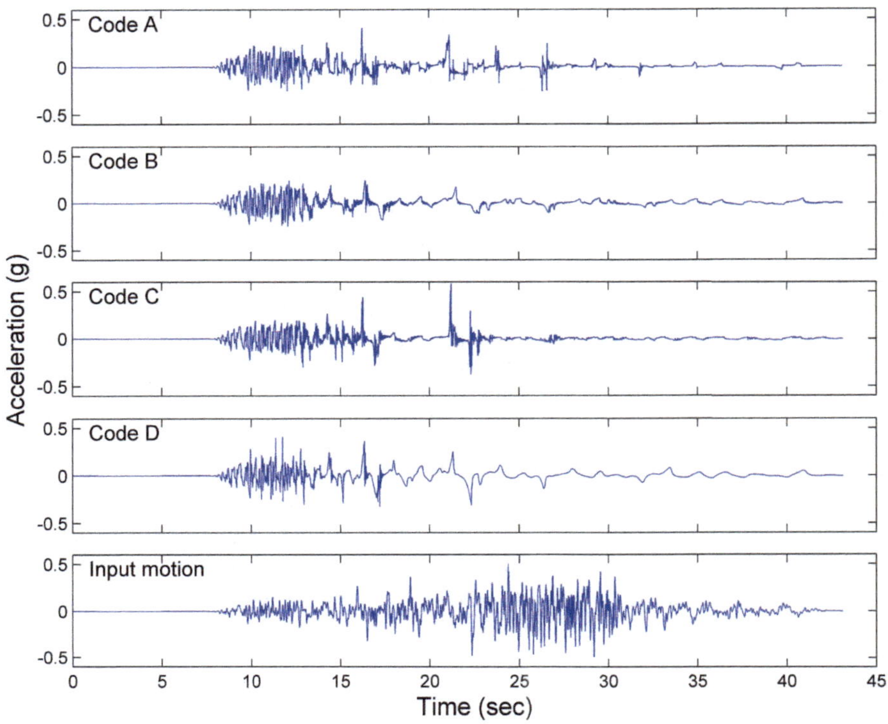

Fig. 19 Computed ground surface accelerograms for 0.50 g rock outcrop motion

for the analyses based on the 0.50 g motion except that the spectral accelerations are higher, the long-period deviations between the spectra are greater and begin at lower periods, and Code A also produces dilation-induced spikes sufficient to affect the low-period portion of the spectrum.

The effects of the post-triggering response of liquefied soil on ground surface response spectra can be expressed in terms of a spectral response ratio,

$$R_L(T) = \frac{[S_a(T)]_{final}}{[S_a(T)]_{trigger}} \tag{4}$$

where $[R_L(T)]_{final}$ is the spectral acceleration computed from a complete acceleration time history and $[R_L(T)]_{trigger}$ is the spectral acceleration computed from the beginning of the record to the time at which liquefaction is triggered. A value of $R_L(T) = 1.0$ indicates that the post-triggering response does not affect the spectral acceleration at period, T, i.e., that the peak response of a oscillator of period, T, occurs before liquefaction is triggered. For weak motions, in which liquefaction is triggered late in the record if at all, $R_L(T)$ values should be at 1.0 or slightly greater. Higher values of $R_L(T)$ should be expected for stronger input motions. Figure 22

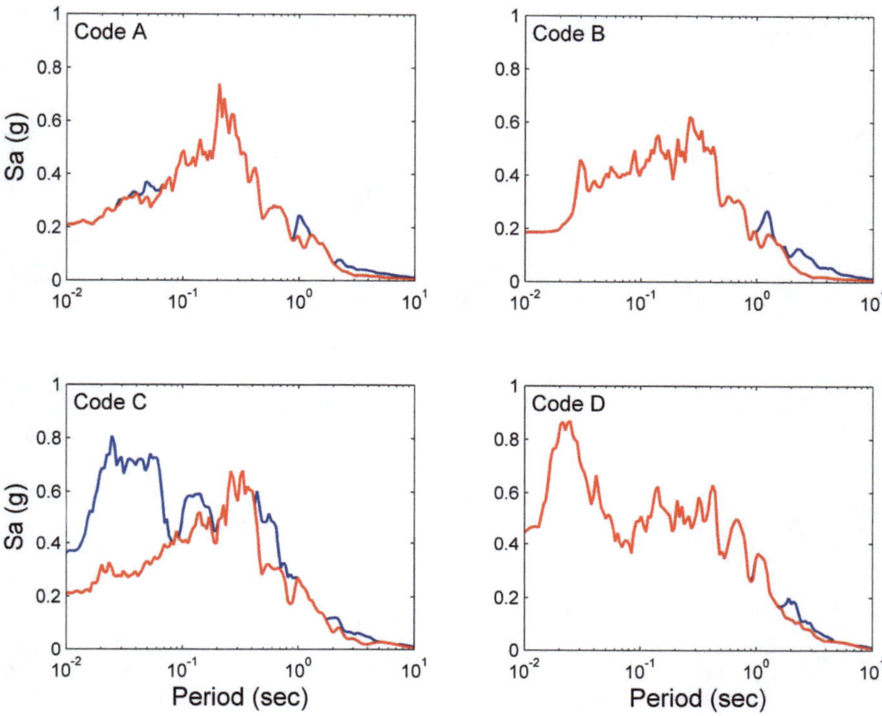

Fig. 20 Computed ground surface response spectra for 0.20 g rock outcrop motion

shows average (over 30 motions) spectral response ratios computed by all four programs for input motion peak accelerations of 0.1, 0.3, and 0.5 g. The spectral response ratios can be seen to be greater than 1.0 at long periods, and by an amount that increases with increasing input motion amplitude. The long-period ratios predicted by the different codes, however, differ noticeably from each other. Code B tends to predict higher long-period ratios, indicating a greater effect of long-period post-triggering response than the other codes. Code D tends to produce lower long-period ratios with Codes A and C being generally consistent. At short periods, the spectral ratios do not rise above 1.0 until the input motion amplitude is higher, and don't reach levels as high as those observed at longer periods. Low-period ratios greater than 1.0 are associated with dilation-induced pulses of acceleration that result from phase transformation behavior. Codes A and C show the strongest tendencies for dilation pulses.

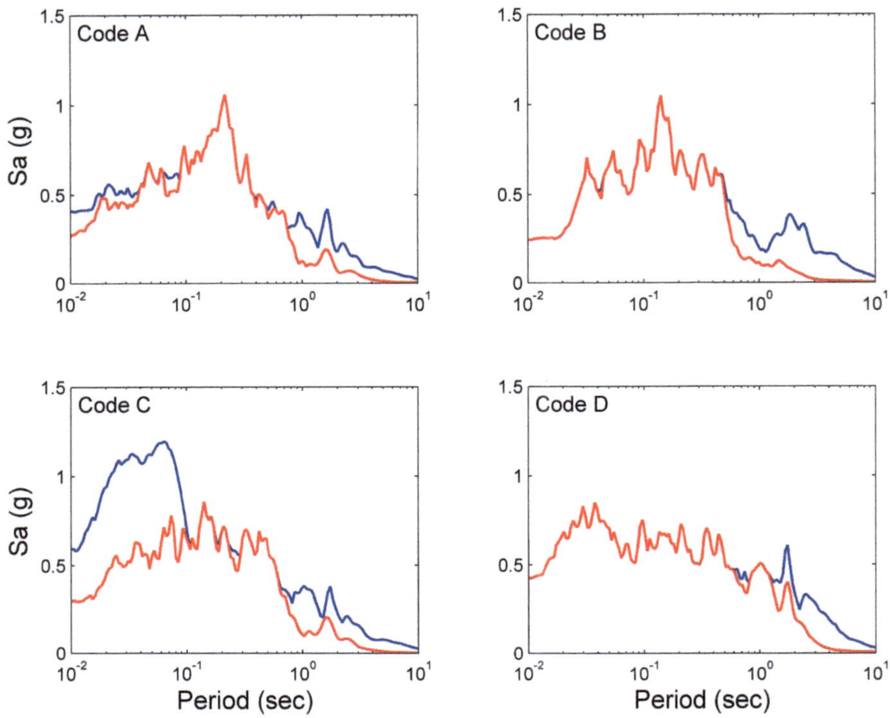

Fig. 21 Computed ground surface response spectra for 0.50 g rock outcrop motion

Summary and Conclusions

Liquefaction is obviously an important geotechnical hazard in earthquakes as evidenced by the many recoded cases of significant lateral and vertical deformations it has produced in past earthquakes. In addition to causing ground failure, however, liquefaction also affects site response and the motions that are transmitted to structures supported on liquefiable soils.

The rapid softening of liquefiable soils as pore pressure approach and reach initial liquefaction causes a soil profile's ability to transmit low and high frequency motions to the surface to increase and decrease, respectively, resulting in a shift of frequency content in the ground surface motion. Phase transformation behavior, which occurs near and after initiation of liquefaction can cause intermittent higher frequency pulses to reach the ground surface even after liquefaction has been triggered.

The complexity of post-liquefaction response suggests that its effects on ground surface motions be isolated and examined. This can be accomplished by identifying the time at which liquefaction is triggered and comparing the response at the time of triggering with the total response. Procedures for identifying the triggering of liq-

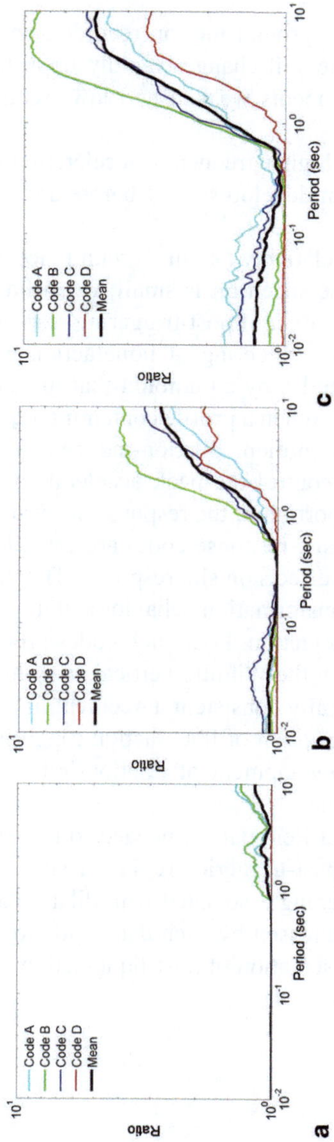

Fig. 22 Spectral response ratios for analyses based on input motions with **(a)** $PGA = 0.10$ g, **(b)** $PGA = 0.30$ g, and **(c)** $PGA = 0.50$ g

uefaction has been developed and applied to recordings of ground motions affected by liquefaction. Numerical techniques capable of representing the most important aspects of liquefiable soil behavior before, at, and after triggering of liquefaction are now available. Examination of recorded earthquake motions and the results of numerical analyses using four such models led to the following conclusions:

1. The frequency content of a ground motion recorded on the surface of a profile underlain by liquefiable soil changes rapidly upon triggering of liquefaction; high frequency components weaken and low frequency components are amplified.
2. Dilation-induced pulses of higher-frequency acceleration associated with phase transformation behavior can develop shortly before and continue after the triggering of liquefaction.
3. When liquefaction occurs relatively late in a ground motion record, its effect on the ground surface response spectrum is small; when liquefaction occurs relatively early in a record, the effect of post-triggering response can be substantial.
4. The response of a site after triggering of liquefaction will generally increase long-period spectral amplitudes by an amount that increases as liquefaction is triggered earlier and earlier. When a profile containing liquefied soil is subjected to continuing strong input motion, dilation-induced acceleration pulses may become strong enough to control the peak acceleration, and thereby strongly influence the short-period portion of the response spectrum.
5. Nonlinear, effective stress site response codes are capable of representing pore pressure generation and its effects on site response. The most advanced of these can also represent phase transformation behavior and its effects of response near and after triggering of liquefaction. Four such codes produced good agreement with the motions recorded at the Wildlife vertical array in the 1987 Superstition Hills earthquake, and generally consistent agreement in numerous analyses of a generic soil profile up to the point of liquefaction triggering. The post-liquefaction behavior, including development of dilation-induced acceleration pulses, however, was highly variable.
6. High-quality laboratory and field data is needed to constrain the calibration of constitutive models in the post-triggering regime. Issues such as fabric degradation and the nature of stiffening associated with dilation above the phase transformation line could be addressed by such data, and would help produce more accurate and uniform representation of post-liquefaction site response.

References

Arefi MJ, Cubrinovsky M, Bradley BA (2012) A model for nonlinear total stress analysis with consistent stiffness and damping variation. Proceedings, 15th World Conference on Earthquake Engineering, Lisboa, Portugal, 10 p

Boulanger RW, Ziotopoulou K (2012) PM4sand (Version 2): a sand plasticity model for earthquake engineering applications, Report no. UCD/CGM-12/01, Center for Geotechnical Modeling, University of California, Davis, 96 p

Earth Technology Corporation (1991) Accuracy of the porewater pressures recoded at Wildlife site during magnitude 6.6 Imperial valley earthquake of 24 November 1987, Unpublished Report to U.S. Geological Survey, Reston, Virginia, 62 p

Elgamal A, Yang Z, Parra E, Ragheb A (2003) Modeling of cyclic mobility in saturated cohesionless soils. Int J Plast 19(6):883–905 (Elsevier Science Ltd., Pergamon)

Hushmand B, Scott RF, Crouse CB (1991) In-situ calibration of USGS piezometer installations. In: Bhatia SK, Blaney GW (Eds) Recent advances in instrumentation, data acquisition, and testing in soil dynamics, vol 29. ASCE Special Publication, pp 49–61

Hushmand B (1992) In-place calibration of USGS pore pressure transducers at wildlife liquefaction site, California, USA, In: Proceeding 10th Earthquake Engineering World Conference, Balkema, Rotterdam, The Netherlands, pp 1263–1268

Iai S, Matsunaga Y, Kameoka T (1992) Strain space plasticity model for cyclic mobility. Soils and Foundations 32(2):1–15

Iai S et al (2011) Dilatancy of granular materials in a strain space multiple mechanism model. Int J Numer Anal Methods Geomech 35:360–392

Idriss IM, Boulanger RW (2008) Soil liquefaction during earthquakes, Monograph MNO-12, Earthq Eng Res Inst

Ishihara K (1985) Stability of natural deposits during earthquakes. Proceedings 11th International Conference on Soil Mechanics and Foundation Engineering, vol 1, pp 321–376

Kostadinov MV, Yamazaki F (2001) Detection of soil liquefaction from strong motion records. Earthquake Eng Struct Dyn 30:173–193

Kutter BL, Wilson DW (1999) De-liquefaction shock waves. Proceedings of the 7th U.S.-Japan workshop on earthquake resistant design of lifeline facilities and countermeasures against soil liquefaction, technical report MCEER-99-0019 (O'Rourke, Bardet, and Hamada eds.), pp 295–310

Manzari MT, Dafalias YF (1997) A critical state two-surface plasticity model for sands. Geotechnique 47(2):255–272

Martin GR, Finn WDL, Seed HB (1975) Fundamentals of liquefaction under cyclic loading. J Geot Eng Div, ASCE 101(GT5):423–438

Masing G (1926). Eigenspannungen und verfestigung beim messing (self stretching and hardening for brass). Proceeding, 2nd Int. Congress for Applied Mechanics, Zurich, Switzerland, pp 332–335

Miyajima M Nozu S, Kitaura M (1997) A detection method for liquefaction using strong motion records. Proceedings of 24th JSCE Earthquake Engineering Symposium, Kobe, 24–26 July. JSCE, pp 265–268 (in Japanese)

Muravskii G (2004) On description of hysteretic behviour of materials. Int J Solids Sturct 42:2625–2644

Phillips C, Hashash YMA (2009) Damping formulation for nonlinear 1D site response analyses. Soil Dyn Earthquake Eng 29:1143–1158

Pyke RM (1979) Nonlinear soil models for irregular cyclic loadings. J Geotech Eng Div, ASCE 105(GT6):715–726

Scott RF, Hushmand B (1995) Discussion of "Piezometer at Wildlife liquefaction site" by T. L. Youd and T. L. Holzer. J Geotech Eng ASCE 121:912–919

Stockwell RG, Mansinha L, Lowe RP (1996) Localization of the complex spectrum: the S transform. IEEE Trans Signal Processing 44:998–1001

Suzuki T, Shimizu Y, Nakayama W (1998) Characteristics of strong motion records at the liquefiable and judgment for liquefaction. 11th European Conference on Earthquake Engineering, Paris, France, 6–11 September

Takada S, Ozaki R (1997) A judgment for liquefaction based on strong ground motion. Proceedings of 24th JSCE Earthquake Engineering Symposium, Kobe, 24–26, July. JSCE, pp 261–264 (in Japanese)

Yang Z, Elgamal A, Parra E (2003) A computational model for liquefaction and associated shear deformation. J Geotech Geoenviron Eng, ASCE 129:1119–1127

Youd TL, Holzer TL (1994) Piezometer performance at Wildlife liquefaction site, California. J Geotech Eng, ASCE 120:975–995

Shaking Model Tests on Liquefaction Mitigation of Embedded Lifeline

Ikuo Towhata, Masahide Otsubo, Taro Uchimura, Masato Shimura, Bangan Liu, Toshihiko Hayashida, Damoun Taeseri and Bertrand Cauvin

Abstract The gigantic earthquake in 2011 caused significant damage in lifeline in the Tokyo Metropolitan area. In particular, the damage was significant in recent artificial islands where liquefaction affected embedded sewage pipelines profoundly. The encountered problem is that the entire subsoil liquefied in addition to loose backfill soils, and that the damage of branch lines introduced liquefied sand into trunk sewage lines, leading to difficult problem of sand clogging. Moreover, the same problem is expected to occur in other areas where strong earthquakes are expected in near future. The present paper addresses the ongoing model tests by which a variety of mitigation measures for sewage pipelines are examined. In the regions where future earthquake is expected, it is not possible to excavate pipes and reconstruct backfills now because of financial limitations. To cope with this situation, less expensive measures such as mechanical constraint, partial injection of grout, or limited installation of drainage measures are studied.

Keywords Earthquake · Liquefaction · Lifeline · Model test · Mitigation

Introduction

Destruction of lifeline facilities caused by the Great East Japan Earthquake on March 11, 2011, was fatal to post-earthquake operation of damaged modern cities, especially near Tokyo Bay. Typical examples of liquefaction-induced damage in embedded lifelines are shown in Fig. 1. Significant floating of manholes and

I. Towhata (✉) · M. Otsubo · T. Uchimura · M. Shimura · B. Liu
Department of Civil Engineering, University of Tokyo, Tokyo, Japan
e-mail: towhata@geot.t.u-tokyo.ac.jp

T. Hayashida
Fudo Tetra Corporation, Tokyo, Japan

D. Taeseri
Department of Civil Engineering, ETH-Zurich, Zurich, Switzerland

B. Cauvin
Department of Civil Engineering, Engineering School ENTPE, Vaulx-en-Velin, France

A. Ansal, M. Sakr (eds.), *Perspectives on Earthquake Geotechnical Engineering*,
Geotechnical, Geological and Earthquake Engineering 37,
DOI 10.1007/978-3-319-10786-8_12

pipelines was observed in a wide area. Moreover, disconnection of pipeline joints was also a serious problem, introducing liquefied backfill sand into the pipes and long delay in resuming operation due to significant efforts to cleaning the clogged pipes.

After the earthquake, discussion has started on the future risk of another gigantic earthquake that may hit other regions of the nation and efforts for damage mitigation are considered very important.

Mitigation Measures

Many studies have been made on the mechanism of liquefaction-induced floating of lifelines and its mitigation (Kitaura et al. 1988; Koseki et al. 1998). Consequently, improvement of backfilling by, for example, the use of cement-mixed soil has been introduced (Miyake et al. 1998; Technical Committee on Earthquake Resistant Design of Sewage Lifelines 2008) and its good performance was validated during the 2007 Niigata-Chuetsu-Oki earthquake (Fig. 2).

Fig. 1 Clogging and floating of sewage pipes. (From website of Urayasu City)

Fig. 2 Prevention of liquefaction damage of lifeline by backfilling by using cement-mixed soil

One of the possible problems of the cement-mixed backfill is the difficulty in re-excavation for maintenance works. Another problem is that the excavation and backfilling along the entire pipelines require time and cost, which are not relevant for urgent earthquake retrofitting in other parts of the nation where the seismic damage has not yet occurred. Consequently, the amount of excavation should be reduced and development of less expensive and easier technology for existing lifeline is expected.

With these in mind, the present study concerns methods which do not require full backfilling and is relevant for lifelines subjected to earthquake problems in near future. In total, four types of mitigative measures were studied (Figs. 3 and 4);

1. Horn-type structure to prevent floating
2. Drainage pipes
3. Chemical grouting
4. Insertion of sheath pipe

Because these mitigations require limited or no excavation, they would be economical. Among the four candidate technologies, installation of a sheath pipe has been proposed to vitalize old deteriorated concrete pipes in which a new pipe is installed. The present study makes use of this and attempts to maintain flow of sewage water even if the outer pipes are affected by disconnection of joints (Fig. 4).

Fig. 3 Mitigation without overall excavation

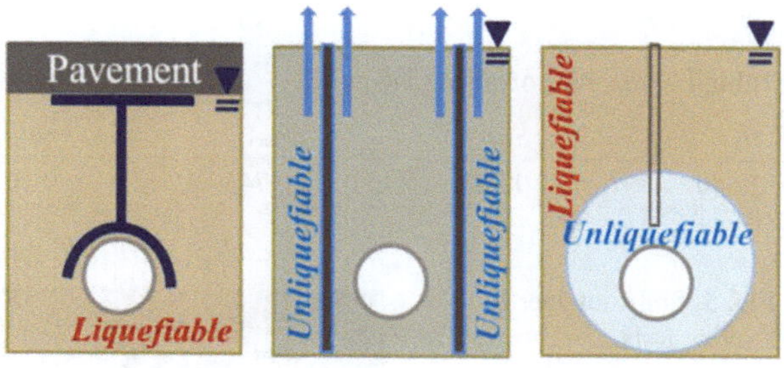

Fig. 4 Insertion of sheath pipe inside an existing pipe

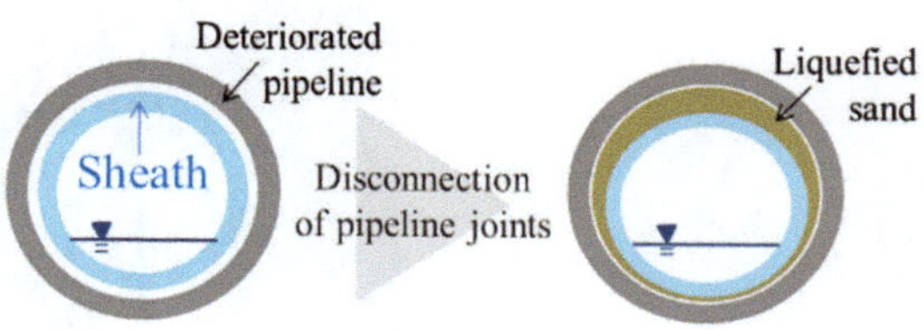

Method of Shaking Model Tests

Tested Cases

Silica sand No. 7 was used for all the experiments reproducing the loose backfill (Dr = 30 %), of which propertiees are listed in Table 1.

Because the size of the model is smaller than the prototype and the effective stress is lower, the time scale was made shorter, running shaking tests at a higher frequency, and the sand was made looser so that dilatancy is similar in both model and prototype (Towhata 2008). Moreover, viscous liquid was used for experiments when the permeability is regarded as the dominant parameter in cases of drainage methods. The viscous liquid was about 5 times more viscous than the normal water.

1-g Shaking Table and Soil Container

The 1-g shaking table and a rigid soil container used in this research are shown in Fig. 5. To reduce the effect of the rigid lateral boundary, cushions were installed at both ends.

Sensors

The sensors were installed to measure excess pore pressure, acceleration and pipeline displacement at positions shown in the section of each method. LVDTs were

Table 1 Properties of silica sand no. 7

G_s	e_{max}	e_{min}	D_{50} (mm)	D_{10} (mm)
2.640	1.243	0.743	0.206	0.115

Fig. 5 Soil container for shaking tests

connected to the buried pipeline to measure its vertical displacement. Besides, strain gauges were pasted along the pipeline to measure the bending strain of buried pipeline during shaking. Note that the bending strain is defined by $(\varepsilon_c - \varepsilon_e)/2$ in which the numerator designates the strain difference between the compression and extension sides of the pipe.

Pipeline Model

A PVC pipe of 6-cm diameter, 256-cm length and 0.39 g/cm^3 of overall density was buried at 30-cm depth at the centre of the soil container. Although damages on manhole structure was observed widely during the gigantic earthquake in 2011, advanced technologies for preventing manhole floating has been developed by many researchers (Japan Sewage Works Association 2006). Besides, flexible joints between manhole and buried pipeline are recommended to use recently. Assuming that manholes keep their position intact during soil liquefaction, the boundary condition was modelled by hinge supports in this research. Figure 6 illustrates schematically an embedded pipeline and its connection with manholes.

Input Acceleration

Figure 7 shows the time history of input horizontal acceleration that was used for all the experiments. The maximum acceleration is around 350 Gal and the duration time is about 25 s with frequency being 10 Hz, which is higher than reality because of the aforementioned similitude reasons.

This input motion was prepared to discuss the post-liquefaction behavior of the buried pipeline and the surrounding ground. The maximum acceleration, around 350 Gal, slightly changed depending on the total weight of the soil used in experiments. This acceleration time history reproduces a Level-2 rare earthquake that occurs once every 500 years or so.

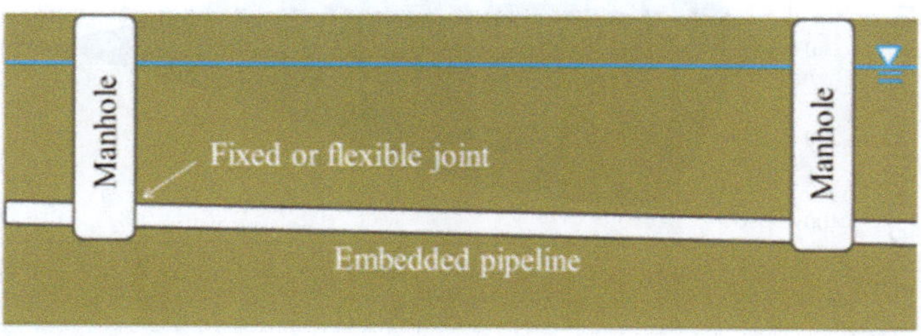

Fig. 6 Schematic illustration of embedded pipeline and manholes connection

Retrofitting of Existing Lifeline Without Backfill Excavation

Seismic retrofitting of embedded sewage pipelines without overall excavation is essentially important in regions where a future gigantic earthquake is feared but a real damage has not yet occurred. This is because financial support is not sufficient in such areas, different from the already damaged areas. Accordingly, the present paper addresses a technology that can improve the liquefaction resistance of pipelines with limited or no excavation of backfill so that costly and time-consuming excavation and backfilling may be reduced. It is aimed that a relatively simple work can achieve reasonable mitigation of uplift of an embedded pipeline. Model studies mentioned in this chapter attempt to validate such technologies while assuming only the backfill is liquefiable and the surrounding ground is stable.

Horn Structure to Prevent Floating of a Pipe

The first idea is a horn-like device that is placed upon a pipe and uses the surface soil and pavement to resist the uplift force (Fig. 8).

As further shown in Fig. 9, the experimental device measures the uplift force by load cells, while the upward punching in the surface soil is resisted by the weight and shear strength of the surface soil and pavement. Hence, D and L in this figure control the mitigative effects of the horn. In this series of experiments, both square and circular shape of plates were employed, and therefore the area of the plate, A, was used as an index in place of L. The tested conditions are tabulated in Table 2.

The experimental set-up of a model is illustrated in Fig. 10 where a model pipe is connected to the end walls of the soil box by hinges, thus somehow reproducing the pipeline connection to manholes. This figure clearly indicates that the width of the

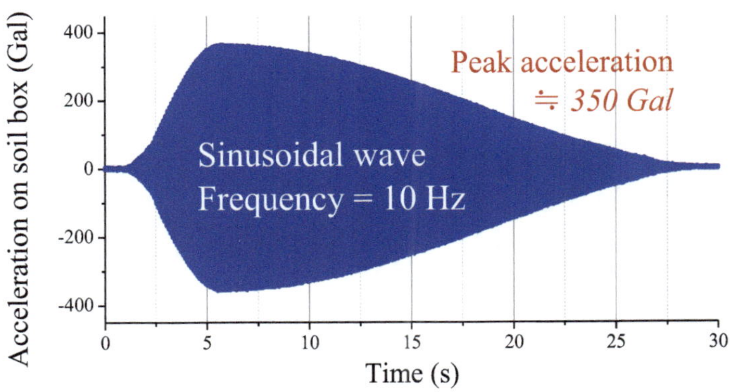

Fig. 7 Time history of input acceleration

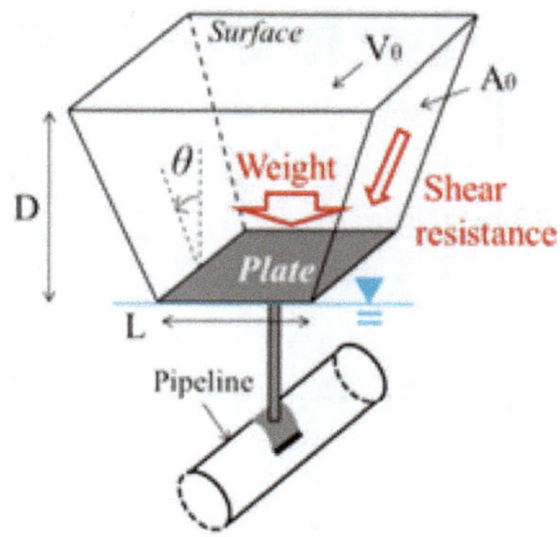

Fig. 8 Schematic illustration of horn

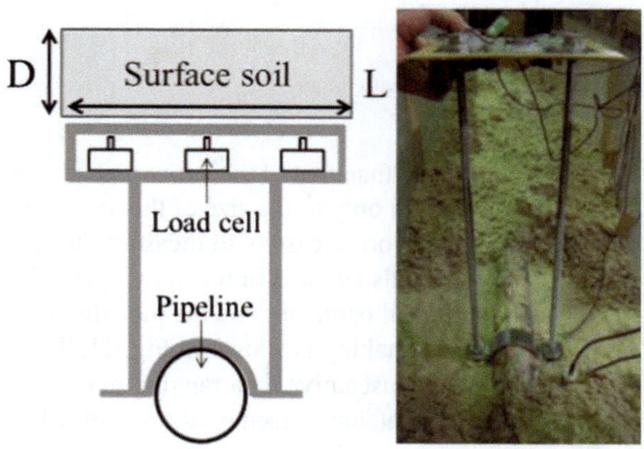

Fig. 9 Structure and photograph of experimental device as horn

Table 2 Specifications of tests on horn

Case	Plate shape	Area of plate: **A** (cm)	Depth of plate: **D** (cm)	Surface layer (density)
Horn.1	Position of horn's plate is fixed to the soil box			
Horn.2	Square	625	10	Gravel (Loose)
Horn.3	Square	225	10	Gravel (Loose)
Horn.4	Square	225	5	Gravel (Loose)
Horn.5	Square	225	10	Sand
Horn.6	Circular	113	7	Gravel (Medium)
Horn.7	Circular	113	10	Gravel (Medium)
Horn.8	Circular	113	10	Gravel (Loose)
Horn.9	Circular	113	15	Gravel (Loose)
Horn.10	Circular	227	7	Gravel (Medium)
Horn.11	Circular	227	12	Gravel (Medium)
Horn.12	Reference case (Without horn structure)			

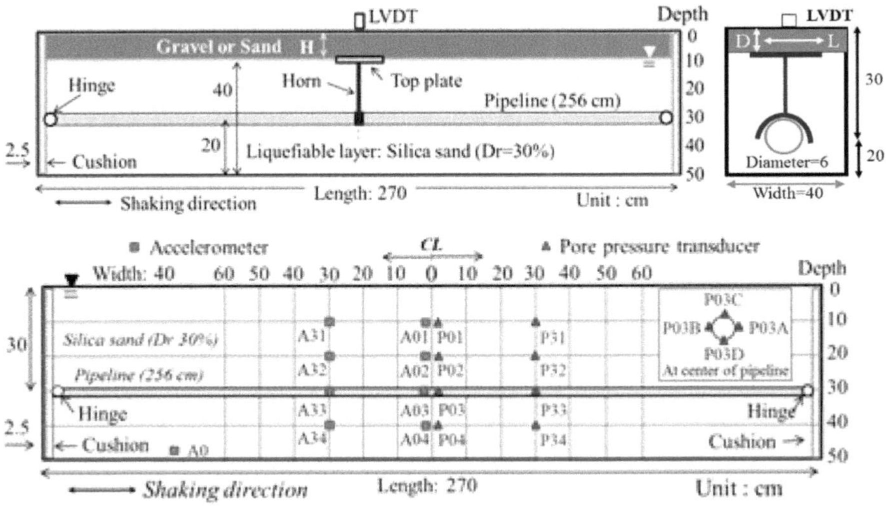

Fig. 10 Model ground with hinge

horn is much shorter than that of the entire pipe length, making the required amount of excavation limited only to the size of the installed horn.

The purpose of Horn.1 case is to measure the applied uplift force during shaking by using load cells situated in the horn's plate. To achieve this goal, the horn's plate was fixed to soil container and the pipe did not move upwards. The measured uplift force during shaking is shown in Fig. 11. The fluctuated data was smoothed by averaging 100 consecutive data ranging over 0.2 s, and the maximum response force, 74.3 N, was obtained based on the smoothed curve.

Figure 12 compares the time history of pipeline displacement in the cases of Horn.2~5 and 12 as examples. As liquefaction took place, the pipeline floated in all cases initially. A possible reason for this is that loosely deposited gravel layer allowed horn's plate to move upwards. After the initial floating, Horn.2 and 3 with 10 cm thickness of the surface loose gravel layer and 10-cm thickness showed subsidence of pipeline during shaking due to the subsidence of the surface gravel layer.

Fig. 11 Response uplift force acting on horn's plate

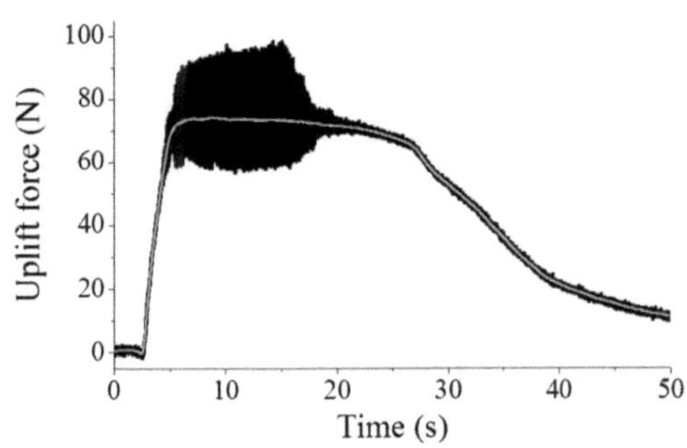

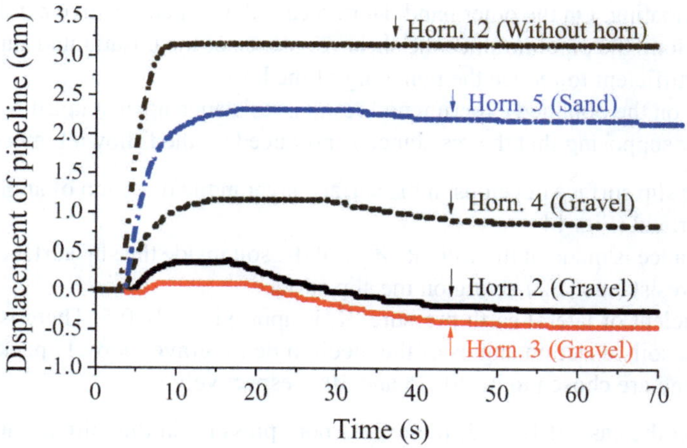

Fig. 12 Examples of time history of pipeline displacement (Horn.2 ~ 5 and 12)

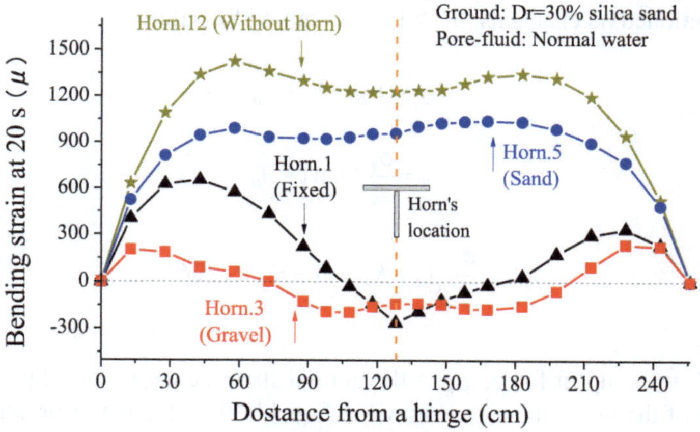

Fig. 13 Examples of bending strain distribution along pipeline at 20sec (Horn.1, 3, 5 and 12)

This would mean that the resistance against uplift force and punching failure was sufficiently high in the 10-cm thick gravel, in spite of its looseness. In contrast, other cases showed floating of pipeline until the end of shaking, and subsidence of pipeline occurred along with consolidation of liquefied ground. This result suggests that the uplift force was greater than the resistance of surface layer because of the small size (A) of the plate (Horn.4) or the reduced shear resistance of the surface sandy layer (Horn.5).

Figure 13 shows the increment of bending strain distribution during shaking along the pipeline at 20 s. in the case of Horn. 1, 3, 5 and 12. Horn.1 case illustrates that this horn-structure system mitigated the floating of the pipe, working as a semi-fixed boundary, because negative bending strain was observed at the horn position. Similarly, Horn.3 case showed negative strain, and consequently reduced the pipeline floating. On the other hand, Horn.5 case developed positive bending strain entirely along the pipeline since the shear resistance of the unsaturated surface sand was not sufficient to reduce the punching of the horn.

Based on the obtained experimental results, resistance against uplift force is discussed by supposing that the resistance is produced by the following mechanisms;

1. Linear slip surface develops in the surface layer in the direction of angle, θ, from the vertical (Fig. 14).
2. Resistance is made of the weight (W_θ) of the soil inside the slip surface (V_θ) and shear resistance (S_θ) acting on the slip surface.
3. Coefficient of lateral earth pressure, K, is approximately 0.5. There is no cohesion in soils. Friction angles of the medium dense gravel, loosely packed gravel and sand are chosen to be 40, 35 and 30°, respectively.

Regarding the case of Horn.5, the excess pore pressure in the initially unsaturated sand layer increased substantially. Therefore, the shear resistance (S_θ) therein was regarded null in the case. Hence, the following equations show the way of calculating the factor of safety (F_s) against punching failure in the surface layer with abovementioned mechanisms. $F_s \geq 1.0$ is required for safety.

$$F_s = (W_\theta + S_\theta)/F_u \tag{1}$$

$$S_\theta = \frac{\sigma_\theta}{2} \cdot \tan\varphi \cdot A_\theta \tag{2}$$

$$\sigma_\theta = \frac{\sigma_v}{2}\{(1+K)-(1-K)\cdot\cos 2\theta\} \tag{3}$$

where F_u = the uplift force, σ_θ = the normal stress acting on the slip surface (at the depth of the horn's plate) as shown in Fig. 14, σ_v = the major principal stress (vertical stress) acting on the horn's plate (at the depth of horn's plate), K = the coefficient of earth pressure, and A_θ = the area of the slip surface.

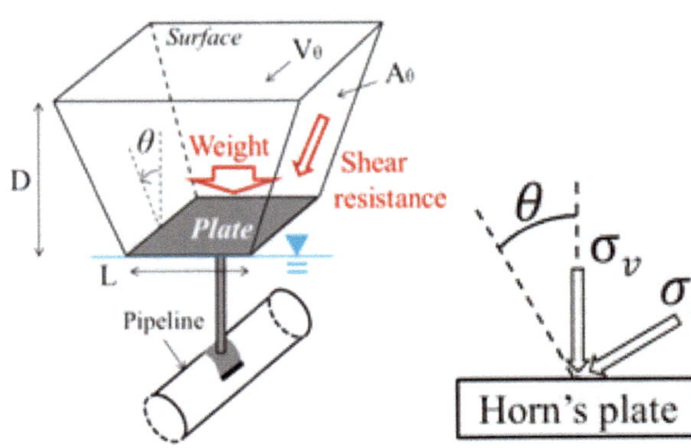

Fig. 14 Assumed slip failure mode of surface layer

Table 3 Residual displacement and factor of safety

Case	Angle of slip surface θ=φ/2 (deg)	Factor of safety, Fs	Residual displacement (cm)
Horn. 2	17.5	1.90	−0.28
Horn. 3	17.5	0.86	−0.49
Horn. 4	17.5	0.34	0.82
Horn. 5	15	0.73	2.15
Horn. 6	20	0.40	0.95
Horn. 7	20	0.73	0.62
Horn. 8	17.5	0.51	1.25
Horn. 9	17.5	1.06	0.56
Horn.10	20	0.66	1.40
Horn.11	20	1.57	0.74
Horn.12	N/A	N/A	3.16

The authors (Towhata et al. 2013) previously suggested using the linear slip surface angle, $\theta=\varphi/2$ by considering the experimental results of Horn.2~5 cases. In this paper, additional 6 tests, where a circular shape plate was utilized, are examined in order to confirm the applicability of the resistance formula against punching failure at the surface.

Table 3 is the summary of the experimental results and the calculated factor of safety when $\theta=\varphi/2$ is selected. For the calculation of the factor of safety, $F_u=74.3$N was selected based on the result of Horn.1. Further, no horn was employed in Horn.12.

Figure 15 shows the relationship between the residual displacement (floating) of the pipe and the factor of safety when $\theta=\varphi/2$ is used. It is obvious that the displacement decreases considerably when the factor of safety becomes greater. The broken curve in Fig. 15 shows a tendency of data except for Horn.5 where dense and unsaturated sand (80 % relative density and separated from the underlying submerged sand by a layer of gravel) was situated upon the liquefiable layer and did not develop the expected shear strength due to submergence during shaking and development of excess pore water pressure. It is inferred that the cases of Horn. 2, 9 and 11 might include a boundary effect due to the presence of front and rear walls of the soil container. As the horn's plate is buried in more compacted soil (greater φ) or at a deeper elevation, or a wider plate was employed, the wall effect became more significant. Thus, the downward arrows in the figure mean that the actual displacement might be smaller than the observation.

Drainage Pipe Method

Similar to the gravel drain that is widely employed to reduce the excess pore water pressure of sandy subsoil during strong shaking, the drainage pipe was intended to

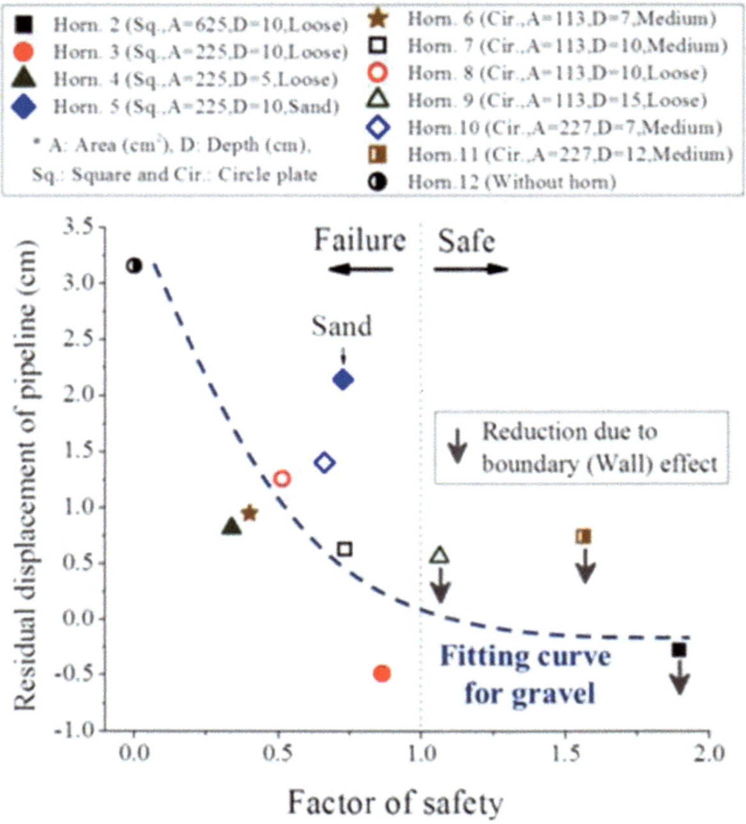

Fig. 15 Relationship between residual displacement of pipeline and factor of safety

reduce the excess pore pressure in the backfill promptly during the shaking. Drainage pipe method has also developed mainly for house foundation (Towhata and Razouli 2013) but not for embedded lifelines. Therefore, in this research, interests regarding this method are as what follows.

1. Effective range of reduction of excess pore pressure by drainage pipes in order to minimize the number of installed drainage pipes.
2. Required spacing between drainage pipes to mitigate floating of a buried pipeline.
3. The effect of unsaturated surface or pavement in reality.

A PVC pipe, having 25 % of its side area being open for drainage, was used in experiments. To prevent soil particles coming into the drainage pipes through the holes, 100 micron sieve was wrapped around the pipes (Fig. 16).

A set of four drainage pipes were installed at every corner of a D × D square as shown in Fig. 17. The bottom of the drainage pipes was placed at the same level (33-cm depth) as the bottom of a buried pipe, which was placed at the centre of the four drainage pipes. Thus, the centre-to-centre spacing between buried pipeline and the nearest drainage pipe was D/2.

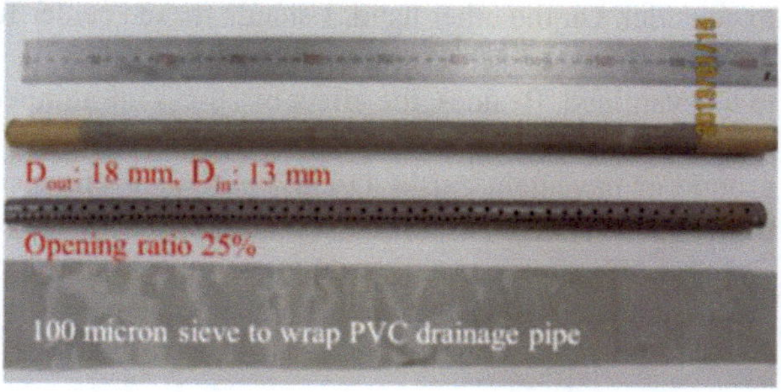

Fig. 16 Drainage pipe used in experiments

Liquefiable ground was prepared with silica sand (Dr = 30 %). The employed pipe, sensors and input acceleration were identical with those in the Horn tests. In the present series of experiments, viscous liquid was used for all the cases to more rigorously satisfy the law of similitude.

Table 4 shows the experimental cases of the drainage pipe method. In this series of experiments, effects of two parameters, **D** and **I**, are examined. First, **D** stands for the center-to-center spacing between drainage pipes as shown in Fig. 17. Since a drainage set consists of four pipes, the spacing between drainage pipes is important

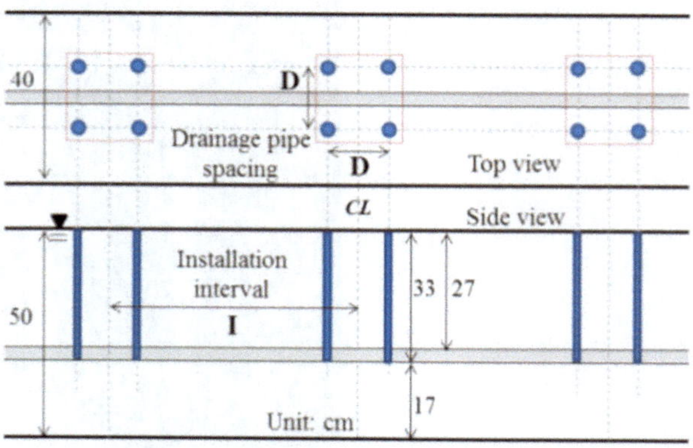

Fig. 17 Positions of installed drainage pipes

Table 4 Specification of tests on drainage pipes

Case	Drainage pipe spacing: **D** (cm)	Installation interval: **I** (cm)
Drain. 1	20	40 (5 sets)
Drain. 2	14	40 (5 sets)
Drain. 3	10	40 (5 sets)
Drain. 4	20	60 (3 sets)
Drain. 5	10	60 (3 sets)
Drain. 6	10	60 (3 sets)
	Unsaturated gravel layer (10 cm)	
Drain. 7	Reference (without drainage pipe)	

to consider. On the other hand, **I** stands for the center-to-center interval of sets of 4 drainage pipes. To achieve economical mitigation, this parameter is very important to be examined. Besides, the effect of 10-cm unsaturated gravel layer at the top was examined in Drain.6 case in order to consider a realistic situation where unsaturated ground or pavement exists. Drain.7 case is a reference and no drainage pipe was used.

Figure 18 shows installed drainage pipes along a pipeline at the bottom. Figure 19 indicates an example of drainage pipe position and prepared ground (silica

Fig. 18 Buried pipeline at the bottom between vertical drainage pipes

sand with Dr = 30 %) for Drain. 5 case. Buried pipeline is located at 30 cm below the surface of the ground.

Figure 20 shows the time history of pipeline displacement when D = 20 cm. Neither Drain. 1 nor 3 showed any mitigative effect, irrespective of the installation interval, **I**.

To understand the performance and capacity of drainage pipes used in the experiments, the relationship between the maximum excess pore pressure ratio and the distance from sensors to the closest drainage pipes was investigated. In Fig. 21, 10 cm can be regarded as the effective distance of drainage pipes. This result is consistent with the D = 20 cm case where the range of effects (10 cm) was half of D and two adjacent drainage pipes could not exert any meaningful mitigation (Fig. 20). It is further obvious herein that I = 40 cm or 60 cm was too long.

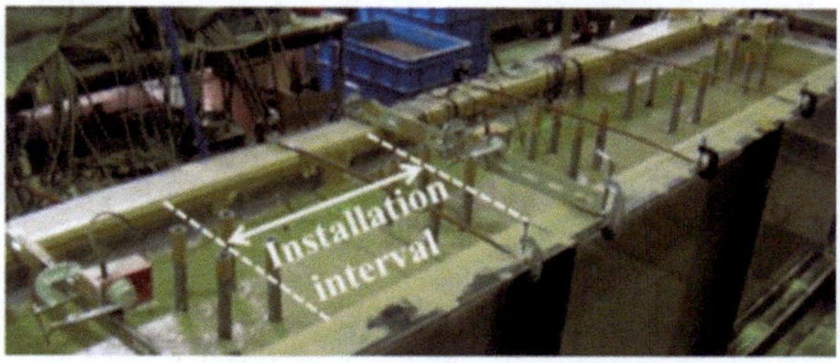

Fig. 19 Preparation of experiments with drainage pipes

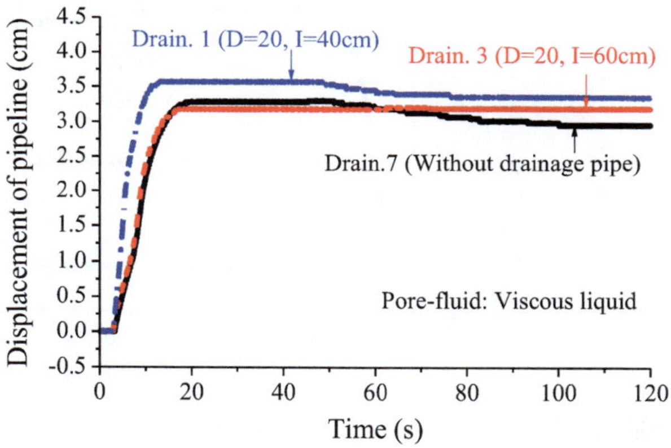

Fig. 20 Time history of pipeline displacement (D = 20 cm)

Figure 22 illustrates the mitigation effect on pipeline displacement developed by drainage pipes of I = 40 cm. It is found that smaller D would lead to more efficient mitigation.

Figure 23 shows the mitigation effect on pipeline displacement developed by drainage pipes of I = 60 cm. In addition, the effect of unsaturated gravel layer situated on the liquefiable sand layer was examined. Although Drain.3 (D = 10, I = 40 cm) case in Fig. 22 showed a satisfactory result, Drain.5 (D = 10, I = 60 cm) case allowed

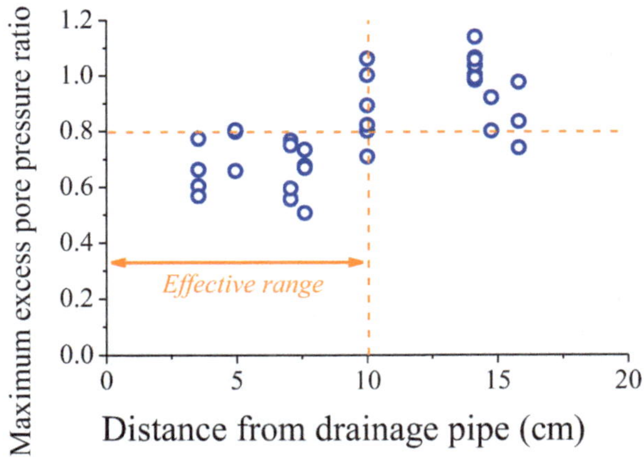

Fig. 21 Relationship between maximum excess pore pressure ratio and distance from drainage pipe

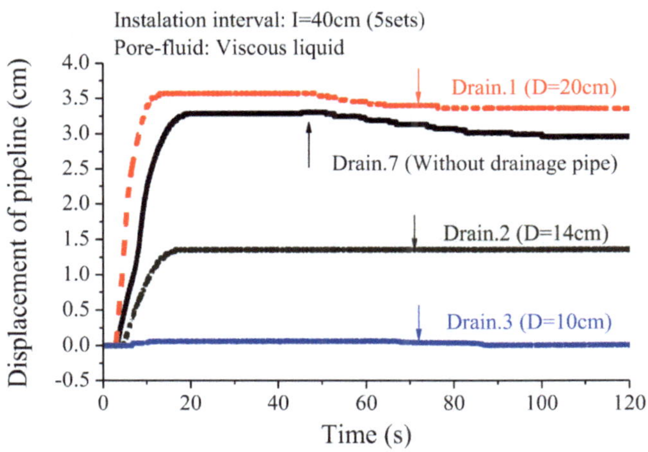

Fig. 22 Time history of pipeline displacement (I – 40 cm)

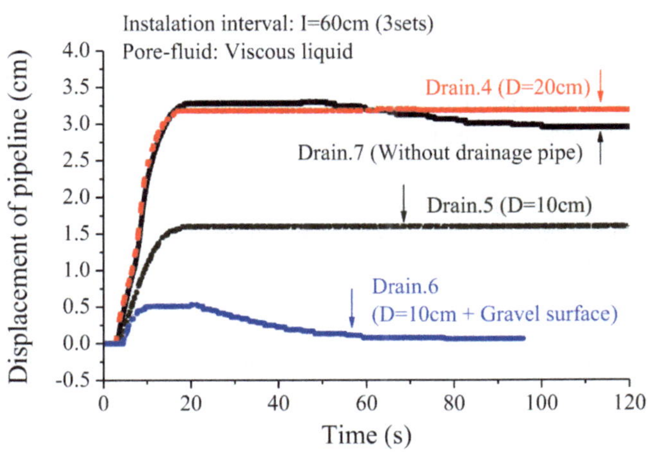

Fig. 23 Time history of pipeline displacement (I – 60 cm)

almost 50 % of the maximum possible floating. On the other hand, Drain.6 case showed only 0.5 cm maximum floating, followed by subsidence after the shaking due to settlement of the surrounding soil.

The floating is reduced by the drainage pipes because the backfill soil columns around the drainage pipes does not liquefy and maintains its rigidity, and thus prevents the upward movement of the pipeline to a certain extent. Moreover, it seems that additional possible mechanism improved the mitigative effects in Drain.6, where a gravel layer was placed at the surface. It is supposed that non-liquefied sand columns above the buried pipeline that was created by the drainage transmitted the weight of the gravel layer to the pipeline and pushed the pipe downwards. This issue is somehow similar to the mechanism of the horn-structure method as stated previously. To examine this surface pavement effect, the excess pore water pressure ratio above and below the centre of the pipeline is shown in Fig. 24 with a schematic illustration of sensor's location. It appears that the excess pore water pressure dissipated quickly above the pipeline and did not allow it to float substantially.

Figure 25 summarizes the effects of D and I as well as the gravelly surface that are exerted on the residual displacement of the buried pipeline. The displacement was measured at the center of the pipeline. Smaller D and I achieved greater mitigation of pipeline floating. However, there is a construction limitation on D because a buried pipeline might be damaged if drainage pipes are installed too close to a buried pipeline. Besides, a greater installation interval is more economical in practice. Therefore, the optimum solution should be proposed by considering required mitigation level and the budget for construction. In addition, existence of unsaturated surface or pavement would enhance the effectiveness of mitigation by drainage method. This effect also should be examined to evaluate the performance of drainage method. Note that the maximum pipeline floating without any mitigation increases if there is no hinge boundary in experiments.

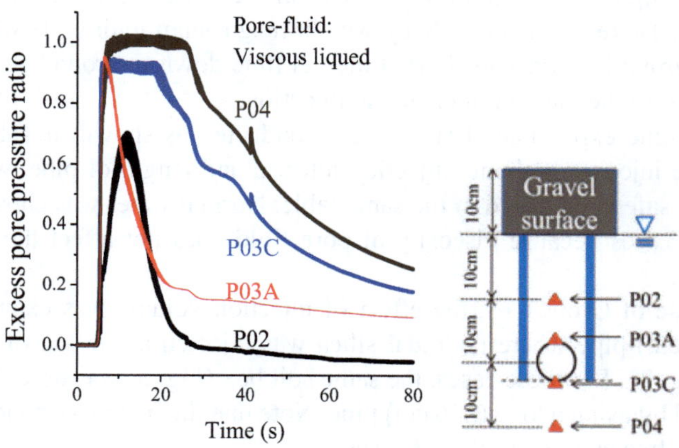

Fig. 24 Time history of excess pore water pressure ratio at the *centre* of pipeline and schematic illustration of sensor's location (Drain.6 case)

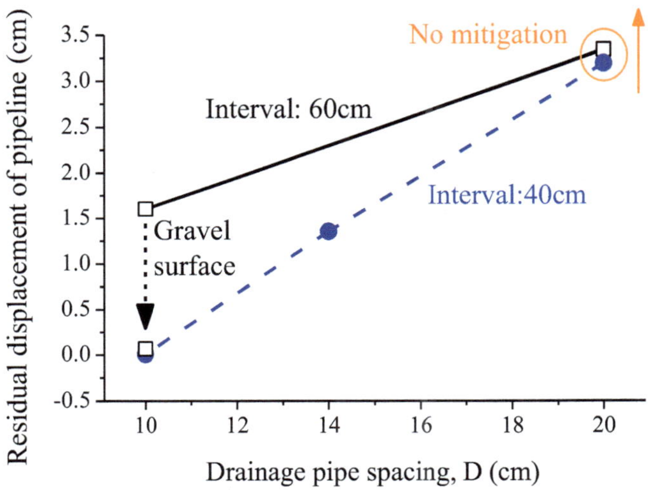

Fig. 25 Residual displacement of pipeline. (Measured at the *centre* of pipeline)

Chemical Grouting Method

Soil solidification by chemical grouting has been used in practice aiming at ground improvement for structure situated upon soft ground. In the present study, application of this method to lifeline protection was examined. The advantage of this method is its quick construction without overall excavation of a backfill. However, problems regarding its high cost still remain. Therefore, effective mitigation with less amount of chemical material (Colloidal silica) was sought for in this study.

Colloidal silica (Towhata and Kabashima 2001) was used as the chemical material to solidify soil surrounding pipeline. The colloidal silica can be injected easily through an inserted small tubes and its chemical composition specified that solidification is completed after more than 1 day in the silica sand used in the current experiments. Therefore, about 3 % by weight magnesium hydroxide was added to silica sand ground in order to shorten the gel time down to around several hours. Figure 26 shows the method of chemical injection.

Tests on nine experimental cases were conducted as shown in Table 5. Main variables are injection volume, injection interval and length of pipe, which affect the factor of safety, Fs, listed in the same table. Normal water was chosen as pore-fluid for all cases because viscosity of pore fluid does not affect the test results significantly.

In the case of Grout.1 ~ 3, the effect of injection volume was examined using a short (35 cm) pipe where colloidal silica was injected at the top of the pipe as shown in Fig. 27. For these cases, the same soil box (Fig. 5) was used. Other cases were studied by using a long (256 cm) pipe. Note that the boundary condition of the pipeline was free end in Grout.1 ~ 3 cases.

Fig. 26 Injection method (Colloidal silica was dyed to be visualized)

Table 5 Experimental cases with chemical grouting

Case	Injected volume (L)	Injection interval (cm)	Pipe length	F.S
Grout. 1	0	–	Short 35 cm	0.24
Grout. 2	0.5	35 cm (1 point)		0.67
Grout. 3	1	35 cm (1 point)		0.80
Grout. 4	0	–	Long 256 cm	0.22
Grout. 5	4	256 cm (1 point)		0.67
Grout. 6	8	256 cm (1 point)		0.82
Grout. 7	8	40 cm (5 points)		0.82
Grout. 8	4	256 cm (compacted sand under pipeline)		0.67
Grout. 9	4	256 cm (10 cm surface replaced by unsaturated gravel)		0.67

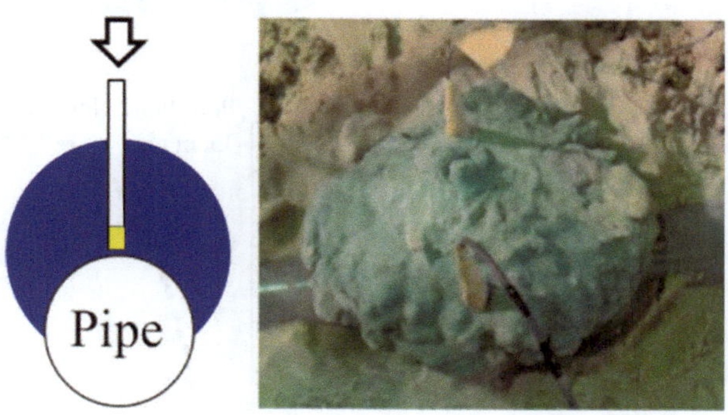

Fig. 27 Example of injection method and products (Grout.2 ~ 3 cases)

In the case of Grout.4 ~ 9, the effect of injection interval and interactions with surrounding compacted sand or pavement were examined. Configuration of soil container is the same as shown in Fig. 5, and a 256-cm pipe was used with the hinged boundary condition (Fig. 28).

The factor of safety against floating (Fs) was calculated by the following equations;

$$F_s = \frac{Total\ weight}{Bouyancy\ force} \tag{4}$$

$$= \frac{\gamma_{pipe} \cdot V_{pipe} + \gamma_{soil} \cdot V_{soil}}{\gamma_{liq} \cdot \left(V_{pipe} + V_{soil}\right)} \tag{5}$$

where γ_{pipe}, γ_{soil} and γ_{liq} are the unit weights of pipe, solidified soil and liquefied soil, respectively. Moreover, V_{pipe}, V_{soil} and V_{liq} designate the volume of the pipe, solidified soil and liquefied soil, respectively.

Figure 29 shows that the pipeline floating is reduced as Fs increases. Thus, Fs affects the velocity of pipeline floating, and consequently residual displacement also changes with Fs. An important point herein is that Fs = 0.8 seems to be sufficient to prevent the pipeline floating.

Figure 30 shows the time history of pipeline displacement with different injection volume and interval. 4 and 8 L of colloidal silica were injected at the centre of the pipeline (injection interval = 256 cm) for the case of Grout.5 and Grout.6. Although the factor of safety was not small compared to the case of Grout.2 and Grout.3, the reduction in displacement was merely 20 % in both cases.

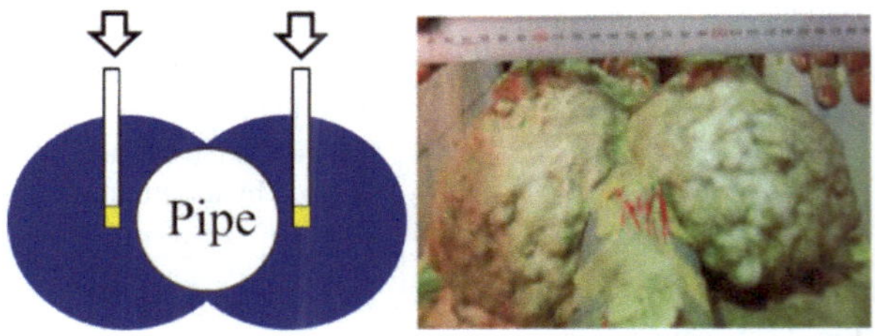

Fig. 28 Example of injection method and products (Grout.5 ~ 9 cases)

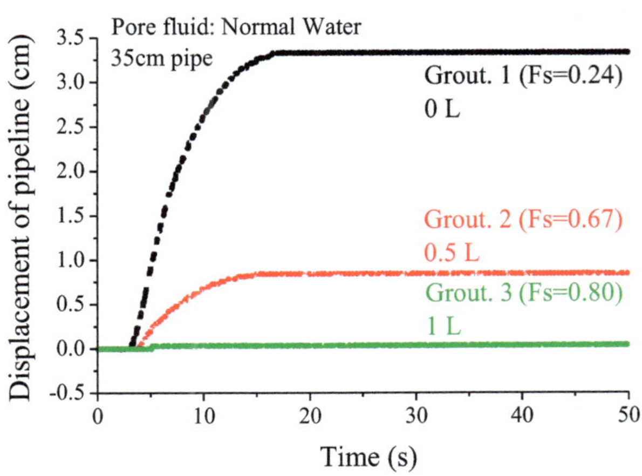

Fig. 29 Time history of pipeline displacement. (The effect of injection volume)

In Grout.7 case, colloidal silica was injected at 5 points (injection interval = 40 cm) to understand the effect of injection interval. The small floating in this test (Fig. 30) indicates that injection interval is one of the key parameters to mitigate the pipeline floating, maintaining a sufficient factor of safety. Figure 31 compares the effect of different injection intervals in Grout. 6 and 7 cases with the same Fs = 0.82. Noteworthy is that the current factor of safety (Eq. 4) cannot account for the efficiency of floating mitigation unless more uniform injection is achieved. In fact, the shortest interval, 35 cm, in Grout.3 case showed no floating with Fs = 0.8, while the longest interval, 256 cm, in Grout.6 case showed significant floating in spite of Fs = 0.82. The reason for this is yet to be known.

An important disadvantage of using the colloidal silica in practice is its high cost as mentioned above. Moreover, a shorter injection interval also leads to a rise of construction budget in spite of its effective mitigation. Therefore, the idea of utilizing a compacted sand layer under the pipeline or unsaturated pavement upon liquefiable layer were examined in the cases of Grout.8 and Grout.9, respectively. Grout.8 intends to connect the pipe to the stable base layer by the solidified soil,

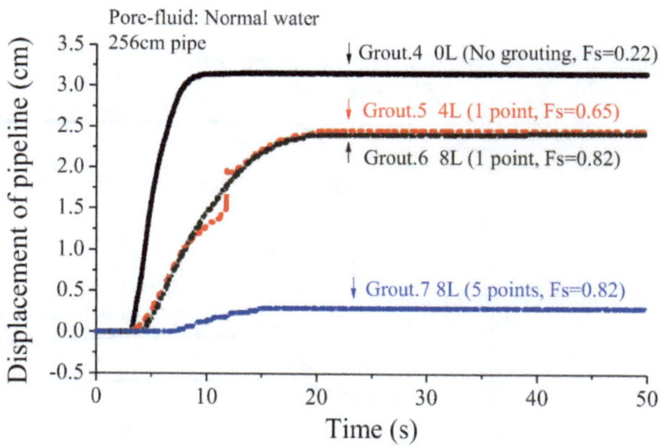

Fig. 30 Time history of pipeline displacement. (The effect of injection volume and interval)

Fig. 31 Comparison on injection interval in the cases of Grout.6 (*Left*) and Grout.7 (*Right*)

while Grout.8 attempts to use the weight of the surface soil to push downwards the pipe. Thus, these methods rely on the structural constraint instead of preventing onset of buoyancy force. The satisfactory mitigation was achieved by these methods as shown in Fig. 32.

Figure 33 shows the detail of a stable structure where the improved soil connects pipeline and compacted sand layer. Thus, the solidified soil worked as an anchorage into the compacted sand during shaking. Consequently, the pipeline floating was mitigated significantly (Fig. 32).

Figure 34 shows the stiff connection between the unsaturated gravel surface (model of pavement) and the embedded pipe. This resistance system is similar to that of the horn-structure method as explained previously. During the injection, colloidal silica flowed into the small gap between sand and gravel layer where the coefficient of permeability was greater than that in the liquefiable sand layer. Consequently, a plate-shape structure was created, which led to a structurally-stable resistance system.

Figure 35 compares the distribution of the bending strain along the pipeline at the time of 20 s in Figs. 30 and 32. Comparison with no mitigation case (Grout.4),

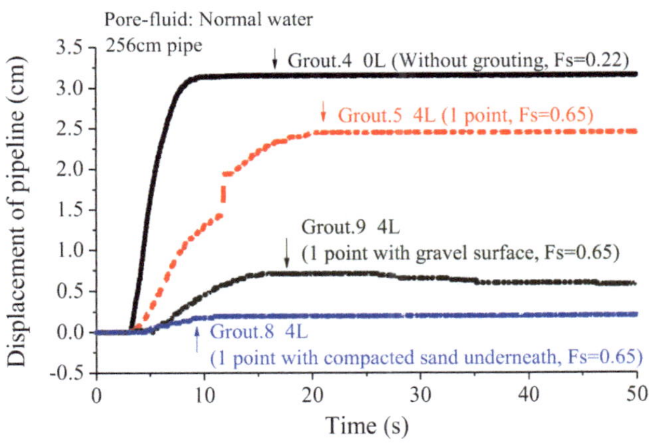

Fig. 32 Time history of pipeline displacement. (The effect of surrounding compacted sand or unsaturated gravel surface)

Fig. 33 Stable structure where the improved soil connects pipeline and compacted sand layer under pipeline (Grout.8)

Grout.5 and 6 show reduced bending strain only at the centre of pipeline, indicating that the effect of chemical grouting was limited only at the injected location. On the other hand, Grout.7 with five injection points shows reduction of bending strain over the entire range. Grout.9 case which was supported by gravel surface also showed nice result. It is noteworthy that Grout.8 case in which the pipe was connected to the compacted base sand at the centre of the pipeline showed negative strain, which clearly indicates that the anchor system worked very effectively.

Figure 36 summarizes the experimental results with chemical grouting method, where the relationship between reduction ratio of pipeline floating and factor of

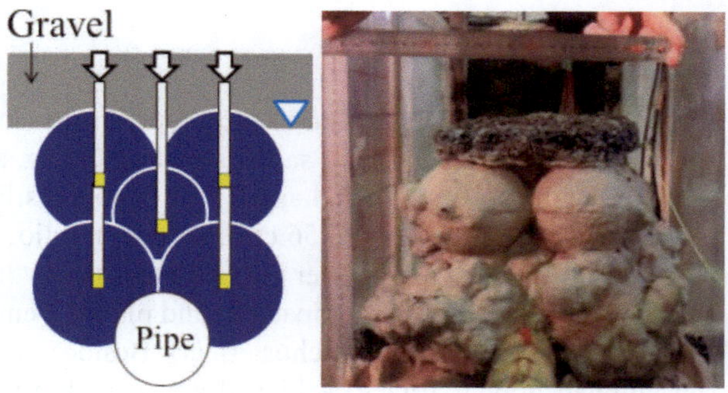

Fig. 34 Connection between unsaturated gravel surface (modelled as pavement) and pipeline (Grout.9)

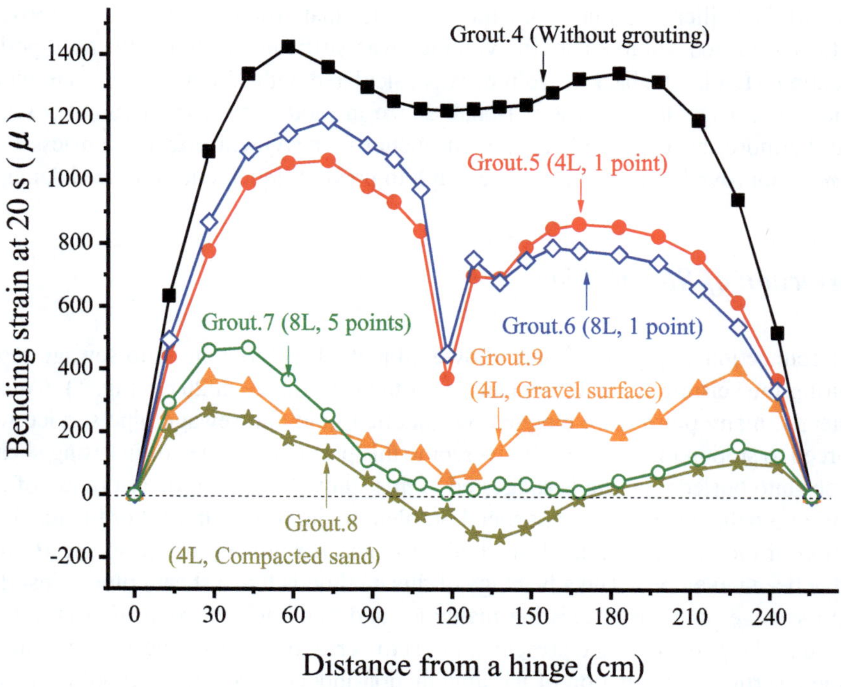

Fig. 35 Comparison on bending strain distribution along pipeline at 20s (Grout.4–9)

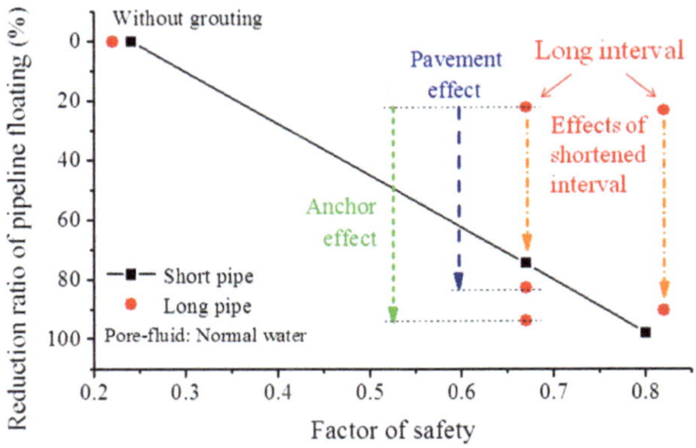

Fig. 36 Relationship between reduction ratio of pipeline floating and factor of safety

safety are discussed. Three data of short pipe results (interval = 35 cm) shows a linear trend between the factor of safety and the floating, and sufficient reduction was achieved with Fs = 0.8. When 4 and 8 L of colloidal silica were injected at the centre of the pipeline (Interval = 256 cm), reduction ratio was clearly smaller than the cases of short pipe, while shorter interval successfully increased the reduction ratio. Moreover, a connection to compacted sand placed 2 cm under pipe prevented pipeline floating significantly (Anchor effect). Besides, a connection to the pavement (unsaturated gravel surface) achieved a remarkable reduction as well (Pavement effect). It should be noted that the volume of solidified sand was 21.4 L when 8 L of colloidal silica was injected, which indicates that around 9.5 % of the improved ratio (solidified volume/backfill volume) was sufficient to mitigate the pipeline floating. Herein, the backfill volume was calculated with 35 cm depth, 25 cm width and 256 cm length which is reasonable size in model test considering 1/5 scale. Furthermore, by taking advantages of anchor or pavement effect, even less than 5 % of improved ratio seems to be enough to prevent liquefaction-induced damage.

Insertion of Sheath Pipe

Disconnection of pipeline joints allowed liquefied soil to come into sewage pipes during the seismic liquefaction caused by the earthquake in 2011 (Fig. 1). Consequently, many pipes were clogged by liquefied sand and sewage pipe service was forced to stop in many cities. To overcome this problem, an idea of inserting sheath pipes into buried pipeline is suggested in this study. Moreover, deterioration of old buried pipeline is an important social problem in the recent times. For the rehabilitation of old and deteriorated pipelines, countermeasures by inserting sheath pipe have been developed. The advantage of this method is that a sheath pipe is inserted into sewage pipes from existing manholes and that backfill excavation is not required. The purpose of the present study is to verify the effectiveness of this sheath pipe insertion in mitigation of the liquefaction-induced problem. Accordingly, two experiments were conducted; one without a sheath pipe (Sheath.1 case), and the other with a sheath pipe (Sheath.2 case). Figures 37 and 38 illustrate the configuration of the tested models with and without a sheath pipe, respectively. The model ground was prepared with Dr = 30 % silica sand entirely. The input acceleration was identical with the one in Fig. 7. Normal water was used as pore fluid.

Two PVC pipes of 128 cm in length were placed close to each other as shown in Fig. 39 (left). The gap between two pipes was around 1 mm, and a waterproof tape was used to connect two pipes as shown in Fig. 39 (right). Because this connection was not strong, it was broken easily when large deformation of buried pipeline occurred upon liquefaction. Figure 40 shows the sheath pipe model used in this research. This sheath is made of polyvinylchloride resin. It is very flexible and deformed substantially more easily than the PVC pipeline.

To visually monitor the situation inside a pipe, a web camera and an LED lamp (light-emitting diode) were installed in the pipe as shown in Fig. 41.

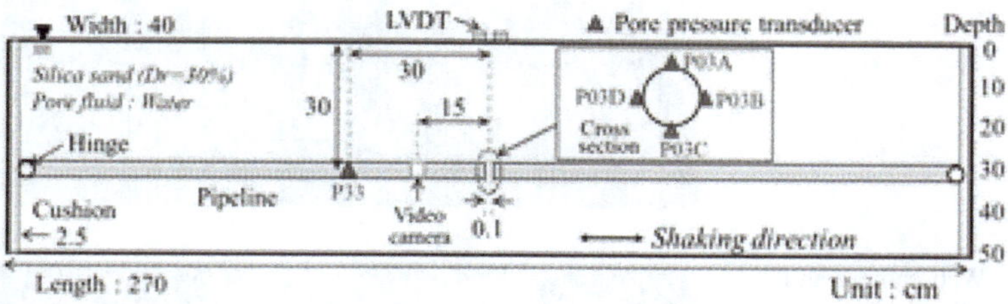

Fig. 37 Configuration of soil container without sheath pipe (Sheath.1 case)

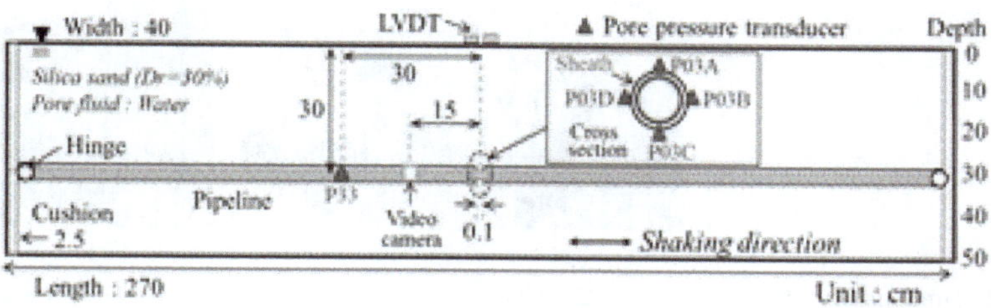

Fig. 38 Configuration of soil container with sheath pipe (Sheath.2 case)

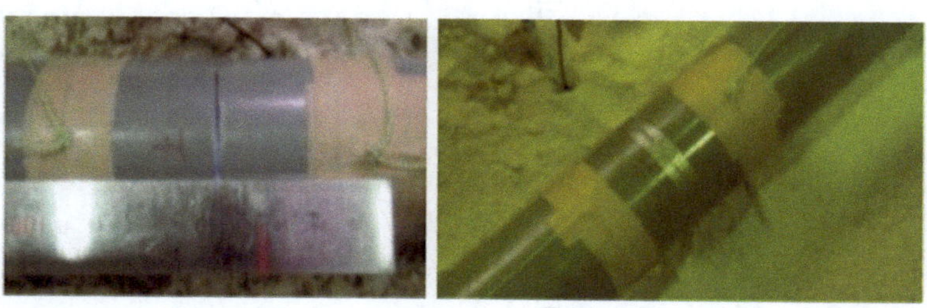

Fig. 39 Connection of two PVC pipes

Figure 42 shows the time history of pipeline displacement with and without a sheath pipe. Apparently, the inserted sheath pipe could not reduce the floating effectively. Figure 43 shows the time history of smoothed excess pore pressure ratio measured at the location of P33 and the connection of two pipelines (average of P03A~D); for their locations, see Figs. 37 and 38. Because the displacement of pipeline and sensor attached on it was significant compared to the overburden depth, a modified formula was used for the calculation of excess pore pressure ratio (E.P.P.R.) as follows;

$$E.P.P.R. \;=\; \frac{U_d + \gamma_w \cdot \delta_{pipe}}{\sigma'_i - \gamma' \cdot \delta_{pipe}} \tag{6}$$

Fig. 40 Sheath pipe inside PVC pipe (Sheath.2 case)

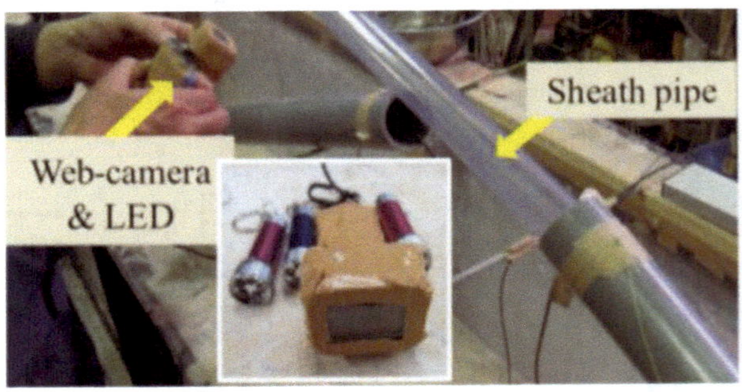

Fig. 41 Web-camera and LED installed inside pipe

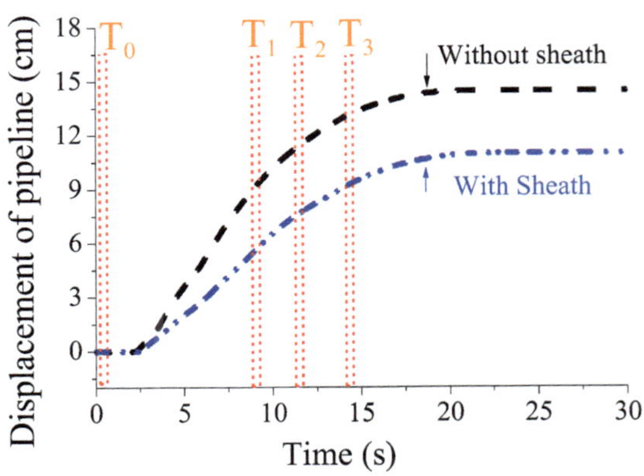

Fig. 42 Time history of pipeline floating. (The effect of sheath pipe insertion)

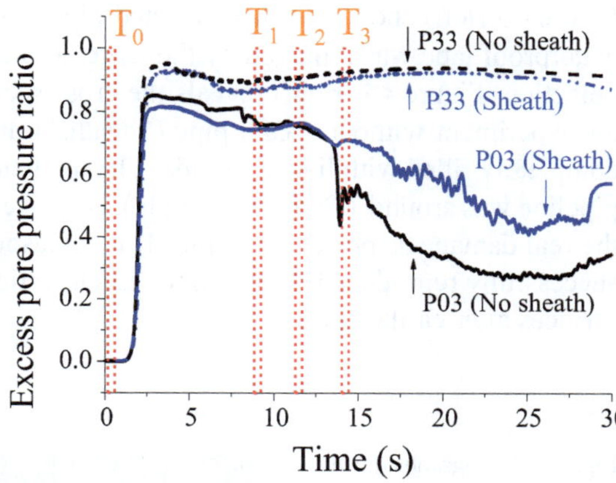

Fig. 43 Time history of excess pore water pressure ratio. (The effect of sheath pipe insertion)

where U_d = measured excess pore pressure, δ_{pipe} = floating displacement of pipeline, σ_i' = initial effective stress at the depth of sensor, and γ_w and γ' = unit weight of water and submerged sand, respectively.

When the excess pore pressure ratio increased close to 1.0 in Fig. 43 (at around 2 s), pipeline started to float in Fig. 42. Because the overall density of the pipeline was increased by 1.9 times when a sheath pipe was inserted, the velocity of pipeline floating became smaller than in the case without a sheath pipe. After the peak value of P03 and P33, only P33 was kept close to 1.0. In contrast, P03 showed a downward trend and significant reduction around the time of T_3. This magnitude of reduction was greater in the case without a sheath pipe. It is most likely that this pore pressure reduction was induced by the flow of liquefied backfill sand into the pipeline through the disconnection. Although E.P.P.R. decreased around the connection of pipelines, pipeline floating continued until the magnitude of input acceleration became small. The possible reason is that the reduction of E.P.P.R. occurred only around the connection of two pipes (total length = 256 cm); therefore, buoyancy force still remained in a significant part of ground.

The photographs taken by the web camera during shaking in Sheath.1 and Sheath.2 cases are shown in Fig. 44. T_1–T_2 in Fig. 44 show the time of photographs which are identical with those shown in Figs. 42 and 43. The camera inside the pipeline was placed at 15 cm away from the connection of two pipes, and the following features are observed. For the case without a sheath pipe (Sheath.1);

1. As the pipeline deformed upward due to liquefaction, a small breakage occurred at T_1.
2. Then, liquefied sand started to flow into the pipeline through the disconnection between two pipelines.
3. As the disconnection became bigger, liquefied sand started to flow into the pipe at T_2.
4. Pipeline was gradually clogged by liquefied sand, and, eventually, the entire cross section was clogged at T_3. This time is almost identical with the moment when the significant reduction in E.P.P.R. was observed in Fig. 43.

Figure 45 (left) shows the disconnected pipe joint without sheath (Sheath.1). The waterproof tape was removed by the large deformation of pipeline during soil liquefaction. Figure 45 (right) reveals the cross section of the excavated pipeline after experiment without sheath pipe (Sheath.1). It is obvious that the pipeline was completely filled with liquefied sand. The total amount of sand that flowed into the pipeline was around 1.3 kg (dry weight). Because this observation is very similar to the real damage as presented in Fig. 1, it is reasonable to state that this experiment successfully reproduced the pipeline disconnections and serious damage in lifeline services after earthquakes.

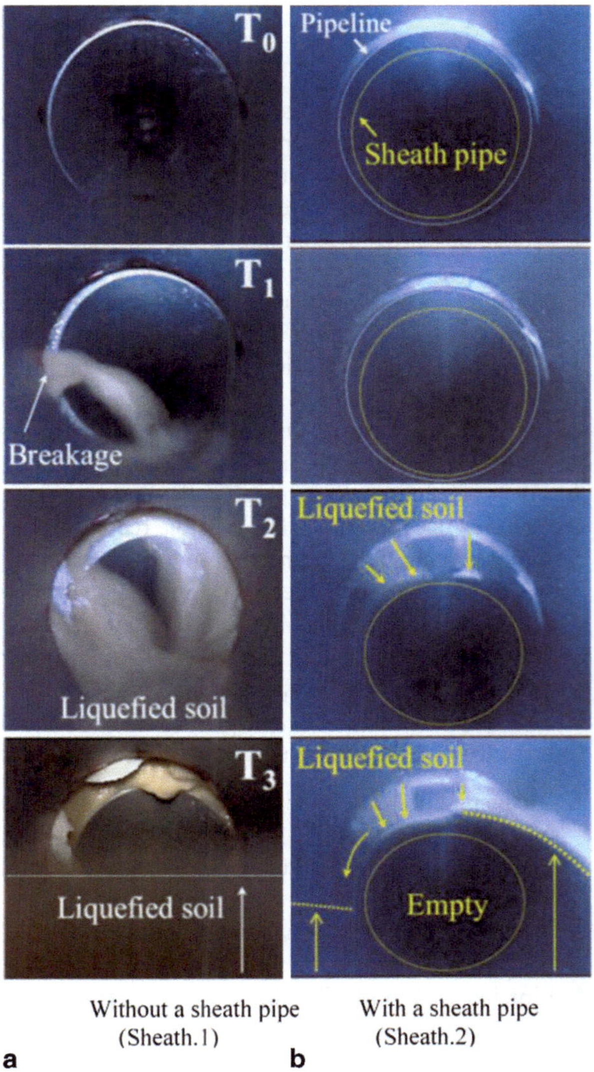

Fig. 44 Photographs during shakings taken by a camera inside pipeline

Fig. 45 Joints disconnection and cross section of broken pipeline without sheath

For the case with a sheath pipe (Sheath.2), the observed features by the camera are summarized as follows.

1. Liquefied sand started to flow into pipeline at around T_2, and then liquefied sand continued to flow between the sheath pipe and the pipeline.
2. After T_3, the liquefied sand gradually moved toward the front side.
3. Eventually, the space between the sheath pipe and the pipeline was clogged by liquefied soil.
4. The sheath pipe successfully maintained the open space inside, allowing the operation of the sewage pipe after subsoil liquefaction.
5. The liquefied sand filled the small space between the sheath pipe and the outer PVC pipe. Because this space was small. the reduction of E.P.P.R. was less significant in Fig. 43.

In comparison with the case of Sheath.1 without a sheath pipe, the time required for pipeline disconnection was longer because the velocity of the pipeline floating was lower than Sheath.1. Because the sheath pipe was very flexible, no damage was observed therein. Thus, a sufficient cross section remained in the pipe for water to flow after the onset of liquefaction. As illustrated in Fig. 46 (left side), no liquefied sand was observed inside the sheath pipe. The total amount of soil that came into

Fig. 46 Joints disconnection of buried pipelines and sheath pipe in it

the space between sheath pipe and pipeline was 0.45 kg (dry weight), which is 35 % of that in the case without sheath pipe (Sheath.1). Thus, it seems possible to continue sewage service after a strong earthquake. It is noteworthy that the sheath pipe could not reduce the floating and the change of the gradient of the pipeline. Because the pipeline gradient is essential for the operation of sewage pipeline, it is recommendable that the sheath pipe is used together with other mitigation technologies as described in this paper.

Conclusion

The seismic resistance of lifelines is one of the key issues in earthquake resistance of modern communities. However, technology has not well developed to protect lifelines from subsoil liquefaction. This situation is particularly significant in embedded sewage pipelines that were seriously damaged during the great earthquake in 2011 in Japan. One of the reasons for this is the substantial length of the lifeline and the other is the limited financial resources. In this regard, the present study conducted shaking model tests in order to validate several new technologies that were supposed to improve the liquefaction resistance of buried pipelines. Consequently, the following conclusions were drawn.

1. Because excavation and backfilling of existing pipes are costly, new technologies that do not require the overall excavation have to be investigated.
2. To avoid overall excavation, the good performance of a horn-shape device, drain pipes, grout injection, and insertion of a sheath pipe was demonstrated.
3. To facilitate practical design of the measures mentioned above, brief ideas of factor of safety against floating and configuration were demonstrated.
4. Through the experiments of a horn-shape device, an estimation formula of punching failure against gravel surface was suggested. This would also be applied to drainage pipes and grout injection method
5. It is interesting that the pavement upon liquefiable layer is utilized as a structural supporter.

Acknowledgments The present study was financially supported by the Ministry of Land, Infrastructure, Transport and Tourism, Japan. Tested colloidal silica was supplied by Kyokado Engineering Company. Those supports are deeply appreciated by the authors.

References

Japan Sewage Works Association (2006) Technical guidance for seismic resistant design of sewage system, pp 140–146 (In Japanese)

Kitaura M, Miyajima M, Yoshida M (1988) Model experiments on damage mitigation of buried pipelines due to liquefaction using gravel piles. Proceeding Faculty of Engineering, University of Kanazawa, Vol 21, No 2, pp 103–109 (In Japanese)

Koseki J, Matsuo O, Tanaka S (1998) Uplift of sewer pipes caused by earthquake-induced liquefaction of surrounding soil. Soils Found 38(3):75–87

Miyake T, Wada M, Miyajima S, Ishida S (1998) A follow-up investigation of back-filled cement treated soils behind quay walls in the earthquake restoration works. Toyo Construction Technical Research Reports, No. 25, pp 17–24 (In Japanese)

Technical Committee on Earthquake Resistant Design of Sewage Lifelines (2008) Earthquake resistant design of sewage lifelines. Committee Report (In Japanese)

Towhata I (2008) Geotechnical earthquake engineering. Springer-Verlag, Berlin, Heidelberg, Sect. 24.14, pp 520–523

Towhata I, Kabashima Y (2001) Mitigation of seismically-induced deformation of loose sandy foundation by uniform permeation grouting. Proceeding Earthquake Geotechnical Engineering Satellite Conference, XVth International Conference on Soil Mechanics and Geotechnical Engineering, Istanbul, Turkey, pp 313–318

Towhata I, Rasouli R (2013) Attempts to protect personal houses from seismic liquefaction problem. Proceeding 4th International Seminar on Forensic Geotechnical Engineering ISFGE, Bangalore, India, pp 191–209

Towhata I, Otsubo M, Uchimura T, Shimura M, Liu B, Hayashida T (2013) Protection of underground lifeline from seismic liquefaction problems. Proceeding 5th International Geotechnical Symposium-Incheon, pp 3–18

Effect of Long Duration of the Main Shock and a Big Aftershock on Liquefaction-Induced Damage During the 2011 Great East Japan Earthquake

Susumu Yasuda, Iwao Suetomi and Keisuke Ishikawa

Abstract The duration of shaking during the 2011 Great East Japan Earthquake was extremely long, and the main shock was soon followed by big aftershocks because the earthquake was a "megathrust earthquake" with extremely large magnitude; $Mw = 9.0$. The unique ground shaking caused the following unusual events: (1) serious liquefaction occurred in a wide area of reclaimed land along Tokyo Bay though seismic intensities in the liquefied zones were not high; (2) some inhabitants testified that boiling did not occur during the main shock but occurred during a big aftershock. The occurrence of liquefaction, the settlement and the inclination of houses must have been affected by the aftershock; (3) shaking continued for a long time after the occurrence of liquefaction. Due to the shaking of the liquefied ground, large horizontal displacement, which is a kind of sloshing of liquefied ground, was induced and caused roads to thrust; (4) the large horizontal displacement of liquefied ground had to have caused the severance of pipe joints and the shear failure of manholes, allowing an influx of muddy water into the pipes and manholes.

Keywords Liquefaction · Megathrust earthquake · Lifeline · House · Road

Introduction

The 2011 Great East Japan Earthquake, a megathrust earthquake with a magnitude of $Mw = 9.0$, occurred in the Pacific Ocean about 130 km off the northeast coast of Japan's main island. Soil liquefaction occurred in the Tohoku region of northeastern Japan and in the Kanto region surrounding Tokyo because the earthquake was huge. The distance between the northernmost and the southernmost liquefied sites was about 600 km. Many houses, roads, lifeline facilities, and river dikes were severely damaged by liquefaction, and some port and harbor facilities, tank yards,

S. Yasuda (✉) · K. Ishikawa
Tokyo Denki University, Tokyo, Japan
e-mail: yasuda@g.dendai.ac.jp

I. Suetomi
Eight-Japan Engineering, Tokyo, Japan

A. Ansal, M. Sakr (eds.), *Perspectives on Earthquake Geotechnical Engineering*,
Geotechnical, Geological and Earthquake Engineering 37,
DOI 10.1007/978-3-319-10786-8_13

electric power stations and tailing dams were also damaged in both the Tohoku and the Kanto regions. In the Tokyo Bay area, the southern district of the liquefied area, though the epicentral distance was very large, about 380–400 km, severe liquefaction occurred in a wide area because many liquefiable reclaimed lands had been constructed along Tokyo Bay. As the 2011 Great East Japan Earthquake was giant in scale, the duration of the main shock was very long, and it was followed by many big aftershocks. These factors caused remarkable, serious liquefaction-induced damage. Many houses settled and tilted due to the main shock and to a big aftershock 29 min later, and a kind of sloshing of liquefied grounds occurred and caused the buckling of roads and the shear failure of sewage manholes. In this paper, the effect of the very long duration of the main shock and of the aftershock on the occurrence of liquefaction and associated damage to houses, roads and sewage facilities are discussed.

Hypocentral Region and Sesimic Waves at Several Sites

The 2011 Great East Japan Earthquake, with a magnitude of $Mw=9.0$, the largest magnitude ever recorded in Japan, occurred in the Pacific Ocean about 130 km off the northeast coast of Japan's main island at 14:46 on March 11, 2011. The rupture plane of this quake was about 450 km in length and 200 km in width. The duration of the failure of the rupture plane was extremely long, about 160 s. Many aftershocks with large magnitudes followed in a wide area immediately after the main shock in and around the rupture zone, as shown in Fig. 1. The biggest aftershock, with a magnitude of $Mw=7.6$, occurred at 15:15, 29 min after the main shock near the southwest boundary of the rupture zone, and the epicenter was only about 110 km from the Tokyo Bay area. Figure 2 shows the distribution of seismic intensities in JMA scale. Very intense shaking of 6–7 in JMA scale, which corresponds to 9–12 in M.M. scale, occurred in many areas from the Tohoku region to the Kanto region. However, the intensity in the Tokyo Bay area, where severe liquefaction occurred, was not high, about 5- 5+in JMA scale, which is around 7–8 in M.M. scale.

Many seismic waves were recorded by several organizations. The ground surface accelerations measured by K-NET (NIED 2011) at six sites are shown in Fig. 3. Two or three peaks, with an interval of about 1 min, hit the Tohoku region and the total duration was very long. In the Kanto region, though there was only one peak, the duration of shaking was very long, i.e., 2–3 min.

Liquefed Sites in Kanto Region

As the liquefaction-induced damage to houses, river dikes, roads, and lifeline facilities was serious, the Kanto Regional Development Bureau of the Ministry of Land, Infrastructure, Transport and Tourism conducted joint research with JGS to

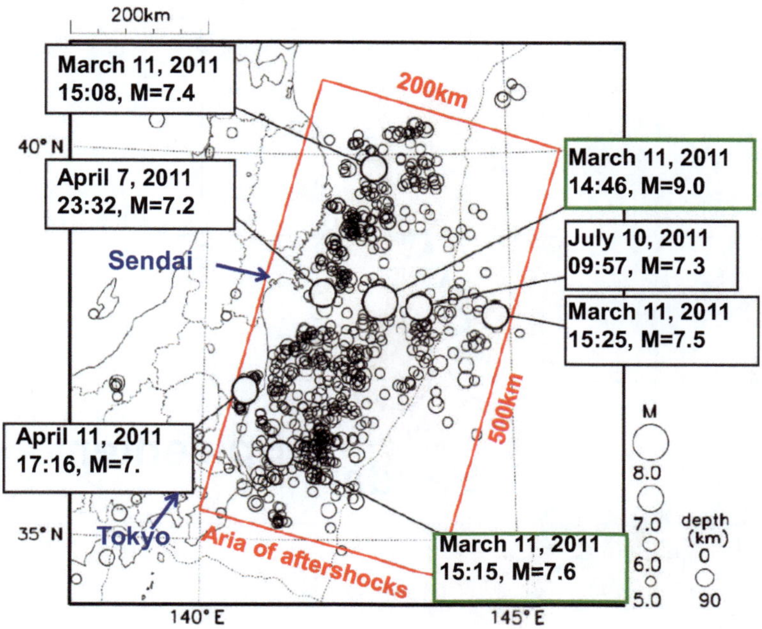

Fig. 1 Locations of the main shock and aftershocks during the 2011 Great East Japan Earthquake. (Partially quoted from JMA)

identify liquefied sites. The results of their joint research were published on the government ministry's website. Figures 4 and 5 are maps of the liquefied sites in the Kanto region and in the Tokyo Bay area, respectively. The farthest liquefied site was Minami-boso City in the Boso Peninsula, where the epicentral distance was 440 km, as measured by Wakamatsu (Wakamatsu 2011). The epicentral distance of the farthest site is plotted against surface magnitude Ms in Fig. 6, assuming $M_w = 9.0$ is equivalent to $Ms = 8.4$. The relationship coincides well with the four relationships proposed by Kuribayashi & Tatsuoka, Ambraseys and Wakamatsu based on past earthquakes.

All of the northern part of Tokyo Bay liquefied, but in the eastern and western parts of Tokyo Bay, liquefaction was observed only in spots (Yasuda et al. 2012, 2013). In the northern part of Tokyo Bay, the ground surface was covered with boiled sands all around the reclaimed lands in Shinkiba in Tokyo, Urayasu City, Ichikawa City, Narashino City and western Chiba City. In contrast, boiled sands were observed only here and there in the reclaimed lands in Odaiba, Shinonome, Tatsumi, Toyosu and Seishin in Tokyo and in eastern Chiba City. The total liquefied area from Odaiba to Chiba City reached about 42 km^2. Many houses, roads, and lifeline facilities were severely damaged in the liquefied zones. The most serious damage was in Urayasu City, where about 85 % of the city foundation soil liquefied. The epicentral distance of the liquefied area in Tokyo Bay was very long, around 400 km. However, the distance from the boundary of the rupture plane was only 110 km because the rupture plane was very wide, as shown in Fig. 1.

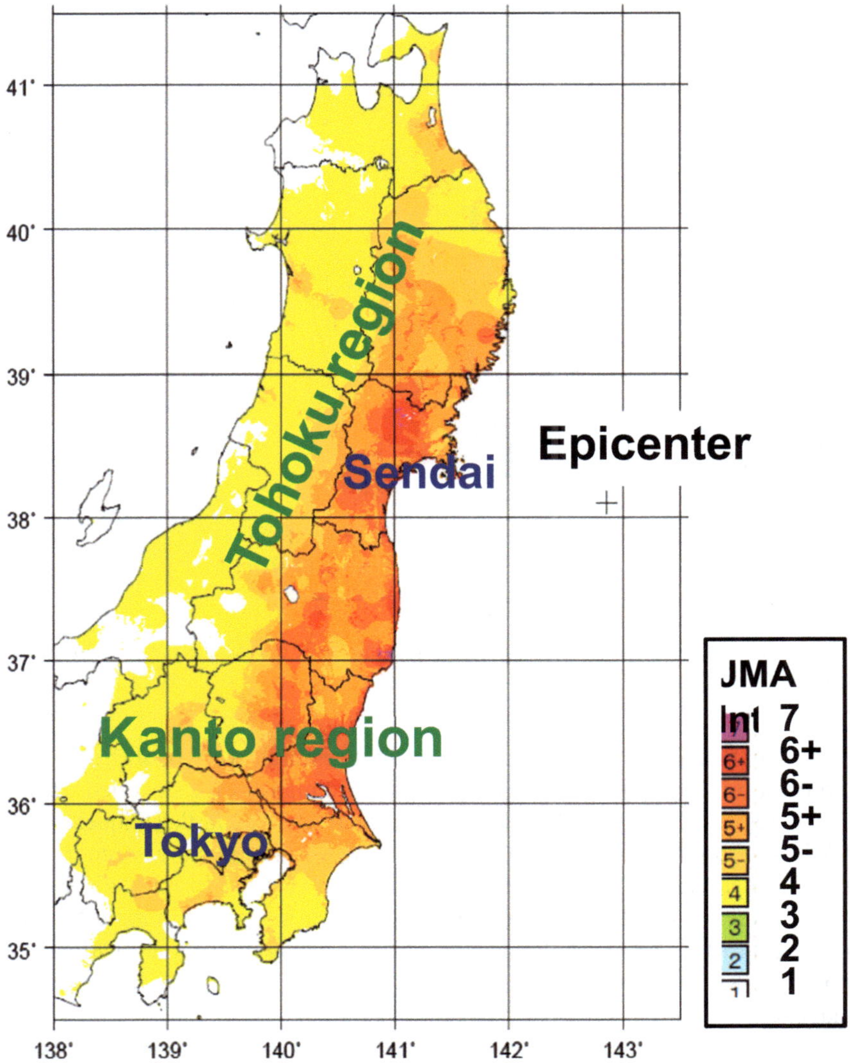

Fig. 2 Distribution of JMA seismic intensities. (Partially quoted from JMA)

Effect of Long Duration and the Main Shock and an Aftershock on Occurrence of Liquefaction

In the Tokyo Bay area, though surface accelerations were not high, about 160–300 cm/s^2, due to the huge scale of the earthquake, the duration of shaking was extremely long and accompanied with strong aftershocks in a short interval. Peak horizontal accelerations composed of NS and EW components during the main shock and a big aftershock 29 min after the main shock at 7 K-NET sites are indicated in Fig. 7. The peak accelerations during the aftershock were almost half those

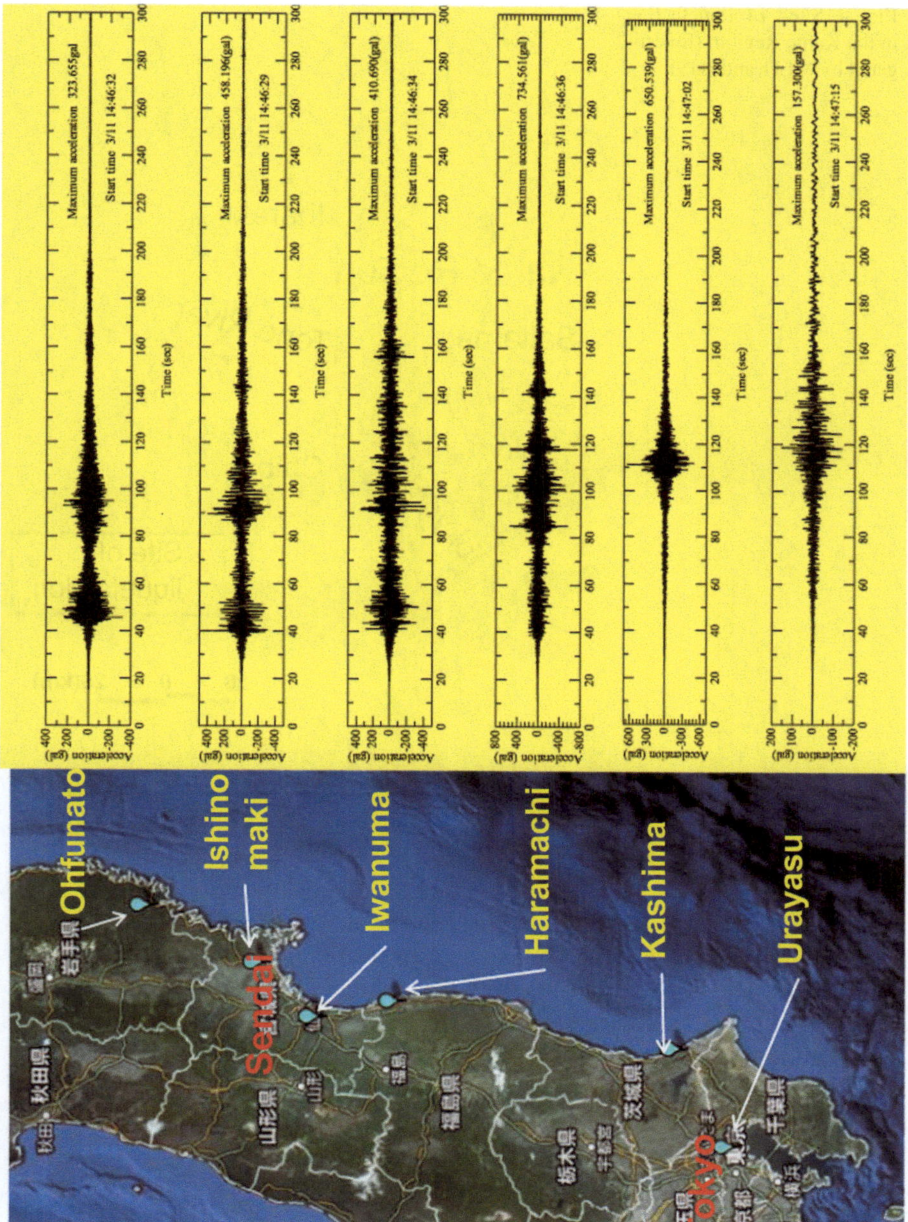

Fig. 3 Seismic waves at six sites recorded by K-net

during the main shock in the Tokyo Bay area. Some inhabitants of Urayasu City observed the boiling of muddy water immediately after the main shock, while others saw spouts of muddy water 5–9 min after the main shock. Surprisingly, some inhabitants testified that boiling did not occur during the main shock but occurred

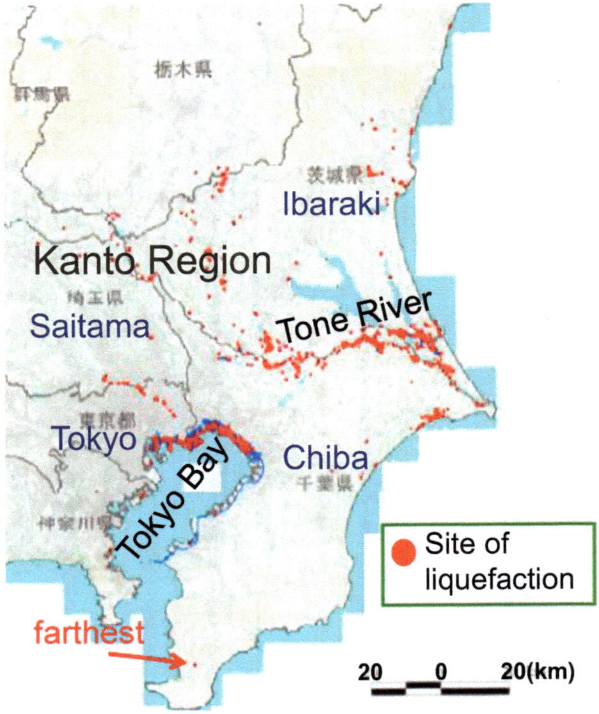

Fig. 4 Sited of liquefaction in the Kanto Region. (Investigated by MLIT and JGS)

during the aftershock. Some excess pore-water pressure may have built up during the main shock, causing complete liquefaction during the aftershock. It is impossible to evaluate the liquefaction time at these sites, but it must have differed by place because of the long duration of the main shock and the subsequent large aftershock.

One of the authors and his colleagues conducted cyclic torsional shear tests to evaluate the effect of the long shaking on the occurrence of liquefaction (Ishikawa et al. 2014). Silty sand taken from the boiled sand in Urayasu City and clean sand called Toyoura sand were used for the cyclic torsional shear tests. The fines content was 36 and 0% for the Urayasu sand and the Toyoura sand, respectively. These sands were poured into a mold with three levels of relative density. Two types of shear wave were applied to the specimens, a sine wave of 20 cycles and the seismic wave recorded during the main shock and the aftershock at Urayasu K-NET. In the case of the seismic wave, excess pore water pressure increased gradually with shear stress, as shown in Fig. 8. Relationships between the stress ratio R (τ_d/σ_c') for the sine wave or Rmax (τ_{max}/σ_c') for the seismic wave and the residual excess pore water pressure ratio u/σc' are plotted in Fig. 9, and the stress ratio to cause liquefaction by 20 cycles of sine wave R_l ($\tau_{d,l}/\sigma_c$') and by the seismic wave Rmax, l ($\tau_{max,l}/\sigma_c$')

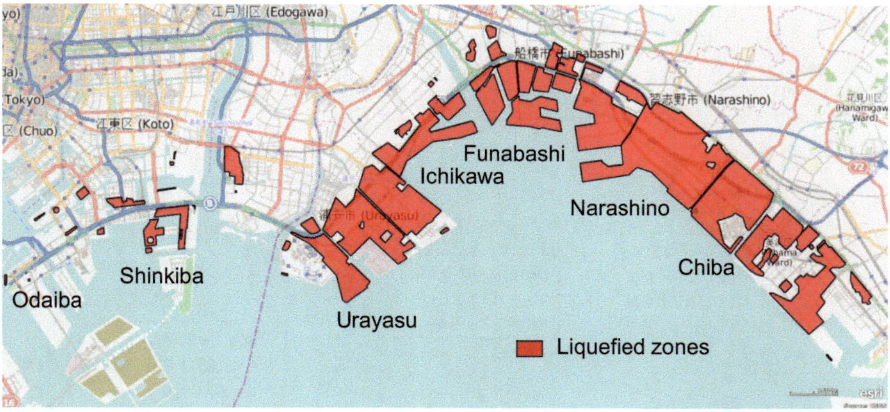

Fig. 5 Areas of liquefaction in the Tokyo Bay area. (Investigated by MLIT and JGS)

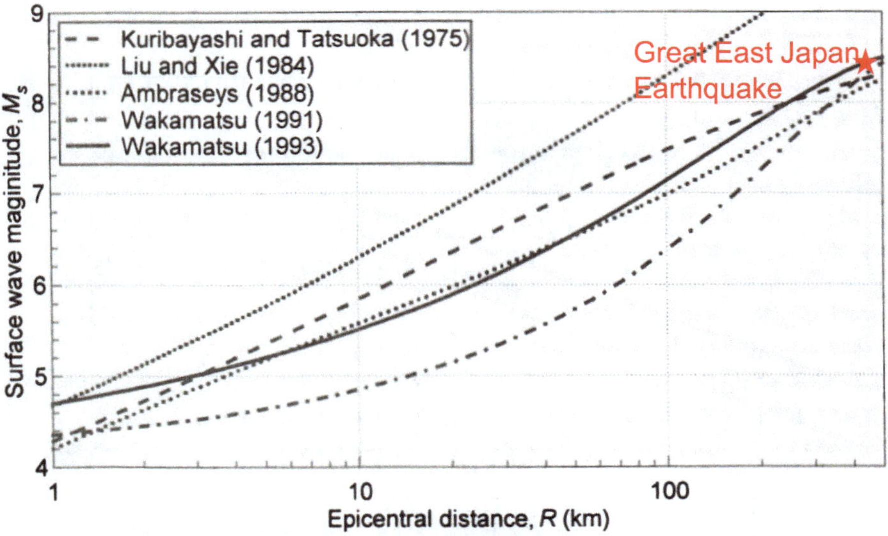

Fig. 6 Relationships between *M*s and maximum epicentral distance of liquefied site

were read at $u/\sigma_c'=0.95$. Then the correction factors $C_2=R\max,1\,(\tau_{\max,1}/\sigma_c')/R_l(\tau_{d,1}/\sigma_c')$ were calculated. Figure 10 shows the relationships between the C_2 thus derived and $R_l(\tau_{d,1}/\sigma_c')$. The C_2 decreased from about 1.5–0.9 with an increase of $R_l(\tau_{d,1}/\sigma_c')$. In the current seismic design codes for highway bridges in Japan (JMR 1996), the correction factor for normal seismic waves is assumed to be around 1.5. However, the test results imply that a smaller C_2 must be used in seismic design for dense sands if the duration of the seismic wave is long.

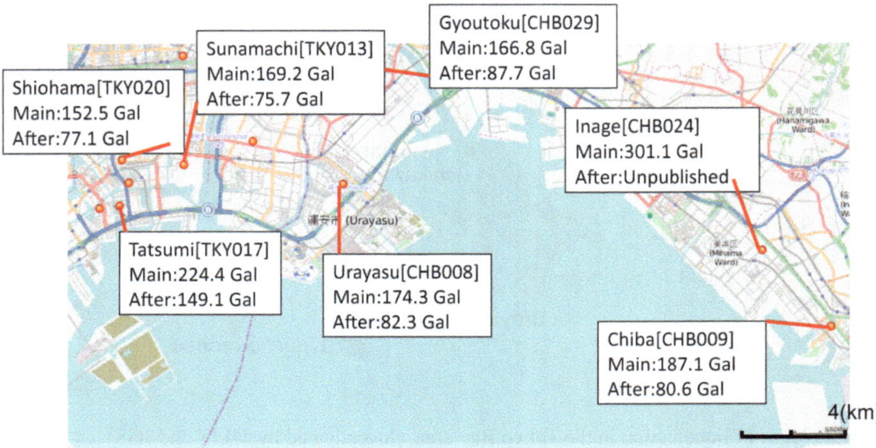

Fig. 7 Surface accelerations during main shock and aftershock 29 min later recorded by K-NET

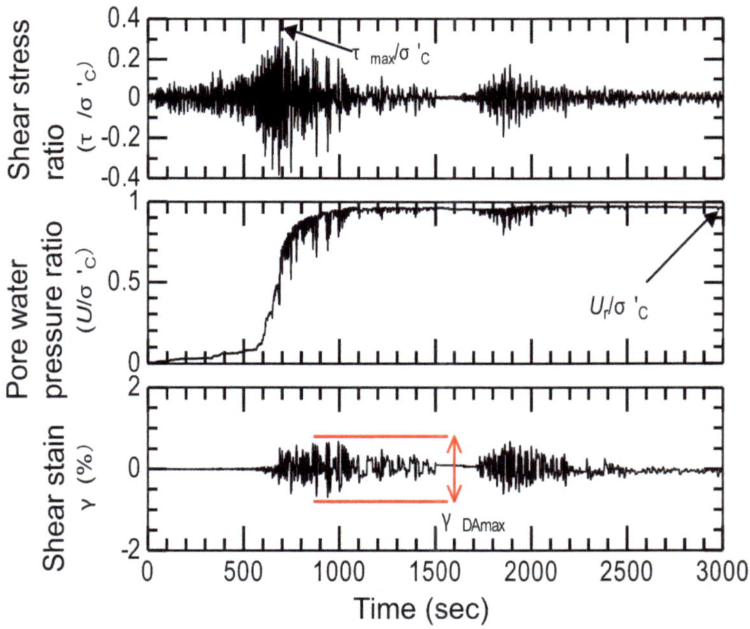

Fig. 8 Results of cyclic torsional shear tests to study correction factor

Fig. 9 Relationships between stress ratio and residual excess pore water pressure ratio

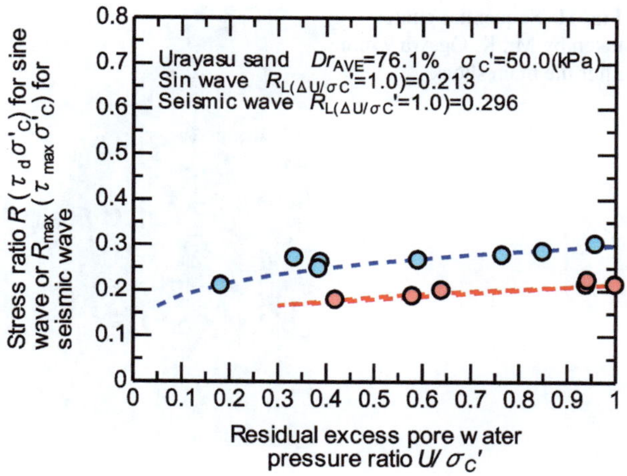

Fig. 10 Relationship between correction factor C_2 for the wave recorded at Urayasu and cyclic stress ratio to cause liquefaction

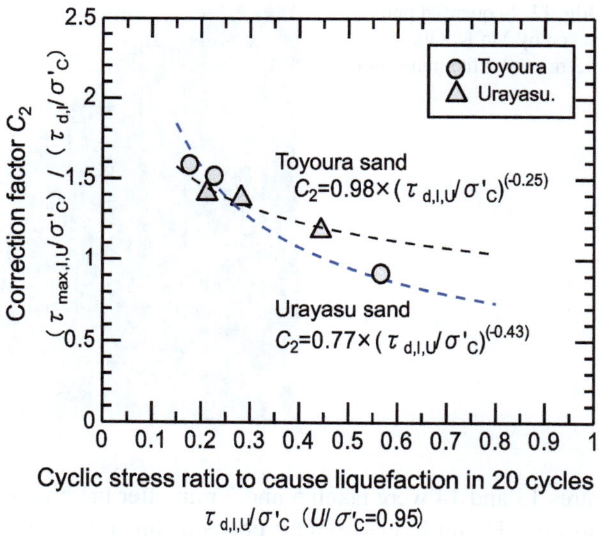

Effect of a Big Aftershock on Liquefaction-Induced Damage to Houses

As the main shock and the big aftershock hit the Tokyo Bay area at 14:47 and 15:16, respectively, many important photos and movies were taken at many sites to study the process and mechanism of liquefaction. A series of photos taken in Urayasu City by Mr. Katsunori Ogawa just after the earthquake is introduced in Figs. 11, 12, 13 and 14 (Yasuda et al. 2012). Figures 11 and 12 were taken 9 and 13 min after the main shock, respectively, before the big aftershock. Muddy water spewed from the ground in a northeast direction several minutes after the main shock subsided. The boiling of muddy water was not intense after the main shock, more like ooze. Fig-

Fig. 11 Sequential photos taken by Mr. K. Ogawa 9 min after the main shock

Fig. 12 Sequential photos taken by Mr. K. Ogawa 13 min after the main shock

ures 13 and 14 were taken 5 and 7 min after the big aftershock at the same avenue as Figs. 11 and 12 but from the opposite direction. The road was covered completely with muddy water, many houses settled and tilted, and cars were submerged in the boiled muddy water.

The authors sent out questionnaires to about 100 inhabitants of Urayasu City to ask the timing of boiling and the height of boiled muddy water. Their answers are summarized in Fig. 15. Only 30 % of the respondents observed the boiling of muddy water immediately after the main shock. However, other respondents recognized spouts of muddy water several minutes or more after the main shock. The height of the muddy water was not high, mainly less than 10 cm. 68 % of the respondents mentioned that the boiling of muddy water continued up to the aftershock, and 86 % of the respondents saw covered water until the aftershock. On the contrary, 79 % observed the spewing of muddy water just after the aftershock at a height slightly greater than the height after the main shock. This means the boiling accelerated due to the aftershock and more severe liquefaction occurred during the aftershock

Fig. 13 Sequential photos taken by Mr. K. Ogawa 5 min after the aftershock

Fig. 14 Sequential photos taken by Mr. K. Ogawa 7 min after the aftershock

at some sites. A rise in the water table after the main shock might have accelerated boiling during the aftershock, or some fissure induced in the ground during the main shock might have accelerated boiling during aftershock.

According to the Ministry of Land, Infrastructure, Transport and Tourism (MLIT), about 27,000 wooden houses in Japan were damaged due to liquefaction by the 2011 Great East Japan Earthquake. In the design of wooden houses, liquefaction has not been considered, whereas the design code for other buildings has considered liquefaction since 1974. This is the main reason such a large number of houses were damaged. In May 2011, the Japanese Cabinet announced a new standard for the evaluation of damage to houses based on two factors, settlement and inclination. A new class of "large-scale half collapsed house" was also introduced, and houses tilted at angles of more than 50/1000, of 50/1000 to 16.7/1000, and of 16.7/1000 to 10/1000 were judged to be totally collapsed, large-scale half collapsed and half collapsed houses, respectively, under the new standard. In Urayasu City,

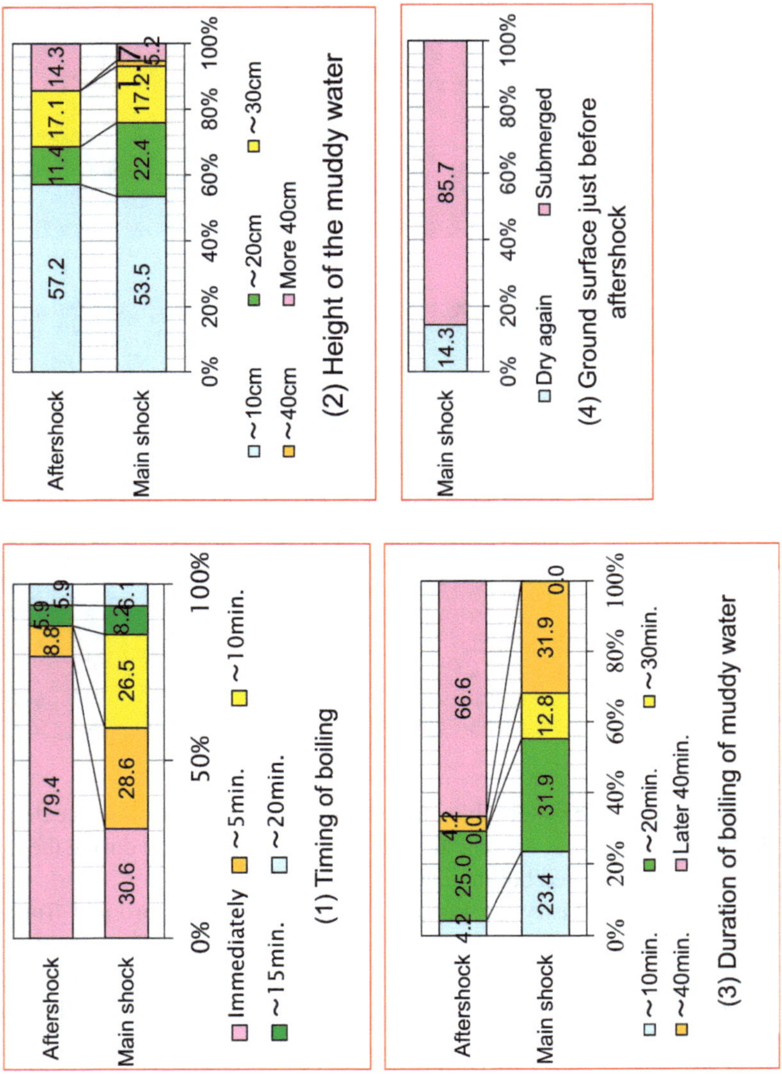

Fig. 15 Responses to questionnaires sent to inhabitants of Urayasu City

the effect of the depth of the water table on the extent of damage to houses was studied and is summarized in Fig. 16. As shown in the figure, the average depths of the water table at sites of completely damaged, large-scale partially damaged, partially destroyed and partially damaged housing lots were 1.38, 1.78, 1.84 and 2.01 m, respectively. Based on this study, one explanation of the effect of the big aftershock on the settlement and tilting is illustrated in Fig. 17. If the depth of the ground water table was medium, such as GL-2 m, houses did not settle and tilt during the main shock though liquefaction occurred in the soil layer under the ground water table, because the ground beneath the footing did not liquefy and was still strong. After the main shock, the water table rose gradually to the ground surface due to the increase of pore water pressure of the liquefied soil layer, and then all the surface soil layers were saturated. During the aftershock, the ground beneath the footing liquefied and caused the settlement and tilting of the house.

Effect of Continued Strong Shaking After the Occurrence of Liquefaction

Many seismic records were obtained in the Tokyo Bay area, as mentioned before. Among them, accelerograms recorded at K-NET Inage in Chiba, shown in Fig. 18, where boiled sand was observed are very important because the liquefaction time can be judged from the recorded waves. Figure 19 shows the accelerograph recorded at Urayasu on ground where liquefaction did not occur. Both records started at almost the same time; 14:46:16 at Inage and 14:46:15 at Urayasu. At Urayasu, the wave frequency did not change drastically after the peak acceleration, which was induced at about 118 s. (14:48:13). On the contrary, at Inage, the wave frequency dropped to a low value after two peaks at 120 s. (14:48:16) and 126 s. (14:48:22) and the amplitude of acceleration decreased suddenly. Therefore, it can be judged that liquefaction occurred at around 14:48:16 to 14:48:22 at Inage. This means many cycles of shear stress, say around 20 cycles over 110 s, might have caused liquefaction at the Inage site.

In Figs. 18 and 19, time histories of velocity, displacement and non-stationary spectra estimated from the recorded acceleration are shown. In Inage's wave, though the amplitude of acceleration decreased suddenly at around 126 s., large amplitudes of velocity and displacement continued after 126 s. for more than 2 min with a predominant period of 3–4 s. Such amplitudes of velocity and displacement were not induced in Urayasu's wave. Shaking continued at a long predominant period for a long time after the occurrence of liquefaction.

The long-period shaking of liquefied ground, which was a kind of sloshing, was recorded by several inhabitants on video. One video taken just after the peak acceleration at Makuhari in Chiba City showed the strange cyclic heaving of a footway at a period of about 4 s. Slow cyclic horizontal movement at an amplitude of about 15 cm was seen in another video taken 2 min after the peak acceleration in the Akemi district of Urayasu City.

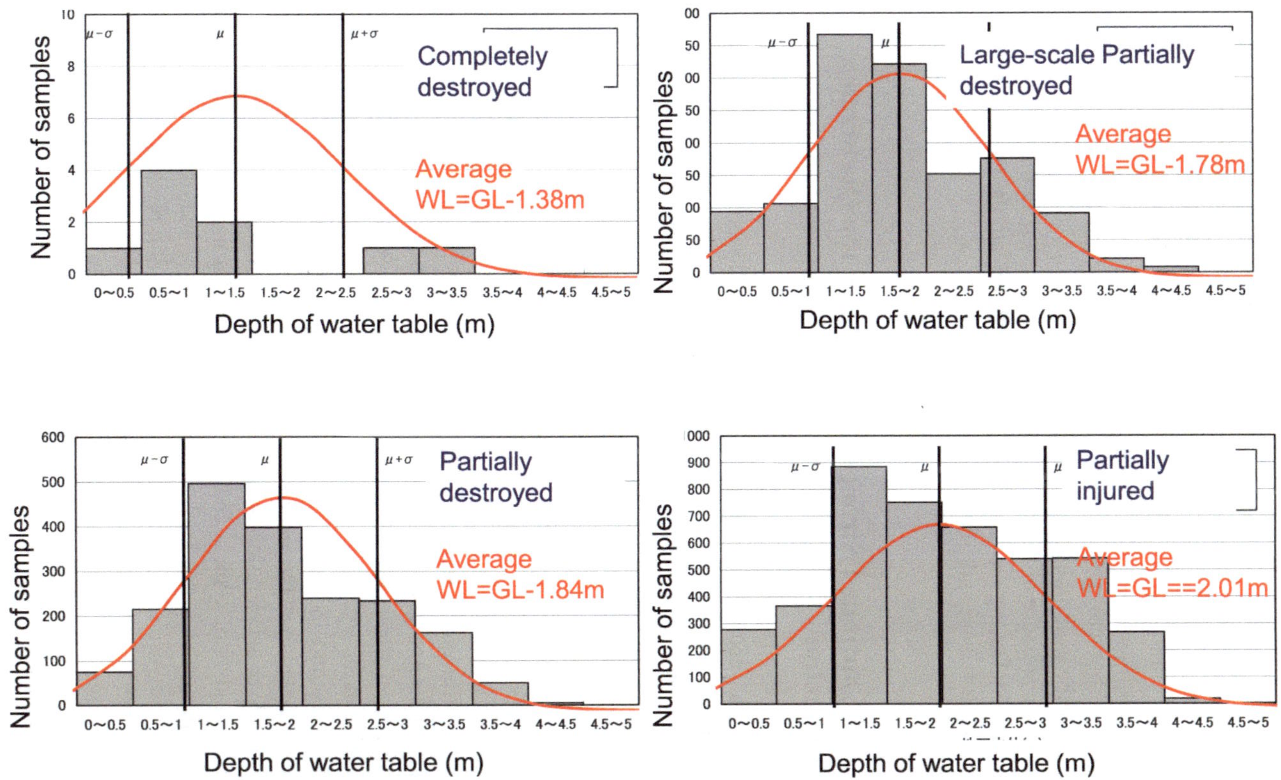

Fig. 16 Depth of water table at completely destroyed, large-scale partially destroyed, partially destroyed and partially damaged houses. (Partially quoted from Technical Committee on Measures Against Liquefaction in Urayasu City 2012)

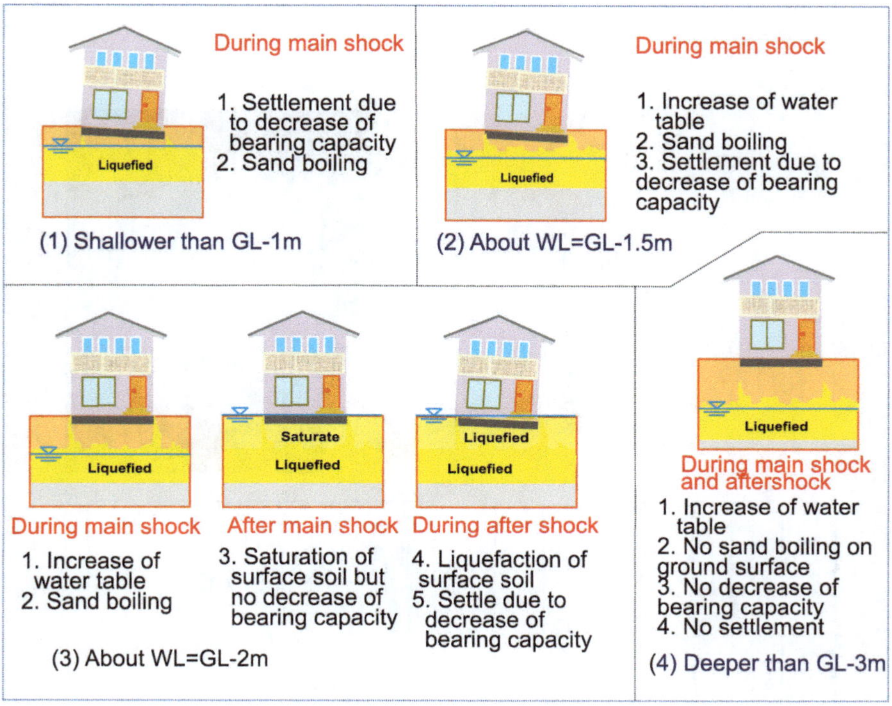

Fig. 17 A possible explanation of effect of the aftershock

Damage to Flat Roads Due to Sloshing of Liquefied Ground

One of the authors visited Takasu and Imagawa in Urayasu on the day after the earthquake and was surprised that a footway heaved and an alley thrust, as shown in Figs. 20 and 21, respectively. The first impression was that the footway heaved due to uplift force of some buried pipes, such as sewage pipes. However, after that, strange thrusts and heavings of footways and alleys were also found in Ichikawa, Chiba and other cities. So, he concluded that some boundaries beside the footways and alleys, such as banks of old sea walls and elevated bridges, caused the thrust or heaving due to a kind of sloshing of liquefied ground, as schematically shown in Fig. 22 (1), because shaking continued for a long time after the occurrence of liquefaction, as mentioned above.

On the other hand, at Maihama in Urayasu, the thrust of an alley was observed though there was no such boundary, as shown in Fig. 23. The locations of heaved footways and alleys in Urayasu City, with and without boundaries, are plotted in Fig. 24. By comparing the locations of heaved footways and alleys without boundaries with the contour lines of the thickness of the filled layer under the groundwater table, it may be said that the heaving occurred at the sites where the bottoms of the

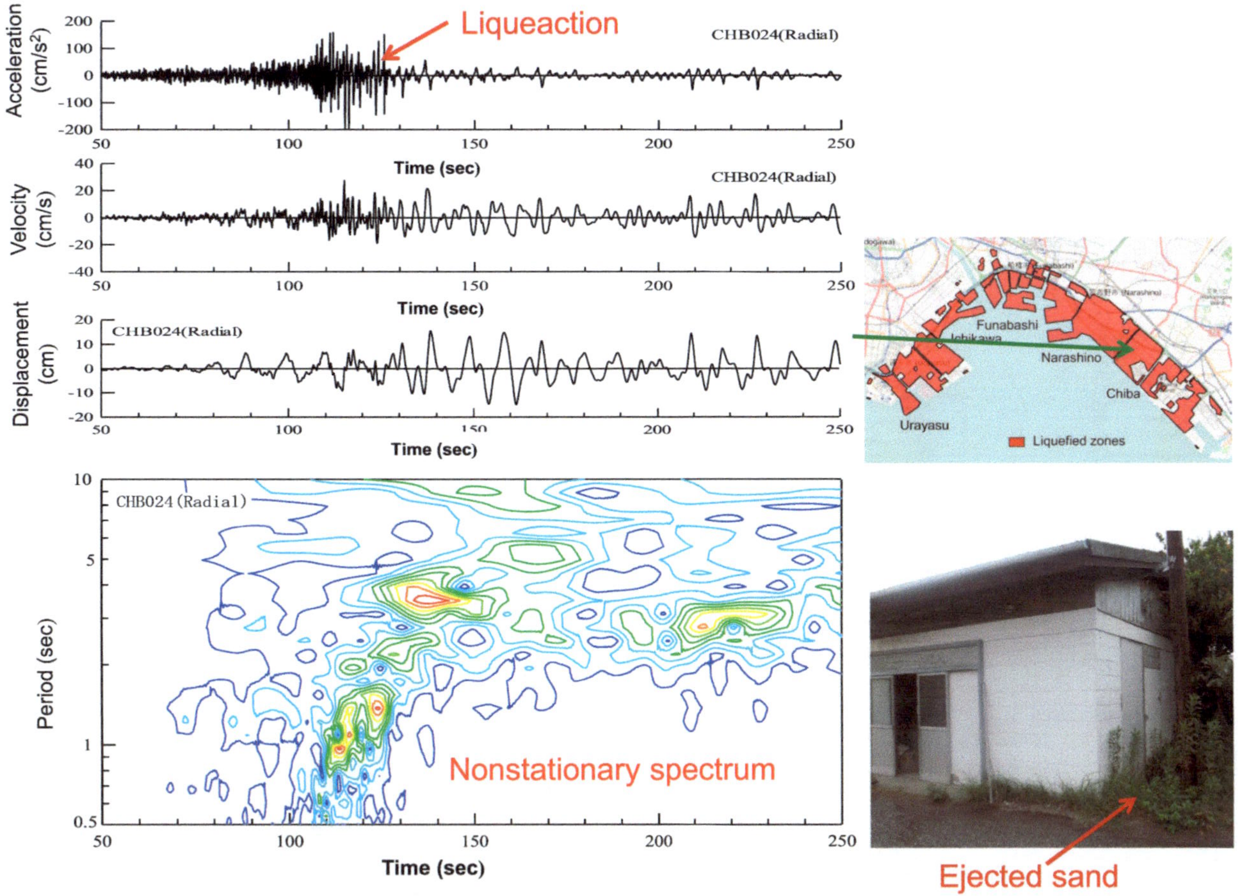

Fig. 18 Acceleration, velocity, displacement and non-stationary spectrum of the record at K-net Inage (site of liquefaction)

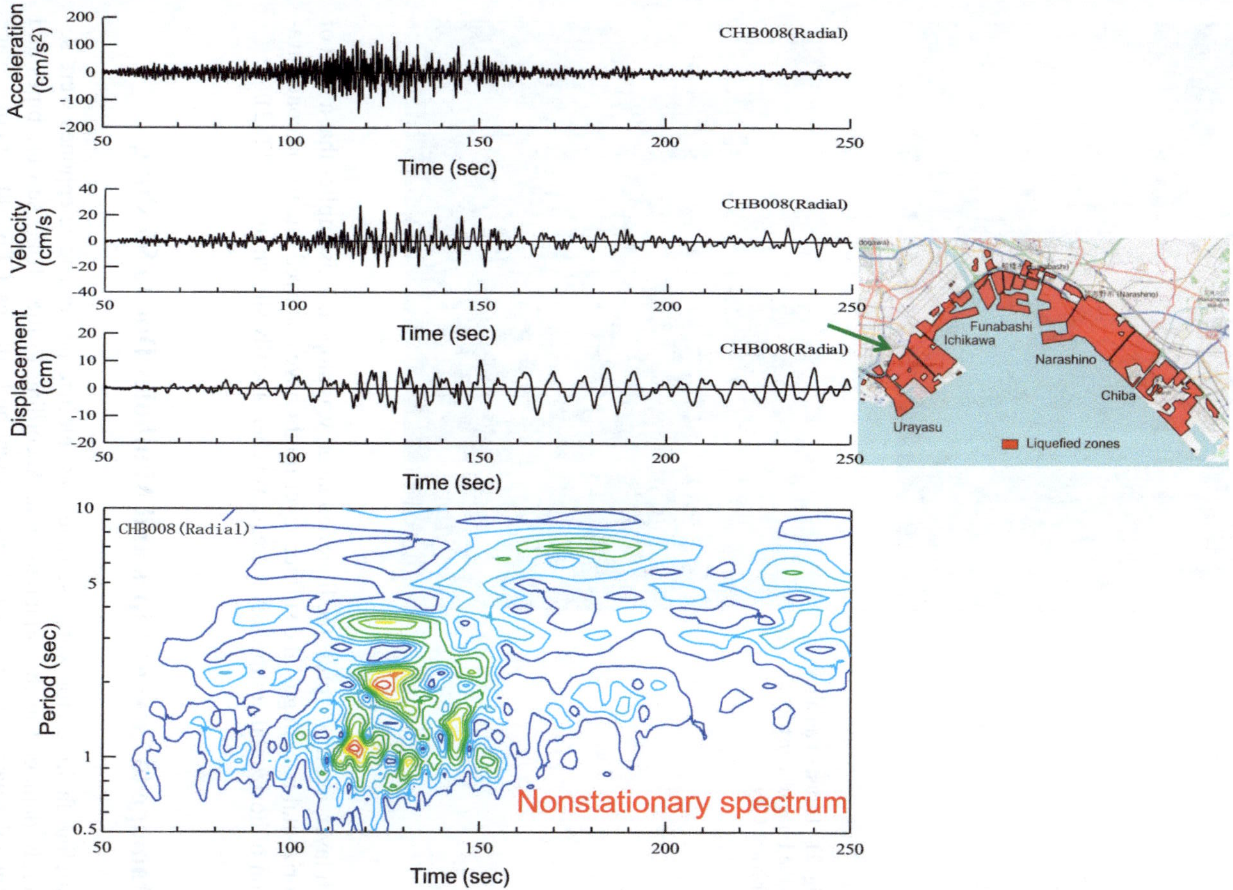

Fig. 19 Acceleration, velocity, displacement and non-stationary spectrum of the record at K-net Urayasu (site of no liquefaction)

Fig. 20 Heaving of a footway in the Takasu district of Urayasu City

Fig. 21 Thrust of an alley in the Imagawa district of Urayasu City

fill layer, in other words, the liquefied layer, was sloped. This implies that a kind of horizontal buckling of the surface layer might have occurred due to the concentration of horizontal compressive stress, as schematically shown in Fig. 22 (2).

Damage to Sewage Pipes and Manholes Due to Sloshing

Lifeline facilities for water, sewage, gas, electric power and telephones were severely damaged in the liquefied area. According to the MLIT, sewage pipes and manholes were damaged in 132 cities, towns and villages. Of 65,001 km of sewage pipes in the cities, towns and villages, 642 km were damaged. As shown in Fig. 25, 65.9 % of the damage to pipes was due to the liquefaction of filled soils and 24.5 % was due to the liquefaction of both filled soils and the surrounding ground. Of the

Fig. 22 Two possible mechanisms of thrust or heaving of the ground

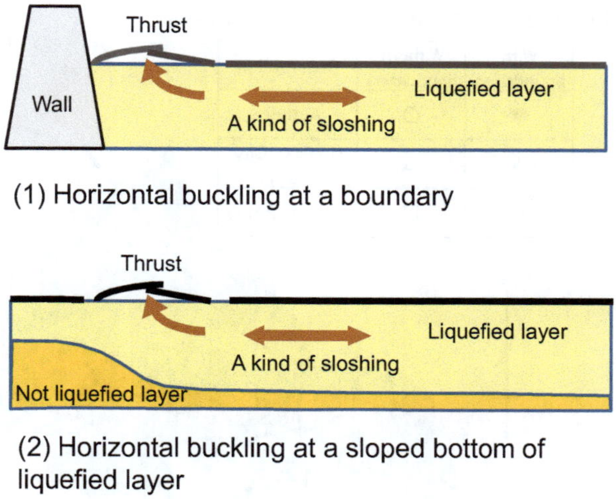

Fig. 23 Thrust of an alley in the Maihama district of Urayasu City. (Photo by Ogawa)

damage to manholes, 41.7% was due to the liquefaction of filled soils and 26.7% was due to the liquefaction of filled soils and the surrounding ground. The damage due to the liquefaction of both filled soils and surrounding ground mainly occurred along Tokyo Bay.

In the Tokyo Bay area, sewage pipes were deformed, cracked, broken and meandered, and joints were sheared or disconnected, as schematically shown in Fig. 26. Muddy water seeped into the damaged pipes and closed pipes completely. Many sewage manholes were cracked and sheared in a horizontal direction and filled with muddy water, as shown in Fig. 27, while a few manholes were lifted or slightly settled. Shear damage to manholes had not occurred during past earthquakes in Japan. During the construction of the manholes and pipes, the ground was excavated to a width of about 2 m. After placing the manholes and pipes, the excavated area

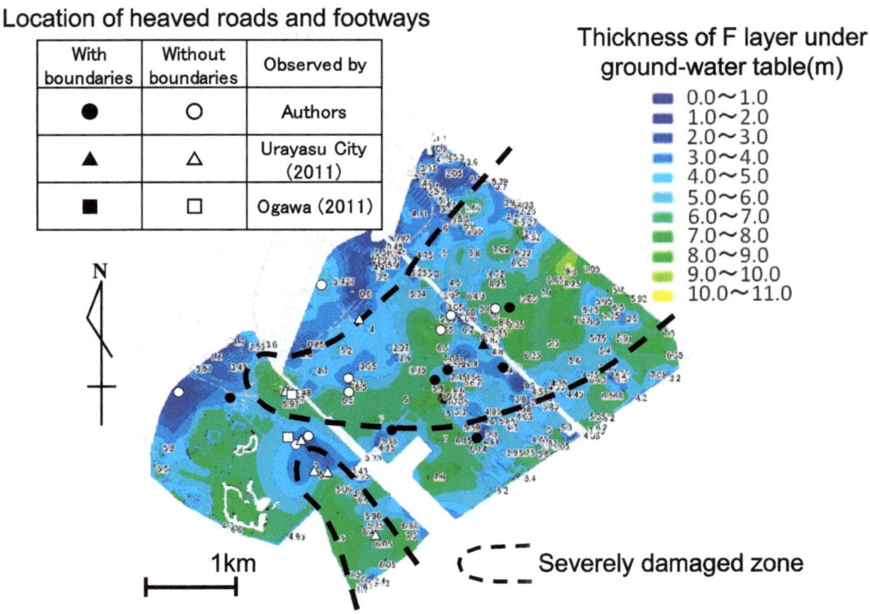

Fig. 24 Estimated thickness of F layer under groundwater table (by Technical Committee on Measures Against Liquefaction in Urayasu City 2012) and locations of thrusted roads and footways. (Yasuda et al. 2012)

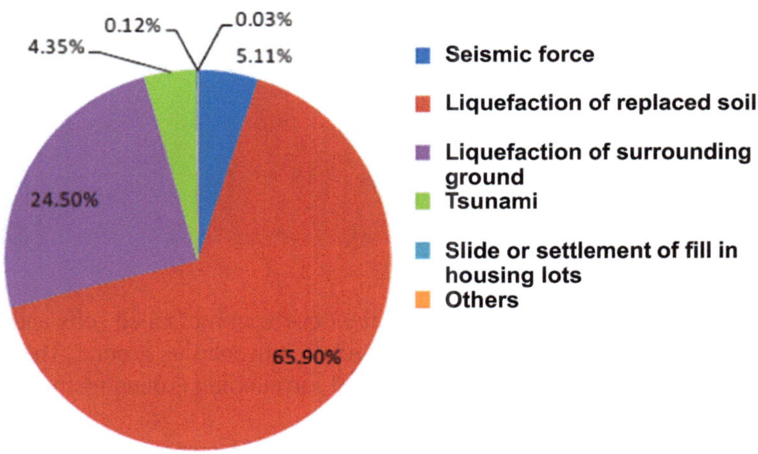

Fig. 25 Triggers to the damage of sewage pipes. (Partially quoted from MLIT)

was filled with sand. During past earthquakes, only the fill soils were liquefied and sewage pipes and manholes uplifted, as schematically shown in Fig. 27. But during the 2011 Great East Japan Earthquake, both the fill soils and the surrounding soils were liquefied and a kind of sloshing of liquefied ground might have occurred due

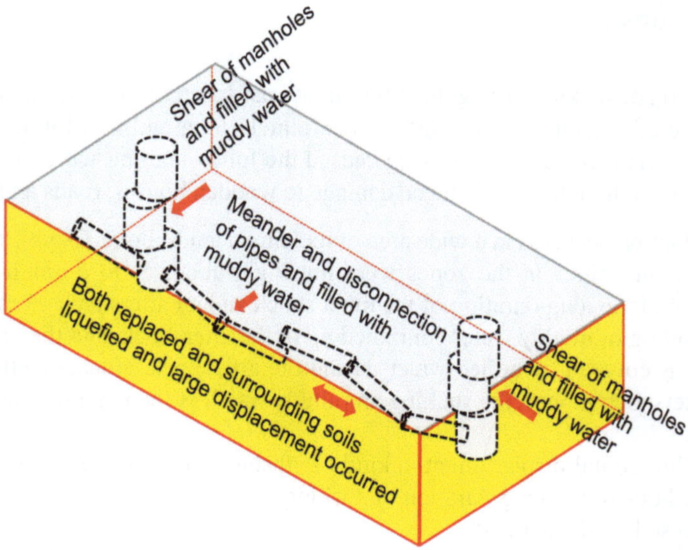

Fig. 26 Damage to sewage pipes and manholes due to the liquefaction of replaced soils and surrounding grounds

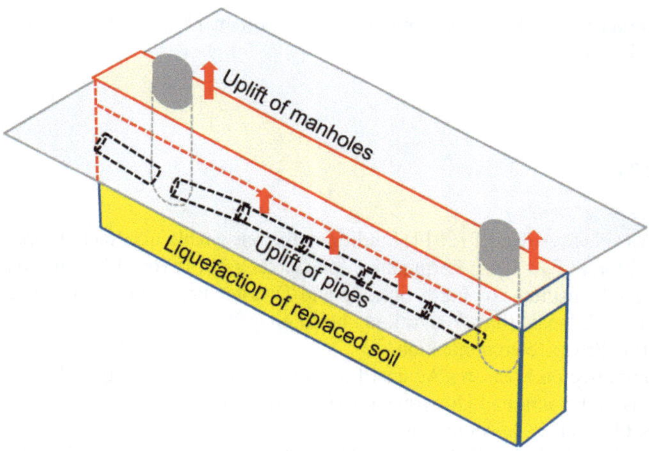

Fig. 27 Damage to sewage pipes and manholes due to the liquefaction of replaced soils

to the long duration of shaking and caused the thrust of roads in the Tokyo Bay area. The large horizontal displacement of liquefied ground had to have caused the disconnection of the pipe joints and the shear failure of the manholes, allowing the influx of muddy water into the pipes and manholes. Fortunately or unfortunately, this muddy water might have prevented uplift.

Conclusions

The duration of shaking during the 2011 Great East Japan Earthquake, a megathrust earthquake of magnitude $Mw = 9.0$, was extremely long and was followed by big aftershocks. This remarkable shaking caused the following unusual liquefaction of the ground and liquefaction-induced damage to wooden houses, roads and lifelines.

1. Liquefaction occurred in a wide area of reclaimed lands along Tokyo Bay though seismic intensities in the zones were not high, about 5- to 5+in JMA scale, because of the long duration of the main shock and an aftershock 29 min later.
2. The boiling of muddy water continued up to the aftershock, and the ground surface was covered by boiled water during the aftershock. Houses settled during the aftershock though the shaking amplitude was less than that during the main shock.
3. Large horizontal displacement, a kind of sloshing of the liquefied ground, was induced due to the long duration of shaking after the occurrence of liquefaction and caused roads to thrust.
4. By the same movement of the ground, pipes were displaced significantly in the horizontal direction, resulting the disconnection or breakage of joints. Manholes were sheared due to horizontal force.

Acknowledgements Mr. K. Ogawa provided us important photos. The authors would like to express their sincere appreciation to him.

References

Ishikawa K, Yasuda S, Aoyagi T (2014) Studies on the reasonable liquefaction-prediction method on the 2011 Great East Japan Earthquake. Jpn Geotech J 9(2):169–183. (In Japanese)
National Research Institute for Earth Science and Disaster Prevention (NIED) (2011) "K-NET WWW service", Japan. (http://www.k-net.bosai.go.jp/)
Ogawa K (2011) Private communication
Technical Committee on Measures Against Liquefaction in Urayasu City (2012) Report on measures against liquefaction in Urayasu City. (In Japanese)
Wakamatsu K (2011) Private communication
Yasuda S, Harada K, Ishikawa K, Kanemaru Y (2012) Characteristics of the liquefaction in Tokyo Bay Area by the 2011 Great East Japan earthquake. Soils and Foundations 52(5):793–810
Yasuda S, Towhata I, Ishi I, Sato S, Uchimura T (2013) Liquefaction-induced damage to structures during the 2011 Great East Japan Earthquake. J JSCE 1:181–193

Liquefaction Observed During the 2010 Chile Earthquake

Ramón Verdugo

Abstract On February 27, 2010, at 3:34 a.m. local time, a large earthquake of Magnitude 8.8 hit the Central-South region of Chile. A significant number of aftershocks followed the initial quake, being the most important of Magnitude 6.2, which occurred 20 min after the main shock. The 2010 Chile Earthquake has been identified as a thrust-faulting type that occurred on the interface between the Nazca and South American plates, at an average depth of 35 km. Seismologists have reported that the rupture zone covered a rectangular area of approximately 550 km by 170 km. High horizontal peak ground accelerations were recorded on soil deposits, being the maximum one 0.94 g at Angol City, located towards South of the rupture zone. To the north of the rupture zone, the horizontal peak ground accelerations (PGA) was recorded in Melipilla City, reaching a value of 0.78 g. Several of the available records show a ground motion that exceeds 2 min of duration, which may explain the significant amount of liquefied sites. Field observations have shown that the earthquake triggered liquefaction in more than 170 different sites, covering a north-south distance of about 950 km, which approximately corresponds to twice of the length of the rupture zone. Liquefaction phenomenon induced damages to the road infrastructure, railroads system, buildings and houses. Liquefaction-induced ground failure displaced and distorted pile foundations of piers impacting seriously the operation of several ports. Especially interesting is the case of Juan Pablo II Bridge, where significant differential settlements were observed, but almost non transversal lateral displacements took place.

Main Features of 2010 Chile Earthquake

On February 27, 2010, at 3:34 a.m. local time, a strong earthquake of Magnitude 8.8 Mw hit the Central-South region of Chile. A series of many aftershocks followed the initial quake; the most important of Magnitude 6.2 occurred 20 min after the main shock. The 2010 Chile earthquake is the second-strongest in the nation's history, only exceeded by the 1960 Valdivia earthquake of Magnitude 9.5. These

R. Verdugo (✉)
CMGI, Santiago, Chile
e-mail: rverdugo@cmgi.cl

A. Ansal, M. Sakr (eds.), *Perspectives on Earthquake Geotechnical Engineering*,
Geotechnical, Geological and Earthquake Engineering 37,
DOI 10.1007/978-3-319-10786-8_14

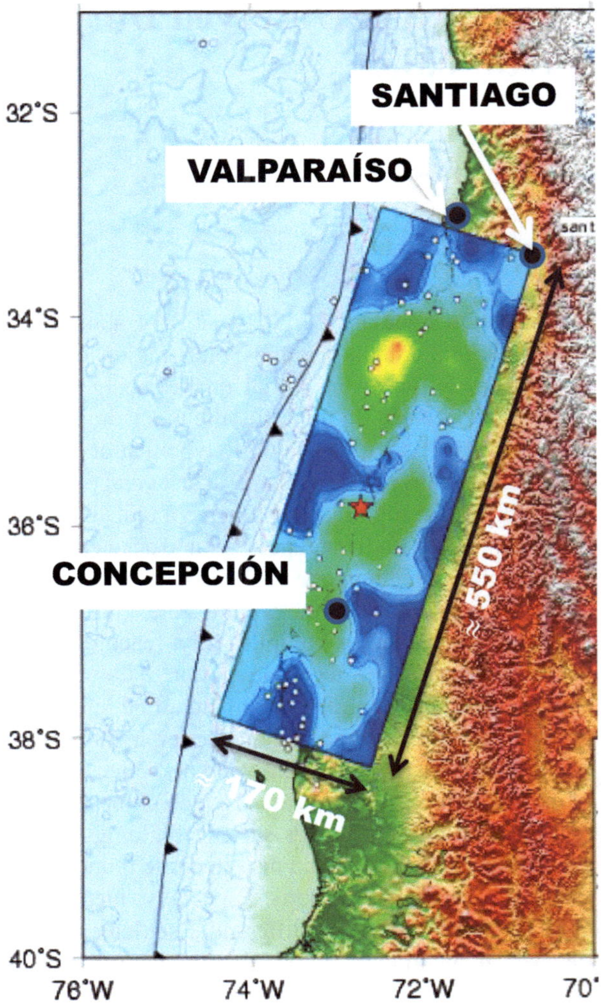

Fig. 1 Rupture zone. (www.tectonics.caltech.edu/slip_history)

earthquakes are associated with the subductive seismic environment generated by the collision between the Nazca and South American tectonic plates, which are converging at an estimated rate of 65–80 mm per year. The Nazca plate subducts below western edge of the South American plate, moving down and landward.

The 2010 Chile Earthquake corresponds to a thrust-faulting type that was originated on the interface between both plates, at an average depth of 35 km. The rupture zone responsible of this quake covered a rectangular area of approximately 550 km by 170 km (Fig. 1). The southern end of the rupture zone overlapped the rupture zone of the 1960 Valdivia earthquake and stopped near the southern end of the 1985 Valparaíso earthquakes of 8.0 Mw (Ruiz et al. 2012).

Considering the large extension of the rupture zone, the concept of hypocenter associated with a single point has to be replaced by the actual physics of the

Fig. 2 Distances from the epicenter to main cities of the affected area

phenomenon, consisting of an approximately planar area from where the seismic energy is emanated according to the evolution of the rupture itself. In the context of earthquakes of large magnitudes, the hypocenter has to be identified as the place of initiation of the rupture, or the starting point of the earthquake. According to USGS, the hypocenter of the 2010 Chile earthquake was located at 35.909°S and 72.733°W at a depth of 35 km, meaning the epicenter on the coast was located at approximately 323 km west of the capital (Santiago) and about 110 km north-northeast of Concepción (Fig. 2). Analysis of GPS stations carried out by Vigny and collaborators have reported that the rupture propagated bilaterally from the hypocenter at about 3.1 km/s, with a slip up to 15 m in the north area of the rupture (Vigny et al. 2011).

The crustal deformations (co-seismic displacements) in the epicentral area were measured by means of the Global Positioning System (GPS) networks previously installed (Vigny et al. 2011). The reported horizontal and vertical displacements are shown in Fig. 3, where the arrows and numbers indicate the direction and magnitude of the displacements in centimeters, respectively. A cross section showing the distribution of the vertical displacements from the trench toward the continent is presented in Fig. 4, where the red circles correspond to near Constitución latitude (35.5°S to 36°S), green circles are near Arauco-Concepción latitude (37°S to 37.5°S), and white circles correspond to intermediate latitudes. The solid line shows the overall trend (Vigny et al. 2011).

The maximum recorded displacements occurred at the tip of Arauco peninsula (60 km south of Concepción city), reaching an uplift of 1.8 m and a horizontal

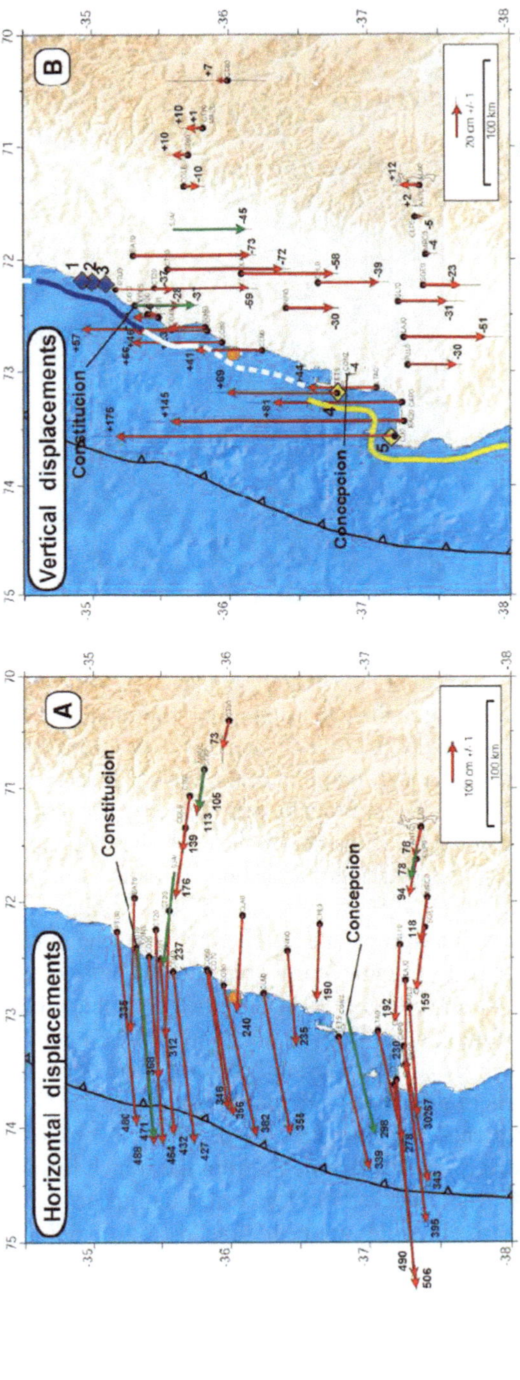

Fig. 3 Co-seismic static displacement field. Survey sites (*red arrows*) and continuous global positioning system stations (*green arrows*). (Vigny et al. 2011)

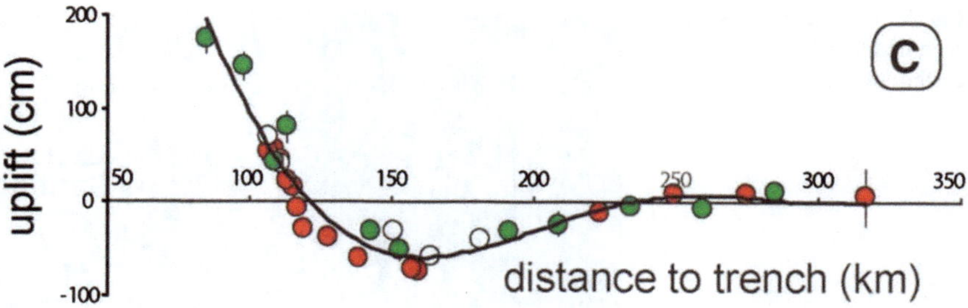

Fig. 4 Land-level changes as a function of distance to the trench. (Vigny et al. 2011)

movement towards the trench of 5.1 m. So far, these significant deformations, commonly observed after major earthquakes, are not considered in any way in engineering designs. Further studies have to be performed in order to understand their potential hazard.

Besides the strong seismic motion that severely disturbed structures, the fault slip distorted the seafloor triggering a tsunami that struck off the Chilean coast, causing a dramatic devastation of many towns located onshore. This phenomenon dramatically increased casualties and the extent of damages, resulting in nearly 600 injured and reported more than 30 billion US dollars economic losses.

A total number of 36 seismic stations located in the most affected area recorded the acceleration time histories on rock outcrops and soil deposits of different geotechnical characteristic (Fig. 5, Tables 1 and 2). The horizontal maximum accelerations recorded are presented in Fig. 6.

It is interesting to observe that horizontal PGA's recorded on rock outcrops are surprisingly moderate; in fact, according to the available data, the maximum recorded value is 0.32 g in Santa Lucía Hill, Santiago. However, it is important to recognize that in rock outcrops there was a clear lack of instruments to cover the most affected area, whereby it is expected that higher values of PGA could have taken place.

On the other hand, the horizontal PGA's recorded on soil deposits were significantly higher, reaching a maximum value of 0.94 g in Angol city, located close to the south end of the rupture zone. The second highest value of PGA was 0.78 g, recorded in the city of Melipilla, located close to the north end of the rupture zone.

Regarding the duration of the earthquake, it is interesting to mention that available records show two general patterns; to the south of the hypocenter, the duration of the ground motion tends to exceed 2 min (i.e. San Pedro, Fig. 7a), but to the north of the hypocenter, the duration is systematically shorter (i.e. Melipilla, Fig. 7b). In any case, the long duration of the ground motions is definitely a characteristic of earthquakes of large Magnitude, which certainly increase both their destructiveness and the liquefaction potential.

Fig. 5 Seismic stations that recorded the 2010 Chile earthquake

Table 1 Stations on soil

Number	City	Location (°)		PGA (g)		
		Latitude	Longitude	H1	H2	V
1	Cabildo	−32.43	−71.07	0.27	0.32	0.13
2	Papudo	−32.51	−71.45	0.30	0.42	0.16
3	Zapallar	−32.55	−71.46	0.18	0.18	0.11
4	Olmué	−33.00	−71.19	0.35	0.25	0.15
5	Viña Centro	−33.03	−71.55	0.22	0.33	0.18
6	Viña El Salto	−33.05	−71.51	0.35	0.34	0.26
7	Valpo. Almendral	−33.05	−71.61	0.22	0.27	0.14
8	Casablanca	−33.32	−71.41	0.29	0.33	0.23
9	Stgo—Las Américas	−33.45	−70.53	0.31	0.23	0.16
10	Stgo—FCFM	−33.46	−70.66	0.17	0.16	0.14
11	Stgo Centro	−33.47	−70.65	0.22	0.31	0.18
12	Stgo—FSR	−33.48	−70.53	0.25	0.24	0.18
13	Stgo—Maipú	−33.51	−70.77	0.56	0.49	0.24
14	Stgo—Peñalolén	−33.50	−70.58	0.30	0.29	0.28
15	Stgo—La Florida	−33.51	−70.61	0.19	0.13	0.10
16	Stgo—Antumapu	−33.57	−70.63	0.23	0.27	0.16
17	Stgo—Puente Alto	−33.58	−70.58	0.27	0.27	0.13
18	San José	−33.64	−70.35	0.47	0.48	0.25
19	Llolleo	−33.62	−71.60	0.33	0.56	0.68
20	Melipilla	−33.69	−71.21	0.57	0.78	0.39
21	Matanzas	−33.96	−71.87	0.34	0.29	0.24
22	Hualañe	−34.98	−71.81	0.38	0.45	0.38
23	Curicó	−34.98	−71.24	0.48	0.41	0.20
24	Constitución	−35.34	−72.41	0.54	0.63	0.35
25	Talca	−35.43	−71.66	0.47	0.42	0.22
26	Concepción	−36.83	−73.05	0.40	0.29	0.37
27	San Pedro	−36.84	−73.11	0.65	0.60	0.58
28	Angol	−37.79	−72.71	0.94	0.70	0.29
29	Valdivia	−39.83	−73.24	0.09	0.14	0.05

General Geological Conditions of Central-South Region

To the South of parallel 33°, the geology of Chile is characterized by the existence of three morphostructural units running parallel one to each other with a north–south orientation. From east to west, these units are Andes Mountain Range, Central Valley (or Intermediate Depression) and Coastal Mountain Range (Fig. 8).

Table 2 Stations on rock

Number	City	Location (°)		PGA (g)		
		Latitude	Longitude	H1	H2	V
1	El Roble	−32.98	−71.02	0.18	0.13	0.10
2	UTFSM	−33.03	−71.60	0.13	0.30	0.07
3	Tórtolas	−33.13	−70.71	0.18	0.17	0.14
4	Santa Lucía	−33.43	−70.63	0.32	0.22	0.22
5	Rapel	−34.03	−71.59	0.20	0.20	0.14
6	Convento Viejo	−34.76	−71.13	0.19	0.15	0.18
7	Melado	−35.76	−71.09	0.14	0.14	0.16

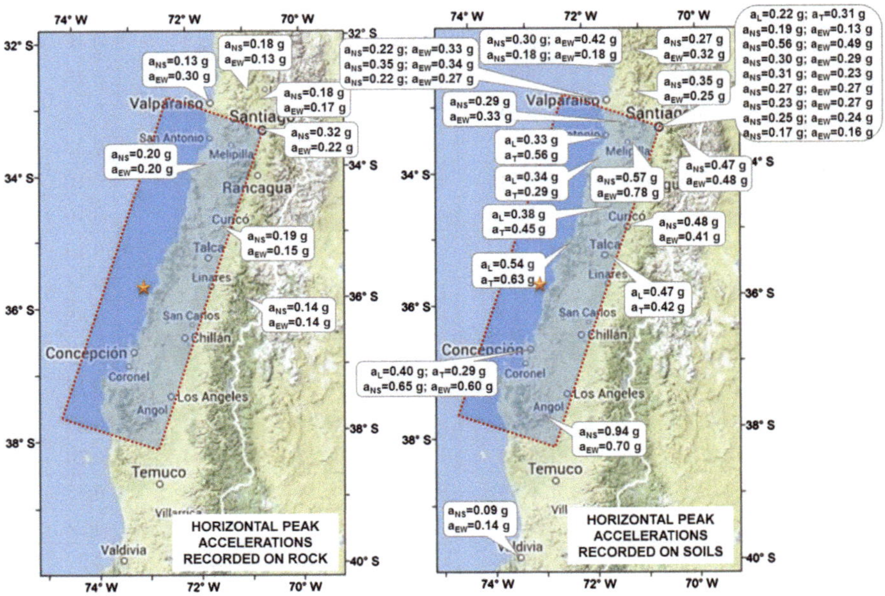

Fig. 6 Horizontal PGA recorded on rock outcrops (*left*) and soil deposits (*right*)

The Andes Mountain Range is located at the east of the country, and it can be seen as a natural barrier that exists along virtually all Chilean territory, disappearing under the sea far south at Cape Horn (it emerges again in the Antarctic). The two tallest mountains are Aconcagua (6959 m.a.s.l.) and Nevado Ojos del Salado (6880 m.a.s.l.). From Santiago to the south, the altitude of the Andes progressively loses height, from 6000 m.a.s.l., in the vicinity of Santiago, to 3000 m.a.s.l. near Chillán (see Fig. 2 to identify these two cities). Along the Andes Range, there is a major volcanic activity which is remarkable towards south. The existence in Chile of more than 40 active volcanic groups is indicative of this singular condition. Eruptions are frequent and usually violent amongst the 25 active volcanoes located in the Southern Andes (Corvalán et al. 1996).

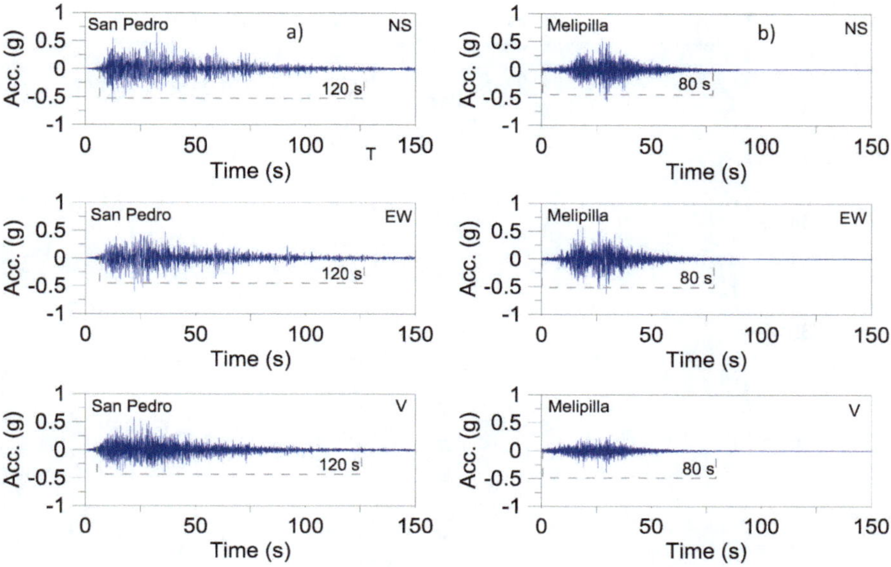

Fig. 7 Acceleration time histories. **a** San Pedro. **b** Melipilla (C.S.N.)

The Coastal Mountain Range extends from Camaraca Mountain, located 20 km south of Arica (north of Chile), to the Taitao peninsula. In the central area, it presents heights that rise above 2000 m.a.s.l. and progressively decreasing southward, being cut by the Biobío River and rising again in Nahuelbuta vicinity, where it reaches an average altitude of 1400 m.a.s.l. Further south, it only reaches 400 m.a.s.l. presenting singular soils of high erosion rates. Several narrow valleys created by rivers flow across the Costal Range, along which sandy soil deposits are quite recurrent. These types of soil deposits are also found along the coast line, where an important number of cities and town are located.

On the other hand, the Central Valley is a depression existing between both the Costal Mountain Range and the Andes Mountains Range, which is shortly interrupted by the transverse valleys and sunk beneath sea level approximately at 42°S. It reappears at north of Santiago, from where the valley widens, although it undergoes several reductions at different locations. In the area affected by the 2010 Chile earthquake, the Intermediate Depression is partly covered with glacio-fluvial sediments from the Andes and soils derived from volcanic ashes. Along the banks of rivers that run from Andes Mountain to the sea, the existence of sandy soil deposits is frequent. Also, in lakes surroundings, sandy soil deposits with a high water table are commonly found.

Therefore, sites composed by saturated sandy liquefiable soil deposits are mainly located in the Intermediate Depression, along the valleys that pass throughout the Coastal Range and along the coast line.

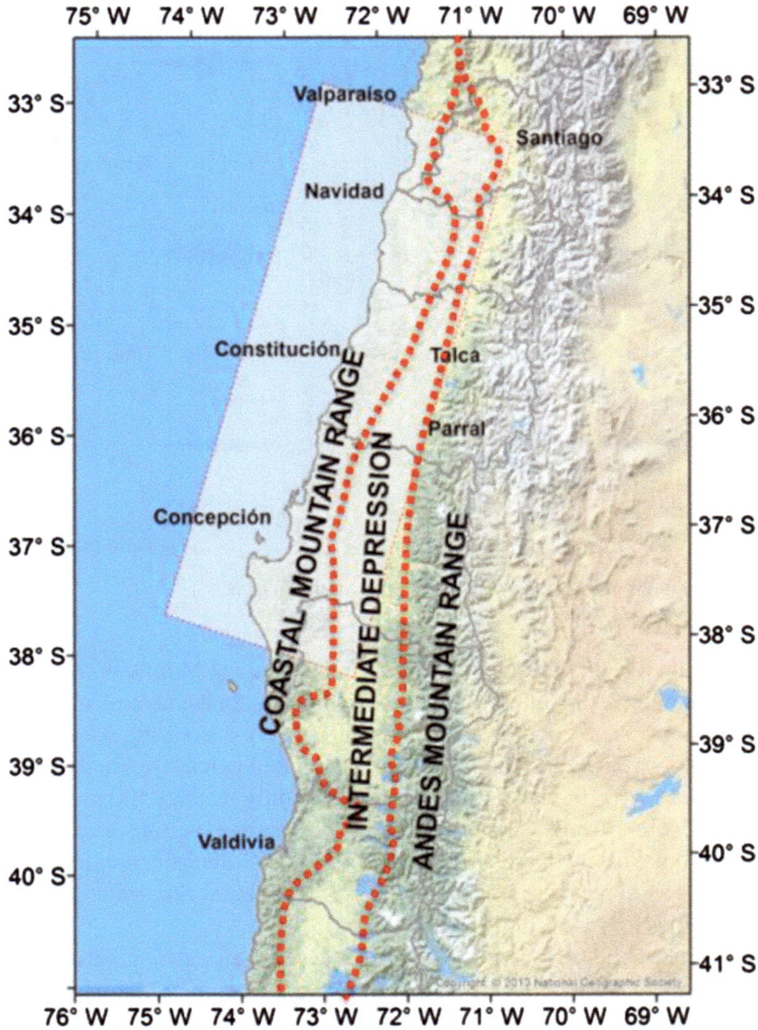

Fig. 8 Main morphostructural units in Central-South Chile

Area Affected by the Liquefaction Phenomenon

After the 2010 Chile Earthquake, the affected area was visited by many national as well as international groups of researchers who were able to identify different sites with evidence of liquefaction. Probably due to the complex situation of the country associated with this natural disaster, a significant number of sites that experienced liquefaction were not reported, so initially they were not visited neither included in the data base. However, later on throughout a careful and systematic review of the web and local newspapers, they could be identified and incorporated to the data

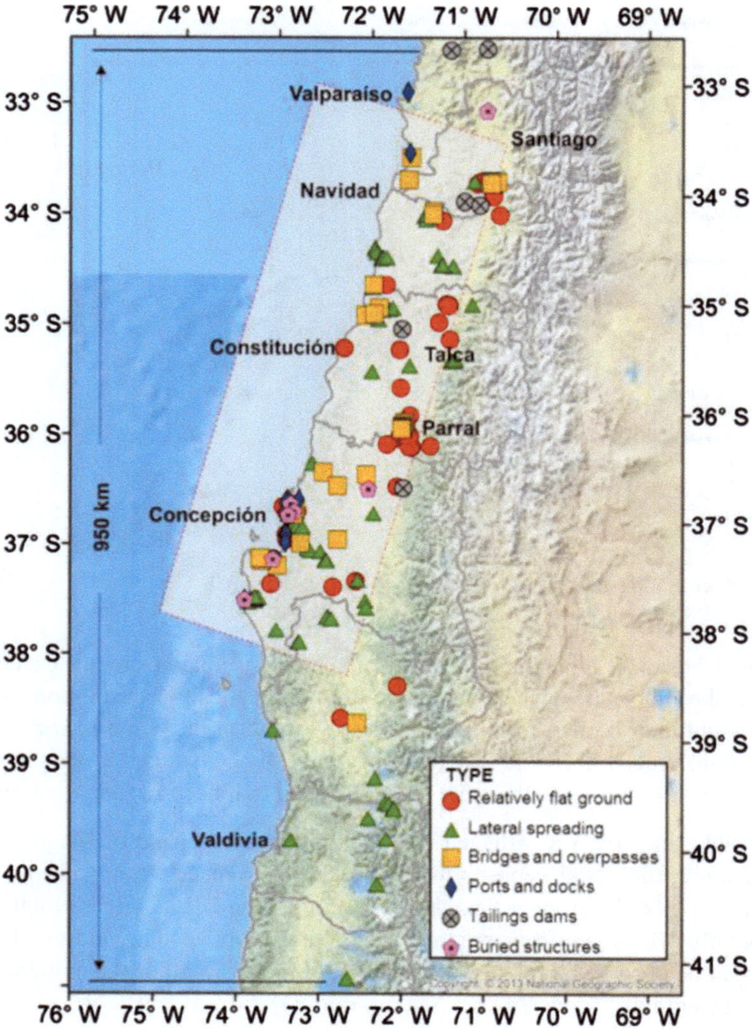

Fig. 9 Sites with evidence of liquefaction

base. As a result, a total number 170 sites with clear evidence of liquefaction have been recognized, which are summarized in Fig. 9.

According to the field observations and integration of the data, there are three important aspects that should be pointed out:

- The area affected by liquefaction has a North-South extension close to 1000 km, which approximately represents twice the length of the rupture zone.
- The liquefaction phenomenon did not extend symmetrically outside the rupture zone, reaching places well south from the south-end of the rupture zone but not far north from the north-end of the rupture zone.
- Liquefaction observed at the south is mainly associated with lateral spreading that affected borders of lakes.

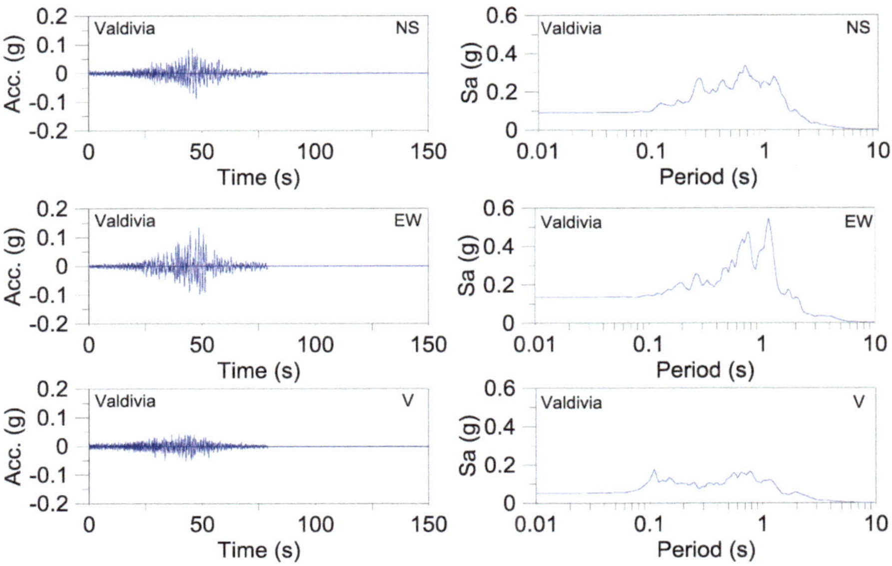

Fig. 10 Acceleration time histories recorded in Valdivia (Renadic)

The extension of liquefaction to the south could be explained by the existence of several lakes in the area, where saturated sandy soil deposits are commonly distributed around their borders. On the contrary, to the north the clime is arid, phreatic surfaces are usually deep and soil deposits are mostly constituted by coarse materials. However, in the south there are two conditions that need further attention:

- In 1960, this area was affected by the Valdivia Earthquake of Magnitude 9.5. Therefore, these soil deposits have already experienced strong seismic disturbances, probably stronger than those induced by the 2010 Chile Earthquake.
- The only available seismic record of this area was obtained in Valdivia (Fig. 10). As can be observed, the PGA reaches 0.14 g, which is a fairly a moderate level of acceleration.

So far, it has not been possible to carry out the necessary field investigation to characterize the Valdivia nearby soil deposits, in order to clarify how these terrains could liquefy even though they are quite far from the epicentral area, where acceleration levels were probably low. In addition, these soil deposits should not be so loose due to the previous strong seismic action, certainly generated by the 1960 Valdivia Earthquake.

Post Liquefaction Settlements

When the liquefied soils are confined by non-liquefiable material, no significant lateral displacements are generally induced, but depending on factors as thickness of the liquefiable strata, initial relative density, level of cyclic shear strains, among

Fig. 11 Post-liquefaction settlements. **a** Costanera route in Concepción. **b** South access to Yali bridge near Navidad. **c** and **d** Railway near Concepción

others, significant settlements may follow the dissipation of the pore water pressure. Four cases of clear evidence of post-liquefaction settlements that induced damages in routes and railways are presented in Fig. 11. Considering that in these cases there is no structures lying on the ground (except pavement or railway), the resulting vertical deformations correspond exclusively to the ground settlements developed after liquefaction.

For the particular geological and geotechnical characteristics of the area affected by liquefaction and from the scrutiny of the different places where post-liquefaction settlements were observed, it is possible to conclude that the maximum settlement in natural ground is limited to around 1.3–1.5 m.

Buildings Affected by Liquefaction of the Ground

Some buildings were damaged due to liquefaction-induced ground failure. Although in some cases the structural damage was important, none of the affected buildings collapsed. In Concepcion, Los Presidentes condominium, constructed after 2006, had four 8-story buildings founded directly on a sandy fill. Evidence of liquefaction of this fill was ostensible by the amount of sand ejected to the surface (Fig. 12a) and three buildings suffered damages due to the liquefaction-induced deformations (Bray et al. 2012). It is worth to mention that the condominium surrounding land did not show any evidence of liquefaction, suggesting that such phenomenon occurred

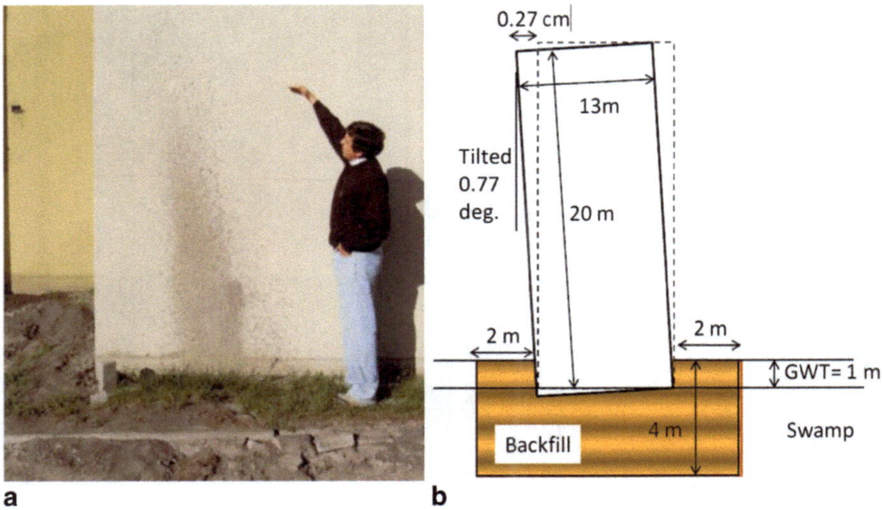

Fig. 12 **a** Trace of ejected sand, **b** General dimensions of tilted building

exclusively within the sandy fill, which compaction probably was not properly performed. The most affected building underwent a tilting of about 1 (Fig. 12b), likely to be the responsible of the significant damages observed.

Another case of liquefaction with a significant loss of bearing capacity is shown in Fig. 13a, where it can be seen that a 2-story residence located in Concepción underwent a serious tilting but did not present evident structural damages.

A church located about 50 km to the south-east of Santiago suffered severe damages in its floor (Fig. 13b). The structural walls were founded on competent ground, while the floors were installed directly on liquefiable sandy soils. Evidence of liquefaction consisting of ejected sandy material was observed in the surroundings of the church. It is possible to conclude that the forces associated with the ejection of sandy soils were high enough to spoil the floor. This type of effect on a non-structural element was systematically observed inside buildings and its importance is associated with the need of serviceability after a strong earthquake.

Seismic Failure of Tailings Dams

In the area affected by the 2010 Chile Earthquake there are seven major tailings deposits owned by Codelco (six) and Anglo American (one). Three of these embankments are made of compacted borrow materials, while the other four are made of cycloned tailings sands. None of these dams presented seismic damages, except for the manifestation of some longitudinal cracks along the crest.

On the other hand, in the same area there exist more than 50 small to medium tailings dams, some of which had limited engineering design. Due to the strong

Fig. 13 **a** Liquefaction-induced failure in Talcahuano, **b** Church located in Abrantes

Fig. 14 Before and after failure of Las Palmas tailings dam. (Bray et al. 2011)

shaking, four of these medium tailings dams developed limited damages and only one of them (Las Palmas) experienced a liquefaction-induced flow failure. The liquefied tailings flowed downstream about 400 m, contaminating and causing four fatalities that were buried under 4 m of tailings inside their home (Fig. 14). Las Palmas tailings deposit consisted of a lower dam and an upper one that were constructed in stages using the upstream construction method (Verdugo et al. 2012a). In situ measurements on the remaining tailings indicated shear wave velocities in the order of 250 m/s, and dynamic cone penetrometer soundings yielded blow counts of about 10 (Bray and Frost 2010).

The present Chilean provision associated with tailings dam projects explicitly indicates that tailings dams constructed by the upper stream method are prohibited due to the abundant seismic experiences of catastrophic flow failures that have left several victims, significant material losses and environmental damages. However, as Las Palmas dam, along the country there is a substantial number of abandoned upper stream dams constructed before the provision, which represent a considerable

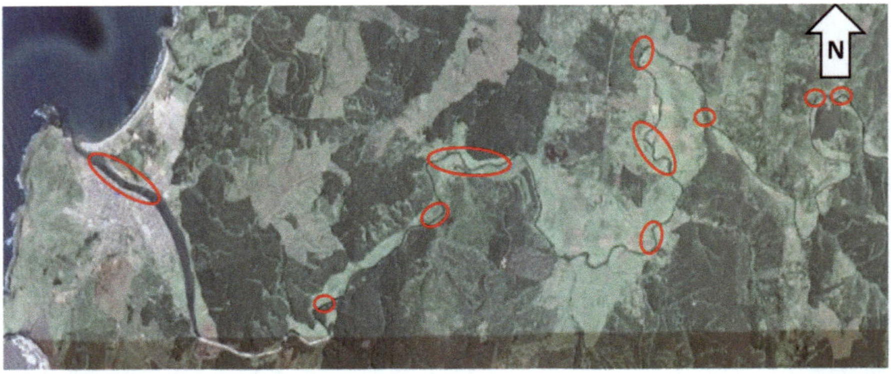

Fig. 15 Lateral spreading observed along Lebu river

risk for the population living within the runoff distance. Although the government is aware of this situation, no adequate counter measurements have been implemented so far. Probably this is a common issue shared with others countries where old mining operations have left unsafe tailings disposals that are difficult to identify as well as to control.

Lateral Spreading

The field investigation of the damages caused directly by the occurrence of liquefaction during the 2010 Chile Earthquake, suggest that those associated with lateral spreading were more extensive and disruptive. After visiting the affected area and the different manifestations of the phenomenon of liquefaction it is possible to point out that lateral spreading is the most systematic manifestation of liquefaction and due to the widespread damages it needs more attention. In some locations, longitudinal cracks associated with the lateral spreading were continuously observed for a couple of kilometers along the banks of rivers (Verdugo et al. 2012b). Manifestations of lateral spreading along the Lebu River are indicated in Fig. 15, proving how extensive this phenomenon can be.

Typical examples of damages generated by lateral spreading on gently natural slopes are shown in Fig. 16, where superficial blocks of dry soil broke up internally, moving downward and "floating" on top of the liquefied layer. In these two cases rather narrow irrigation channels collapsed allowing the lateral displacements.

More critical and economically serious damages caused by liquefaction-induced lateral spreading took place in several ports as shown in Fig. 17. In these cases it seems that the piles were not large enough, so they did not penetrate in non-liquefiable layers. Thus the piles were horizontally dragged by the lateral movement of the ground compromised by the lateral spreading, i.e. the crust layer and the liquefied materials.

Fig. 16 Lateral spreading observed in. **a** Apalta route and. **b** Nancagua Town

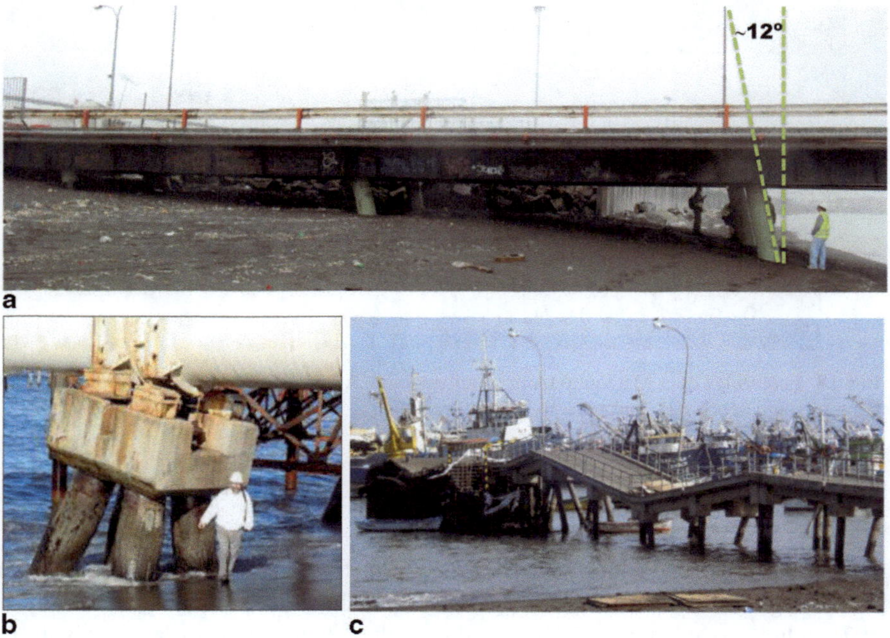

Fig. 17 Lateral spreading. **a** Coronel Port. **b** Bocamina Port and. **c** fishermen's Port of Coronel

Bridges were also affected by liquefaction-induced lateral spreading, being normally affected the accesses and the abutments of these structures. A footbridge over Mataquito River (pasarela Escalones), located to the west of Licantén Town, underwent severe damages on its south access sub-structure as shown in Fig. 18.

In this location the cracks observed on the ground penetrated around 63 m from the river and near 100 m in the immediate vicinity. As can be observed in Fig. 18d, the principal cracks presented a width of about 3 m. Therefore, in this site the maximum lateral displacement of the ground can be estimated in more than 10 m, considering that the images suggest at least three sets of cracks.

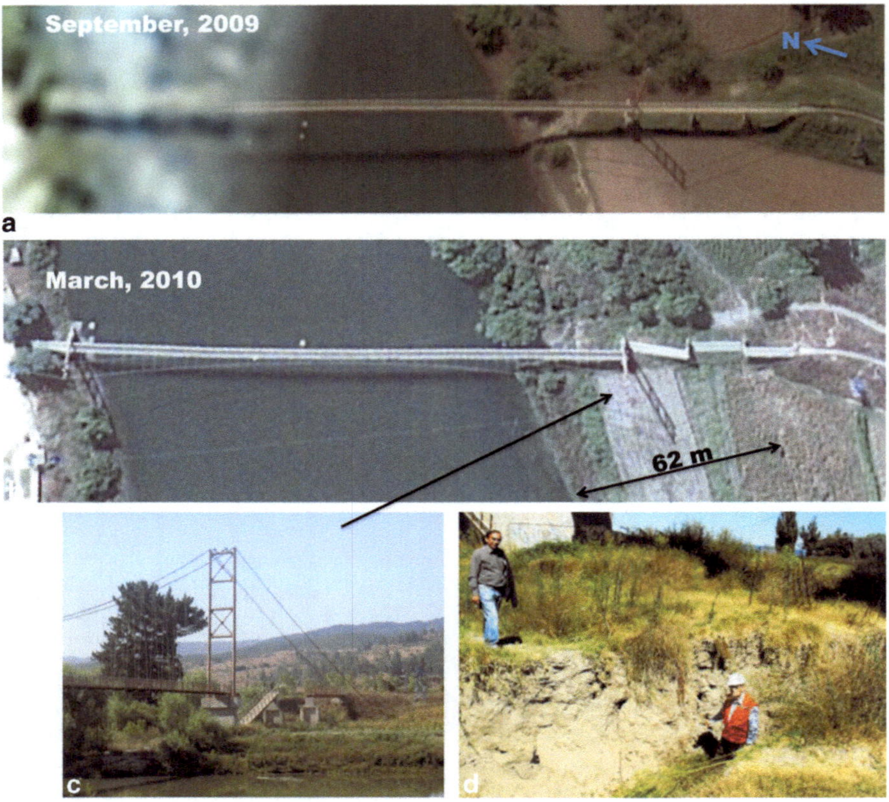

Fig. 18 Footbridge over Mataquito river. **a** Before and. **b** After earthquake (Google-Earth). **c** Collapse of south access. **d** Cracks due to lateral spreading

Juan Pablo II Bridge

Juan Pablo II Bridge is located in the city of Concepción and it crosses the BioBio River (Fig. 19). Its construction began in September 1968, being opened in April 1974. It has a total length of 2310 m and a width of 21.9 m with six lanes. It is structured by means of 70 spans, each composed of seven reinforced concrete girders and a concrete deck. Each span is supported by reinforced concrete columns, connected to driven cylinder piers of 2.5 m in diameters and 16 m in length (Fig. 20).

This bridge underwent seismic damages due to the 2010 Chile Earthquake. Clear evidence of liquefaction-induced lateral spreading in the Northern abutment (access to Concepción) was observed (Ledezma et al. 2012). However, the piers located into the river bed did not present any evidence of lateral displacements, but important settlements due to the occurrence of liquefaction took place. This condition can be seen in photos of Fig. 21.

After the earthquake a topographic survey along the Juan Pablo II Bridge was conducted by the Ministry of Public Work which results, in terms of settlements, are shown in Fig. 22.

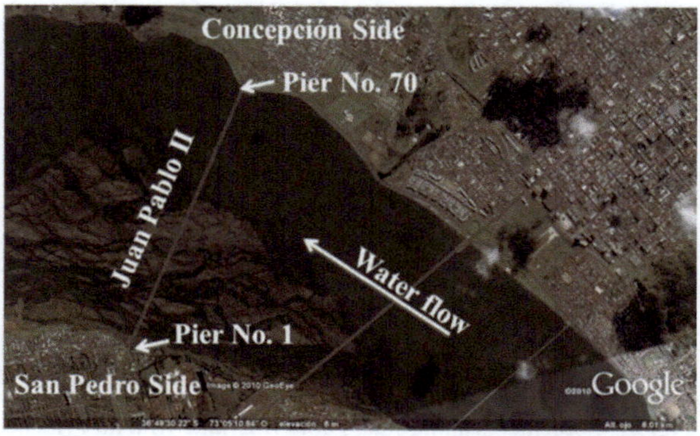

Fig. 19 Location of Juan Pablo II bridge

Fig. 20 Supporting structure

According to the available information, in general, the piers along the left side (upstream) of the bridge developed greater settlement than the ones located along the right side (downstream). Larger settlements were developed at the Concepción abutment bank. The largest settlement is practically 1.8 m, followed by settlements of 1.1 and 0.9 m.

After the 2010 Earthquake, a series of 16 boreholes with N-SPT measurements were carried out along Juan Pablo II Bridge. An energy ratio of 60 % was assumed considering that the Chilean procedure follows the U.S. testing practice, consequently, a correction factor $C_E = 1.0$ was adopted. The resulting values of N_1 (normalized to a vertical pressure of 1 kg/cm^2) are plotted in Fig. 23.

It has been estimated that values of N_1-SPT lower than 25 blows per foot are associated with liquefiable sandy soil layers. Accordingly, those sandy soil layers that should liquefy are indicated in red in Fig. 23. On the contrary, those sandy soil layers with N_1-SPT greater than 25 blows per foot (non-liquefiable) are not coloured.

Fig. 21 Settlements along Juan Pablo II Bridge with no evidence of horizontal displacements

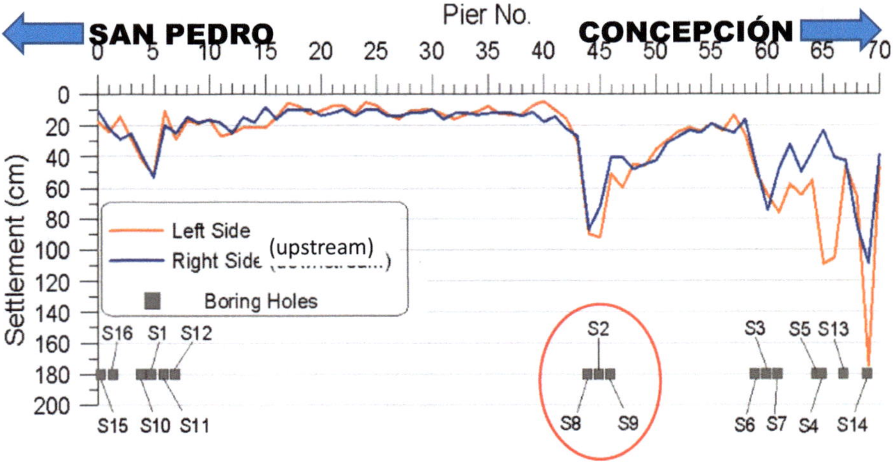

Fig. 22 Measured settlements along the Juan Pablo II Bridge

The layers of silty materials are marked in green, being not clear whether they may or not liquefy, because they are mainly non-plastic silts.

According to the available information of the original project, the piers had to be pushed down 16 m below the river bed. Therefore, the piers 45, 46, 67 and 69 would be resting on silty soft materials and piers 60, 64 and 65 would be in contact with liquefiable sandy soils. Except by pier 44, there is a good agreement between the observed settlements and the soil conditions established from the N-SPT data. These results suggest that the non-plastic silt layers liquefied under the strong seismic load imposed by the 2010 Chile Earthquake of Magnitude 8.8.

It is interesting to realize that although important settlements occurred on several piers, no permanent horizontal displacements were observed along the bridge. This fact necessarily implies that non-liquefiable layers were able to provide sufficient lateral stiffness and strength to ensure the lateral stability of the bridge. From a point of view of the structural design, this is a crucial issue because it considerably

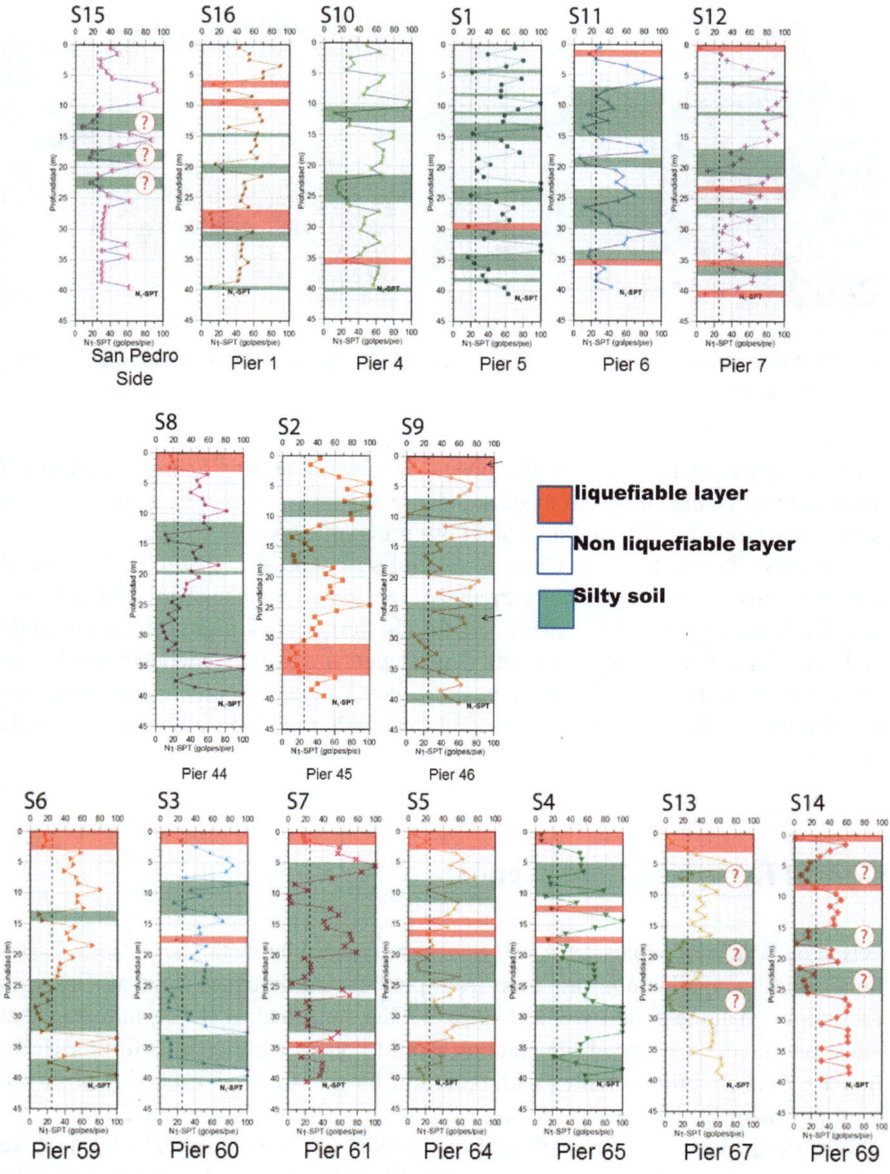

Fig. 23 Normalized N-SPT values at Juan Pablo II Bridge

reduces the sub-structure transverse displacements of the bridge, guaranteeing an adequate lateral stability.

In the light of available background, it is feasible to postulate that, in high seismic environments as the case of Chile, the seismic history of a site is necessarily associated with liquefaction of those initially loose sandy soil deposits. These episodes have the singularity of modifying their structure and to likely transform

Fig. 24 Liquefaction in Muelle Schuster at Valdivia city: a 1960 Valdivia earthquake. b 2010 Chile earthquake

them in highly heterogeneous soil deposits, in terms of density. Therefore, a layered structure that would consist of loose and dense soil layers is expected, which means a stratified structure of liquefiable and non-liquefiable layers.

Actually, the existence in Chile of natural soil deposits constituted by several meters of just loose sandy soil is extremely rare. Actually the author is not aware of any single soil deposit of this characteristic. On the contrary, highly stratified sandy soil deposits constituted by loose and dense layers are usually encountered. Therefore, it is likely to find only two types of sandy soil deposits: dense, without any problem of liquefaction; and strongly stratified, with liquefiable and non-liquefiable soil layers.

Sites of Recurrent Liquefaction

Based on both a reconnaissance survey of the area shaken by the 2010 Chile Earthquake and the available information from previous earthquakes, several sites with recurrence of liquefaction were identified. This empirical evidence indicates that sites that have experienced liquefaction would liquefy again due to further earthquakes. This condition of sites with recurrent liquefaction has been also observed after the Great East Japan Earthquake of March 11, 2011 (Wakamatsu 2012).

The north part of the area affected by liquefaction due to the 2010 Earthquake has been previously subjected to a long history of large ground motions, being the most recent ones the 1985 Earthquake, of Magnitude 7.8, and the 1906 Valparaíso Earthquake, of Magnitude 8.2. The south part has also been exposed to high seismicity, being the most important in the recent past the 1960 Valdivia Earthquake of Magnitude 9.5.

One example is presented in Fig. 24 where the same site (Muelle Schuster), located at Valdivia City, was affected by important liquefaction-induced damages during the 1960 Valdivia Earthquake (Fig. 24a) and then by the 2010 Chile Earthquake (Fig. 24b).

Fig. 25 Liquefaction-induced damages in Lo Gallardo Bridge: **a** Collapse after the 1985 earthquake. **b** Misaligned of new bridge after the 2010 Chile earthquake

Another case occurred in the mouth of the Maipo River, approximately 90 km south-west of Santiago. During the 1985 earthquake, part of Lo Gallardo Bridge collapsed due to the occurrence of liquefaction (Fig. 25a). Later on, due to the 2010 Earthquake again it underwent important lateral deformations, as can be observed in Fig. 25b, even though it had deeper piles.

The Hospital overpass, located approximately at 45 km to the south of Santiago, presented some degree of damages after the 1985 Earthquake, suggesting that a limited liquefaction could have happened. This was definitely confirmed by the severe damages resulting from liquefaction triggered by the 2010 Earthquake. Permanent horizontal displacements of 87 cm were reported at the north abutment, being the responsible for the fall of the bridge deck (Fig. 26). The stratigraphy of the ground is shown in Fig. 27, indicating the existence of loose layers of sandy soils enclosed by fines soils. In this case, the previous earthquake of 1985 (M = 7.8) only induced a limited liquefaction and its recurrence was activated by the 2010 Earthquake of Magnitude 8.8. It is possible to speculate that the 1906 Valparaíso Earthquake (M = 8.2) could generate a sort of threshold cyclic stress that was not exceeded by the 1985 Earthquake.

Concluding Remarks

On February 27, 2010, at 3:34 a.m. local time, a giant earthquake of Magnitude 8.8 hit the Central-South region of Chile, being the second-strongest in the nation's history, only exceeded by 1960 Valdivia earthquake of Magnitude 9.5. It corresponds to a thrust-faulting type event originated on the interface between Nazca and Continental plates, at an average depth of 35 km. The rupture zone covered a rectangular area of approximately 550 km by 170 km, overlapping the rupture zone of the 1960 Valdivia earthquake towards south and, at some extent, limited by the 1985 Valparaíso earthquake rupture zone northwards.

Fig. 26 Collapse of hospital overpass

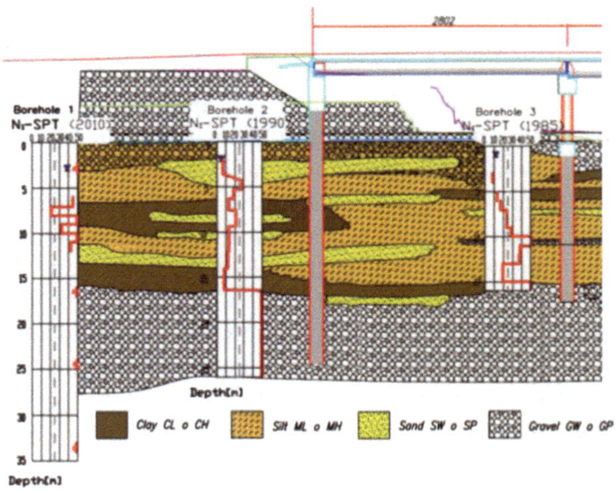

Fig. 27 Stratigraphy North Abutment

Maximum crustal deformations were recorded at the tip of Arauco peninsula, with an uplift of 1.8 m and a horizontal movement towards the trench of 5.1 m. These significant deformations have been observed after major earthquakes and they should be studied in order to understand their hazard potential.

Regarding the phenomenon of liquefaction, it affected a sector close to 1000 km (north–south direction), which approximately represents twice the length of the rupture zone. The extension of liquefaction reached places well south from the south-end of the rupture zone, where lateral spreading affected borders of lakes.

According to the reconnaissance and the particular geological and geotechnical characteristic of the area, it is possible to indicate that the maximum settlement in natural ground is limited to 1.3–1.5 m.

Due to the catastrophic experience left by the seismic failures of tailings dams built using the upstream construction method, they are prohibited in Chile. However, an abandoned tailings dam constructed by this method (Las Palmas) experienced a flow failure, killing four persons. In the country there are a substantial number of mid height abandoned tailings dams that were constructed by the upstream method before the prohibition, which present a considerable risk to the people and environment. This situation might also be the reality of others countries, where old mining operations have left unsafe tailings disposals that are difficult to identify as well as to control.

The field investigation of the damages caused directly by the occurrence of liquefaction during the 2010 Chile Earthquake, suggests that those associated with lateral spreading were more extensive and disruptive. Accordingly, it is possible to point out that lateral spreading is the most systematic manifestation of liquefaction and due to the widespread damages it needs more attention from a design point of view.

According to the field observations it is postulated that in high seismic environments, as the case of Chile, the seismic history is necessarily associated with the occurrence of liquefaction in initially loose sandy soil deposits. These episodes should modify the initial structure generating highly heterogeneous soil deposits in terms of density. Therefore, a layered structure consisting of loose and dense soil layers is expected, which means a stratified structure of liquefiable and non-liquefiable layers. Hence, there can be recognized only two types of sandy soil deposits: dense, without any problem of liquefaction, and strongly stratified, with liquefiable and non-liquefiable soil layers.

Based on a reconnaissance survey of the area shaken by the 2010 Chile Earthquake and the available information from previous earthquakes, several sites with recurrence of liquefaction were identified. This empirical evidence confirms what have been observed in others earthquakes all around the world.

Acknowledgements The author wants to acknowledge the very useful comments done by my colleagues Abelardo Julio and Javiera Gonzalez. Also, thanks are due to CMGI Ltda. and Department of Civil Engineering of University of Chile for the support provided during the reconnaissance carried out after the earthquake.

References

Bray JD, Frost JD (eds) (2010) Geo-Engineering reconnaissance of the 2010 Maule, Chile earthquake. A report of the NSF-sponsored GEER Association Team, primary authors: Arduino et al. http://www.geerassociation.org/

Bray JD, Frost JD, Rathje EM (September 2011) Turning disaster into knowledge. ASCE G-I Geo-Strata, pp 18–26

Bray J, Rollins K, Hutchinson T, Verdugo R, Ledezma C, Mylonakis G, Assimaki D, Montalva G, Arduino P, Olson S, Kayen R, Hashash Y, Candia G (2012) Effects of ground failure on buildings, ports, and industrial facilities. Earthq Spectr 28(S1):S97–S118

Corvalan J, Herve F (1996) "Geological origin of Chile," large dams in Chile. CIGB—ICOLD, Santiago

C.S.N. Centro Sismológico Nacional. Universidad de Chile. http://www.sismologia.cl/

Ledezma C, Hutchinson T, Ashford S, Moss R, Arduino P, Bray J, Olson S, Hashash Y, Verdugo R, Frost D, Kayen R, Rollins K (2012) Effects of ground failure on bridges, roads, and railroads. Earthq Spectra 28(S1):S119–S143

Renadic: Red Nacional de Acelerógrafos Departamento de Ingeniería Civil. Universidad de Chile. http://www.terremotosuchile.cl/

Ruiz S, Madariaga R, Astroza M, Saragoni R, Lancieri M, Vigny C, Campos J (2012) Short-Period rupture process of the 2010 Mw 8.8 Maule earthquake in Chile. Earthq Spectr 28:S1–S18

Symposium on Engineering Lessons Learned from the Giant Earthquake. One year after 2011 Great East Japan Earthquake. Tokyo, Japan

Verdugo R (2012) Comparing liquefaction phenomena observed during the 2010 Maule Chile Earthquake and 2011 Great East Japan earthquake. In: Proceedings of the International Symposium on Engineering Lessons Learned from the 2011 Great East Japan Earthquake

Verdugo R, Sitar N, Frost D, Bray J, Candia G, Eldridge T, Hashash Y, Olson S, Urzua A (2012a) Seismic performance of earth structures during the February 2010 Maule, Chile earthquake: dams, levees, tailings dams, and retaining walls. Earthq Spectra 28(S1):S75–S96

Verdugo R, González J, González V, Torres A (2012b) Características y efectos del fenómeno de licuefacción. Mw=8.8 Terremoto en Chile, 27 de Febrero 2010, Capítulo 3, 63–105. Edited by Department of Civil Engineering, University of Chile

Vigny C, Socquet A, Peyrat S, Ruegg JC, Metois M, Madariaga R, Morvan S, Lancieri M, Lacassin R, Campos J, Carrizo D, Bejar-Pizarro M, Barrientos S, Armijo R, Aranda C, Valderas MC, Ortega I, Bondoux F, Baize S, Lyon-Caen H, Pavez A, Vilotte JP, Bevis M, Brooks B, Smalley R, Parra H, Baez JC, Blanco M, Cimbaro S, Kendrick E (2011) The 2010 Mw 8.8 Maule Mega-Thrust earthquake of Central Chile, monitored by GPS. Science 331:1417–1421

Wakamatsu K (2012) Recurrent liquefaction induced by the 2011 Great East Japan earthquake compared with the 1987 earthquake. In: Proceedings of the International Symposium on Engineering Lessons Learned from the 2011 Great East Japan Earthquake, Japan

Yasuda S (2011) Lessons learned from most recent large earthquakes. 5th International Conference on Earthquake Geotechnical Engineering. Santiago, Chile

A Case Study on Silty Sand Liquefaction—2010 Hsin Hwa Liquefaction in Taiwan

Wei F. Lee, C. C. Chen, M. H. Chang and Louis Y. N. Ge

Abstract Silty sand liquefaction has been of great interest of research in geotechnical earthquake engineering, especially for the silty sand containing a large amount of non-plastic fines. This article presents a case study on non-plastic silty sand liquefaction. A new sampling technique, Gel-Push sampler, that could retrieve undisturbed non-plastic silty sand and results of series dynamic triaxial tests those investigated influences of fines contents, void ratio, as well as sampling disturbance are reported. Research progress presented here is hoped to be helpful in understanding mechanism as well as consequences of non-plastic silty sand liquefaction.

Keywords Fines content · Gel-Push sampler · Silty sand · Soil liquefaction · Undisturbed sampling technique · Void ratio

Introduction

Non-plastic silty sand liquefaction has been of great research interests in geotechnical earthquake engineering. During the 1999 Chi-Chi earthquake, serious soil liquefaction damages were observed in central Taiwan including Wu Feng, Nan Tou, and Yuen Lin areas. Post-earthquake reconnaissance indicated that most soil liquefactions were taken place in silty sand deposits with high fines content. Christchurch and its vicinity of New Zealand have also suffered from severe liquefaction damages during series of earthquakes from 2010 to 2011. Non-plastic silty sand

W. F. Lee (✉)
MICE Engineering Consultants, 1F, No. 11, Lane 295,
Sec. 1 DunHua S. Road, Taipei, Taiwan
e-mail: wflee0206@gmail.com

C. C. Chen · M. H. Chang
Department of Civil Engineering, National Chen Kung University, No. 1 University Road, Tainan, Taiwan 701

L. Y. N. Ge
Department of Civil Engineering, National Taiwan University, No. 1, Sec 4, Roosevelt Road, Taipei, Taiwan, 10617

A. Ansal, M. Sakr (eds.), *Perspectives on Earthquake Geotechnical Engineering*,
Geotechnical, Geological and Earthquake Engineering 37,
DOI 10.1007/978-3-319-10786-8_15

Fig. 1 Non-plastic silty sand liquefaction observed in 1999 Taiwan Chi-Chi earthquake, 2010~2011 New Zealand Christchurch earethquakes, and 2011 Japan Tohoku earthquake

has been recognized as the major sources of soil liquefaction. Moreover, Tokyo bay area and Chiba perfect was suffered from serious soil liquefaction damages during the 2011 Great East Japan earthquake. Preliminary reconnaissance also concludes that majority liquefaction occurred in the reclaimed silty sand deposits (Fig. 1; Lee et al. 2012).

Seed and Idriss (1976) indicated that uniformly-graded clean sand and silty sand are two types of soil susceptible to liquefaction mostly, and the plasticity of fines is a major concern. Ishihara et al. (1980) preformed a series of cyclic triaxial tests on remolded specimens and concluded that silty sand has lower cyclic stress ratio (liquefaction resistance) than clean sand. Seed and Idriss (1982) proposed SPT-N modification on plastic fines content for a simplified liquefaction evaluation with consideration of D_{50}. Ishihara (1985) indicated that "Whether or not a soil containing such a large percentage of fines can develop liquefaction would be a subject of concern requiring further investigations. However, it would appear that the liquefiability of soils with high fines contents will depend on the physical characteristics of the fines themselves." Seed et al. (1985) proposed modification on the simplified liquefaction evaluation method, based on field observation on case histories, reminders of characteristics of fines. Ishihara (1985, 1993) concluded that characteristics of fines such as plasticity, mineral contents, particle shape, size, and gradation are rather important than percentage of fines on silty sand liquefaction. Figure 2 summarizes the research time table of silty sand liquefaction accounting for the influence of fines

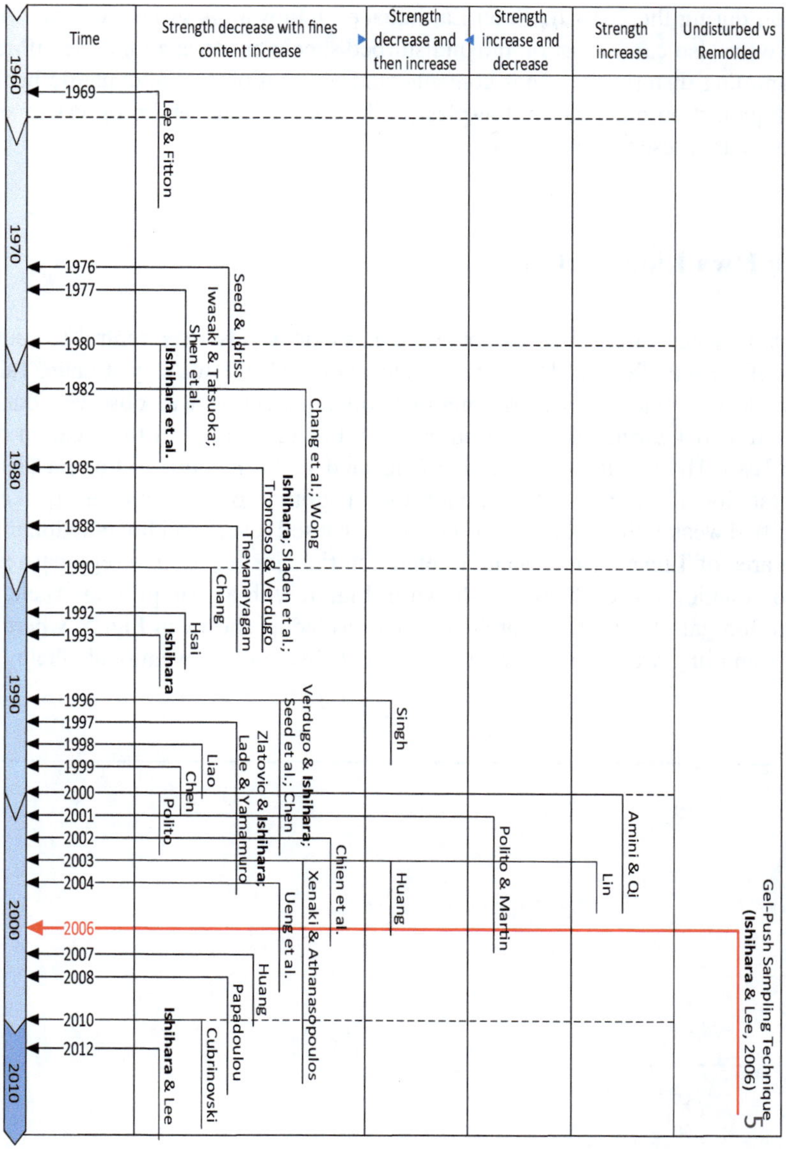

Fig. 2 Summarized research time table on silty sand liquefaction, with emphasis on influence of fines content.

content. However, it has been difficult to retrieve undisturbed samples of silty sand. Influences of fine contents, void ratio as well as sampling disturbance on liquefaction potential of silty sand containing non-plastic fines were not well studied in the past. Ishihara and Lee introduced the Gel-Push sampler to retrieve undisturbed samples from a drilling project in Hsin Hwa, where was suffered from soil liquefaction during the 2010 Jia Shan Earthquake in Taiwan. A series dynamic triaxial tests were also carried out on the undisturbed specimens to examine the influences of sampling disturbance, fines contents, and void ratios on the dynamic behavior and liquefaction potential of non-plastic silty sand. Details and progress of such a research are presented as follow.

Hsin Hwa Liquefaction

The silty sand with high fines content exists extensively over central to southern parts of western Taiwan. The test site is located in Hsin Hwa City, Tainan, Taiwan. This site was selected because massive soil liquefaction was observed during a magnitude 6.4 earthquake occurred in 2010. Figure 3 indicates the location of the Hsin Hwa (HH01) site, the epicenter is located 17.1 km south-east of Jia Sian station (station ID: KAU047). Although the magnitude of Jia Sian earthquake was classified weaker than Chi-Chi earthquake, it caused severe soil liquefaction in Hsin Hwa area of Tainan. The ground acceleration time history from a regional seismic station (station ID: CHY063) is shown in Fig. 4, with a peak ground acceleration about 385 gal. A total four boreholes were drilled as shown in Fig. 5, where Gel-push sampling was conducted in three boreholes, and conventional Shelby tube

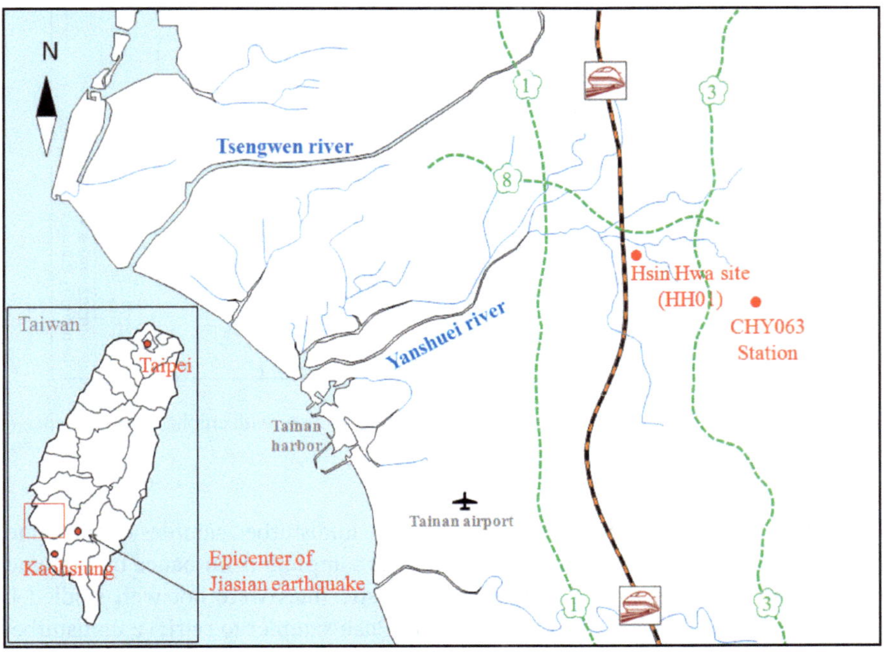

Fig. 3 Geographical location of Jia Sian earthquake

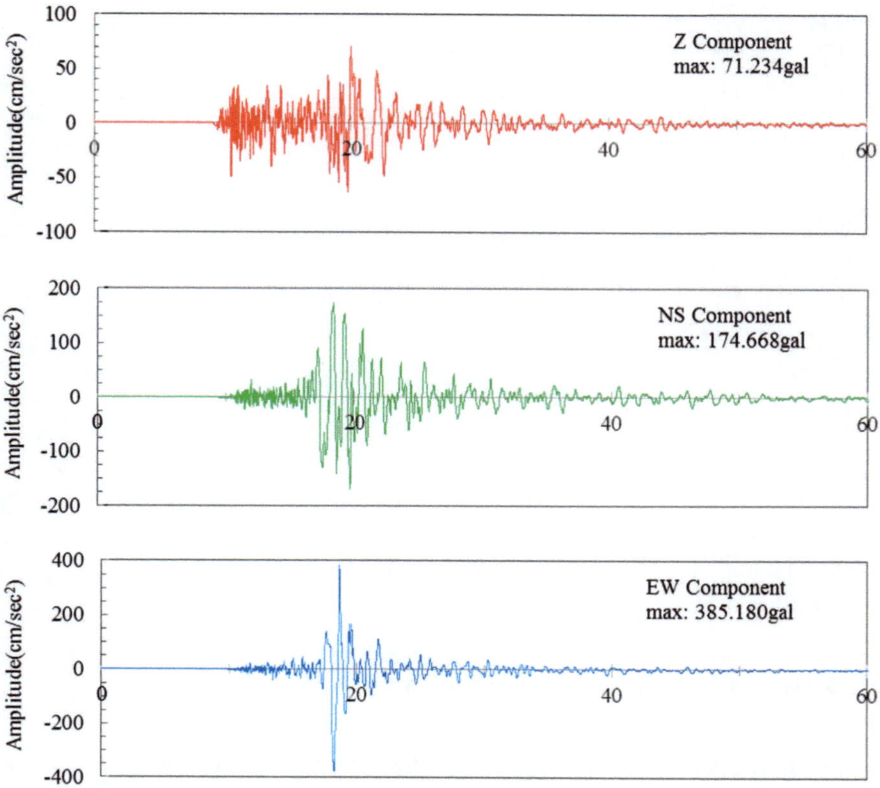

Fig. 4 Ground acceleration time history of Jia Sian earthquake on CHY063

Fig. 5 Soil drilling program at Hsin Hwa study site

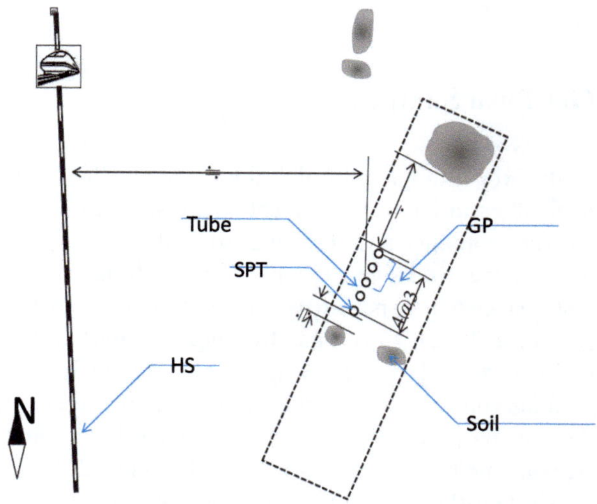

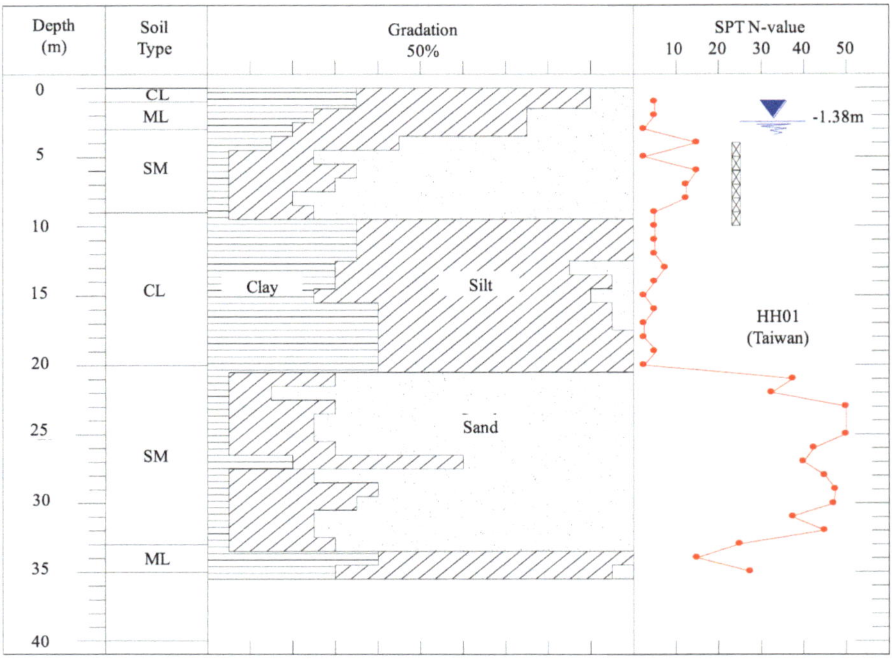

Fig. 6 Summarized soil profile at Hsin Hwa study site

sampling was performed in the fourth one for the purpose of comparison. The soil profile including the SPT-N values of the test site is given in Fig. 6. As depicted in the figure, a silty sand layer which is located between 2 and 10 m below the ground surface contains high fines content ranging from 10 % to more than 50 %.

Gel-Push Sampler

Undisturbed sampling of high fines content silty sand was facing several technical difficulties in the past. Conventional Osterberg's Shelby tube sampling technique has shortcomings in retrieving good quality high fines content silty sand specimens because the excessive friction generated during penetration tends to cause serious disturbance to the specimens and results in incomplete soil sample and poor quality (Lee et al. 2012). The ground freezing technique or tube freezing process those generally were used for preserving sampled soil quality in the past would cause drifting of fines content and disturbance on sensitive micro structure during freezing and de-freezing process. Serious fines content loss would probably occur when such freezing methodologies are adapted (Lee et al. 2012).

The Gel-Push sampling technique was first developed to retrieve gravel material as an alternative replacing ground freezing method in Japan in 2004. The Gel-Push sampler was then introduced to Taiwan by Lee and Ishihara in 2006 in an effort to obtain undisturbed silty sand with non-plastic fines content. It was modified to accommodate the thin wall tube inside the sampler to become a triple tube system. The newly developed sampler was designed to allow polymer lubricant to seep into the tube wall while the tube was penetrated into the soil by hydraulic pressure. Figure 7 shows configuration of the Gel-Push sampler.

Figure 8 shows the schematic drawings of the Gel-Push sampler at different stages of sampling processes. As shown in the figure, the outer tube is designed to secure the borehole and to keep the penetration rod and piston fixed in alignment

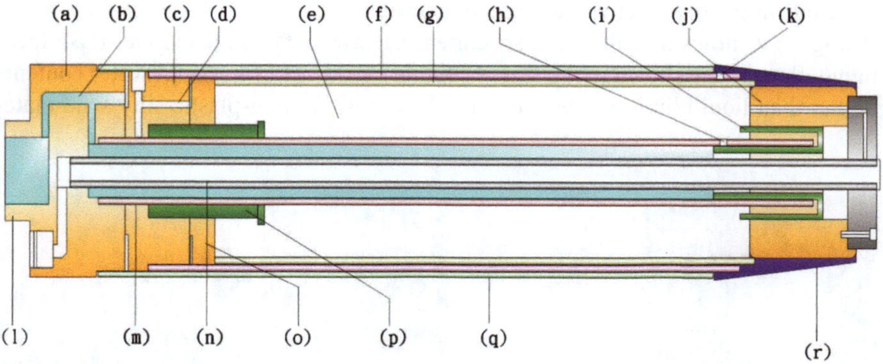

Fig. 7 The configuration of the Gel-Push sampler

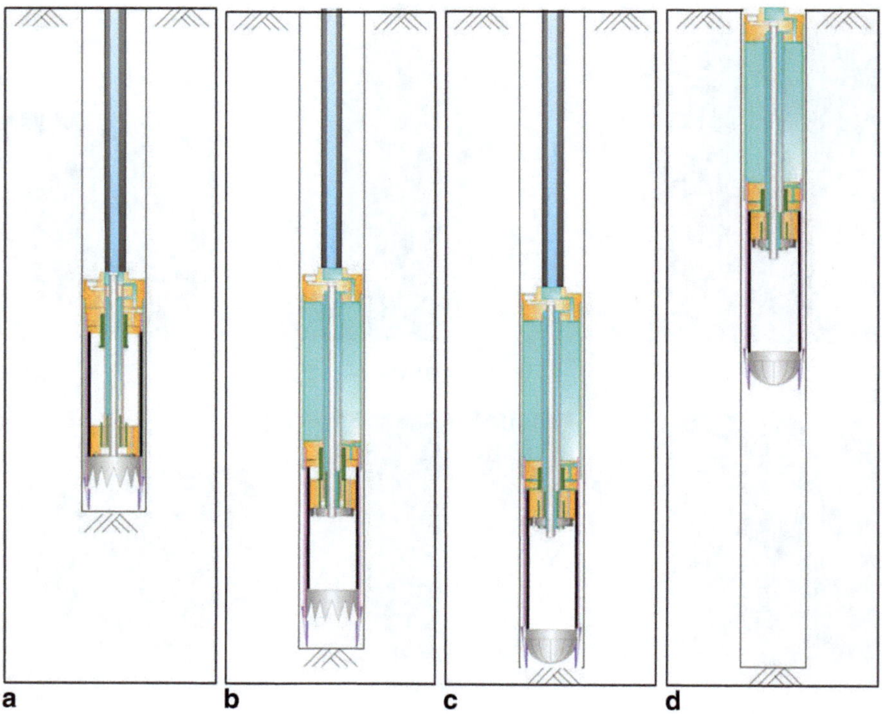

Fig. 8 Schematic drawings of Gel-Push sampling technique. **a** Fc=7.5%. **b** Fc=15.6%. **c** Fc=33%

during penetration. The middle tube acts as the guiding tube to push sampler into soil. While sampling process starts, polymer gel is squeezed out of the chamber and seep into both outer side of the guiding tube and inner side of the thin wall tube. The sampler is also designed with a cutter attaching to the guiding tube to allow smooth penetration, and a catcher fixed at bottom of the thin wall tube to hold soil specimen from falling out during uplifting. The polymer gel would contaminate limited superficial portion of the specimen because very small amount of polymer gel is applied. However, it could effectively reduce the wall friction so as to allow sensitive silty sand specimen to be recovered in good quality.

Figure 9 shows the silty sand specimen that was obtained using Gel-Push technique. Specimens shown in the figure contains 5–33 % of fines with water contents higher than liquid limit. It was recognized as sensitive non-plastic silty sand mate-

Fig. 9 Undisturbed non-plastic silty sand specimens retrieved using Gel-Push sampling technique (before testing, after testing, and cross section)

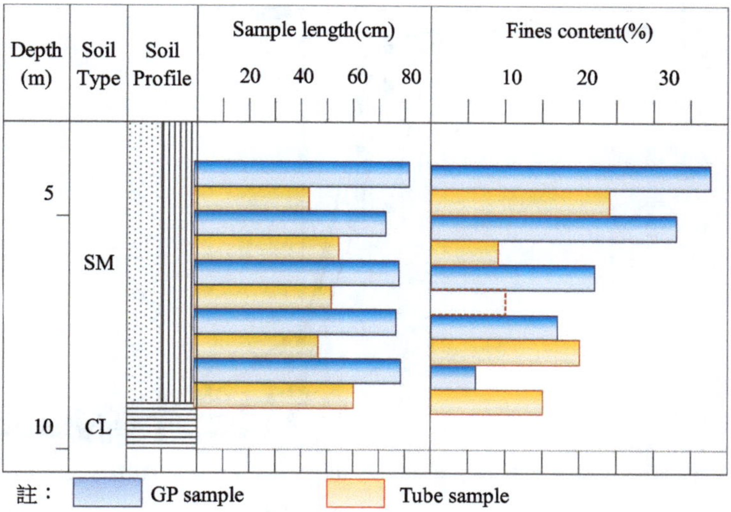

Fig. 10 Comparison of sampling results between conventional tube sampler and Gel-Push sampler

rial that was difficult to be retrieved in the past by using conventional Shelby tube sampler. As shown in the figure, specimen retrieved using Gel-Push sampler could self-stand without freezing preservation. Figure 10 show the sampling results of both conventional Shelby tube sampler and Gel-Push sampler. The sampling depth is from 4 to 9 m below ground surface where most silty sand deposits locate. As depicted in the figure, Gel-push sampler had successfully recovered more sample than the tube sampler. For completed sampling length, Gel-Push sampler also preserved more fines contents than tube sampler did. Tube sampler could only preserve clayey portion or coarse sand portion by comparing to soil samples retrieved using Gel-Push sampler from the laboratory observation.

Dynamic Properties of Silty Sand

The non-plastic silty sand specimens used in the cyclic triaxial testing program were retrieved by the Gel-Push sampling technique at the study site (HH01). Figure 11 shows the grain size distribution curve of HH01 specimens. As shown in the figure, the average grain size (D_{50}) is about 0.1~0.19 mm, and fines content (F_C) is 9.0~29.1. The soil type of HH01 specimens belongs to SM or ML in USCS system. Figure 12 shows the Scanning Electron Microscope (SEM) image of fines particles obtained from the test site. Fines particles of such silty sand material are in angular to sub-angular shapes as shown in the figure. This evidence clearly indicates that almost no plasticity could be possibly exerted within such soils.

Testing results of two representative specimen groups were selected for comparative illustration. The groups were categorized by fines contents near 10 and 28 % respectively. Each group contains undisturbed specimens and remolded specimens.

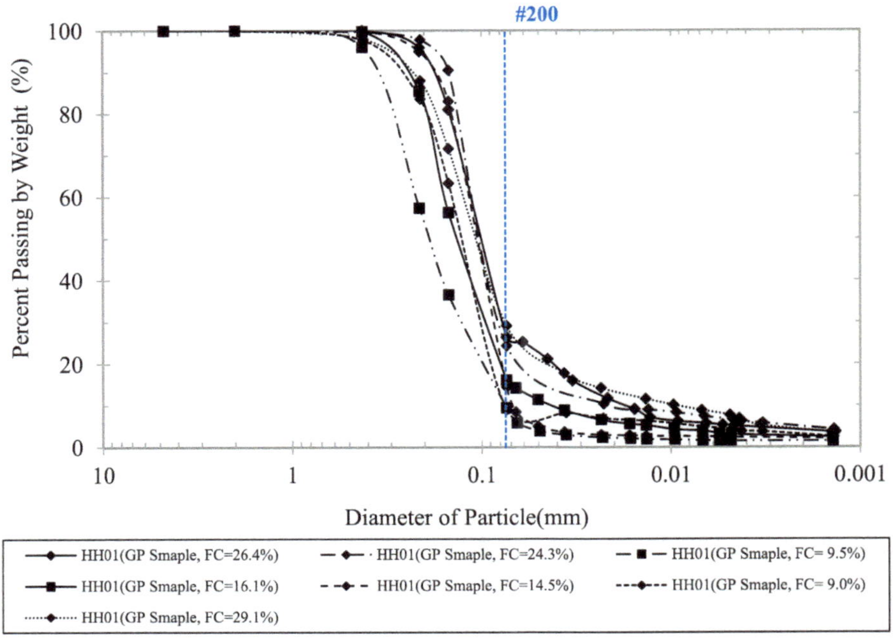

Fig. 11 The grain size distribution curve of HH01 specimens

Fig. 12 Fines particles SEM images of studied silty sand

Both undisturbed and remolded specimens were tested under similar void ratios and confining pressure control for comparison. The remolded specimen were prepared using wet-damping methodology to effectively control the void ratio; and the undisturbed specimens were prepare from the Gel-Push sampler with minor trimming. Figure 13 shows the axial strain and pore water pressure excitation histories of two selected groups. As shown in the figure, the failure pattern of undisturbed specimens was similar to plastic deformation when the excess pore water pressure increased

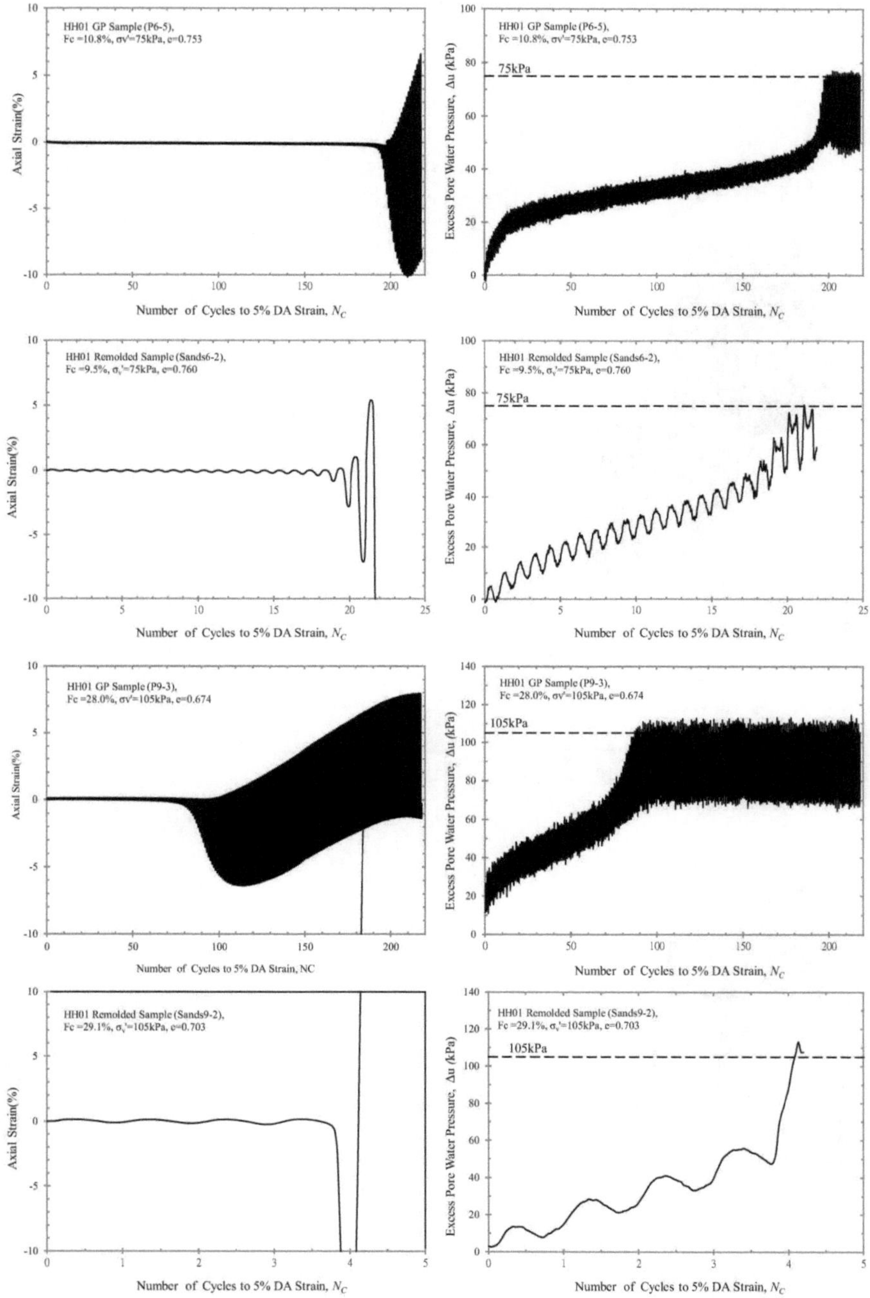

Fig. 13 Axial strain and excess pore water pressure histories of cyclic triaxial tests

slowly and incrementally. For remolded specimens, the excess pore water pressure increased quickly, and brittle failure has occurred under the same testing condition.

Figure 14 shows the strain-stress behaviors of undisturbed and remolded specimens respectively. As shown in the figure, the undisturbed specimen has stronger

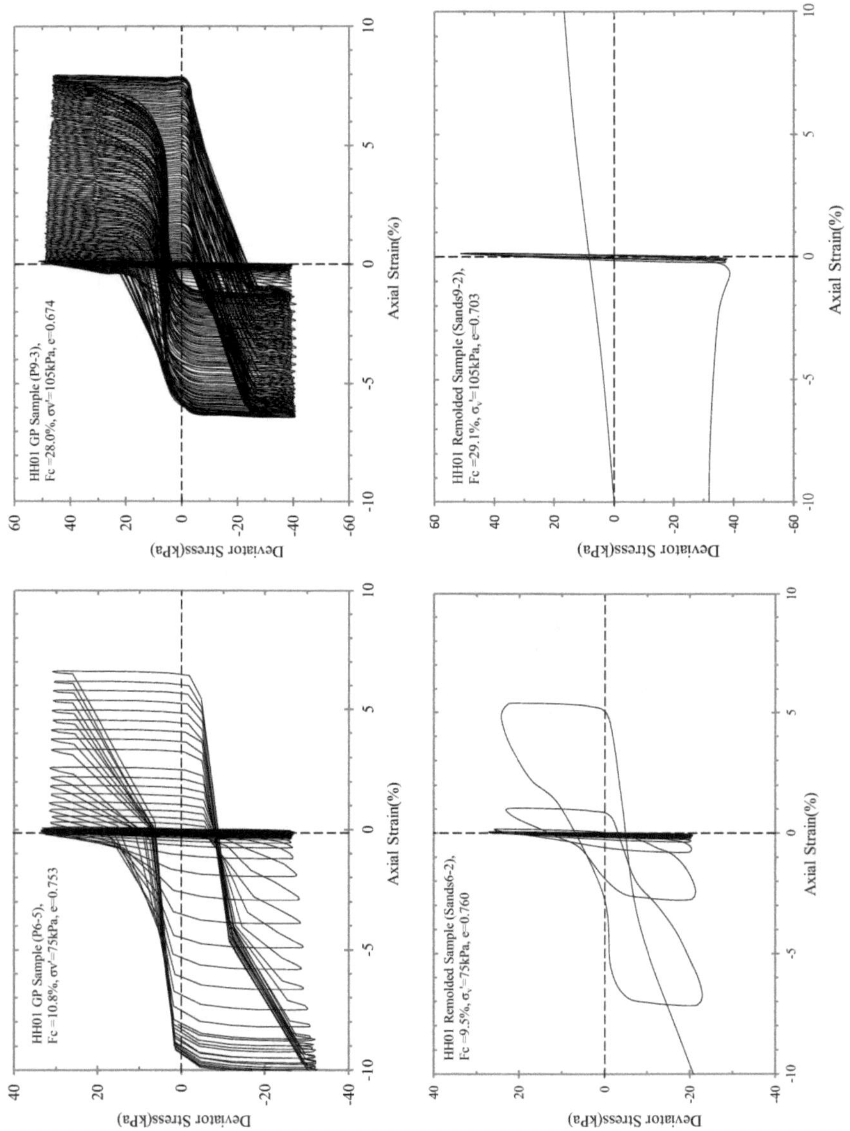

Fig. 14 Strain-stress behaviors of cyclic triaxial tests

cyclic resistance and possesses a larger yielding strain than the remolded specimen with a similar density and deviator stress. To make further comparison, the close-up first 5 cycles of strain-stress history are plotted in Fig. 15. For specimens with similar fines contents, the remolded ones appear to soften-up (yield) quickly while the undisturbed specimens still exert plastic behavior. For undisturbed specimens with different fines contents, specimens with 28 % fines deform less than those with 10 %, indicating that fines content plays an important role in stabilizing the micro structure of non-plastic silty sand. However, for remolded specimens, specimen with less fines content appear to yield much less than those with higher fines content.

Figure 16 shows the effective stress paths of undisturbed and remolded specimens. As shown in the figure, undisturbed specimens could bear a larger cyclic shear stress or load cycles, while the remolded specimens yield much more quickly before liquefaction was observed. This phenomenon become clearer on specimens with higher fines content.

Discussion on Engineering Practice

Cyclic Stress Ratio

Figure 17 summarizes the relationships between the cyclic stress ratio (CSR) and the number of load cycles (Nc) at liquefaction of all tests conducted. As illustrated in the figure, undisturbed specimens have higher cyclic strengths than those of remolded specimens with the same fines contents and similar void ratios. It indicates that sampling disturbance could definitely influence the soil liquefaction resistance of non-plastic silty sand. Under the same void ratio condition, specimens with higher fines contents tend to have lower cyclic strengths. Under the similar fines content condition, specimens with higher void ratios tend to have lower cyclic strength. This behavior becomes more noticeable for the remolded specimens.

The proposed regression curves of the CSR versus Nc relationships are shown in Figs. 18 and 19. In these figures, when the fines content and relative density of specimens are similar, the cyclic stress ratio of undisturbed specimens is clearly higher than it of remolded specimens. The cyclic stress ratio decreases when the relative density decreases. The cyclic stress ratio also decreases as the fines content increases under the same relative density condition.

Influence of Fines Content and Void Ratio

Figure 20 shows the influence of sampling disturbance to dynamic properties of non-plastic silty sand with various fines contents. In the figure, the vertical axis is the ratio of CSRs between remolded specimens and undisturbed specimens at the

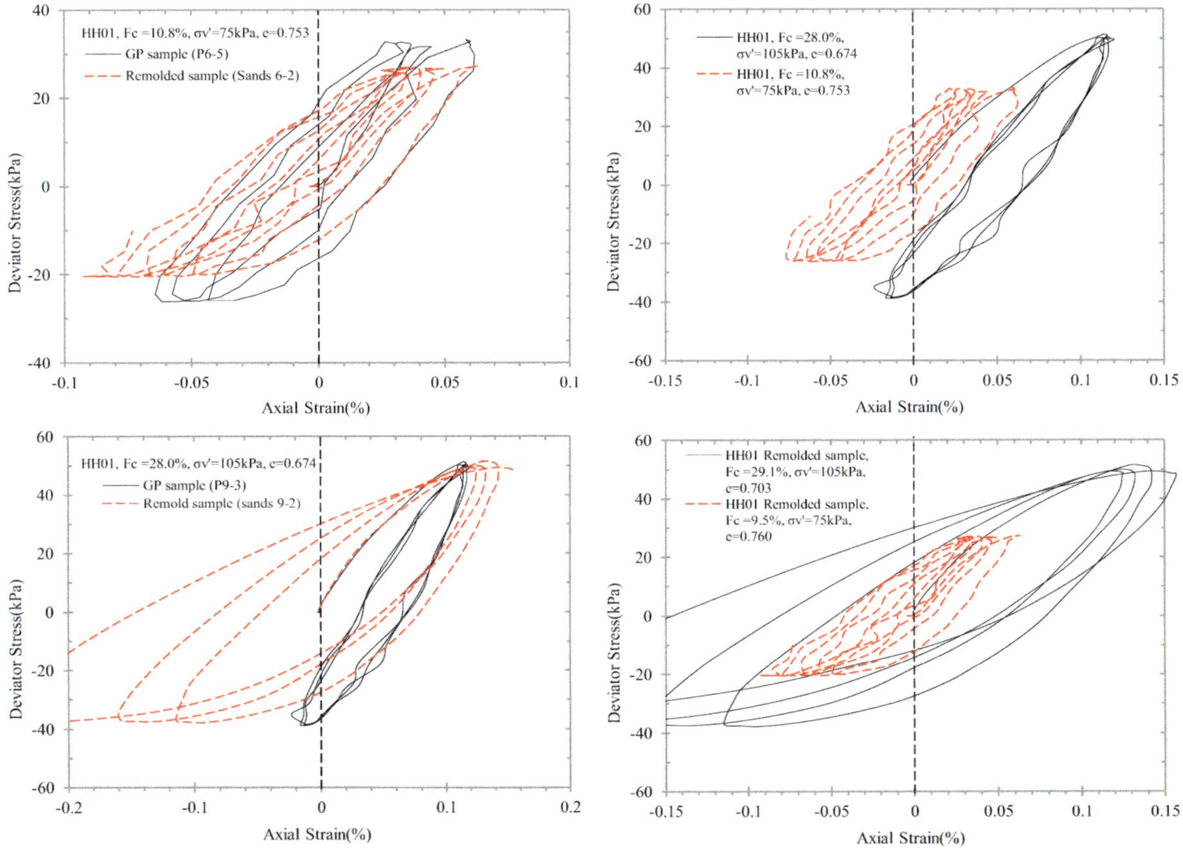

Fig. 15 The close-up first 5 cycle of strain-stress histories

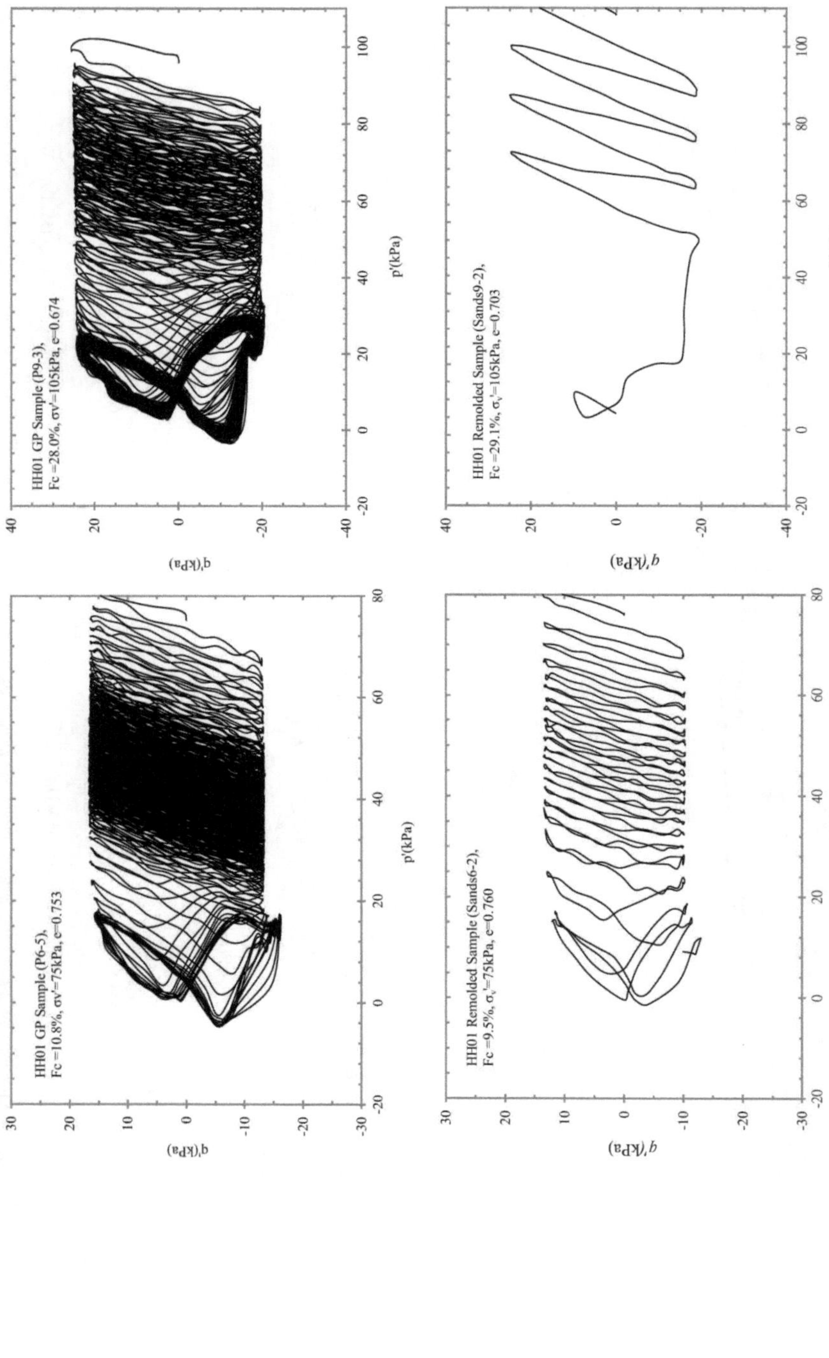

Fig. 16 Effective stress paths of the cyclic triaxial tests

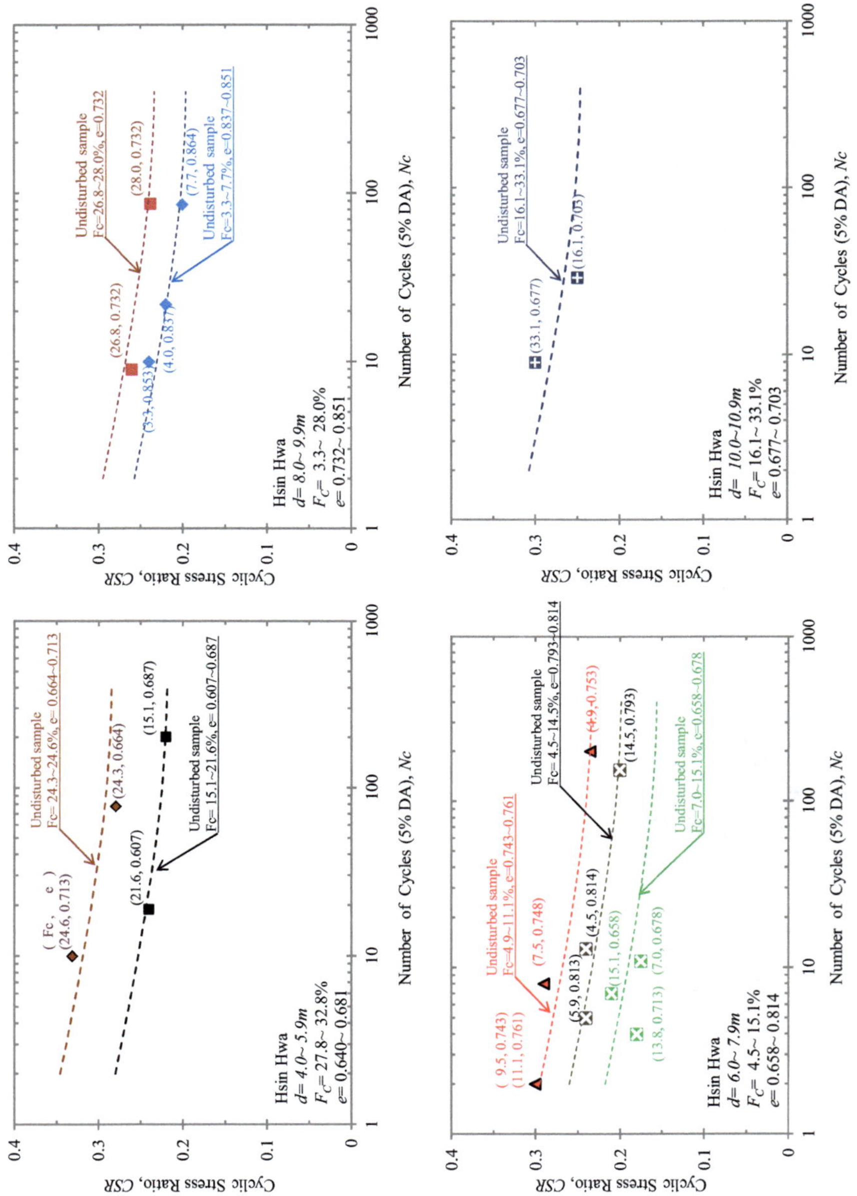

Fig. 17 Summary of results of conducted cyclic triaxial tests

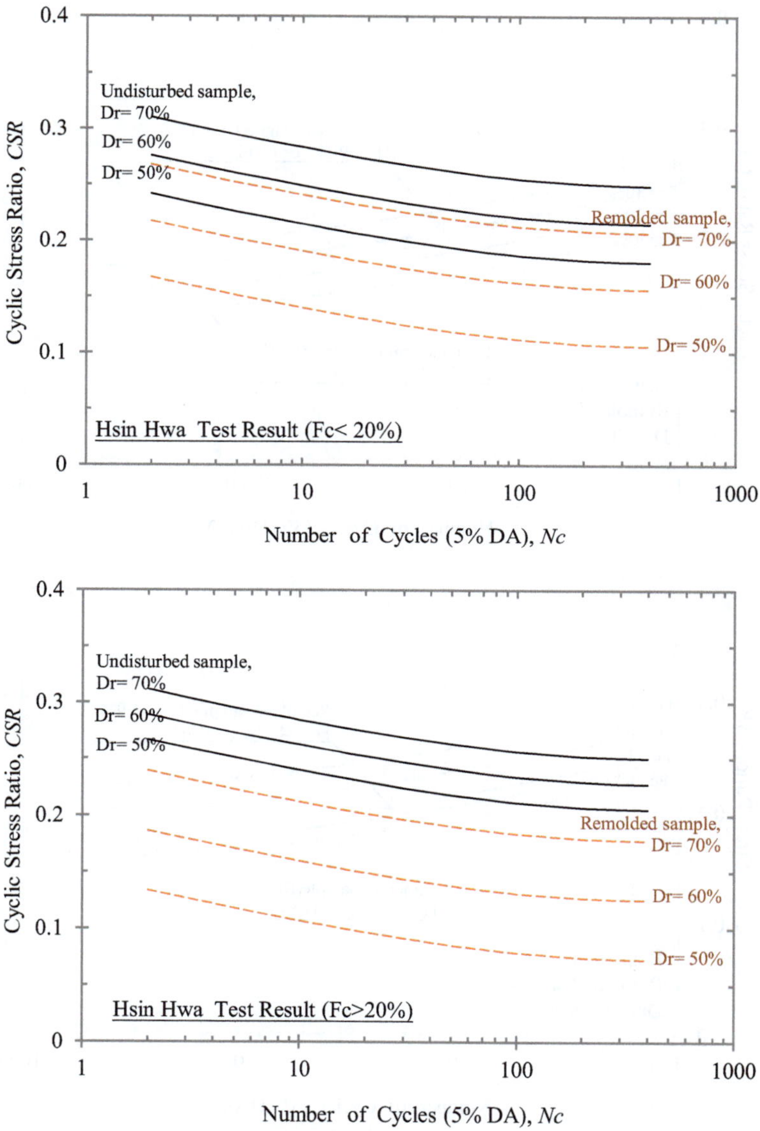

Fig. 18 The proposed regression lines of cyclic triaxial test results with different specimen conditions, fines contents, and void ratios

5th, 15th, 20th load cycle (Nc). As shown in the figure, the more fines content the non-plastic silty sand has, more reduction in cyclic strength are observed. When the fines content of soil specimens was 30 %, there was a 40 % reduction in its cyclic strength. Regardless what void ratio the silty sand specimen has, the cyclic strength

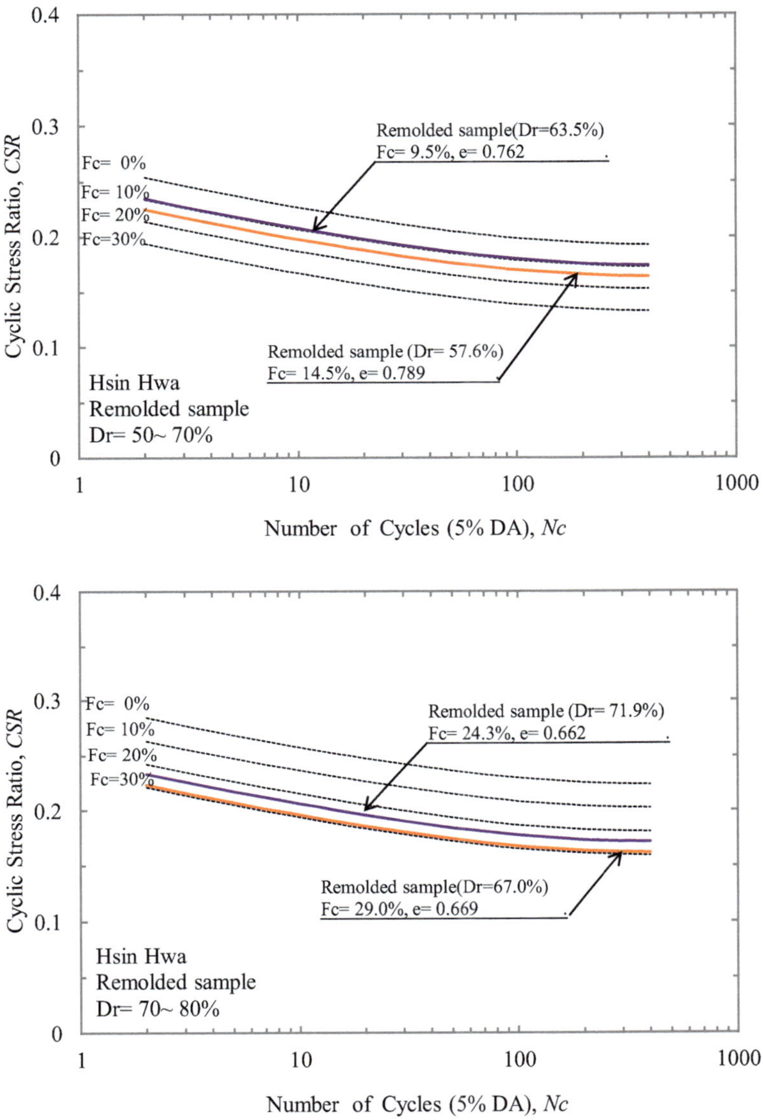

Fig. 19 Proposed regression lines of remolded cyclic triaxial test results at different relative densities

decreases when fines content increase as shown in Fig. 21. This phenomenon could be related to the cohesion of natural deposition and compaction process. Generally, under natural condition, the void ratio decrease when fines content increase, and the particle structure was stronger. Hence, fines content and void ratio are two important factors to the liquefaction resistance of non-plastic silty sand.

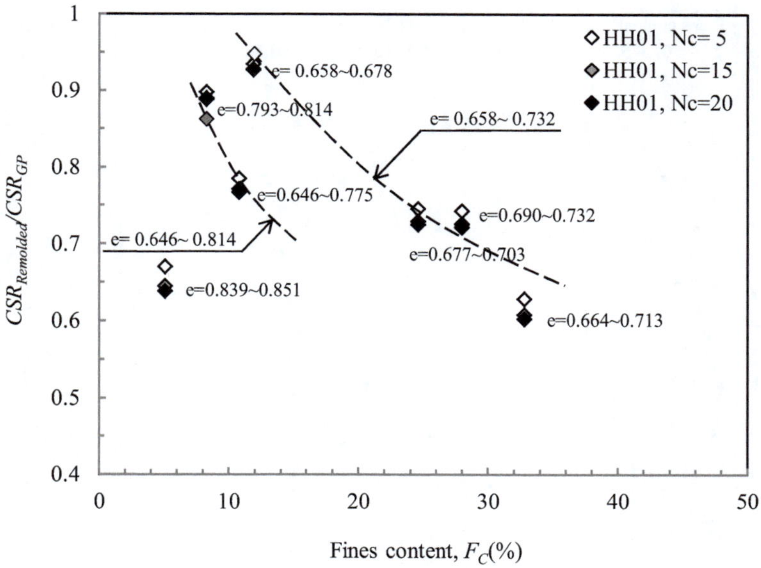

Fig. 20 Influence of sampling disturbance to dynamic properties of non-plastic silty sand with various fines contents

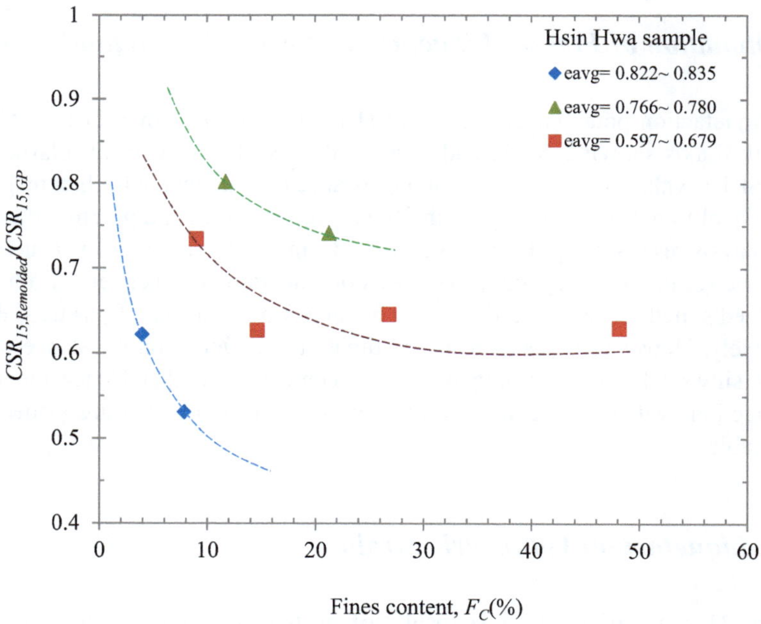

Fig. 21 Summary of fines content influence and disturbance effects on cyclic stress ratio of non-plastic silty sand

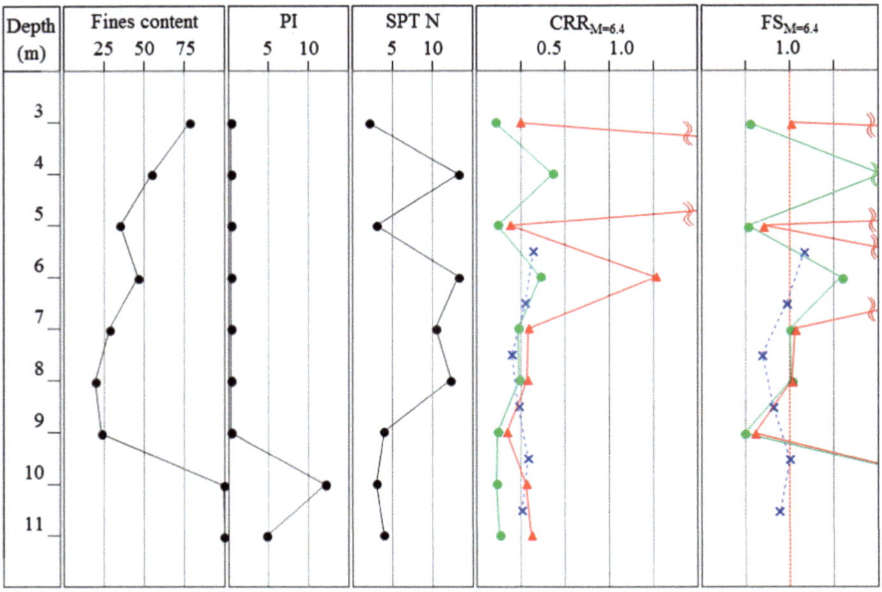

Fig. 22 Comparison of different liquefaction potential evaluations of non-plastic silty sand

Examination of Present Liquefaction Evaluation Method

The liquefaction potential evaluation of Hsin-Hwa site is shown in Fig. 22, with a vertical axis showing depth, and horizontal axes of fines content, plastic index (PI), SPT-N value, and the liquefaction resistance ratio under the Richter magnitude (Mr) equal to 6.4, respectively. In the figure, red triangles and green circles show the analysis results using the NJRA and Seed's methods, respectively. Blue x indicates the results of the cyclic triaxial tests on undisturbed specimens. Both NJRA and Seed's methods would estimate the liquefaction resistance of plastic soil layer accurately. However, it seems to overestimate the liquefaction resistance in non-plastic silty sand layer, especially when fines content was higher. Hence, the present liquefaction evaluation methods in a non-plastic silty sand layer are inadequately applicable.

Post-Liquefaction Volumetric Strain

Figure 23 summarizes the test results of post-liquefaction volumetric strains, where the post liquefaction volume change was measured from the dissipation of excess pore water after liquefaction was identified in each cyclic triaxial test. As shown in the figure, remolded specimens clearly possess larger volumetric strains than undisturbed ones. Post-liquefaction volumetric strains of remolded

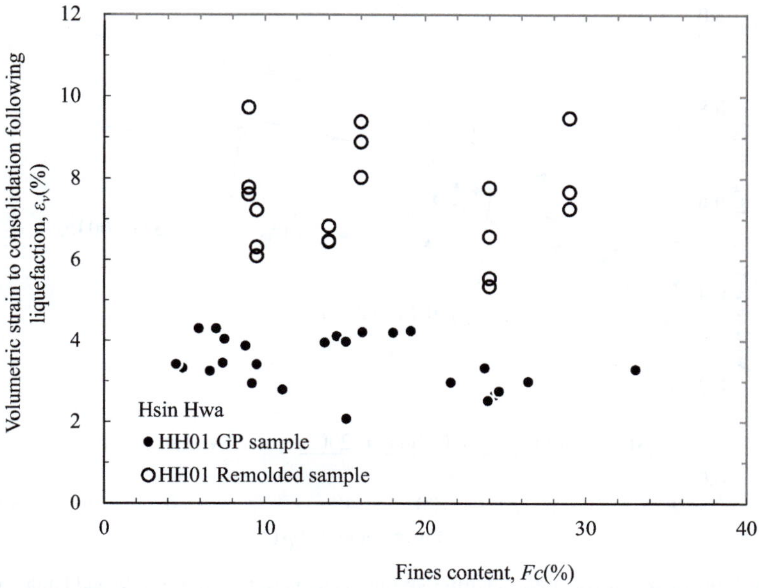

Fig. 23 Test results of post-liquefaction volumetric strains according to various fines contents

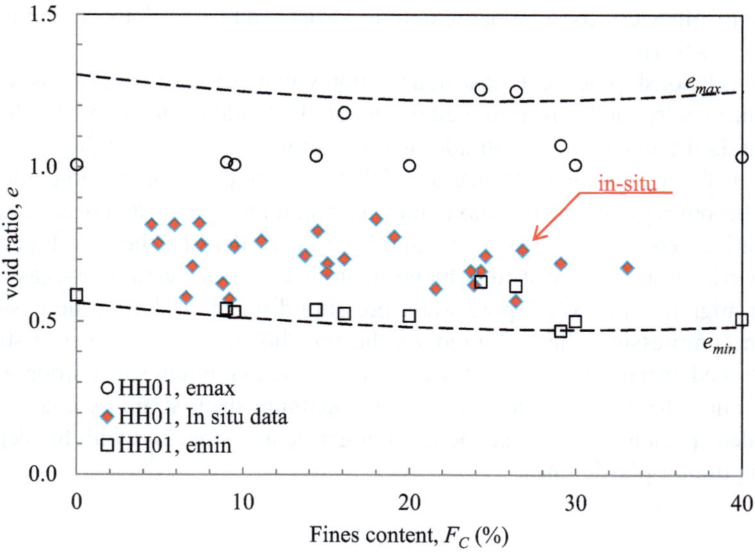

Fig. 24 Void ratios of tested material

specimens would be as high as 8 to 10 %, whereas those of undisturbed specimens remain between 2 and 5 %. It was found that scale of post liquefaction volumetric strain depends very much on the void ratio range, i.e. $e_{max} - e_{min}$. Figure 24 summarizes the void ratios of tested specimens at different fines content. Void ra-

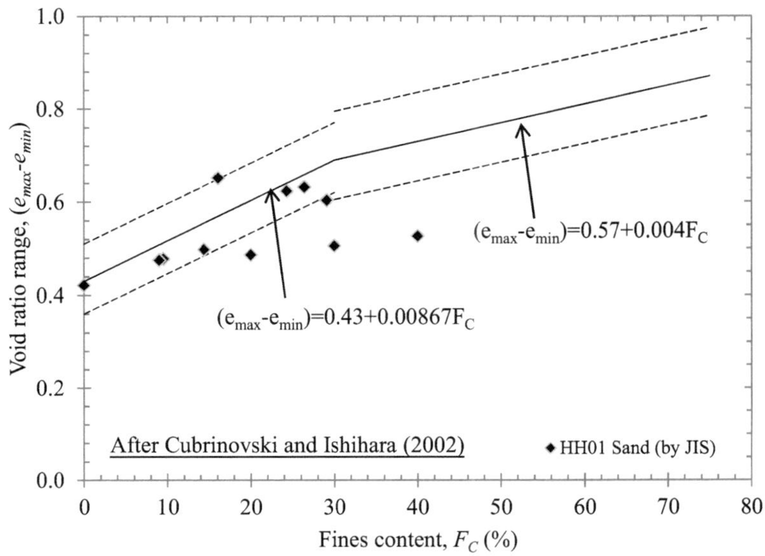

Fig. 25 Comparison of void ratios at different fines content (after Cubrinovski and Ishihara 2002). **a** Undisturbed. **b** Remolded

tio test results were also compared to those proposed by Cubrinovski and Ishihara (2002) as shown in Fig. 25.

The relationship between post-liquefaction volumetric strain and shear strain of non-plastic silty sand was presented in Fig. 26. It should be noted that the horizontal axis is the maximum amplitude of shear strain (γ_{max}), $\gamma_{max} = 1.5\varepsilon_{(d,\ max)}$, where $\varepsilon_{(d,\ max)}$ is the maximum amplitude of axial strain. The post-liquefaction volumetric strain becomes greater as the maximum shear strain increases, and decreases when its relative density increases. For a similar relative density, the post-liquefaction volumetric strain of silty sand is higher than that of clean sand. Non-plastic fines would migrate with excess pore water pressure dissipation during the post-liquefaction compression phase. Moreover, the post-liquefaction volumetric strain of undisturbed specimen tends to stay constant as the maximum shear strain exceeds 8 %, while it tends to stay constant as the maximum shear strain exceeds 11 % for remolded specimens. Those phenomena might be in connection with the degree of drifting of non-plastic fines.

Conclusions

Gel-Push appears to be a capable tool for acquiring intact sample of non-plastic silty sand under a high ground water table (GWT) and loose deposit. Results of laboratory tests indicate that higher the void ratio and fines content of the non-plastic silty sand would have a larger cyclic strength reduction when dis-

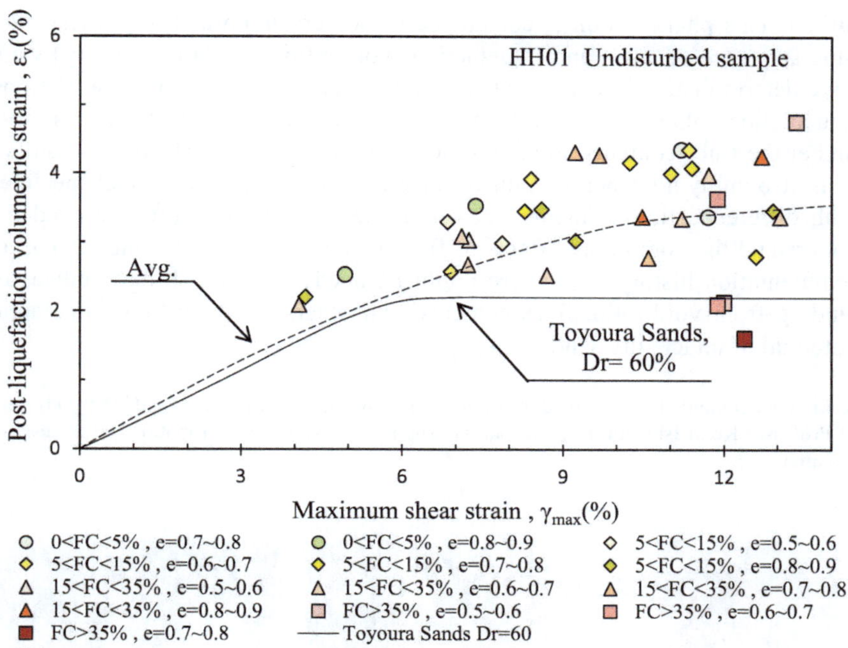

a

Undisturbed

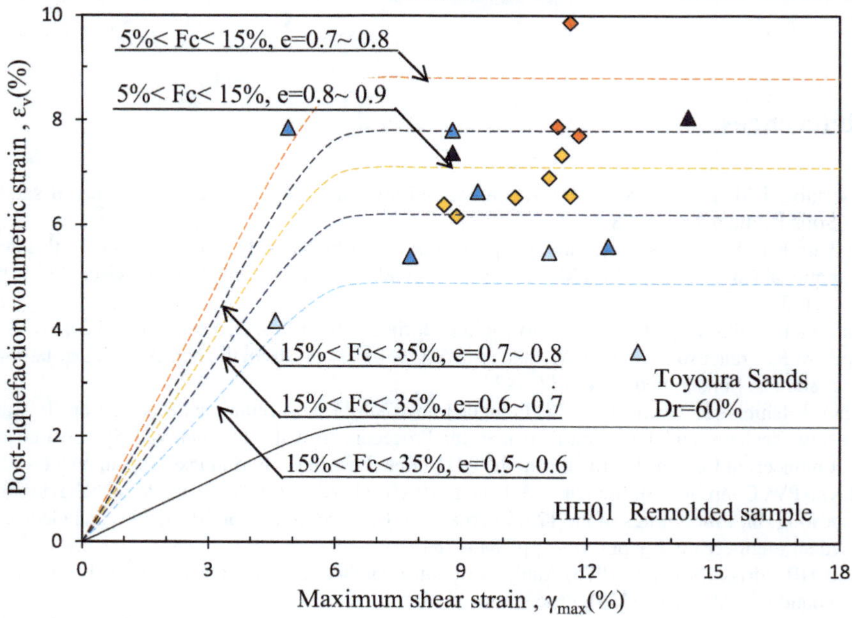

b

Remolded

Fig. 26 Summary relationships between volumetric strain and shear strain of non-plastic silty sand turbed. This phenomenon is considered to be very important in understanding time sequence of silty sand liquefaction. Void ratio (or relative density) would have deterministic effect on the silty sand liquefaction potential, so as the post-liquefaction volumetric strain. Results of tests on remolded specimens also show higher the fines contents, lower the liquefaction resistance. However, results of tests also imply no general trends of liquefaction resistance for intact specimens with difference fines contents. It is suggested that more efforts are needed to understand the combining effects of fines content, particle packing, aging, and sedimentation history, etc. There might be needs to establish new indices extending from void ratio, fines contents, and others for describing liquefaction potential of intact silty sand.

Acknowledgement The presented research progress was under inspirations and kindly advisories of Professor Kenji Ishihara. His wisdom and supports make great contribution to the described research.

a b c

References

Cubrinivski M, Ishihara K (2002) Maximum and minimum void ratio characteristics of sands. Soils Found 42(6):65–78

Ishihara K (1985) Stability of natural deposits during earthquakes. In: Proceedings of 11th International Conference on Soil Mechanics and Foundation Engineering, San Francisco, Vol 1, pp 321–376

Ishihara K (1993) Liquefaction and flow failure during earthquakes. Geotechnique 43(3):351–415

Ishihara K, Trancoso J, Kawase Y, Takahashi Y (1980) Cyclic strength characteristics of tailings materials. Soils Found 20(4):127–142

Lee WF, Ishihara K, Chen CC (2012) Liquefaction of Silty Sand-Preliminary Studies from Taiwan, New Zealand, and Japan Earthquakes. In: Proceedings of the International Symposium on Engineering Lessons Learned from the 2011 Great East Japan Earthquake, Japan, pp 747–758

Mayne PW, Coop MR, Springman SM, Huang AB, Zornberg JG (2009) Geomaterial behavior and testing. In: Proceedings of the 17th International Conference on Soil Mechanics and Geotechnical Engineering, Egypt, Vol 4, pp 2844–2845

Seed HB, Idriss IM (May, 1976) Analysis of soil liquefaction: Niigata earthquake. J Soil Mech Found Div ASCE 93(SM3), Proc. Paper 4233

Seed HB, Idriss IM (1982) Ground motion and soil liquefaction during earthquakes. Earthquake Engineering Research Institute Monograph, Oakland

Seed HB, Tokimatsu K, Harder LF, Chung RM (1985) Influence of SPT procedures in soil liquefaction resistance evaluations. J Geotechnical Eng 4SCE 111(12):1425–1445

Post-Earthquake Analysis for a Seismic Retrofitting: The Case History of a Piled Foundation in Augusta (Italy)

Michele Maugeri and Francesco Castelli

Abstract The case history of a seismic retrofitting of a piled foundation is reported. To analyze and to empathize the importance of the different phases of the seismic design process, in the paper the approach adopted for the structural upgrading of a six storey reinforced concrete frame resisting building in the Saline district of Augusta (Italy), damaged by the Sicilian earthquake ($M_L = 5.4$) of 13th December 1990, is described and discussed. For evaluating the possibility to repair the building, an investigation on soil, structure and foundation was carried out. As building was founded on piles, for the seismic retrofitting of the structure, the response of the piles subjected to earthquake loading has been studied. Vertical and horizontal loading tests on single pile and pile groups were conducted to test the integrity of the piles after the seismic event. To extrapolate the behaviour of the piled foundation under the design seismic actions, a numerical analysis of inertial and kinematic effects have been performed. The site was well investigated by in-situ and laboratory tests. To estimate dynamic stress-strain geotechnical characteristics of soils resonant column tests were also performed. Loading tests showed that the seismic event of December 13th 1990 have not damaged the effectiveness of the soil-pile system. Nevertheless, as the numerical analysis showed that the existing foundation was enable to carry on design seismic actions, the structural upgrading and seismic retrofitting of the piled foundation required the enlargement of the foundation and new bored piles.

Keywords Seismic analysis · Piled foundation · Inertial effect · Kinematic effect

---

M. Maugeri (✉)
University of Catania, Catania, Italy
e-mail: mmaugeri@dica.unict.it

F. Castelli
University of Enna "*Kore*", Enna, Italy
e-mail: francesco.castelli@unikore.it

A. Ansal, M. Sakr (eds.), *Perspectives on Earthquake Geotechnical Engineering*,
Geotechnical, Geological and Earthquake Engineering 37,
DOI 10.1007/978-3-319-10786-8_16

Introduction

In the third millennium the seismic risk mitigation is one of the greatest challenges of the Civil Engineering and an important contribution toward this challenge can be given by the Geotechnical Earthquake Engineering. In this field the researches are essentially devoted to two different goals: the first, which is the most traditional, deals with the modelling of the behaviour of the single geotechnical construction and/or soil-structure interaction.

The second deals with the stability of physical environment. This last field involves, more than the first, different scientific sectors, because the Geotechnical Earthquake Engineering interacts with seismology, geology and structural engineering. The common goal is the evaluation of the design seismic action.

Generally the seismic action is defined by the National Regulations. Nevertheless, a "soil factor" due to possible amplification phenomena given, for example, on the basis of medium values of soil properties (Vs, 30), can be excessive for particular soil profile conditions. Thus, for relevant and strategic structures, it is appropriated to evaluate the seismic site response, by means of 1-D, 2-D and/or 3-D codes, taking into account the soil profile, up to a significant depth, and the nonlinear soil behaviour.

For strategic structures, monumental and historical buildings the deterministic evaluation of the seismic action could be more appropriated. The deterministic evaluation of the seismic action can be achieved by means of the use of recorded accelerograms, conveniently scaled in amplitude and frequency, in order to simulate the scenario earthquake, or by means of synthetic accelerograms developed from the modelling of the source mechanisms, related to the scenario earthquake, of medium or high magnitude.

An accurate geotechnical characterization is fundamental for evaluating the seismic local site response, as well as for modelling the soil behaviour and for analyzing the stability of the site where the structure foundations rest on.

The soil investigations must be planned in relation to the specific design. The geotechnical characterization must include an estimation of the static and dynamic soil properties, and for strategic structures it is necessary to perform specific dynamic in-situ and laboratory tests. For common structures the dynamic geotechnical characterization can be performed through empirical correlations, which join the dynamic soil properties to the static ones.

As reported in Eurocode 8 Eurocode 8—Part 5 (2003), with the geotechnical characterization it is necessary to investigate the overall stability of the site where the structures rest on. In this context, the most important phenomena to be analyzed are the slope stability, both in static that in dynamic conditions, and the liquefaction. Evaluated the overall stability of the site, it is possible to investigate the behaviour of shallow and deep foundations in relation to the ultimate limit state and to the damage limitation state in static and/or dynamic conditions.

To analyze and empathize the importance of the different phases of a seismic design process, in the paper the approach adopted for the seismic retrofitting of a group of reinforced concrete buildings founded on piles (Fig. 1), located on the southern part of eastern Sicily (Augusta, Italy), damaged by the Sicilian earthquake of 13th December 1990 ($M_L = 5.4$), is described (Fig. 2) and discussed.

One of the most seriously damaged buildings was chosen as a case study (Fig. 3) for structural upgrading and seismic retrofitting.

Fig. 1 View of building. **a** And damages. **b** Caused by earthquake

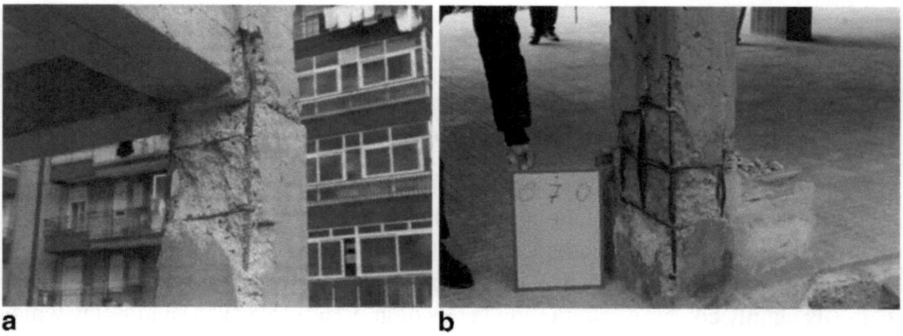

Fig. 2 Typical damages occurred on buildings. **a** Damages on beam and. **b** Damages on column

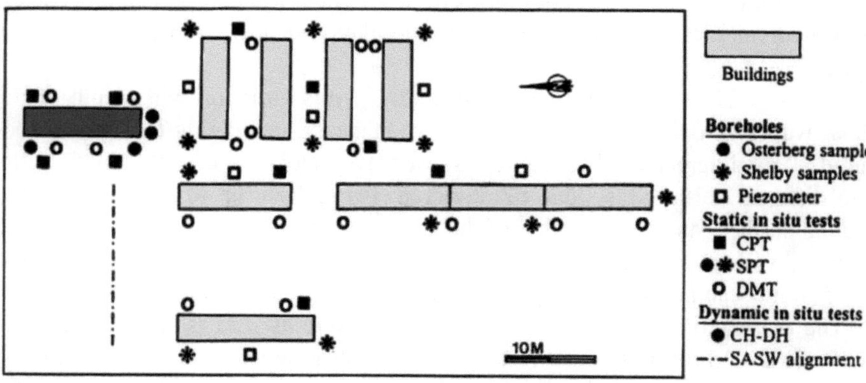

Fig. 3 Layout of the site with the location of buildings and in situ tests

The site was well investigated by means of laboratory and in-situ tests including seismic tests (Fig. 3). Seismic action has been evaluated by a deterministic approach (Frenna and Maugeri 1995; Maugeri and Frenna 1995) based on accelerograms registered during the 13th December 1990 earthquake, scaled to simulate the effects of the 1693 scenario earthquake.

For evaluating the possibility to repair the buildings, an investigation on soil, structures and foundations was carried out. As buildings are founded on piles, vertical and horizontal loading tests on two single pile tests were conducted.

Also tests on a three-piles group were carried out. Even if the loading tests showed that the seismic action have not damaged the effectiveness of the soil-pile system, to extrapolate the behaviour of piles under design seismic actions, a numerical simulation of inertial and kinematic effects have been performed. As this numerical analysis showed that the foundation was enable to carry on this design actions, the seismic retrofitting of the piled foundation required the enlargement of the foundation and new bored piles.

The procedure discussed and analysed permit to define a state of practice in the field of structural upgrading and/or seismic retrofitting of pile foundations.

Experimental Investigation

Soil Geotechnical Characterization

The investigated area has plane dimensions of 4100 m^2. A detailed description of the soil properties is reported by Castelli and Maugeri (2004).

The thickness of the deposits varies from 50 up to 300 m. The upper part of these deposits mainly consist of alternating layers of grey-silty-clay and sandy-clay which is locally called "*Augusta Clay*". Layers of sand were found at depths of between 9 and 12 m. The water table is located at around −0.80 m from the ground surface.

The index properties and the mechanical characteristics of the Augusta clay were derived from in situ and laboratory tests (Fig. 4). The lower deposits mainly consist of a medium stiff, over-consolidated marine clay with low to medium plasticity index.

By Cross-Hole (*CH*) and Down-Hole (*DH*) tests variation with depth of the shear waves velocity *Vs* was determined. In this case, comparing the values measured, a good agreement between the shear wave velocities V_s determined by the two different seismic tests can be observed. Particularly up to −10 m from the ground surface, an average value of the shear waves velocity V_s=320 m/s can be derived, while an average value of the shear waves velocity approximately equal to V_s=530 m/s can be estimated up to −30 m from the ground surface.

The Marchetti's flat dilatometer tests (DMT) were used to assess OCR and the coefficient of earth pressure at rest k_o (Fig. 5). For a depth up to −10 m from the ground surface, DMT results show an OCR ranging between 1 to 3 (k_o=0.5÷1.0),

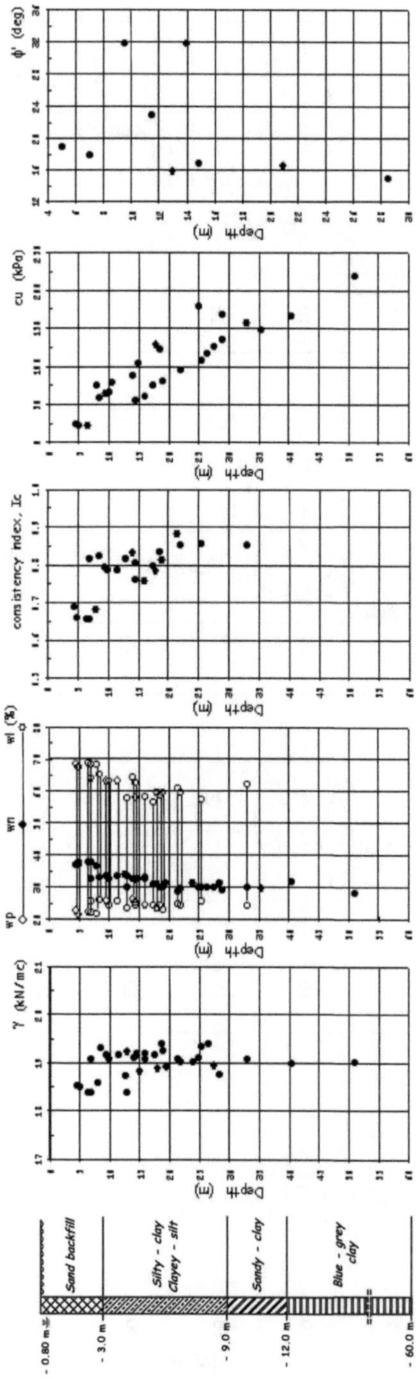

Fig. 4 Soil properties derived by laboratory tests

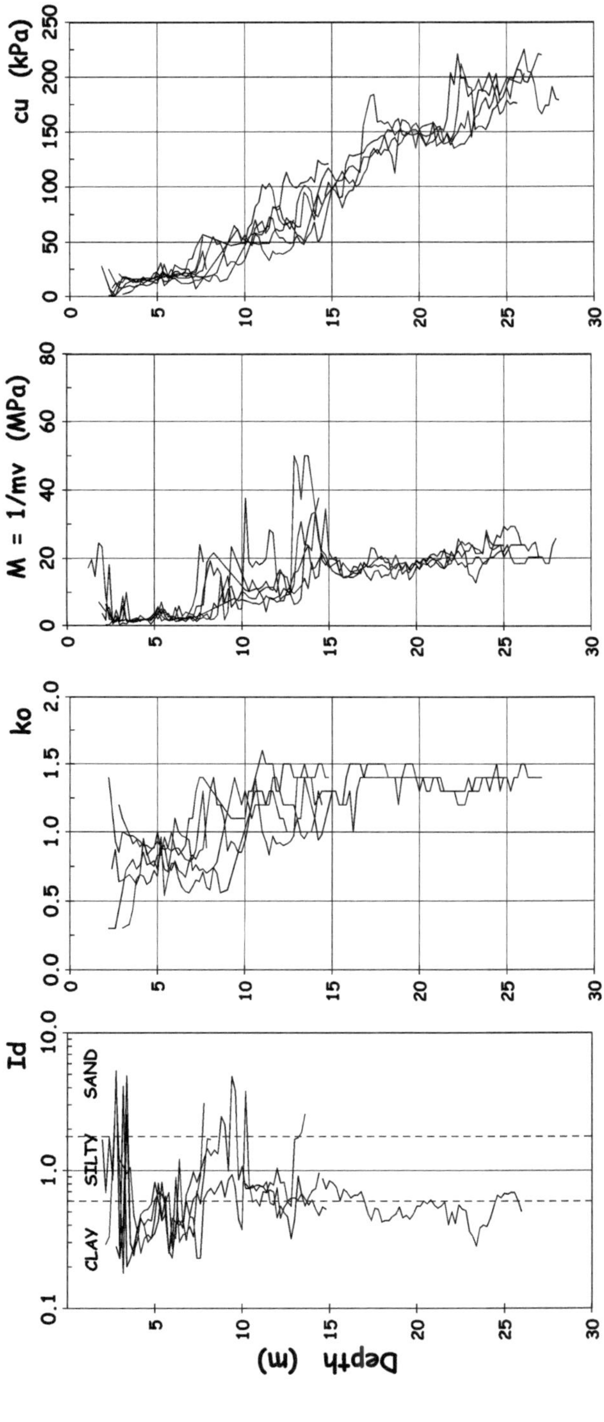

Fig. 5 Marchetti's flat dilatometer test results

showing that the upper clay deposit can be considered normally consolidated or lightly over-consolidated. For the lower clay deposit, the OCR values obtained by DMT range from 5 to 7 (k_o=1.0÷1.5) with an average value equal to 6 up to a depth of about −30 m.

For estimation of the soil response and to characterize the soil behaviour for dynamic loading conditions, values of the small strain shear modulus G_o were determined by in-situ and laboratory tests. In laboratory the shear modulus and the damping ratio (*D*) was determined by Resonant Column tests (*RCT*) and cyclic loading torsional shear tests (*CLTST*). It was also assessed by empirical correlations based on in-situ test results. By laboratory tests, the values of G_o were determined at shear strain levels less than 0.001 %.

The values of G_o for the upper silty-clay steadily increase from 20 to 80 MPa with depth. In the transition zone, where stiff sand layers are present, G_o increases up to 110 MPa. In the lower blue-grey clay G_o values range in the range 80 up to 120 MPa (Cavallaro and Maugeri 1996). A small difference between the values of shear modulus G_o determined by this last procedure and the values derived by resonant column tests can be observed in the first 10 m.

Loading Tests on Piles

The site under consideration, the Saline of Augusta, is one of the most seismically active areas of Italy. The city of Augusta has been destroyed by three disastrous earthquakes with an MKS intensity ranging from IX up to XI (Postpischl 1985). Even if 1990 earthquake was of moderate magnitude (M_L=5.4 and MKS intensity equal to VII), it caused 19 victims and severe damages to buildings and infrastructures.

The reinforced concrete buildings under consideration are founded on bored piles having a length of 20 m and a diameter of 500 mm. To check the integrity of the piled foundations after the earthquake, the piles were inspected and an experimental loading tests program was carried out. Vertical and horizontal loading tests on two single pile and on a three-piles group were carried out (Fig. 6).

The piles have been disjointed from the foundation by the removal of a terminal pile segment (Fig. 7). The connection between the piles and the cap was restored after the loading tests (Fig. 8). A detailed description of the loading tests on piles is reported by Castelli and Maugeri (2004) and Maugeri and Castelli (2007).

Two different vertical loading and un-loading cycles have been carried out. The piles were subjected to a maintained load and during each step of loading the pile head settlements were measured. The vertical load was applied in different stages up to around 150 % of the working load, estimated equal to 650 kN (Castelli and Maugeri 2004).

The experimental results shows a marked non-linear behaviour of the soil-pile interaction. The load-settlement curves seem to follow a hyperbolic law, then the loading curve may be analyzed by transforming the data and plotting to yield a

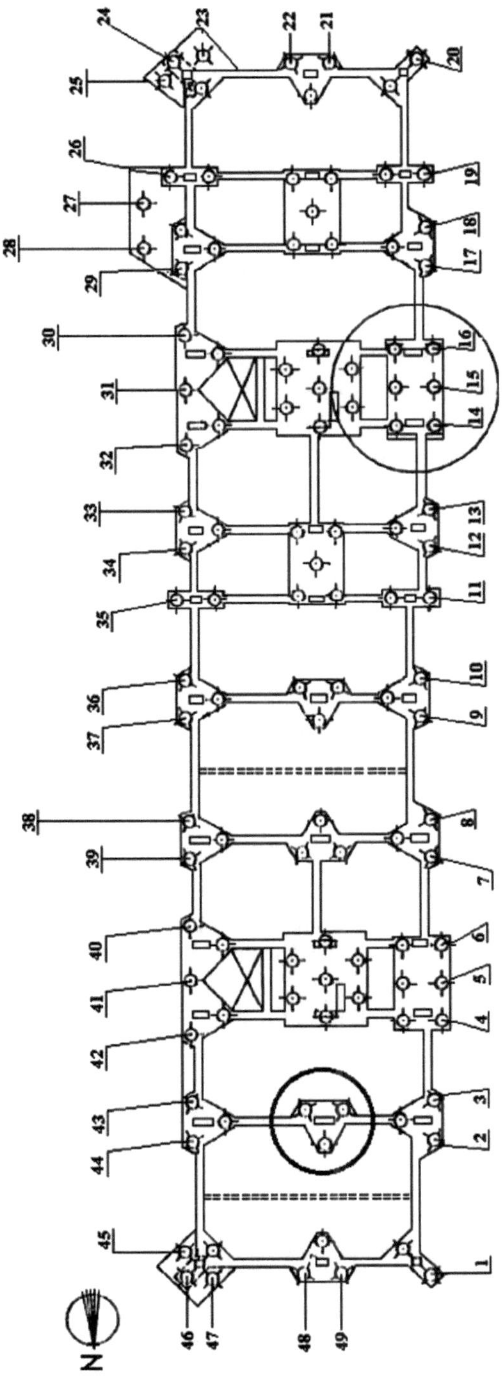

Fig. 6 Piled foundation with the location of the tested piles

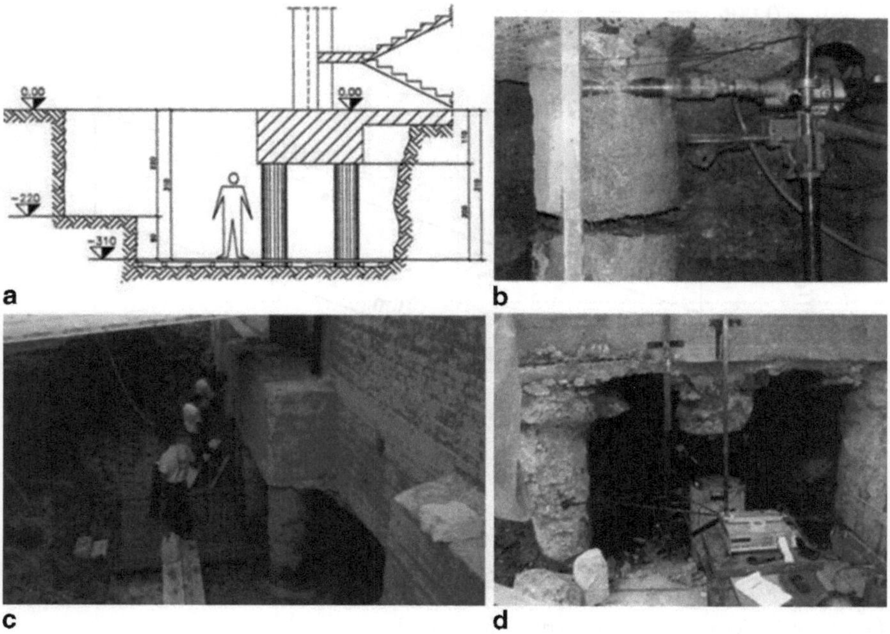

Fig. 7 Piled foundation. **a** section. **b** Pile cutting. **c** View of the excavation. **d** Loading test on pile

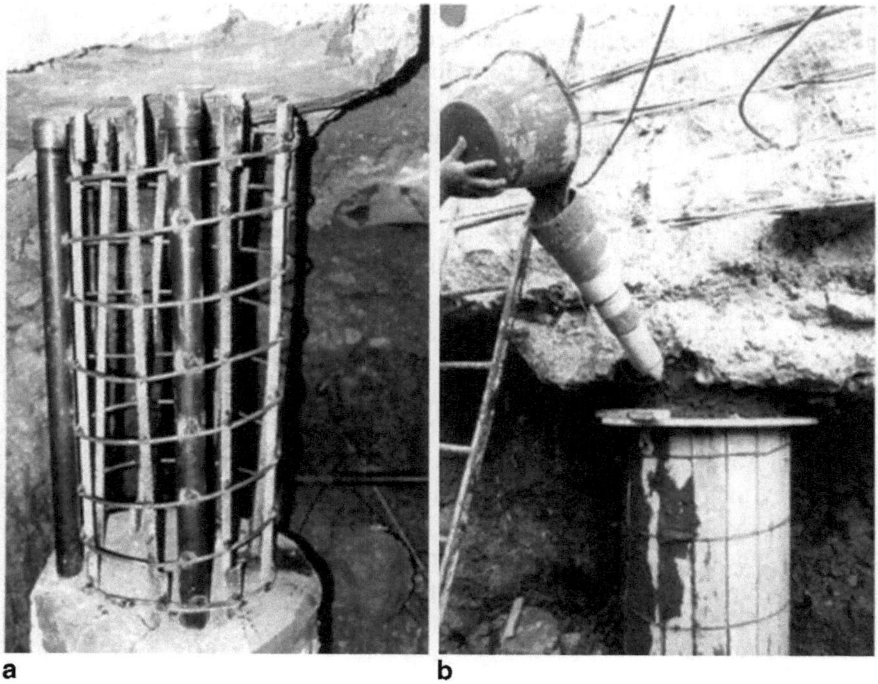

Fig. 8 Pile restoration. **a** Steel bars. **b** Concrete placement

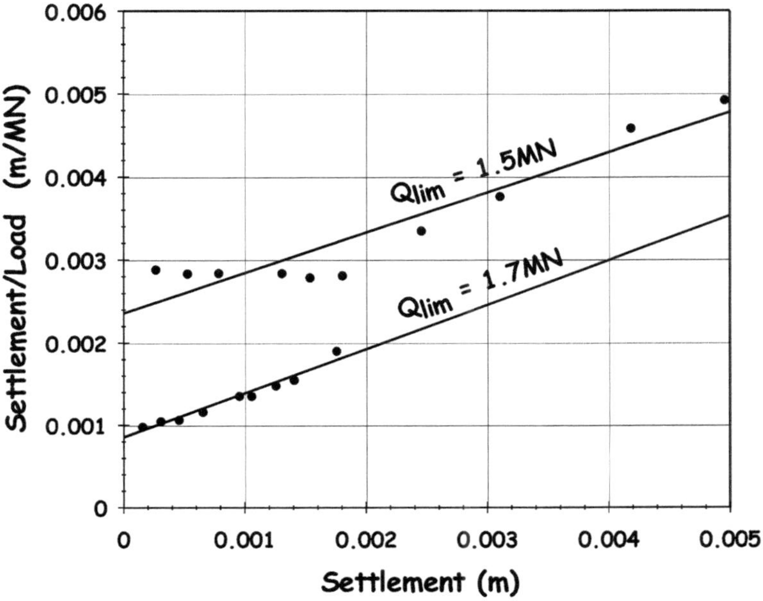

Fig. 9 Evaluation of the ultimate vertical failure load

straight-line relationship in a plane (*settlement, settlement/load*), the slope of which allows to estimate the ultimate failure load, as suggested by Chin (1970). The ultimate pile capacity is assumed generally equal to 90 % of the value so determined.

According to this procedure, the ultimate vertical load derived by the loading tests was ranging between 1.5 and 1.7 MN (Fig. 9), in a good agreement with the value of 1.53 MN computed in undrained conditions by the well-known Vesic's (1977) relationships. This allowed the validation of the loading tests performed at the time of construction and the predictions of the ultimate load derived from the geotechnical soil properties. The effectiveness of the piles was considered sufficient to sustain vertical loads, instead of the damages occurred on building.

To evaluate the pile performance due to lateral loads, two lateral loading tests were also carried out. Figure 10a and Fig. 10b report the experimental load-deflection curve derived by the loading tests named *Test no.1* and *Test no.2* respectively.

The experimental results show that the measured lateral deflection for the maximum load applied was ranging between 1.5 and 4.0 mm, while the residual lateral deflection was ranging between 0.45 and 1.3 mm. The ultimate failure lateral load derived by the hyperbolic representation of the load-deflection curve is equal to around 19 kN.

Finally, the horizontal loading test carried on a three-pile group (Fig. 11a) gives the results reported in Fig. 11c in terms of load-deflection curve.

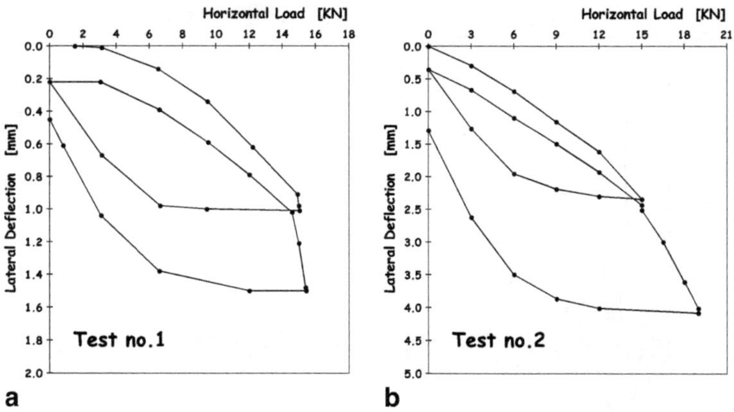

Fig. 10 Single pile horizontal loading test results no. 1 (**a**) and no. 2 (**b**)

Fig. 11 Horizontal loading test on pile group. **a** Piles scheme. **b** View. **c** Loading test results

Seismic Soil Response

The analysis of damages caused by the 13th December 1990 earthquake in the "*Saline of Augusta*" site (Sicily, Italy), has emphasized the phenomenon of the amplification effects, which caused serious damages to concrete frame resisting buildings.

The seismic response of the soil, schematized as a one-dimensional system with several degrees of freedom and lumped masses, has been evaluated with a hysteretical-simplified model, by the integration of the system of differential equations of the dynamic equilibrium with variable coefficients, using the Newmark's numerical integration method step-by-step.

The model is capable to take into account the decay of the elastic shear modulus with strain (Maugeri and Frenna 1987).

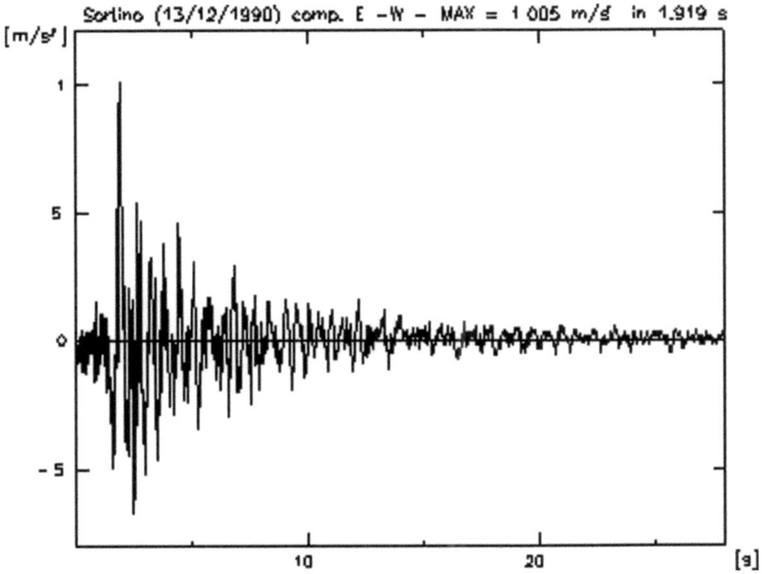

Fig. 12 Accelerogram recorded at Sortino E-W (Italy)

The input excitation at the base of the model were the accelerograms of the 13th December, 1990 earthquake, in particular that recorded at *Sortino E-W* (Sicily, Italy) which is one of the most representative among those on rock (Fig. 12).

As it has been emphasized, related to the seismic characteristics of the area, to analyze the possibility that even more intense earthquake could occur, the accelerations and the frequencies of the recordings at *Sortino E-W* have been scaled (Maugeri and Frenna 1995; Frenna and Maugeri 1995). The recording was scaled using the maximum acceleration 0.3 g and the predominant period of earthquake T=0.5 s.

The model for the evaluation of the seismic soil response requires the knowledge of the depth of the conventional bedrock. In this case it has been placed 80 m below the ground surface.

The numerical analysis provides the time-history response in terms of displacement, velocity and acceleration profiles versus depth (Fig. 13).

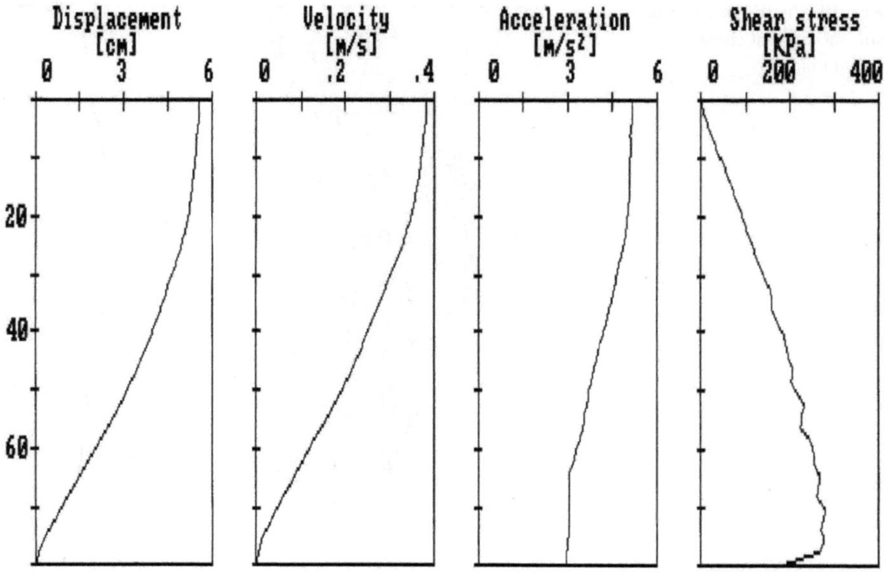

Fig. 13 Results of the seismic soil response

Numerical Simulation

In the last years several simplified approaches for the analysis of single piles or pile groups have been developed that can be used with little computational effort. These methods have given results that are often in remarkable agreement with the mathematical models (Novak 1974; Markis and Gazetas 1992; Abghari and Chai 1995; Tabesh and Poulos 2001).

Pseudo-static approaches, as example, for the seismic analysis and design of foundations are attractive for design engineers because they are simple, when compared to difficult and more complex dynamic analyses.

During an earthquake there are two sources of loading on piles: "*inertial*" loading of the pile head, caused by the lateral forces imposed on the structure by the earthquake and which are then imposed on the piles, and "*kinematic*" loading along the length of the pile, imposed by the lateral ground movements developed by the earthquake (Fig. 14).

Generally in the design of piles only the inertial loading are taken considerations, even if kinematic loading is also very important (Gazetas 1984; Tazoh et al. 1988; Masayuki and Shoichi 1991; Kavvadas and Gazetas 1993).

While there is ample experience of carrying out the equivalent static analysis for the inertial loading, no specific method or procedures are available to predict deformations and bending moment from the kinematic loading.

To take into account these two sources of loading, Abghari and Chai (1995) proposed an analysis in which the pile are subjected to the *free-field* soil displacements

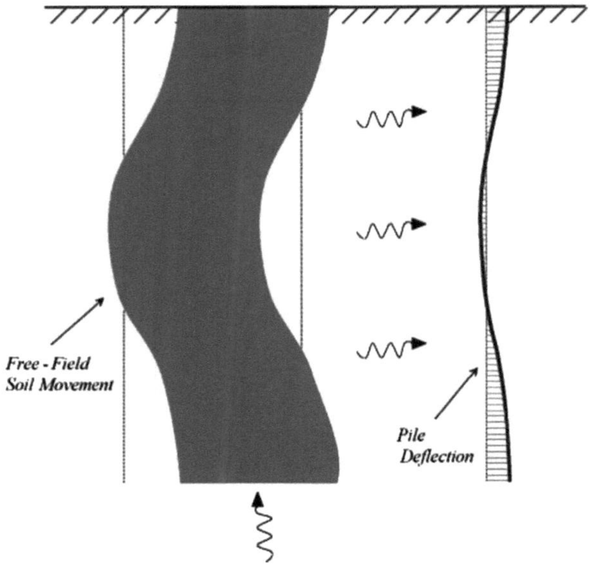

Fig. 14 Profile of external soil movement caused by earthquake

at each node along its length. These displacements are obtained from a separate *free-field* site response analysis. The inertial forces acting on the pile are obtained from the product of superstructure mass and spectral acceleration. These forces are applied to the heads of the piles as static forces.

Similarly Tabesh and Poulos (2001) proposed a simple methodology for estimating the maximum internal forces of piles subjected to lateral seismic excitation. The method requires the evaluation of the maximum *free-field* soil movements caused by earthquake computed, for example, by the well known computer program SHAKE (Schnalbel et al. 1972) or similar computer code and the analysis of the response of the pile to the maximum free-field soil static movements plus a static loading at the pile head (Poulos 2006), which depends on the computed maximum spectral acceleration of the structure being supported.

According to this approach, a pseudo-static push over analysis using a "*p-y method*" is adopted to simulate the behaviour of a single pile and/or an individual pile of a group subjected to lateral soil movements (kinematic loading) and static loadings at the pile head (inertial loading).

The approach involves two main steps: first a *free-field* site response analysis is performed to obtain the maximum soil displacements along the pile. Next a static load analysis is carried out for the pile, subjected to the envelope of the *free-field* soil displacements and the static loading at the pile head (Fig. 15) based on the maximum ground surface acceleration (Castelli et al. 2008; Castelli and Maugeri 2009).

The response of the complete system constituted by soil, foundation and superstructure is obtained by the superposition of the effects, even if the validity of the principle of the superposition of the effects is correct only in the hypothesis of linear behaviour of all components (Kausel and Roesset 1974; Gazetas and Mylonakis 1998).

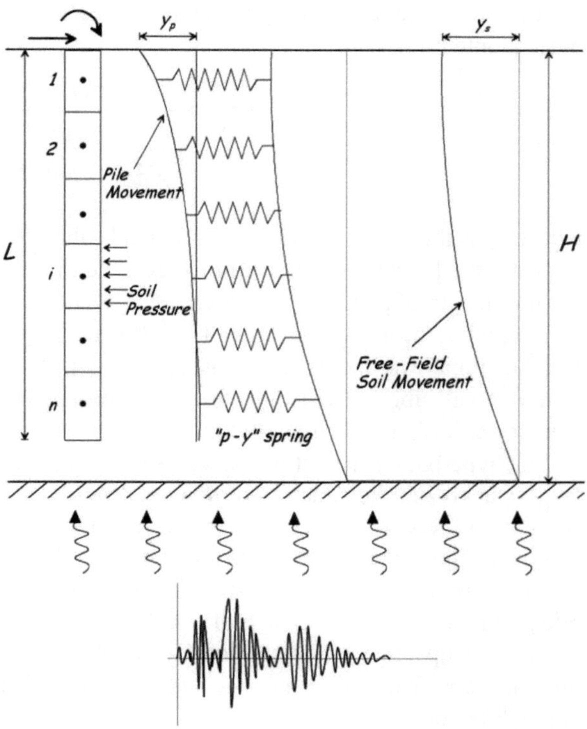

Fig. 15 Pile model for pseudo-static seismic analysis

Nevertheless, the superposition of the effects can be considered acceptable for systems which are moderately non linear (Mylonakis et al. 1997).

Inertial Loading

Several approaches are available to analyze the behaviour of piles subjected to lateral load ranging from complex models, as non linear dynamic analysis by 2D or 3D finite element methods, to the use of simplified approaches as limit equilibrium (LE) approach and *p-y* analysis approach.

The "*p-y method*" is still widespread in practice and based on literature review it appears to be one of the most attractive methods for civil engineers. The "*p-y approach*" for analyzing the response of laterally loaded piles is essentially a modification of the basic Winkler model, where *p* is the soil pressure per unit length of pile and *y* is the pile deflection. The soil is represented by a series of nonlinear *p-y* curves along the pile length.

Researches based on results of field tests on full scale piles, both in cohesive than in cohesionless soil, suggests to employ non linear "*p-y relationship*" (Reese et al. 1974; Reese and Welch 1975; Reese & Van Impe 2001; Juirnarongrit and Ashford

2006), thus in the proposed approach the following hyperbolic *p-y* relationship has been adopted:

$$p(z)=\frac{y_p(z)}{\dfrac{1}{E_{si}(z)}+\dfrac{y_p(z)}{p_{lim}(z)}} \tag{1}$$

where E_{si} is the initial modulus of horizontal subgrade reaction, y_p is the pile lateral deflection [L], p and p_{lim} are the mobilized and the ultimate horizontal soil reaction per unit length of pile [FL^{-1}] respectively.

Therefore, the units of E_{si} are in FL^{-2}. If p and p_{lim} are expressed as the mobilized and the ultimate horizontal soil resistance [FL^{-2}], then $k_{si}=E_{si}/D$ is the coefficient of horizontal subgrade reaction (force per unit volume [FL^{-3}]) being D the pile diameter (Prakash and Kumar 1996).

The hyperbolic *p-y* relationship (1) is defined by the two parameters p_{lim} and E_{si}. By this approach the situation of a layered soil deposit can be easily analyzed assuming for these parameters different values along the pile length.

To take into account the group effect due to pile-soil-pile interaction, load-transfer functions must be modified to correlate the behaviour of a single pile to that of a pile group. Similar to the design of axially loaded piles using an efficiency method, in a pile group subjected to horizontal load, the *p-y* relationships for the individual piles are modified to take into account the group effects by "*stretching*" the curve in the direction of deflection.

Brown et al. (1998), as example, defined the concept of *p*-multiplier f_m, a multiplier of the limiting values p_{lim} capable of stretching the *p-y* curve for the single pile to account for the interaction among the piles in a group. The multiplier has values in the range 0–1.

Similarly, it should be expected that the resulting initial modulus of horizontal subgrade reaction E_{si} of a pile group is softer than the value of an isolated pile (Ashour et al. 2004). Thus to adapt the hyperbolic *p-y* relationships defined for a single pile to the case of a pile group, the multiplier ζ_m has been introduced also for the initial modulus of horizontal subgrade reaction E_{si} (Castelli and Maugeri 2009). Consequently we can assume for the pile group:

$$p(z)=\frac{y_p(z)}{\dfrac{1}{\zeta_m E_{si}(z)}+\dfrac{y_p(z)}{f_m p_{\lim}(z)}} \tag{2}$$

The multiplier ζ_m has obviously values in the range 0–1.

Field experiments have been carried out in the last decade for the determination of the *p-y* multipliers. The value of the group reduction factors seems not to depend on the soil type, pile type and load level, thus it is reasonable to suggest values only in terms of number of piles, pile spacing and position in the group (McVay et al. 1998; Rollins et al. 1998; Zhang et al. 1999; Mandolini et al. 2005; Rollins et al. 2006).

The advantage of the approach is that the load-deflection response of the pile group can be calculated using solutions for the response of a single pile. Naturally, the approach is based on an estimation of only the average head deflection of the pile group.

The numerical simulation is based on an iterative procedure taking into account the decrease of the model stiffness with the increasing of the applied horizontal load (Castelli et al. 1995b; Castelli and Maugeri 1999; Castelli 2002, 2006).

Kinematic Loading

The passage of seismic waves through the soil surrounding a pile imposes lateral movements and curvatures on the pile, generating kinematic bending moments even in absence of a superstructure.

The simplified numerical approach adopted in the paper is based on the idea to couple and to evaluate the effects due to the applied load at the pile head and lateral movements along the pile length by a series of independent "*p-y curves*", relating soil reaction and relative soil-pile movements (Castelli and Maugeri 2009). Since these relations are non linear, the equivalent linear procedure using secant modulus is normally used to establish a law relating a set of discrete interaction forces at the soil/pile interfaces to a corresponding set of discrete relative motions between piles and *free-field* soil.

The implementation of the method involves imposing a known *free-field* soil movement profile. When the expected *free-field* movement is large enough to cause the ultimate pressure of laterally spreading soils to be fully mobilized, the ultimate pressure, instead of *free-field* soil movement, may be used. This soil movement will in turn displace the pile by a certain amount depending on the relative stiffness between the pile and the soil. The soil loading must be considered by taking into account the relative movement between the soil and the pile.

If the soil mass moves and the pile movement y_p is less than the soil movement y_s, the soil exerts a driving force on the pile. However, if the pile movement y_p is greater than the soil movement y_s, the soil provides the resistance force p_{lim} to the pile. The response of the pile can then obtained by solving the following governing differential equation:

$$EI\frac{d^4y}{dz^4} - p(y_p - y_s) = 0 \tag{3}$$

where EI = pile stiffness, p = soil reaction per unit pile length and z = depth.

Equation (3) can be solved by a numerical procedure based on a finite element discretization, in which the pile load (force per unit area [FL^{-2}]) due to relative pile-soil movement ($y_p - y_s$), can be represented by a series of p-y curves on both sides of the pile shaft along its length (Fig. 16).

As a first approximation an idealized elastic p-y relationship could be used:

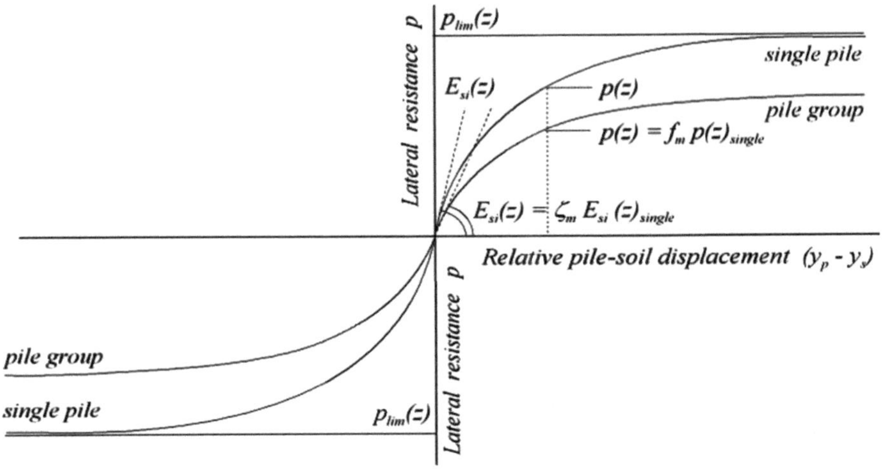

Fig. 16 *p-y* curves to model loading due to lateral movements

$$p(z) = \frac{E_{si}(z)}{D}[y_p(z) - y_s(z)] \tag{4}$$

where D is the pile diameter.

To take into account that the lateral pile response to static or dynamic loading is non linear, the following p-y relationship could be adopted (Castelli and Maugeri 2009):

$$p(z) = \frac{[y_p(z) - y_s(z)]}{\dfrac{1}{E_{si}(z)} + \dfrac{|\,y_p(z) - y_s(z)\,|}{p_{lim}(z)}} \tag{5}$$

in which E_{si}, y_p, p and p_{lim} are defined as in Eq. (1).

Comparison with field measurements, shows that this procedure can give lateral deflection, bending moment and shear force distributions on piles which agree with experimental ones (Castelli and Maugeri 2007, 2009).

When an individual pile of a group is taken into consideration, in Eq. (5) the empirical reduction factors f_m and ζ_m could be adopted for the pseudo-static loading at the pile head (*inertial effect*) and for the kinematic loading (*kinematic effect*) induced by the ground movements developed during an earthquake.

It must be considered, that the *p*-multipliers reported in the literature were developed from the analysis of static or cyclic load tests on single piles and pile groups. These tests do not represent the dynamic loading conditions during an earthquake event.

Being *s* the pile spacing, it can be noted (NCHRP 2001) that the main factors that affect the dynamic *p*-multipliers are the spacing ratio, *s/D*, and the pile-head displacement ratio, *y/D*. The *p*-multipliers increase as *s/D* increases, meaning that the group effect decreases. The *p*-multipliers also increase as the *y/D* increases.

This means that during a dynamic loading event, which is characterized by large pile head displacement, the pile-soil-pile interaction decreases, and the piles tend to behave as individual piles. It can be seen, also, that the *p*-multipliers are less dependent on frequency than on pile deformation (NCHRP 2001).

It must be emphasized that these considerations are based on limited results and further investigations should be done before any conclusions drawn from these observations can be generally accepted.

Behavior Of Piles Under Lateral Loading

To simulate the behaviour of piles subjected to lateral loading the model proposed by Castelli et al. (1995a) and Castelli and Maugeri (2009), and implemented in an original computer code, has been used. The code allows the assessment of the lateral response (deflection, moment and shear force distribution) of an isolated pile and a pile group including the *p-y* curve along the length of the isolated pile or the individual pile in the group.

Validation of the proposed approach has been carried out comparing the numerical results with those measured in loading tests (Castelli et al. 1995a), both on single pile and piles group Castelli and Maugeri (2009). Model parameters have been assessed to obtain numerical results close to those measured.

A difficulty of applying the proposed *p-y* curve approach is the appropriate evaluation of the functions parameters for a realistic estimation of single pile and/or pile group performance. The adoption of hyperbolic *p-y* curves requires the determination of the ultimate horizontal soil resistance p_{lim} and initial modulus of horizontal sub-grade reaction E_{si}.

In a single pile-soil interaction, the ultimate horizontal soil resistance p_{lim} can be evaluated according to the formulas existing in literature (Matlock 1970) both for cohesive (Broms 1964a) than for cohesionless soils (Broms 1964b). As concern the initial modulus of horizontal sub-grade reaction E_{si}, the relationships proposed by Matlock (1970), Welch and Reese (1972), Robertson et al. (1989) for hyperbolic *p-y* curves can be used.

Taking into account the profile of the values of the undrained shear strength derived from in situ and laboratory tests (Fig. 2), the numerical analysis has been carried out with values of p_{lim} ranging between 0.024 MN/m^2 from the ground surface up to 1.33 MN/m^2 at the pile base.

The initial modulus was: $E_{si} = (E_{soi} + k_i z)$, being z the depth, k_i equal to 19.75 N/cm^3 the gradient of the initial modulus of horizontal sub-grade reaction and E_{soi} the initial modulus at the ground surface, that in the analysis was assumed equal to 37 MN/m^2. The tested single piles were in a *free-head* condition. As the horizontal load during the loading test was applied at a distance from the ground surface of 43 cm (Fig. 17a), the numerical analysis has been carried out considering the pile loaded with a horizontal force and a moment at the head.

The comparison between measured and computed single pile head lateral deflection for *Test no.1* (Fig. 17b) shows a good agreement between numerical and experimental results. Loading tests showed that the seismic actions have not damaged the effective-

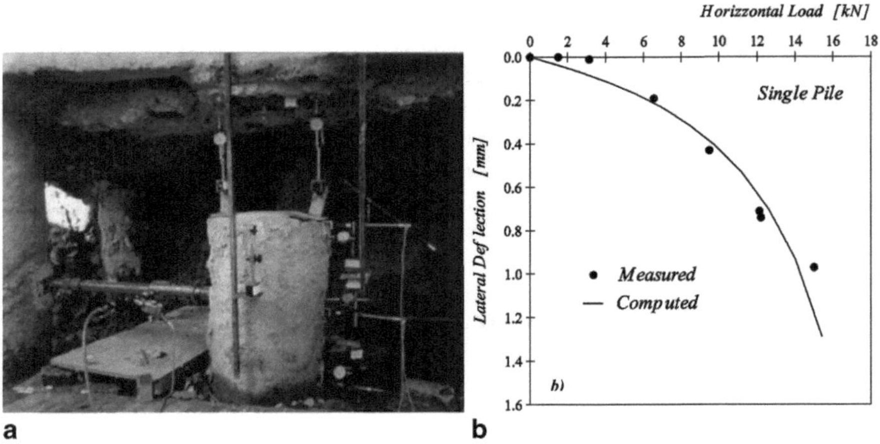

Fig. 17 **a** Horizontal loading test on pile. **b** Comparison between experimental and computed load-deflection curve

ness of the soil-pile system, as the bearing capacity of the piles appeared unchanged in spite of the additional loads applied on the structures during the earthquake.

To reduce the computed load-carrying capacity of the single pile to the pile group rows, according to the design curve reported by Rollins et al. (1998), a *p*-multiplier f_m equal to 0.4 was assumed. The initial modulus of horizontal sub-grade reaction of the individual pile of the group was reduced assuming for the empirical factor ζ_m the same value of f_m.

A good agreement between numerical results and experimental evidences can be observed also for the case of the individual pile of the group (Fig. 18).

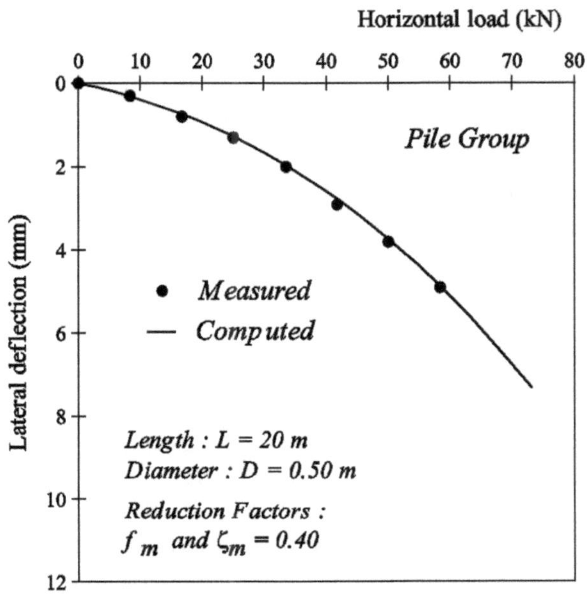

Fig. 18 Comparison between experimental and computed load-deflection curve for the individual pile of the three-pile group

Evaluation of Inertial and Kinematic Effects

The behaviour of the building during the 13th December 1990 earthquake has shown some resistance to horizontal actions (Oliveto and Decanini 1998). However, the severe damages observed have shown that the structure reached its resistance limit. Fortunately the collapse mechanism was not activated for the short duration of the earthquake and its low energy input in spite of a relatively high level of maximum acceleration.

Nevertheless, the 13th December 1990 earthquake represents a moderate event, not comparable with the design earthquake associated with the seismicity of the region. Thus, an additional structural system capable of absorbing all the seismic design actions was considered essential for the seismic retrofitting of the structure.

The structural upgrading of the piled foundation was referred exclusively to the inertial effects, the only prescribed by the Italian seismic regulation when remedial works have been designed and executed. A pseudo-static numerical analysis via the procedure proposed by Castelli et al. (1995b) and Castelli and Maugeri (1999) has been carried out.

Numerical simulation provides the lateral deflection (Fig. 19a) and the bending moment distribution (Fig. 19b) along the pile length for lateral seismic design loading ranging from 12.2 to 15 kN (*inertial effects only*).

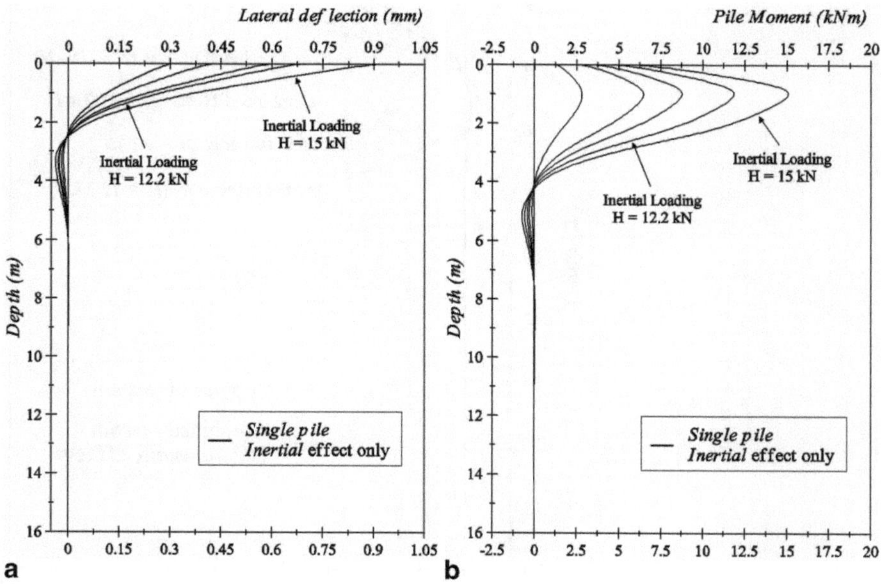

Fig. 19 Seismic design loading. **a** Computed lateral deflection. **b** Computed bending moment

Yet, to estimate the kinematic effects on piles related to seismic design actions, the displacement profile of the *free-field* soil movements evaluated by the seismic soil response analysis (Fig. 13) can be considered. So that, numerical simulation provides bending moment distribution along the pile length related to lateral inertial loading and kinematic ground movements included in the analysis (combined effects).

The numerical analyses relative to this last case of combined effects (*inertial* and *kinematic*) have been performed according to *p-y* relationship defined by Eq. (5). The results obtained for the single pile are summarized in Fig. 20.

Analyzing these results it is possible to observe that, as expected, if *kinematic effects* (lateral movements) are ignored and only *inertial effects* (lateral load applied at the pile head) are considered, the maximum moments along the pile length can be underestimated.

When combined effects are taken into consideration, an increasing of the maximum bending moment can be observed, with a considerable reduction of the safety margin respect to the estimated plasticity moment of the single pile equal to $M_y = 120$ kNm.

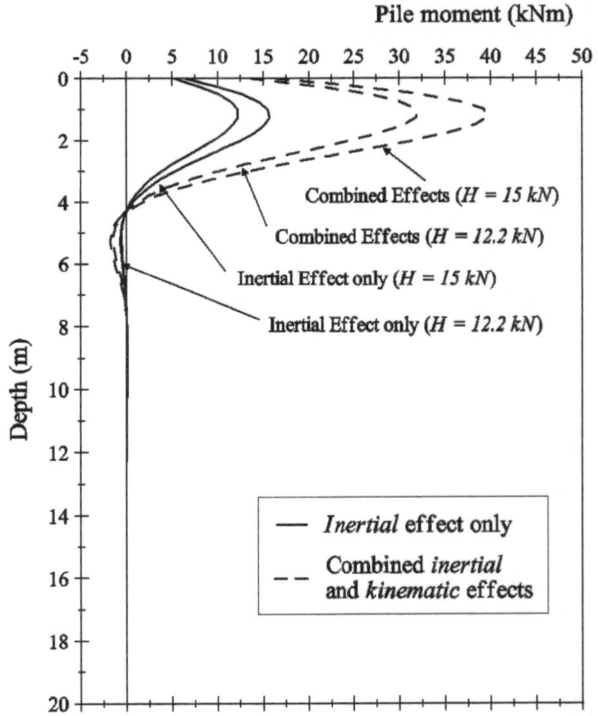

Fig. 20 Numerical results for combined effects on single pile (inertial and kinematic)

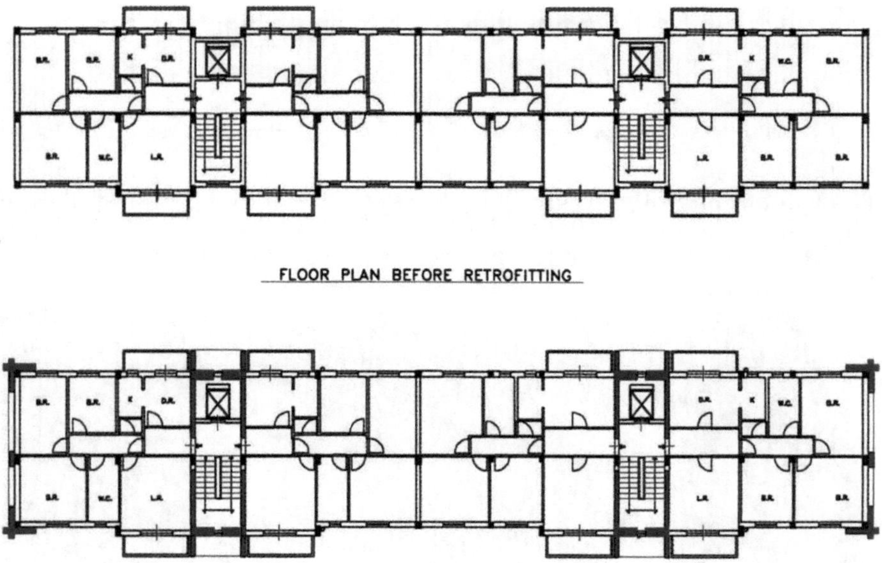

Fig. 21 Selected retrofitting system. RC stiffening cores and corner elements. (Oliveto and Decanini 1998)

In any case, the vertical and horizontal load-bearing capacities of the existing foundation were not sufficient for the design seismic actions. A strengthening of the foundation was necessary.

The strengthening of the existing structural system was realized by several kinds of seismic resistant reinforced concrete walls, by the participation to the resistance in the longitudinal direction also of the walls in the transversal direction. This was achieved through the solidarization of longitudinal and transversal adjacent walls, thus obtaining two stiffening reinforced concrete cores in each of the staircases (Fig. 21).

The stiffening cores involve a concentration of vertical forces in the pile groups around the foot. Seventy-six new bored piles of 50 cm in diameter and 25 m in length were disposed as in Figure (Fig. 22a).

This concentration could lead to the overcoming of the vertical load-bearing capacity, and even of the ultimate load, in those piles while leaving the remaining ones understressed (Fig. 22b). The connection of the new to the old foundation footings was accomplished by means of post-tensioned high strength steel bars (Fig. 22c).

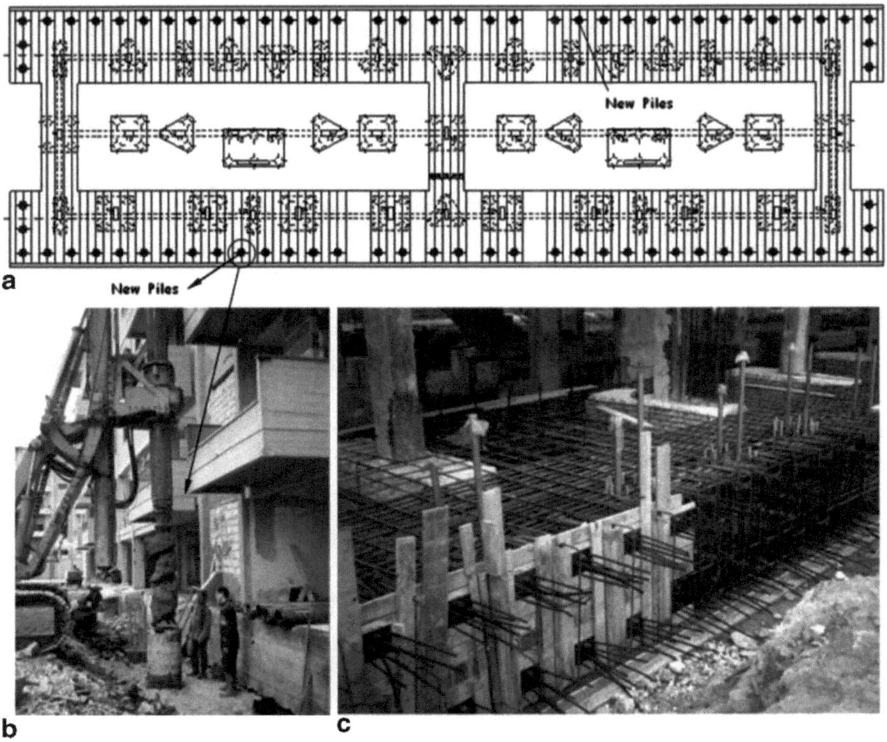

Fig. 22 Piled foundation with new piles. **a** Plan view. **b** Pile drilling. **c** Foundation enlargement

Conclusions

The design of single pile and pile groups for axial and lateral capacity has tended to rely on semi-empirical design methods that make it difficult for practical designers to select the appropriate procedure. It would, therefore, be preferred to define more reliable approaches and relevant soil parameters that can be adopted in the design methods. With this aim, the various phases of the study for the seismic retrofitting of a six storey reinforced concrete frame resisting building in the Saline district of Augusta (Italy) have been described.

To evaluate the possibility to repair a building damaged by the Sicilian earthquake of 13th December, 1990, an investigation on soil, superstructures and foundations was carried out. As buildings are founded on piles, vertical and horizontal loading tests on single piles and a three-pile group were executed. At the end, the connection between the piles and the cap was restored.

The loading tests showed that the seismic action have not damaged the effectiveness of the soil-pile system. For the seismic retrofitting of the structures, the internal response of the piles subjected to earthquake loading has been studied.

Although the seismic upgrade in this case did not consider kinematic ground movement effects, the predicted moment under the combined effects of inertial and kinematic loading was fortuitously still less than the ultimate load capacity. The numerical analysis showed that the foundation was able to carry on the design actions, the structural upgrading and seismic retrofitting of the piled foundation required the enlargement of the foundation and new bored piles.

Acknowledgements This paper is one of the last works of professor Michele Maugeri, who died on 1 November 2014 after a long battle with illness. He was a scientific "ambassador" of geotechnical earthquake engineering and his career was full of countless tasks of primary importance. He was a nice colleague and a friend. He will be greatly missed.

References

Abghari A, Chai J (1995) Modelling of soil-pile superstructure interaction for bridge foundations. In: Turner JP (ed) Performance of deep foundations under seismic loading. ASCE, New York, pp 45–59

Ashour M, Pilling P, Norris G (2004) Lateral behavior of pile groups in layered soils. J Geotech Geoenviron Eng ASCE 130(6):580–592

Broms BB (1964a) Lateral resistance of piles in cohesive soils. J Soil Mech Found Eng Div ASCE XC(SM-2):27–63

Broms BB (1964b) Lateral resistance of piles in cohesionless soils. J Soil Mech Found Eng Div ASCE XCI(SM-3):123–156

Brown DA, Morrison C, Reese LC (1998) Lateral load behaviour of pile group in sand. J Geotech Geoenviron Eng ASCE 114(11):1261–1276

Castelli F (2002) Discussion on "Response of laterally loaded large-diameter bored pile groups", by CWW Ng, L Zhang & DCN Nip. J Geotech Geoenviron Eng ASCE 128(11):963–965

Castelli F (2006) Non linear evaluation of pile groups lateral deflection. In: Proceedings International Conference on Piling and Deep Foundations, Amsterdam, 31 May—2 June, 2006, pp 127–134

Castelli F, Maugeri M (1999) Nonlinear soil effects in the design of pile group in pseudo-static conditions. In: Proceedings IX National Conference "Ingegneria Sismica in Italia", Torino, Italy, 10 p (in Italian)</bib>

Castelli F, Maugeri M (2004) Analysis of the behaviour of the piled foundations of a group of earthquake damaged buildings. In: Proceedings V International Conference on Case Histories in Geotechnical Engineering, New York, 13–17 April, 2004, paper no. 1. 45, 6 p

Castelli F, Maugeri M (2007) A pseudo-static analysis for the evaluation of the lateral behaviour of pile groups. In: Proceedings of 4th International Conference on Earthquake Geotechnical Engineering, Thessaloniki, Greece, Paper no. 1399

Castelli F, Maugeri M (2009) Simplified approach for the seismic response of a pile foundation. J Geotech Geoenviron Eng ASCE 135(10):1440–1451

Castelli F, Maugeri M, Motta E (1995a) Modelling of a horizontal loading test on a pile beneath a building damaged due to earthquake. In: Proceedings II National Conference on Earth. Engineering, Siena. Italy I:185–194 (in Italian)

Castelli F, Maugeri M, Motta E (1995b) Non linear analysis of the deflection of a pile subjected to horizontal loads. Italian Geotech J XXIX(4):289–303

Castelli F, Maugeri M, Mylonakis G (2008) Numerical analysis of kinematic soil-pile interaction. In: Proceedings MERCEA 2008, International Seismic Engineering Conf. Commemorating

the 1908 Messina and Reggio Calabria Earthquake, Reggio Calabria and Messina. July 8–11, 1:618–625
Cavallaro A, Maugeri M (1996) Comportamento tensionale e deformativo dell'argilla di Augusta sottoposta a carichi ciclici. Ing Sismica XIII:30–40 (in Italian)
Chin FK (1970) Estimation of ultimate load of piles not carried to failure. In: Proceedings 2nd SE Asian Conference on Soil Engineering, Singapore, pp 81–92
Eurocode 8—Part 5 (2003) Design of structures for earthquake resistance—Part 5: foundations, retaining structures and geotechnical aspects. European Prestandard, ENV1998, European Commission for Standardization, Brussels, Belgium
Frenna SM, Maugeri M (1995) Microzoning of the saline site of Augusta town (Sicily, Italy). In: Proceedings 10th International Conference on Earthquake Engineering. Vienna I:43–49
Gazetas G (1984) Seismic response of end-bearing single piles. Soil Dyn Earthq Eng 3(2):82–93
Gazetas G, Mylonakis G (1998) Seismic soil-structure interaction: new evidence and emerging issue. In: Dakoulas P, Yegian MK, Holtz RD (eds) Geotechnical earthquake engineering and soil dynamics III. ASCE, II, pp 1119–1174
Juirnarongrit T, Ashford SA (2006) Soil-Pile response to blast-induced lateral spreading. II: analysis and assessment of the p-y method. J Geotech Geoenviron Eng ASCE 132(2):163–172
Kausel E, Roesset JM (1974) Soil structure interaction for nuclear containment structures. In: Proceedings ASCE, Power Division Specialty Conference, Boulder, Colorado
Kavvadas M, Gazetas G (1993) Kinematic seismic response and bending of free-head piles in layered soil. Géotechnique 43(2):207–222
Mandolini A, Russo G, Viggiani C (2005) Pile foundations: experimental investigations, analysis and design. In: Proceedings 16th International Conference on Soil Mechanics and Geotechnical Engineering (16ICSMGE). Osaka 1:177–213
Markis N, Gazetas G (1992) Dynamic pile-soil-pile interaction. Part II: lateral and seismic response. Earthq Eng Struct Dyn 21:145–162
Masayuki H, Shoichi N (1991) A study on pile forces of a pile group in layered soil under seismic loading. In: Proceedings 2nd International Conference on recent advances in geotechnical earthquake engineering and soil dynamics, St Louis, p 3
Matlock H (1970) Correlations for design of laterally loaded piles in soft clay. In: Proceedings II Offshore Technical Conference, Houston. Texas 1:577–594
Maugeri M, Castelli F (2007) Seismic retrofitting of the piled foundation of a reinforced concrete building. In: Proceedings 2nd Greece—Japan Workshop on Seismic Design, Observation and Retrofit of Foundations, Shinjukuku, Tokyo. April 3–4 2007, pp 320–334
Maugeri M, Frenna SM (1987) Modello isteretico semplificato per la determinazione della risposta dei terreni in campo non lineare. Proceedings III National Conference "Ingegneria Sismica in Italia" II:169–288 (in Italian)
Maugeri M, Frenna SM (1995) Soil-response analyses for the 1990 South-East Sicily earthquake. In: Proceedings 3rd International Conference on recent advances in geotechnical earthquake Eng. and soil dynamics, St. Louis, Vol.II, no. 9. 17, pp 653–658
McVay M, Zhang L, Molnit T, Lai P (1998) Centrifuge testing of large laterally loaded pile groups in sands. J Geotech Geoenviron Eng ASCE 124(10):1016–1026
Mylonakis G, Nikolaou A, Gazetas G (1997) Soil-pile-bridge seismic interaction: kinematic and inertial effects. Part I: soft soil. Earthq Eng Struct Dyn 27(3):337–359
National Cooperative Highway Research Program—NCHRP (2001) Static and dynamic lateral loading of pile groups, Report 461
Novak M (1974) Dynamic stiffness and damping of piles. Can Geotech J Ott 11:574–598
Oliveto G, Decanini LD (1998) Repair and retrofit of a six storey reinforced concrete building damaged by the earthquake in south-east Sicily on the 13th December 1990. Soil Dyn Earthq Eng 17:57–71
Postpischl D (1985) Atlas of isosismal maps of Italian Earthquake. Italian National Research Council (CNR), Geodynamical Project, Roma

Poulos HG (2006) Ground movements—A hidden source of loading on deep foundations. In: Proceedings International Conf. on "Piling and Deep Foundations", Amsterdam, pp 2–19 (John Mitchell Lecture)

Prakash S, Kumar S (1996) Nonlinear lateral pile deflection prediction in sands. J Geotech Geoenviron Eng ASCE 122(2):130–138

Reese LC, Van Impe WF (2001) Single piles and pile groups under lateral loading. Balkema, Rotterdam, The Netherlands

Reese LC, Welch RC (1975) Lateral loading of deep foundations in stiff clay. Geotech Test J, GTJODJ XII(1):30–38

Reese LC, Cox WR, Koop FD (1974) Analysis of laterally loaded piles in sand. In: Proceedings VI Offshore Technology Conference, Houston, Texas, pp 473–483

Robertson PK, Davies MP, Campanella RG (1989) Design of laterally loaded driven piles using the flat dilatometer. Geotech Test J GTJODJ 12(1):30–38

Rollins KM, Peterson KT, Weaver TJ (1998) Lateral load behavior of full-scale pile group in clay. J Geotech Geoenviron Eng ASCE 124(6):468–478

Rollins KM, Olsen KG, Jensen DH, Garrett BH, Olsen RJ, Egbert JJ (2006) Pile spacing effects on lateral pile group behavior: analysis. J Geotech Geoenviron Eng ASCE 132(10):1272–1283

Schnalbel B, Lysmer J, Seed HB (1972) SHAKE, a computer program for earthquake response analysis of horizontally layered sites. Rep. no. EERC 72-12, Earthquake Engineering Res., Ctr., National Information System for Earthquake Engineering

Tabesh A, Poulos HG (2001) Pseudo-static approach for seismic analysis of single piles. J Geotech Geoenviron Eng ASCE 127(9):757–765

Tazoh T, Wakahara T, Shimizu K, Matsuzaki M (1988) Effective motion of group pile foundations. In: Proceedings 9th World Conference Earthquake Eng. Tokyo 3:587–592

Vesic AS (1977) Design of pile foundations. Transport Res. Board. Report no 42:68

Welch RC, Reese LC (1972) Laterally loaded behaviour of drilled shafts. Research Report no. 3-5-65-89, Center for Highway Research, University of Texas

Zhang L, Mc Vay M, Lai P (1999) Numerical analysis of laterally loaded 3×3 to 7×3 pile groups in sands. J Geotech Geoenviron Eng ASCE 125(11):936–946

Pile Design in Laterally Spreading Soil: Feedback from Numerical Predictions and Model Test Results

George D. Bouckovalas and Yannis K. Chaloulos

Abstract The main findings are summarized of a systematic research effort regarding the response of pile foundations in laterally spreading soils. The incentive of the research were findings from properly scaled (with regard to the pore fluid) centrifuge experiments suggesting that severe soil dilation may occur at the upper part of the pile, as a result of large soil-pile relative movement, causing soil pressures to significantly increase. In essence, this observation negates current design practice which univocally accepts that soil liquefaction drastically reduces seismic demands. To further explore this new evidence, a 3D nonlinear numerical methodology was developed and tested against the aforementioned experiments. On this basis valuable feedback was first gained with regard to current numerical techniques, with most important the need of a new type of tied-node boundaries for the simulation of submerged infinite slopes. Comparative analyses, with the old and the new boundaries, revealed that the former (which also reflect the kinematic response of the laminar box containers employed in model tests) can significantly underestimate soil pressures imposed to the foundation. In the sequel, the numerical methodology was applied parametrically (for various soil, pile and excitation characteristics) and a new set of multivariable relationships was statistically established for the practical estimation of ultimate soil pressures applied to the pile. Compared to existing empirical relationships, the new ones can capture the aforementioned dilation phenomena, which may develop at the upper part of the pile in the case of relatively low permeability soils (e.g. fine-grained, silty sands) and have an overall detrimental effect on pile response.

Introduction

Lateral spreading of liquefied soils is a common source of damage to infrastructure during earthquakes. The phenomenon was first brought into the attention of the engineering community during the Niigata 1964 earthquake in Japan, and has become a subject of extensive investigation then after. In the case of pile foundations, large

Y. K. Chaloulos (✉) · G. D. Bouckovalas
National Technical University of Athens, Athens, Greece
e-mail: ioannischaloulos@gmail.com

A. Ansal, M. Sakr (eds.), *Perspectives on Earthquake Geotechnical Engineering*, Geotechnical, Geological and Earthquake Engineering 37,
DOI 10.1007/978-3-319-10786-8_17

ground displacements induced by lateral spreading result in large residual kinematic loads, which may cause significant bending moments and pile head movement. The analysis of these effects is mainly performed numerically, by adopting a simplified "Beam on Winkler Foundation" model, in which the pile is treated as a beam, while soil is replaced by Winkler springs that obey a non-linear load-deformation law (p-y curve). Free field ground displacements are applied at the fixed-end of the soil springs.

It is evident, that the efficiency of the above simulation largely depends upon the realistic definition of the nonlinear p-y relationship for the lateral springs. Therefore, it is no surprise that a large number of research studies, mainly experimental, have been dedicated to this topic, and have given valuable insight to the parameters that affect the p-y curves. In general, the various empirical methodologies for the estimation of p-y curves are based on the corresponding relations for firm soil, while the effects of liquefaction are taken into account by reducing the associated ultimate resistance and the initial subgrade modulus. Modification of the curves for firm soils is performed either by applying appropriate reduction factors (e.g. Brandenberg et al. (2005)) or by considering empirical relations for the residual strength of liquefied soil (e.g. Cubrinovski and Ishihara 2007). In any case, the pursued reduction is related only to the relative density of the sand.

Nevertheless, recent experimental studies (e.g. Tokimatsu and Suzuki (2009)) acknowledge that a number of additional parameters (e.g. soil permeability, excitation characteristics, pile properties e.t.c.) may be also of importance. Along the same line, it is of special interest to refer to González et al. (2009) who have shown that significant negative excess pore pressures may develop near the pile head, for values of soil permeability commonly encountered in the field, thus increasing instead of decreasing the soil pressures compared to the non-liquefied case. Such dilation effects are masked in small scale model tests using water (instead of a more viscous) pore fluid, and have been consequently overlooked in the majority of experimental studies.

In view of the above, the interaction of piles with laterally spreading sand deposits was revisited with the aid of non-linear 3D numerical analyses. To ensure the realistic prediction of pile and soil response, the numerical analyses were first calibrated against the aforementioned experiments of Gonzalez et al. and subsequently they were utilized parametrically in order to investigate the effect of various problem parameters. Thus, feedback was obtained and conclusions could be drawn on the following aspects of practical interest:

a. *Numerical simulation of piles in laterally spreading ground*, with emphasis on the simulation of the pile-soil interface, the appropriate boundary conditions for a horizontally shaken submerged infinite slope, as well as the permeability of the soil during seismic shaking and liquefaction.
b. *Dilation effects on the subgrade reaction of liquefied soils*, with emphasis upon the mechanisms which control the liquefied soil response around the pile and the associated ultimate soil pressures, as well as the identification of the important problem (soil, excitation and pile) parameters and the quantification of their effects within a new design framework.

The following presentation is an overview of the basic assumptions and the main findings of this study, and it is largely based on the recently published papers by Chaloulos et al. (2013) and Chaloulos et al. (2014) which address in more detailed issues (a) and (b) above.

Numerical Methodology Calibration

Centrifuge Test Setup

The set-up of the centrifuge test by González et al. (2009), used to calibrate the 3D numerical methodology, is shown in Fig. 1a. Soil consists of a 6 m thick layer of Nevada sand, placed at a relative density of $D_r=40\%$, over a 2 m thick non-liquefiable layer of slightly cemented sand. The laminar box container has a mild inclination of five degrees relative to the horizontal. The pile has a diameter D=0.60 m and a bending stiffness EI=9000 kNm^2, while a sinusoidal excitation of 30 cycles with period T=0.5 s and peak acceleration $a_{max}=0.30$ g is applied horizontally, at the base of the container.

The test has been performed under 50 g centrifugal acceleration, and the system response has been monitored by densely placed instrumentation consisting of: strain gauges, pore pressure transducers, accelerometers and linear variable differential transducers (LVDTs). The soil is saturated with metulose whose viscosity is 50 times that of water so that, under the 50 g centrifugal acceleration, the test reproduces the drainage conditions met in the field.

Figure 1b shows the variation of recorded lateral displacements with depth, at various instants during the test, and compares it to the maximum lateral displacement that is allowed by the laminar box container (heavy dashed line). It can be observed that the ultimate displacement capacity of the box has been reached at the early stages of shaking, i.e. at t=3 s for depth z=5.5 m and at t=9 s for depth z=4 m. It is thus speculated that all test measurements at later test stages have been affected by this artificial constraint and should be evaluated with caution.

Numerical Simulation

The Numerical Model The centrifuge test described above was simulated using the Finite Difference Code FLAC3D (v.4.0). A major advantage of this code is that it allows coupling between pore water flow and dynamic loading, while it makes use of an explicit integration algorithm, which is more efficient for highly non-linear dynamic problems, such as earthquake-induced soil liquefaction. The mesh built to simulate the experiment is shown in Fig. 2. Its dimensions are those of the the laminar box (i.e. 35.5 m long, 9 m wide and 8 m tall). Following a number of mesh sensitivity analyses, the height was divided into 16 zones of 0.5 m each,

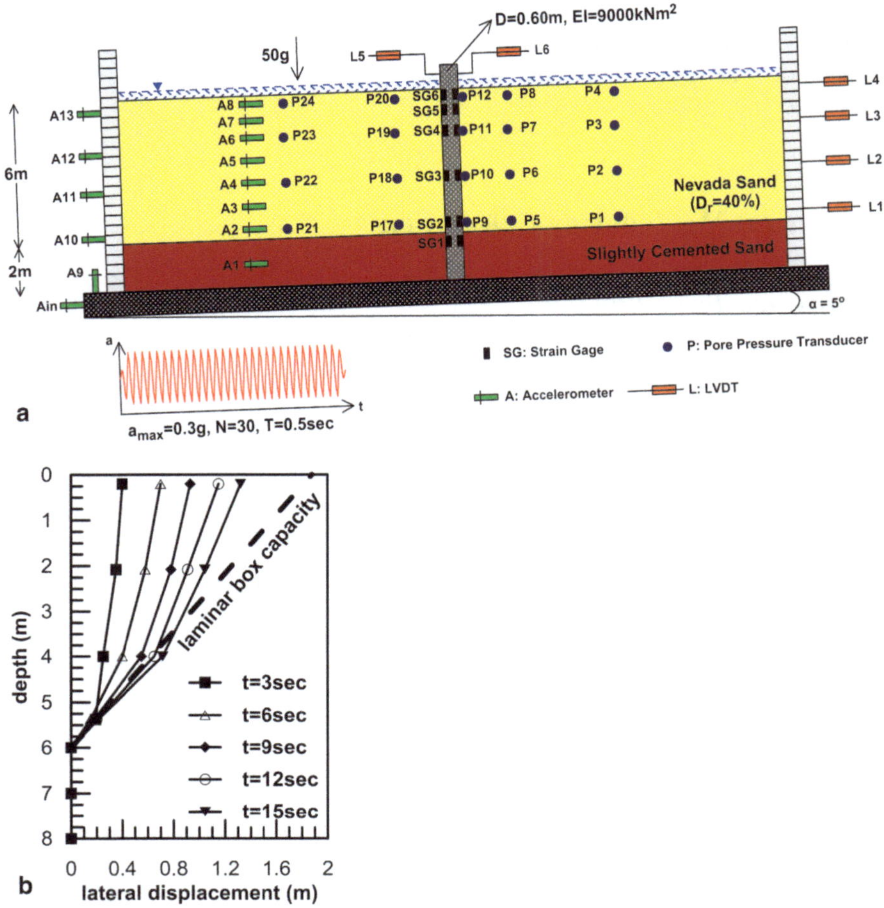

Fig. 1 **a** Setup and instrumentation used in the experimental model (prototype units). **b** Variation of lateral displacements with depth measured at various test stages. (Chaloulos et al. 2013)

while the width of the zones starts from a minimum length of 0.1 m at the pile-soil interface and increases progressively with distance away from the pile. Finally, the ground surface inclination is modeled by the superposition of a horizontal gravitational component.

Soil Conditions The liquefiable sand response is simulated with the NTUA Sand model (Andrianopoulos et al. 2010; Papadimitriou and Bouckovalas 2002) implemented to FLAC3D by Karamitros (2010), and calibrated against monotonic and cyclic liquefaction element tests on Nevada sand. It incorporates the Critical State Theory of Soil Mechanics, so that the effects of initial state (relative density and mean effective stress) are simulated in terms of a uniquely defined state parameter (Been and Jefferies 1985). In parallel, the hysteretic Ramberg-Osgood formulation is adopted for small strain increments, allowing for the accurate simulation of the

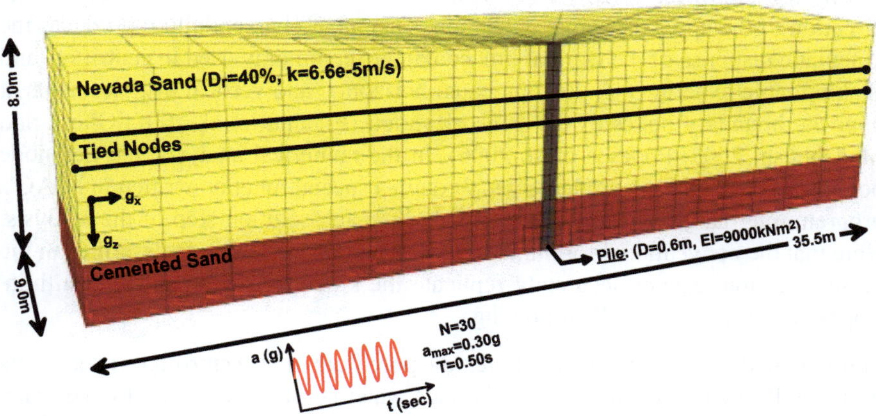

Fig. 2 3D Finite Difference mesh used to simulate the test by Gonzalez et al.

non-linear hysteretic response of sands (decrease of shear modulus and increase of hysteretic damping with increasing cyclic shear strain amplitude) over a wide range of strain amplitude levels. Shake-down effects and liquefaction-induced softening during cyclic loading are properly simulated by quantifying the effect of fabric evolution on plastic strain increments.

It is important to note here that, the use of such advanced constitutive models (instead of simpler uncoupled formulations), although it increases the complexity of the analyses, is essential to capture key mechanisms of the pile-soil interaction such as liquefaction related cyclic mobility at the free field, as well as, dilation and negative excess pore pressure build-up near the pile (Chaloulos et al. 2013).

The cemented sand is assumed to behave as linearly elastic, with Poisson's ratio $v=0.33$ and elastic (small strain) shear modulus $G_o=120000$ kPa. The G_o value corresponds to the elastic stiffness of clean Nevada sand, at the same Relative Density and confining stress levels, multiplied by a factor of two. This approach for the modulus definition of the cemented soil is based on various studies (Acar and El-Tahir 1986; Saxena et al. 1988; Schnaid et al. 2001; Sharma and Fahey 2003) which suggest that cementation may (on average) double the stiffness relative to the same un-cemented soil.

Pile & Pile-Soil Interface The pile is considered elastic, as no plastic hinges have been developed during the test, while the elastic constants are selected so as to yield the reported flexural stiffness $EI=9000$ kNm^2. *Interface elements* are placed between the pile and the soil in order to allow slippage development, which affects dilation and negative excess pore pressure build-up. Comparative analyses demonstrate that without the use of such interfaces, ultimate soil pressures can be significantly overestimated (Chaloulos et al. 2013). Finally, taking into account that sand grains were glued to the outer part of the pile, the interface is assigned zero cohesion and friction angle equal to that of the sand.

Boundary Conditions The well-known tied-node method was incorporated in the direction of the applied seismic excitation. This method essentially reproduces the kinematic response of the laminar box containers commonly used in centrifuge and shaking table experiments, yielding equal horizontal and vertical displacements at opposite boundary nodes of the same elevation (Elgamal et al. 2005; Ghosh and Madabhushi 2003; Popescu et al. 2006). In the numerical analysis, the tied-node boundary conditions are implemented through a special function written in FLAC's programming language FISH, executed during each solution step of the analysis. Note that the conventional formulation of tied-node boundaries, as described in the present section, cannot accurately replicate the kinematics of submerged infinite slopes and is revisited in the following.

Soil Permeability As described earlier, the pore fluid in the centrifuge experiments was metulose with the intension to simulate water permeability in prototype scale, under static steady state conditions. Still, for earthquake-induced liquefaction, the actual value of soil permeability in the field is an unresolved issue. Namely, while Darcy's coefficient has been established to quantify flow under static conditions and through the pores of a stable soil skeleton, the present application involves dynamic loading that may cause alternating direction of the pore fluid flow, while excess pore pressure build-up and liquefaction leads to a rather unstable soil skeleton.

Hence, the following three different scenarios are considered herein with regard to the permeability coefficient of the sand, based on relevant recommendations from the literature:

a. *"Static" Permeability Coefficient* $k = 6.6 \times 10^{-5}$ m/s, determined from constant head permeability tests under 1 g gravitational acceleration for $D_r = 40\,\%$ (Arulmoli et al. 1992)
b. *"Dynamic" Permeability Coefficient* $k = 2.1 \times 10^{-5}$ m/sec, as proposed by Liu and Dobry (1997) in order to take into account the alternating flow direction within the pores of the sand skeleton during shaking.
c. *Variable Permeability Coefficient,*

$$k_b = k_{ini} \cdot \left[1 + (a-1) r_u^{\beta}\right] \tag{1}$$

as proposed by Shahir et al. (2012), in order to take into account that during liquefaction soil particles gradually loose contact, leading to the creation of additional flow channels within the sand skeleton.

Note that Eq. (1) was implemented in the analysis through a FISH function, which constantly updates the value of permeability at each zone and at each time step. The initial value of the permeability coefficient in Eq. 1 was set to $k_{ini} = 2.1 \times 10^{-5}$ m/s, while, based on Shahir et al. it was further assumed that $\alpha = 10$ and $\beta = 1$.

All above scenarios were evaluated through comparison with the experimental data. An optimum overall fit of the test results was achieved for $k = 6.6 \times 10^{-5}$ *m/s*. Note that this value is approximately equal to the logarithmic mean of the $k = f(r_u)$ scenario for $r_u = 0$ and $r_u = 1.0$. [e.g. Eq. (1)]. In the following, numerical results and corresponding comparison with experimental data are shown only for the optimum

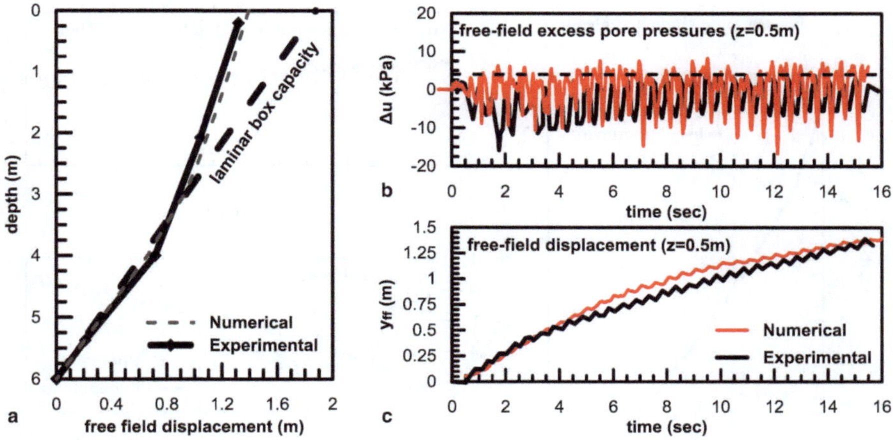

Fig. 3 Comparison between experimental data and numerical predictions. **a** Free-field displacements at the end of shaking. **b** Excess pore pressures at the free field (P4 in Fig. 1a). **c** Free-field displacements near the ground surface (L4 in Fig. 1a)

case of $k = 6.6 \times 10^{-5}$ *m/s*, while more details and comparisons for the calibration of permeability can be found in Chaloulos et al. (2013).

Evaluation of Numerical Predictions

Free Field Response Predictions of free field response are evaluated Fig. 3. Namely, Fig. 3a shows the variation of free field displacements with depth, at the end of shaking, while Fig. 3c presents their evolution with time at depth z = 0.5 m from ground surface. Note that, in order to take into account the lateral displacement limits of the laminar box, marked with black dashed line in Fig. 3a, the numerical predictions $y_{ff}(h)$ have been corrected as follows:

$$y_{ff}(h) = \int_0^h \gamma(h) \cdot dh \tag{2}$$

where h is the height above the bottom of the liquefiable layer and $\gamma(h)$ is the numerically computed shear strain at h, which is not allowed to exceed the maximum shear strain of the laminar box, i.e. $\gamma(h) \leq \gamma_{ult} \approx 33\%$. In both figures corrected numerical displacements are in fairly good agreement with the experimental data.

In addition, Fig. 3b compares the evolution of excess pore pressures, at z = 0.5 m depth (Transducer P4 in Fig. 1a). Observe that excess pore pressures become almost equal to the initial effective stress, denoted with a dashed line in this figure, signifying soil liquefaction. However, all time histories exhibit significant dilation spikes which are typical for soils undergoing lateral spreading.

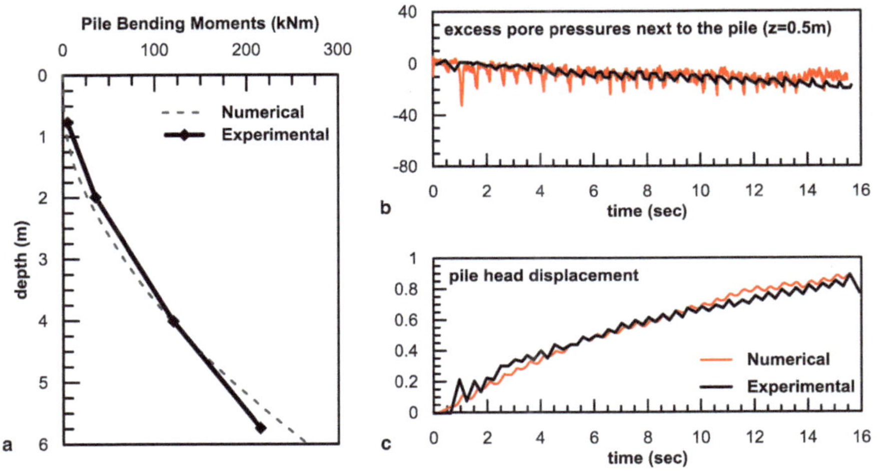

Fig. 4 Comparison between experimental data and numerical predictions. **a** Pile bending moments. **b** Excess pore pressures next to the pile (P12 in Fig. 1a). **c** Pile head displacements (L4 in Fig. 1a)

Pile and Surrounding Soil Response Using a similar presentation format, Fig. 4 summarizes the response of the pile and the adjacent soil. Namely, Fig. 4a, and Fig 4c compare numerically predicted and experimentally recorded pile bending moments and pile head displacements respectively. Note that, due to the previously discussed laminar box limitations, bending moments are not compared at the end of shaking, but at $t = 6\ s$ when the constraints imposed by the lateral boundaries of the box have not yet been fully mobilized. On the other hand, pile head displacements are shown in a time history format, for the entire duration of testing, although emphasis is given again to the first 6 sec of shaking. In this case also, the numerical methodology simulates the experimental results with fair accuracy.

Furthermore, Fig. 4b compares experimental measurements and numerical predictions of excess pore pressures at transducer P12 placed next to the pile head (Fig. 1a). It is observed that the numerical predictions replicate with quantitative accuracy the experimental trends, namely that excess pore pressure become systematically negative, despite the partially drained conditions which prevail in this area. Note that this is a rather peculiar response pattern, indicating a tendency for strong dilation due to the combined effects of low confining pressures and large shear strains induced by the flow of the liquefied soil around the pile head.

The accuracy in predicting dilation-induced negative excess pore pressure build up is further evaluated in Fig. 5a, 5b which compare numerical and experimental contours of excess pore pressure ratio at the end of shaking. To avoid confusion due to the different color scales of the two figures, a black dotted line is used to mark the contour of zero excess pore pressure ratio. Observe that the numerical analysis predicts consistently the experimental results, namely that partial or complete liquefaction (i.e. $r_u \approx 0.50 \div 1.00$) takes place at large depths along the pile and at

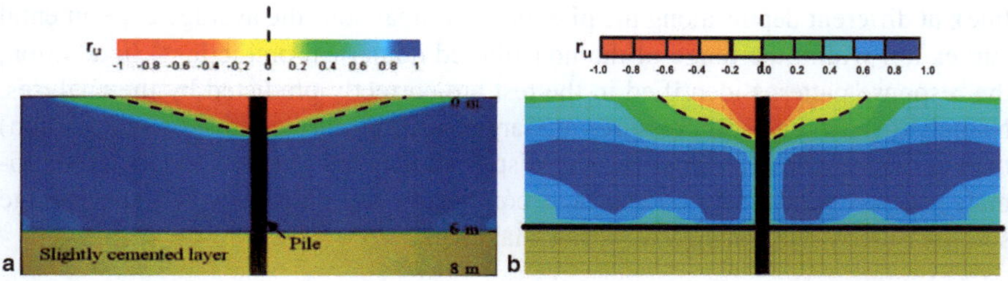

Fig. 5 Comparison between excess pore pressure ratio contours at the end of shaking. **a** Experimental recordings. **b** numerical predictions

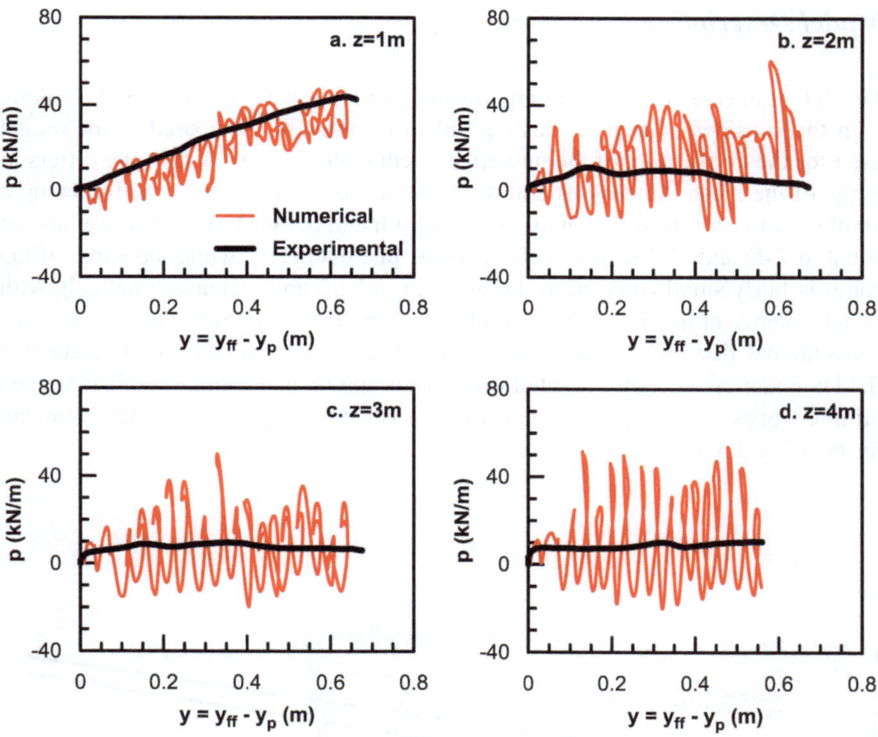

Fig. 6 Comparison between experimental and numerical p-y predictions

the free field, while significant dilation (i.e. $r_u < -0.50$) occurs near the pile head, leading to the formation of an inverted cone of high negative excess pore pressures.

P-y Relations Finally, Fig. 6 compares the average experimental p-y curves (black line), back-calculated from the recorded bending moments after the cyclic component was filtered out, with the respective non-filtered numerical predictions (red line), at different depths along the pile. In all comparisons the average experimental curves fall within the range of the non-filtered numerical predictions. In addition, the response patterns identified in the test are correctly predicted by the analyses. Namely: (a) average soil pressures are larger near the surface (due to soil dilation) than at depth, while (b) large relative displacements are required to reach the ultimate pressure near the ground surface, contrary to larger depths where the average soil pressure is stabilized early during shaking.

Parametric Analyses

Model Description

The 3D mesh created for the parametric analyses is shown is shown in Fig. 7, along with the considered range of basic problem parameters. Compared to the model used for the verification of the numerical methodology (Fig. 2), this one differs in terms of the mesh dimensions and the boundary conditions. Namely, following a number of mesh sensitivity analyses, the length and the width of the model are set equal to 74D and 17D respectively (D is the pile diameter), while the width of the zones is fairly small adjacent to the pile (e.g. 0.1 m) and increases gradually with radial distance at the free field lateral boundaries (e.g. up to 5D). Furthermore, the conventional tied-node formulation used in the previous section to simulate free field boundaries was now revisited in order to capture the kinematics of submerged infinite slopes. The modified concept is described in more detail in the following Sects. 3.2 and 3.3.

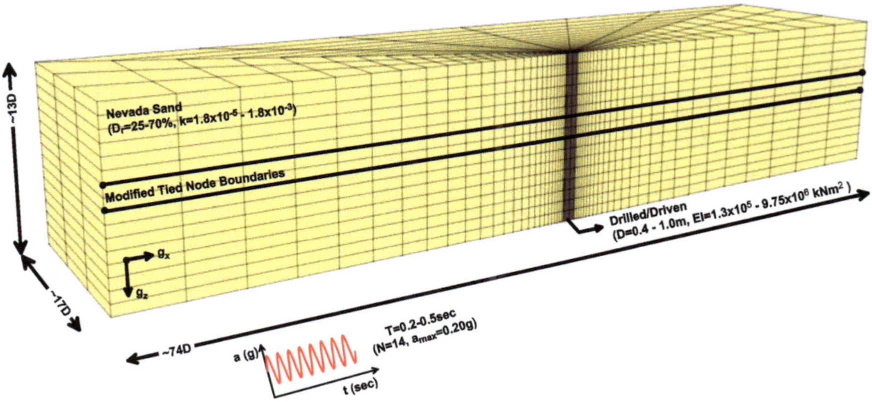

Fig. 7 3D mesh and parameters considered in the parametric investigation

In total, 26 parametric analyses were performed for the following soil, pile and excitation characteristics: relative density $D_r = 25$ to 70 %, permeability coefficient $k = 1.8 \times 10^{-5}$ to 1.8×10^{-3} m/s, pile diameter D = 0.4–1.0 m, bending stiffness EI = 1.3e5 to 9.75e6 kNm^2, pile type: drilled or driven and harmonic excitation period T = 0.20 to 0.50 s. Note that, when one of the above problem parameters is varied, all other parameters are kept equal to the reference values $D_r = 50$ %, $k = 6.1 \times 10^{-5}$ m/s, D = 0.60 m, EI = 1.3e6 kNm^2 and T = 0.30 s. In addition, analyses for different pile diameter are performed by uniform scaling of all mesh dimensions in Fig. 7.

Modified Tied-Node Boundaries

Numerical analyses of infinite slopes should replicate the actual field conditions, namely that effective stresses and displacements in between any two lateral boundaries are uniform along the free ground surface and vary only with depth. For instance, for uniform soil conditions and a Cartesian coordinate system *xz* fitted to the free ground surface of an infinite slope (*x* is parallel to the ground surface), the static effective stresses vary linearly with *z*, as:

$$
\begin{aligned}
\sigma_z' &= (\gamma' \cos\theta)\, z \\
\sigma_X' &= k_o \sigma_z' \\
\tau_{XZ} &= (\gamma' \sin\theta)\, z
\end{aligned}
\qquad (3)
$$

where θ is the ground slope inclination relative to the horizontal, γ' is the effective unit weight of the soil, and k_o is the geostatic earth pressure coefficient.

As described previously, the numerical analyses are performed with a horizontal mesh, while soil inclination is simulated by applying a horizontal gravitational component. For the common case of submerged slopes, considered in this paper, this implies that the phreatic surface is not horizontal anymore but it is rotated by an angle θ so that it becomes perpendicular to the inclined unit gravitational force γ'.

This transformation of the problem geometry is schematically explained in Fig. 8. As a result, although effective stresses at the two lateral boundaries remain

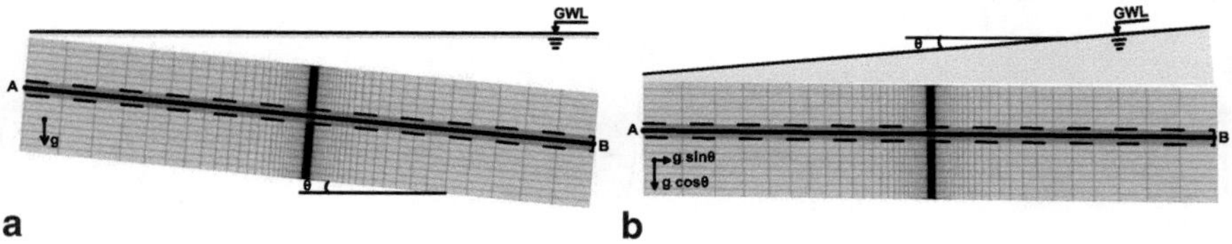

Fig. 8 **a** Physical Problem of the submerged slope. **b** Numerical interpretation with horizontal gravitational component and inclined phreatic surface

Fig. 9 Equilibrium of the infinitesimal slice of soil around grid points A and B

equal, hydrostatic pore pressures and total stresses are higher at the downslope than at the upslope boundary. This condition is perfectly acceptable in the free field, where no boundary constraints are required in order to achieve equilibrium. However, in a numerical analysis, where the slope of infinite length has to be replaced by a model of finite length, the inclined phreatic surface and the associated total stress difference induce an unbalanced force in the direction of the slope gradient, in excess of the corresponding unit gravitational component $\gamma' \sin\theta$. Thus, the static stress field, as well as, the response during seismic excitation is significantly distorted.

The above destabilization mechanism is further explained with reference to the soil slice between tied nodes A and B in Fig. 8b, and the associated body force diagram of Fig. 9. Namely, due to the inclined phreatic surface, hydrostatic pore pressures at grid point B are larger than at point A (i.e. $u_B > u_A$). Assuming further that the infinite slope is initially at equilibrium, it comes out that horizontal effective stresses at A and B are equal (i.e. $\sigma'_{X,B} = \sigma'_{X,A}$) while total stresses at B are larger than at A (i.e. $\sigma_{X,B} > \sigma_{X,A}$).

Assume now that tied-node boundaries are applied and the system is allowed to stabilize. For the above stress values, a solution cycle in FLAC 3D will proceed as follows:

Step 1 Calculation of the total horizontal force, F, at grid points A and B from the average *total* horizontal stresses. For positive downslope forces, accelerations and velocities, these forces are computed as:

$$\begin{aligned} F_A &= -\sigma_{XX,A} \cdot A_s \\ F_B &= \sigma_{XX,B} \cdot A_s \end{aligned} \tag{4}$$

where A_s is the area corresponding to grid points A and B.

Step 2 Calculation of acceleration, a, and velocity, v, for each grid point from the corresponding total forces:

$$\begin{aligned} a_A &= F_A / m \\ v_A &= a_A \cdot dt = \left(-\sigma_{XX,A} \cdot A_s / m\right) \cdot dt \end{aligned} \tag{5}$$

and

$$\begin{aligned} a_B &= F_B / m \\ v_B &= a_B \cdot dt = \left(\sigma_{XX,B} \cdot A_5 / m\right) \cdot dt \end{aligned} \tag{6}$$

where *m* is the mass corresponding to each grid point and *dt* is the time step of the analysis.

Step 3 At that point of the regular solution algorithm, the tied-node function, programmed in FISH, is called in order to modify the velocities at the boundary grid points by assigning to them an average common velocity. According to the conventional formulation the latter is obtained as follows:

$$v_{av} = \frac{v_A + v_B}{2} \tag{7}$$

Note that, due to the total stress difference at grid points *A* and *B*, the velocity magnitude at *B* is larger than that at *A* (i.e. $|v_B| > |v_A|$) and consequently the average velocity v_{av} is not zero, as under dry slope conditions, but takes a positive value pointing in the down slope direction.

Step 4 The solution cycle closes with the calculation of strains (based on v_{av}) and new effective stresses (based on the adopted constitutive law).

Steps 1–4 are repeated until v_{av} from Eq. (7) becomes zero. Since this velocity is the result of the difference in total stresses between corresponding points of the same elevation, equilibrium can only be achieved when the total stresses at the two boundaries become equal. However, at that state, down slope effective stresses will be less than those upslope, violating the actual free-field conditions.

Having understood the mechanisms triggered by the tied-node boundary conditions of submerged slopes, it is now feasible to modify their formulation so that the effective stress and deformation conditions within the infinite slope remain uniform along the ground surface. The modified tied-node formulation proposed herein is based on the idea that an upslope velocity should be applied at the lateral boundaries of the model, so that the average velocity v_{av} computed from Eq. (7) is zero despite the total stress difference between points A and B of the lateral boundaries. In other words, equation should be re-written as:

$$v_{av} = \frac{v_A + v_B + v_{up}}{2} \tag{8}$$

where

$$v_{up} = -(v_B - v_A) \tag{9}$$

Furthermore, combining equations (4) and (5), and taking also into account that $\sigma'_{xxA} = \sigma'_{xxB}$, Eq. (9) yields:

$$v_{up} = \frac{A_s \cdot dt}{m}\left(\sigma_{xx,A} - \sigma_{xx,B}\right) = \frac{A_s \cdot dt}{m}\left(u_A - u_B\right) \tag{10}$$

Note that all quantities in Eq. (10) can be estimated based on the geometry of the problem at hand, so that v_{up} can be evaluated at the beginning of the analysis, prior to any stepping initiation.

Evaluation of Lateral Boundary Effects

The preceding analysis revealed that kinematic constraints imposed by conventional tied-node boundaries, and subsequently by laminar box containers in centrifuge and shaking table experiments, are not consistent with submerged infinite slope conditions. To evaluate in more detail how this inconsistency may affect the predicted pile-soil system response, the analysis of the centrifuge test by (González et al. 2009) has been repeated with the use of the modified, instead of the conventional tied node boundaries.

Figures 10 and 11 show the contours of horizontal effective stresses σ'_{xx}, after static equilibrium as well as the deformed model shape and the corresponding contours of excess pore pressures Δu after shaking respectively. Observe the smooth shape of the deformed model, as well as, the σ'_{xx} and Δu contours obtained by the modified tied node formulation. Non-uniform conditions develop only at the vicinity of the pile, while uniform infinite slope conditions are gradually restored with distance away from the pile.

For a quantitative assessment of boundary formulation effects, Fig. 12 compares horizontal free field and pile displacements, as well as pile bending moments predicted with the two types of lateral boundaries. Free field displacements are somewhat over-predicted by the conventional tied-node formulation, with the maximum differences observed near the bottom of the container and at the ground surface. This trend is reversed for the lateral pile displacements and bending moments. In this case, the conventional formulation proves significantly un-conservative, with

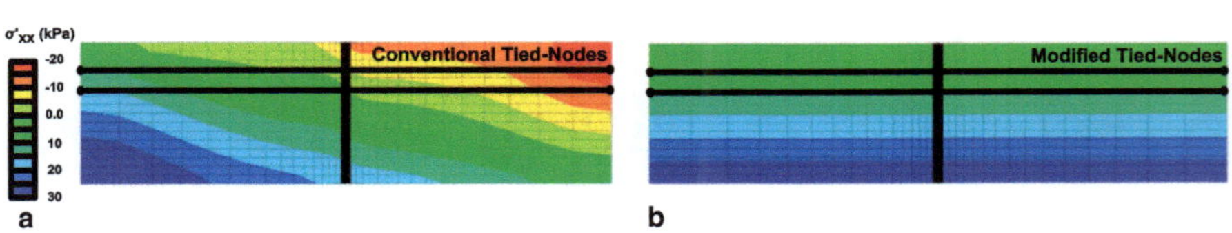

Fig. 10 Comparison of horizontal effective stresses after static equilibrium using. **a** The conventional. **b** The modified tied-node formulation

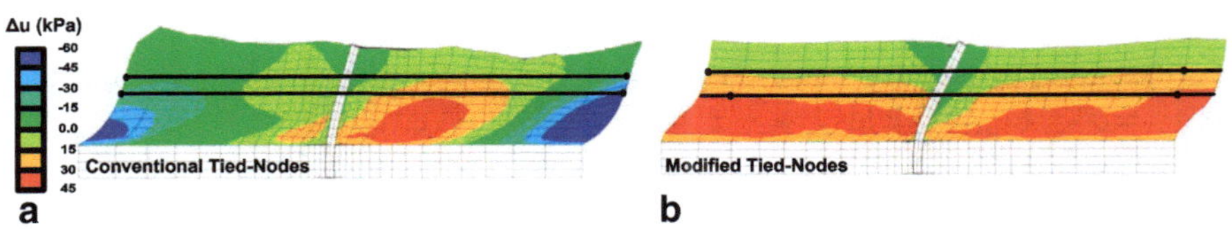

Fig. 11 Comparison of deformed shape and excess pore pressures at the end of shaking using. **a** The conventional. **b** The modified tied-node formulation

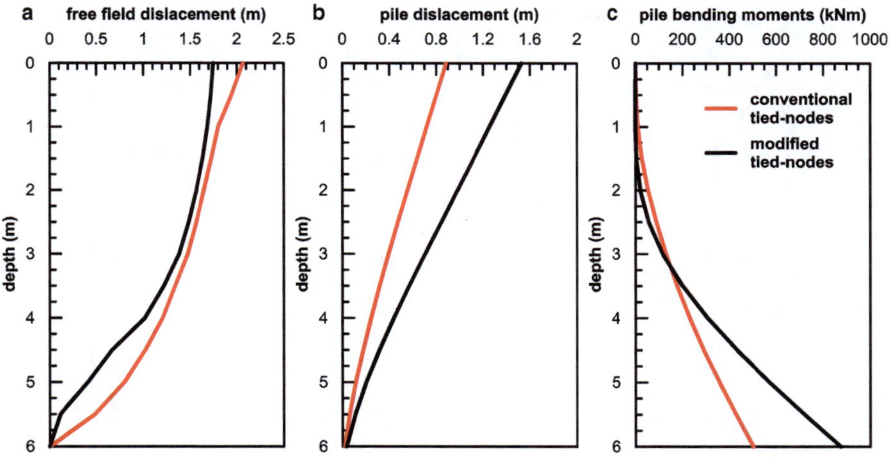

Fig. 12 Comparison of **a** Free field displacements. **b** Pile displacements. **c** Pile bending moments with the conventional and the modified tied-node formulation

the maximum pile (head) displacements and the maximum (base) moments under-predicted by 40–45 %.

Finally, Fig. 13 compares predicted p-y curves at four different elevations along the pile. The conventional formulation proves again un-conservative, yielding systematically lower values of the ultimate soil pressures applied to the pile. Differences become more pronounced at medium depths (z=2–3 m) where ultimate soil pressures are underestimated by 60–80 %. The above differences can be attributed to the dilative response of the soil near the pile, confirming the role of the latter as the key mechanism for the development of subgrade reaction.

Empirical p-y Relations

Interpretation of Numerical Predictions

Figure 14a, 14b show the p-y curves, obtained from a typical parametric analysis, for two depths along the pile (red line). Observe that, although the p-y response within each loading cycle exhibits hardening, the average p-y response exhibits softening. Based on this observation, and following previous studies for dry sands (e.g. Det Norske Veritas 1980; Georgiadis et al. 1992) as well as for laterally spreading soils (e.g. Brandenberg et al. 2007; Cubrinovski and Ishihara 2007; Tokimatsu and Suzuki 2009), it is assumed that the average p-y curves can be fitted by a hyperbolic function expressed analytically as follows (black line in Fig. 14a, 14b):

$$p = \frac{y_{rel}}{\dfrac{1}{k_{ini,liq} z} + \dfrac{y_{rel}}{p_{ult,liq}}} \tag{11}$$

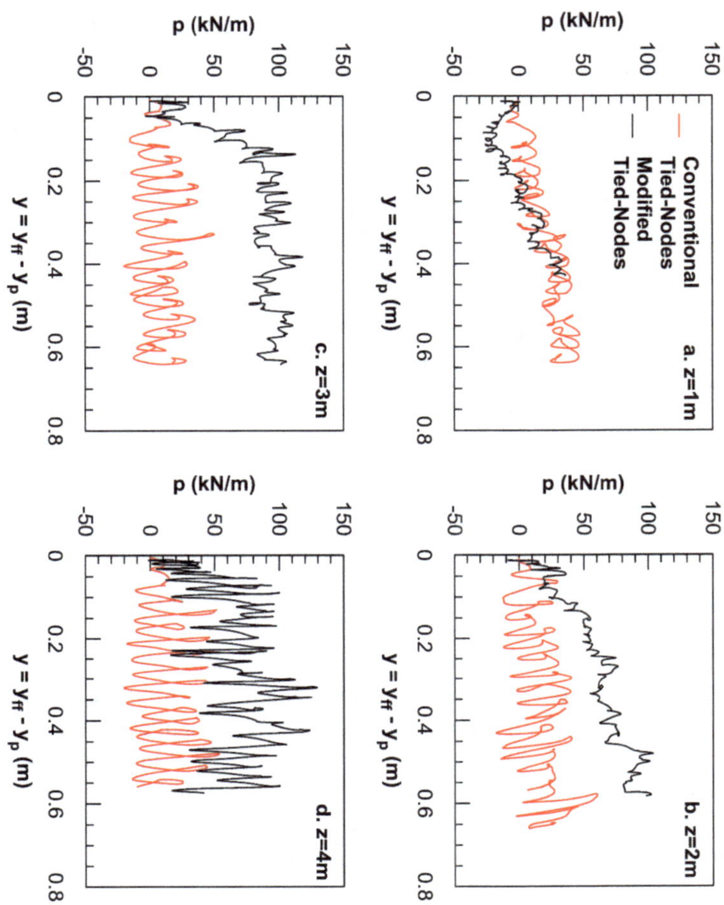

Fig. 13 Comparison of p–y curves at four different depths with the conventional and the modified tied-node formulation

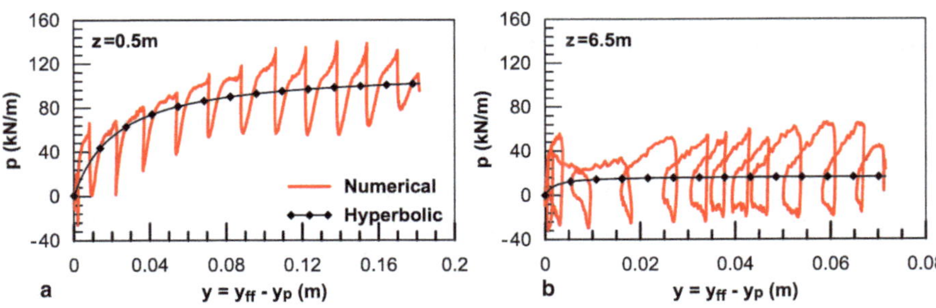

Fig. 14 Adjustment of a hyperbolic curve on the numerically obtained p-y curve for the estimation of ultimate soil pressure

where p is the soil pressure [F/L], y_{rel} is the relative displacement between the pile and the free field soil [L], $k_{ini,liq}$ is the gradient with depth of the initial subgrade modulus of the liquefied soil [F/L^3] and $p_{ult,liq}$ is the associated ultimate soil pressure [F/L]. In the above, F and L stand for force and length units respectively.

In the sequel, following a trial-and-error optimization, the best fit $k_{ini,liq}$ and $p_{ult,liq}$ values were obtained. This procedure was consequently applied to all 26 analyses of this study, with $k_{ini,liq}$ and $p_{ult,liq}$ values obtained for eight different depths along the pile, thus leading to a set of 208 pairs of values.

Initial Subgrade Reaction Parameter $k_{ini,liq}$

The most consistent average fit of the numerically predicted p-y curves is obtained assuming that $k_{ini,liq}$ is approximately equal to one half of the initial stiffness, for the non-liquefied soil, i.e.:

$$k_{ini,\,liq} = {}^{1}\!/_{2} \cdot k_{ini,\,firm} \tag{12}$$

Note that $k_{ini,liq}$ does not represent the small strain elastic stiffness of the soil, but the secant stiffness at very small relative pile-soil displacements (i.e. in the order of a few mm).

It is acknowledged that this is a rather simplifying assumption. However, it is adopted here following the observation that the ultimate soil pressure of the liquefied soil $p_{ult,liq}$, which is the focus of this study, is generally reached early during shaking, so that the initial stiffness of the p-y curve becomes of secondary importance. It is noteworthy that Dobry et al. 2003 use a similar argument and recommend to simply use $p_{ult,liq,}$ rather than perform a complete p-y analysis, for assessing the lateral pile response.

Ultimate Soil Pressure, $p_{ult,liq}$

One of the main findings of this study is that ultimate pressures are uniquely related to the excess pore pressure ratio along the pile (Chaloulos et al. 2014). This relation is demonstrated in Fig. 15a where the normalized ultimate soil pressures, $p_{ult,liq}/(\sigma'_{vo}D)$, are plotted versus the excess pore pressure ratio near the pile, $1\text{-}r_{u,pile}$. The latter approaches unity at relatively large depths and exhibits large negative values close to the surface. As a result $p_{ult,liq}$ is large at the upper part of the pile and decreases with depth. Graphically, this relation is demonstrated in Fig. 15b, which shows that the normalized ultimate soil pressure along the pile increases log-linearly with decreasing depth, until it reaches a constant peak value near the pile head. Namely, $p_{ult,liq}/(\sigma'_{vo}D)$ can be calculated as follows:

$$\frac{p_{ult,\,liq}}{\sigma'_{vo}\,D} = A\left(\frac{z}{L}\right)^{-B} \leq C \tag{13}$$

where z/L is the normalized depth over the pile length and A, B and C are problem specific coefficients.

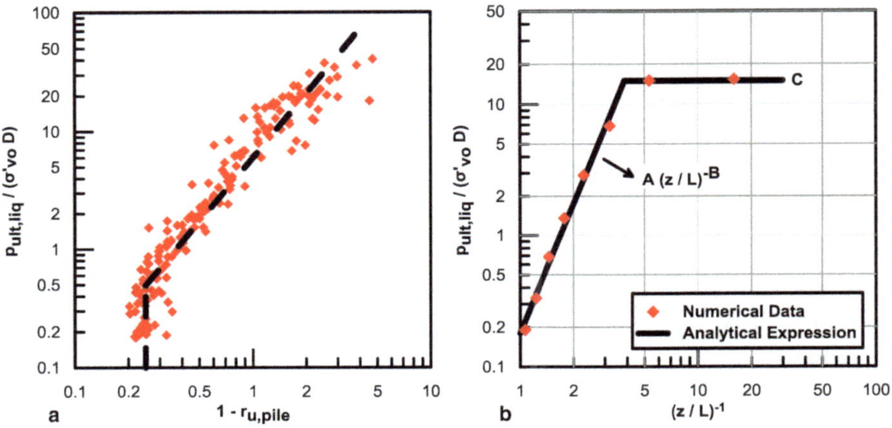

Fig. 15 Relation between $p_{ult,\,liq}\,/\,(\sigma'_{vo}D)$ and **a** 1-ru,pile and (b) (z/L) for the whole set of analyses. (Chaloulos et al. 2014)

From a physical point of view, coefficient A corresponds to the normalized pressure near the tip of the pile (z=L), i.e at relatively large depth where soil response is contractive and the excess pore pressure ratio approaches unity. On the other hand, coefficient C represents the normalized soil pressure at small depths, where dilation is significant and leads to large negative excess pore pressures. Finally, coefficient B represents the rate of transition with depth, from small soil pressures at large depths, to large soil pressures near the pile head.

A multivariable regression analysis of the numerically predicted ultimate soil pressures (Chaloulos et al. 2014), finally led to the following relationships for the estimation of coefficients A, B and C in terms of the basic problem parameters T(s), D(m), k(m/s), Dr(%) and EI(kNm^2):

$$A = 0.045 \cdot D_r(\%)^{0.9} \cdot T^{1.55} \cdot \left(k \cdot 10^4\right)^{0.06} \tag{14}$$

$$B = 1.35 \cdot \exp\left[-0.023 \cdot D_r\right] \cdot T^{-0.95} \cdot \left(k \cdot 10^4\right)^{-0.11} \cdot \exp\left[1.3 \cdot D\right] \tag{15}$$

$$C = 0.063 \cdot \exp\left[-0.025 \cdot D_r\right] \cdot \min\left[\left(EI \cdot 10^{-5}\right)^{0.3},\ 8.6 \cdot \left(EI \cdot 10^{-5}\right)^{-0.45}\right] \cdot \left(\frac{k \cdot T}{D}\right)^{-0.6} \tag{16}$$

The effects of pile head constraint and installation method were found to be minor and have not been included to the above empirical relations.

The overall accuracy of Eqs. 11–16 is evaluated in Fig. 16a where all analytical predictions are one-to-one compared with the corresponding numerical results. Furthermore, Fig. 16b presents the associated relative error, defined as the ratio of numerical over analytical predictions of $p_{ult,liq}\,/\,(\sigma'_{vo}D)$. In both figures, different

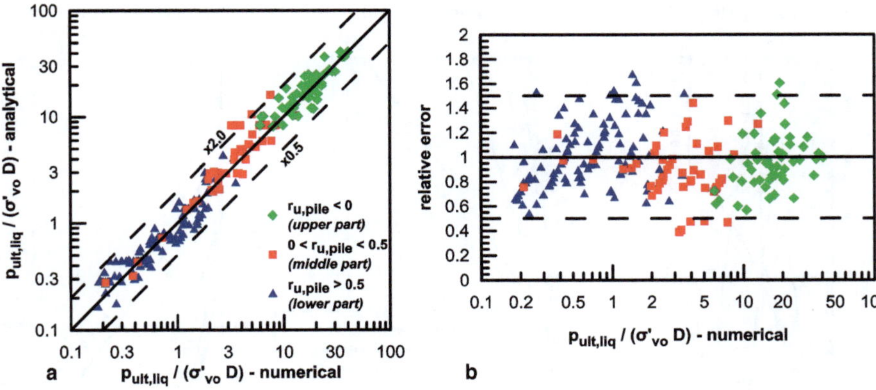

Fig. 16 Evaluation of analytical relationships for the estimation of ultimate soil pressure. **a** One-to-one comparison. **b** Relative error

symbols are used for the upper part of the pile with $r_{u,pile} < 0.0$, the middle part with $r_{u,pile} = 0.0–0.5$ and the lower part where $r_{u,pile} > 0.5$. The agreement between empirical and numerical predictions is consistent for the entire range of data, while the relative error for approximately 95 % of the data points is less than 50 %. No bias is observed with regard to the different pile segments (upper, middle or lower).

Dilation Effects on Pile Performance

The main novelty of the proposed relationships is that they can capture the effects of soil dilation along the upper part of the pile, and the associated increase of ultimate pressures. To identify the cases of practical significance for this new feature, and also to built confidence on the proposed relationships, they are compared to the recently proposed and widely used relationships by Brandenberg et al. (2007), Cubrinovski and Ishihara (2004, 2007) and Tokimatsu and Suzuki (2009). These relationships draw upon extensively documented experimental results from centrifuge and large shaking table tests, under presumably highly permeable soil conditions, and consequently they do not make explicit reference to the dilation effects discussed in this paper.

The comparison is shown in Fig. 17, for a representative (L = 8 m, D = 0.6 m, EI = EI = 190,000 kNm^2) concrete pile which is installed in an 8 m thick uniform Nevada sand layer with $D_r = 50\,\%$ and is rigidly supported at its base. The excitation period was taken as T = 0.3 s, while the friction angle and the buoyant weight of the soil were equal to $\varphi = 33$ and $\gamma' = 9.81$ kN/m^3. Based on the empirical relation by Tokimatsu and Seed (1987), the equivalent SPT blow count for this soil was estimated as $(N_1)_{60-cs} \approx 11$.

To explore the effects of dilation at the upper part of the pile, predictions are shown for two different permeability coefficients that may be encountered in the field: $k = 6 \times 10^{-5}$ m/s, a typical low value for fine sands or silt-sand mixtures, and $k = 3 \times 10^{-3}$ m/s, a typical high value for coarse sands or sand-gravel mixtures. Note

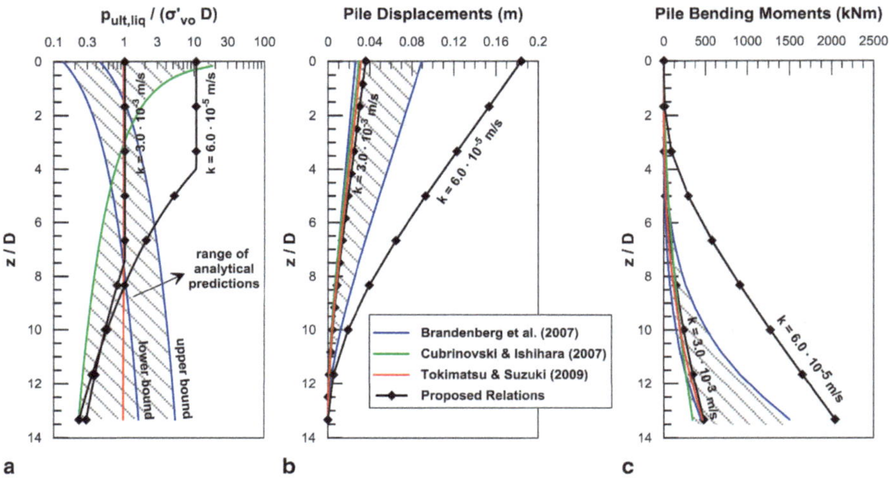

Fig. 17 Effect of permeability and dilation on. **a** Ultimate soil pressures. **b** Pile displacements. **c** Pile bending moments. (Chaloulos et al. 2014)

that existing methodologies do not account for permeability effects and consequently the associated predictions remain the same for both values of k. Furthermore, the predictions according to Brandenberg et al. are drawn as a range, defined by the limit values of the p-y curve multiplier $m_p = 0.050$ and 0.165, while the predictions by Tokimatsu and Suzuki are drawn for a reduction multiplier value equal to $\beta \approx 0.10$ (for $z < 10$ m).

Focusing first upon the variation with depth of the normalized ultimate soil pressures, inspection of Fig. 17a reveals that:

- When dilation effects are not pronounced, i.e. at medium and large depths and for large permeability coefficients, the proposed relationships fall within the range of the existing ones. This finding is positive, as the latter are supported by an extensive set of credible experimental data.
- On the contrary, when dilation effects become significant, i.e. at shallow depths and low permeability coefficients, ultimate soil pressures from the proposed relationships increase drastically relatively to predictions from the existing empirical methods.

The above observations come in support of the basic argument of González et al. (2009), namely that dilation effects may have been masked in scaled model experiments, used to establish current empirical methods, due to saturation of the liquefiable sand with water instead of a higher viscosity fluid. Furthermore, they imply that current empirical methods perform reasonably well for high permeability coarse sands and sand-gravel mixtures, but may underestimate significantly the liquefied soil pressures, in the case of low permeability fine and silty-sand deposits.

The associated predictions of pile displacements and bending moments are compared in Fig. 17b, 17c. The trends identified previously for the ultimate soil pressure distribution with depth, apply here as well. In addition, the opportunity is given to appreciate the potentially detrimental effect of dilation-induced negative pore pressures on pile response: pile head displacements are increased by 5–6 times, while bending moments are increased by about 4 times.

Conclusions

A comparative evaluation of 3D numerical predictions and experimental data has been presented aimed to explore the potential effects of soil dilation on piles undergoing large kinematic loads due to ground liquefaction and lateral spreading. The main feedback and conclusions drawn from this study are the following:

Numerical simulation of piles in laterally spreading ground

a. Accurate simulation of pile response requires incorporation of special features such as:
   - An advanced, effective stress constitutive model which can replicate the extremely non-linear liquefied soil response, as well as soil dilation due to pronounced shearing.
   - Slip and separation interface elements, to simulate the flow of liquefied ground around the pile head.
   - Drainage parallel to dynamic shaking and excess pore pressure build-up.
   - Consistent lateral boundaries which can replicate accurately the free-field conditions of submerged infinite slopes.

b. An optimum fit of the test measurements was obtained for the permeability coefficient k obtained from constant head permeability tests under 1 g gravitational acceleration. Lower k values, reflecting the effects of dynamic loading, distorted the numerical predictions. Furthermore, variable k values, increasing with the excess pore pressure ratio r_u, required considerably more computational time, without improving the accuracy of predictions.
c. Conventional tied node boundaries, developed for seismic loading of level ground or dry slopes, are not appropriate for submerged infinite slopes. Hence, the tied node formulation had to be modified in order to account for the hydrostatic pressure surplus at the down slope lateral boundary.
d. Laminar box containers, commonly used in centrifuge and shaking table simulators, are compatible with the conventional tied node boundaries used in numerical analyses. Hence, their application in problems related to liquefaction-induced lateral spreading is subject to the same limitations described in (c) above. For the case analyzed herein, laminar box containers in model tests, as well as conventional tied node boundaries in numerical analyses, may underestimate pile head displacements and bending moments by as much as 50 %.

Dilation effects on the subgrade reaction of liquefied soils

e. The ultimate pressure of the liquefied soil is uniquely related to the average excess pore pressure ratio in the vicinity of the pile ($r_{u,pile}$) and not in the free field ($r_{u,ff} \approx 1.0$).

f. Near the pile tip, relative soil-to-pile displacements are limited and consequently dilation is too weak to prevent positive excess pore pressure build up during seismic shaking. As a result, $r_{u,pile}$ approaches the free field value $r_{u,ff} \approx 1.0$ and the ultimate soil pressure decreases well below the reference value for non-liquefied soil.
g. At the upper part of the pile, soil-to-pile relative displacements are large and induce considerable dilation of the surrounding soil. Hence, excess pore pressures may become negative and the associated soil pressures may increase far beyond the corresponding values near the pile tip. This response is confirmed by centrifuge tests using high viscosity (i.e. metulose) pore fluid, but it is masked when water is used, leading to quick dissipation of the dilation-induced negative excess pore pressures.
h. The proposed multivariable relationships capture the aforementioned dilation effects, which become prominent at the upper part of the pile, for relatively low permeability soils (fine grained sands, silty sands, etc.). At the lower parts of the pile and for more permeable soils (coarse sands, sand and gravel mixtures, etc.), dilation effects are diminished and the proposed relations become comparable with existing ones which are based on data from centrifuge and shaking table experiments with water saturated sand.
i. The eventual development of negative excess pore pressures at the upper part of the pile, and the subsequent increase in the associated soil pressures, have a detrimental overall effect on pile performance and should not be overlooked.

Note that all above findings refer to the case of liquefiable soil layers which extend to the ground surface, i.e. without a high strength non-liquefiable cap. The presence of such a cap deserves further study, as it may alter pile-soil interaction effects on dilation-induced excess pore pressures and ultimate soil pressures (e.g. by reducing the relative pile-soil displacements, as well as, by cutting off drainage towards the ground surface). In addition, it is felt that more experiments, similar to that used to calibrate the numerical methodology, are required in order to identify independently the pile-soil interaction mechanisms described herein and to quantify their effects.

References

Acar YB, El-Tahir E-TA (1986) Low strain dynamic properties of artificially cemented sand. J Geotech Eng (Louisiana State Univ, Baton Rouge, LA, USA, Louisiana State Univ, Baton Rouge, LA. USA) 112(11):1001–1015

Andrianopoulos KI, Papadimitriou AG, Bouckovalas GD (2010) Bounding surface plasticity model for the seismic liquefaction analysis of geostructures. Soil Dyn Earthq Eng 30(10):895–911 (Department of Geotechnical Engineering, School of Civil Engineering, National Technical University of Athens, Greece)

Arulmoli K, Muraleetharan KK, Hossain MM, Fruth LS (1992) VELACS: verification of liquefaction analyses by centrifuge studies; Laboratory Testing Program—Soil Data Report.Research Report, The Earth. Technology Corporation

Been K, Jefferies MG (1985) A state parameter for sands. Géotechnique 35(2):99–112

Brandenberg SJ, Boulanger RW, Kutter BL, Chang D (2005) Behavior of pile foundations in laterally spreading ground during centrifuge tests. J Geotech Geoenviron Eng 131(11):1378–1391 (Dept. of Civil and Environmental Engineering, Univ. of California, Davis, CA 95616 United States)

Brandenberg SJ, Boulanger RW, Kutter BL, Chang D (2007) Static pushover analyses of pile groups in liquefied and laterally spreading ground in centrifuge tests. J Geotech Geoenviron Eng 133(9):1055–1066 (Dept. of Civil and Environmental Engineering, Univ. of California at Los Angeles, Los Angeles, CA 90095. United States)

Chaloulos YK, Bouckovalas GD, Karamitros DK (2013) Pile response in submerged lateral spreads: common pitfalls of numerical and physical modeling techniques. Soil Dyn Earthq Eng 55:275–287

Chaloulos YK, Bouckovalas GD, Karamitros DK (2014) Analysis of liquefaction effects on ultimate pile reaction to lateral spreading. J Geotech Geoenviron Eng (accepted for publication)

Cubrinovski M, Ishihara K (2004) Simplified method for analysis of piles undergoing lateral spreading in liquefied soils. Soils and Foundations 44(5)119–133

Cubrinovski M, Ishihara K (2007) Simplified analysis of piles subjected to lateral spreading: parameters and uncertainties. In: Pittilakis K (ed) 4th International Conference on Earthquake Geotechnical Engineering. Thessaloniki, Greece

Det NV (1980) Rules for the design, construction and inspection of offshore structures. Appendix F: Foundations. Hovik, Norway

Dobry R, Abdoun T, O'Rourke TD, Goh SH (2003) Single piles in lateral spreads: Field bending moment evaluation. J Geotech Geoenviron Eng (Dept. of Civil Engineering, Rensselaer Polytechnic Institute, Troy, NY 12180-3590. United States) 129(10):879–889

Elgamal A, Lu J, Yang Z (2005) Liquefaction-induced settlement of shallow foundations and remediation: 3D numerical simulation. J Earthq Eng (Department of Structural Engineering. University of California, San Diego, San Diego, CA 92093, United States, pp 17–45. 9(SPEC ISS)

Georgiadis M, Anagnostopoulos C, Saflekou S (1992) Centrifugal testing of laterally loaded piles in sand. Can Geotech J (NRC Research Press) 29(2):208–216

Ghosh B, Madabhushi SPG (2003) A numerical investigation into effects of single and multiple frequency earthquake motions. Soil Dyn Earthq Eng 23(8):691–704 (Geotechnical Research Group, Department of Engineering, University of Cambridge, Cambridge. United Kingdom)

González L, Abdoun T, Dobry R (2009) Effect of soil permeability on centrifuge modeling of pile response to lateral spreading. J Geotech Geoenviron 135(1):62–73 (Eng Dept. of Civil Engineering, Univ. of Chile, Avda. Blanco Encalada 2002, Santiago. Chile)

Karamitros DK (2010) Development of a numerical algorithm for the dynamic elastoplastic analysis of geotechnical structures in two and three dimensions. PhD Thesis, Dept of Civil Engineering, NTUA, Athens

Liu L, Dobry R (1997) Seismic response of shallow foundation on liquefiable sand. J Geotech Geoenviron ASCE 123(6):557–566

Papadimitriou AG, Bouckovalas GD (2002) Plasticity model for sand under small and large cyclic strains: a multiaxial formulation. Soil Dyn Earthq Eng 22(3):191–204 (Department of Geotechnical Engineering, Faculty of Civil Engineering, National Technical University of Athens, 42 Patission Street, 10682 Athens, Greece.)

Popescu R, Prevost JH, Deodatis G, Chakrabortty P (2006) Dynamics of nonlinear porous media with applications to soil liquefaction. Soil Dyn Earthq Eng 26(6–7):648–665 (Faculty of Engineering and Applied Science, Memorial University, St. John's, Nfld. A1B 3X5. Canada)

Saxena SK, Avramidis AS, Reddy KR (1988) Dynamic moduli and damping ratios for cemented sands at low strains. Can Geotech J 25(2):353–368 (Department of Civil Engineering, Illinois Institute of Technology, Chicago, IL 60616, USA)

Schnaid F, Prietto PDM, Consoli NC (2001) Characterization of cemented sand in triaxial compression. J Geotech Geoenviron Eng 127(10):857–868 (Dept. of Civ. Engrg., Fed. Univ. of Rio Grande do Sul, Av. Osvaldo Aranha, 99, 3. andar, Rio Grande do Sul 90035-190. Brazil)

Shahir H, Pak A, Taiebat M, Jeremić B (2012) Evaluation of variation of permeability in liquefiable soil under earthquake loading. Computers and Geotechnics, Department of Engineering, Tarbiat Moallem University, P.O. Box 31979-37551, Karaj, Iran, 40, pp 74–88

Sharma SS, Fahey M (2003) Degradation of stiffness of cemented calcareous soil in cyclic triaxial tests. J Geotech Geoenviron Eng 129(7):619–629 (Ctr. for Offshore Foundation Systems, The Univ. of Western Australia, 35 Stirling Highway, Crawley, WA 6009. Australia)

Tokimatsu K, Seed HB (1987) Evaluation of settlement in sands due to earthquake shaking. J Geotech Eng 113(8):861–878 (Tokyo Inst of Technology, Tokyo, Jpn, Tokyo Inst of Technology, Tokyo. Jpn)

Tokimatsu K, Suzuki H (2009) Seismic soil-pile-structure interaction based on large shaking table tests. In: Kokusho T, Tsukamoto Y, Yoshimine M (eds) Performance-based design in earthquake geotechnical engineering. Taylor & Francis

Integrating use of Swedish Weight Sounding Tests for Earthquake Reconnaissance Investigations

Yoshimichi Tsukamoto

Abstract The use of Swedish weight sounding tests has been examined by the authors for a wide range of stability problems of natural and reclaimed soil deposits during earthquakes. The present study is aimed at integrating these studies into one framework and at facilitating the use of Swedish weight sounding tests for future earthquake reconnaissance investigations. The example case history is given to prove the usefulness of subsurface soil profiling by means of Swedish weight sounding tests, and the procedures for determining the undrained shear strength, liquefaction strength and post-liquefaction settlement are examined and described in detail. Swedish weight sounding tests prove to be useful in analyzing the field performance of natural and reclaimed soil deposits during earthquakes.

Keywords Swedish weight sounding · Liquefaction-related evaluation

Introduction

It is quite decent and sound for geotechnical engineers and researchers residing in earthquake-prone areas to take a stance and an approach to conduct earthquake reconnaissance investigations and to gain as many lessons as possible. Among various field investigation tools, Swedish weight sounding tests have been increasingly used for post-earthquake subsurface soil profiling during recent devastating earthquakes, (Orense et al. 2002; Towhata et al. 2002, and others).

The use of Swedish weight sounding tests can be twofold. One is obviously to use them for postearthquake reconnaissance investigations. The other one is to evaluate the allowable shear strength of soils for residential house constructions. Swedish weight sounding tests were officially recommended as an investigation tool for this purpose by the Ministry of Land, Infrastructure and Transport in Japan. Since the great earthquake struck the eastern mainland of Japanese artipelago on March 11, 2011, and caused hugely wide-spread geotechncial hazards associated with soil liquefaction, including settlement and tilt of over 10,000 of residential

Y. Tsukamoto (✉)
Department of Civil Engineering, Tokyo University of Science, Chiba, Japan
e-mail: ytsoil@rs.noda.tus.ac.jp

A. Ansal, M. Sakr (eds.), *Perspectives on Earthquake Geotechnical Engineering*,
Geotechnical, Geological and Earthquake Engineering 37,
DOI 10.1007/978-3-319-10786-8_18

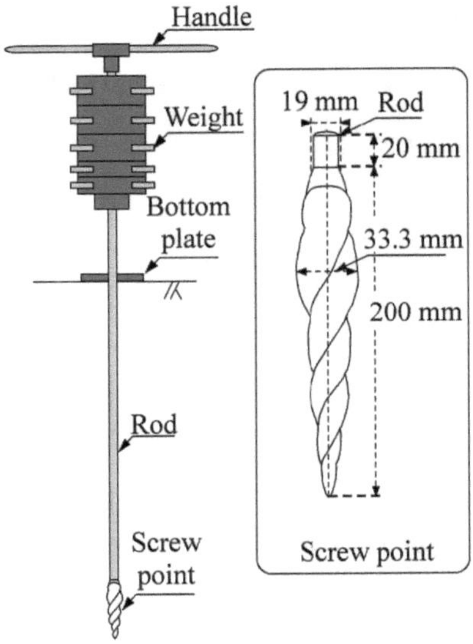

Fig. 1 Swedish weight sounding test equipment

houses located on the reclaimed lands at the water-front areas around Tokyo Bay, it has become important to pursue more advanced uses of geotechnical investigation tools such as Swedish weight sounding tests.

Swedish weight sounding tests are known to be portable and robust as shown in Fig. 1. In addition to the testing equipment shown in Fig. 1, it is recommended and usually necessary to bring a commercially available small kit for pulling out the rod penetrated into the soil deposits. The testing procedure and the interpretation of test results are described by Tsukamoto et al. (2004a). This field test is comprised of two phases, static penetration and rotational penetration. In the phase of static penetration, the screw-shaped point attached to the tip of the rod weighing 49 N (5 kg) is statically penetrated by putting several weights (10, 10, 25, 25, 25 kg) stepwise in increments until the total load becomes equal to 980 N (100 kg). At each load increment, the depth of static penetration is measured and the total weight is denoted as W_{sw} (kN). When static penetration is ceased, the rotational penetration is conducted. The horizontal handle attached to the top of the rod is rotated, and the number of half a turn necessary to penetrate the rod through 25 cm is denoted as N_a. The values of N_a are then multiplied by 4, and are converted to the number of half a turn per metre, N_{sw} (ht/m). From the experiences gained from the earthquake reconnaisance investigations conducted by the authors, Swedish weight sounding tests are relatively good at detecting thin weak soil layers. It is sometimes argued that it is not directly possible to determine depths of groundwater levels and types of soils and soil strata encountered during penetration, though it is optionally viable to tell whether the rod surface is dry or wet and also to classify soil particles adhered to the surface of the penetrated rod during its extraction. In addition, whenever it is possible to acquire any borehole data from standard penetration tests nearby, it is tractable to extrapolate types of soils and soil strata from such nearby SPT data.

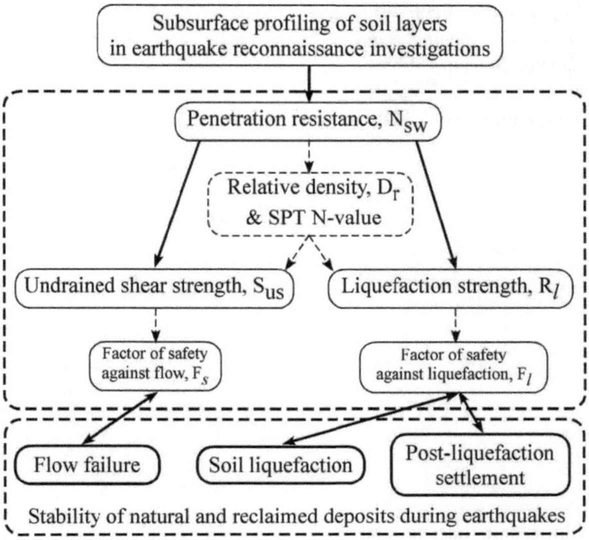

Fig. 2 Procedure suggested for use of Swedish weight sounding tests

The procedures for determining the engineering properties of soils have been studied by making use of Swedish weight sounding tests, such as the estimation of relative density, converting to SPT *N*-value, the evaluation of undrained shear strength, and predicting post-liquefaction settlement, (Tsukamoto 2009; Tsukamoto and Ishihara 2010, 2011; Tsukamoto et al. 2004a, b, 2009a, b, c). The present study is aimed at integrating these studies into one framework, and an advanced approach is proposed for the use of Swedish weight sounding tests during earthquake reconnaissance investigations, as shown in Fig. 2. The proposed framework covers a wide range of geotechnical interest, including soil liquefaction, and associated flow failure as well as post-liquefaction settlement. To investigate these phenomena observed in fields, the subsurface soil profiling, the procedures for determining undrained shear strength and liquefaction strength and for estimating post-liquefaction settlement are presented in the present study that follows.

Evaluation of Relative Density and Conversion to SPT N-Value

The conversion of the field penetration resistance to these engineering parameters requires the states of the soil to be determined properly. Herein, the relative density D_r was chosen as a key parameter that plays an important role for converting N_{sw}-values, (Tsukamoto et al. 2004a). Figure 3 shows the plots of N'_{sw1}-values against the relative density D_r for clean sand and various silty sands with non-plastic fines, which were obtained from the calibration chamber tests. Herein, the value of N'_{sw1} is defined as follows,

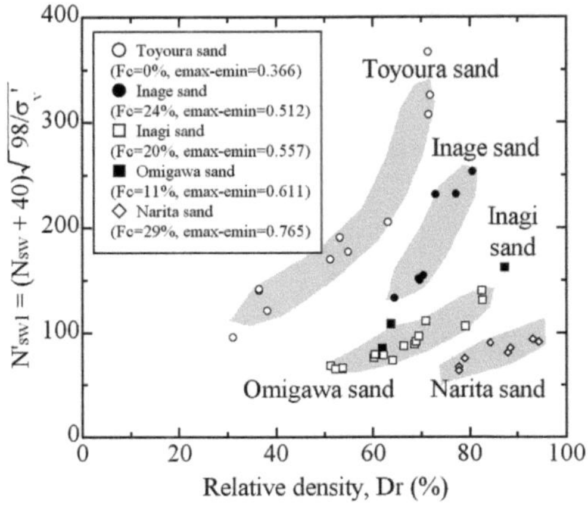

Fig. 3 Plots of converted values of N'_{sw1} against relative density D_r from calibration chamber tests. (Data after Tsukamoto et al. 2004a)

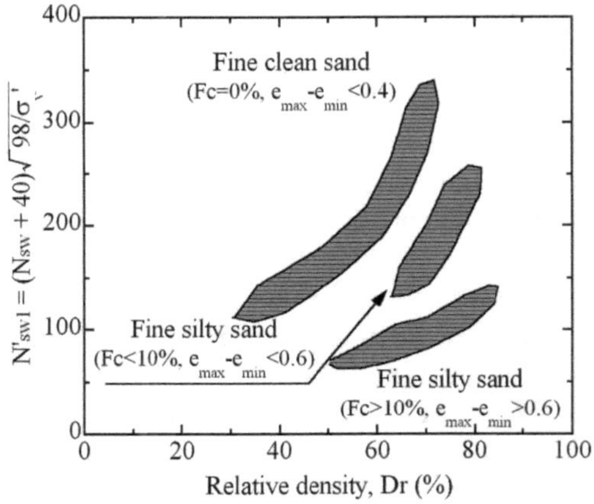

Fig. 4 Summary relations between converted values of N'_{sw1} against relative density D_r from calibration chamber tests

$$N'_{sw1} = (N_{sw} + 40)\sqrt{98/\sigma'_v} \tag{1}$$

where σ'_v is the effective vertical stress in kPa. The above expression takes into consideration the effect of static penetration, where the effect of static penetration is represented by the equivalent N_{sw}-value of 40. It is found that the silty sand with more non-plastic fines tends to locate its relation in a right-hand side of the diagram. Based on Fig. 3, the summary relations between N'_{sw1}-values against relative density D_r are illustrated in Fig. 4.

The values of penetration resistance thus measured by Swedish weight sounding tests can be converted to SPT *N*-values by employing the empirical expressions. There are several empirical correlations proposed in the past literature. Inada (1960) proposed the following correlations,

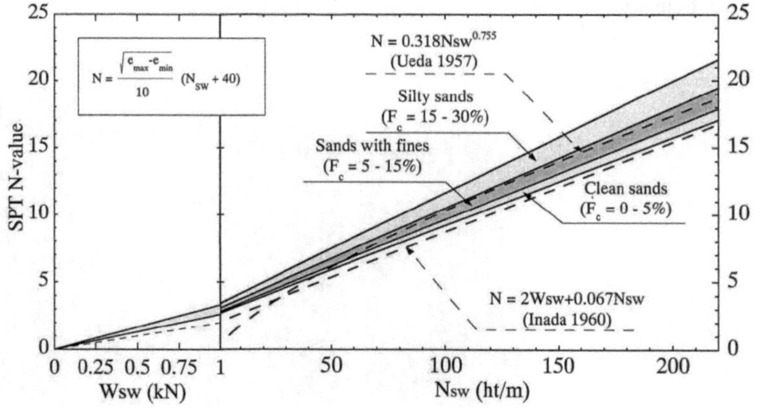

Fig. 5 Relations between Swedish W_{sw} and N_{sw}-values and SPT *N*-value for sands and silts with reference to fines content F_c

$$N = 0.002\, W_{sw} + 0.067 N_{sw} \quad \text{for sandy soils} \tag{2}$$

$$N = 0.003 W_{sw} + 0.050 N_{sw} \quad \text{for clayey soils} \tag{3}$$

where W_{sw} is expressed in the unit of N. The most recently proposed empirical expression for non-plastic silty sands is found in Tsukamoto et al. (2004a) as follows,

$$N = \frac{\sqrt{e_{\max} - e_{\min}}}{10}(N_{sw} + 40) \tag{4}$$

It is noticed that this correlation depends upon the grain characteristics represented by the void ratio range, $e_{max} - e_{min}$. From the practical point of view, it is more useful to indicate this relation with reference to the value of fines content, F_c. Cubrinovski and Ishihara (2004) examined extensively the relations between void ratio range, $e_{max} - e_{min}$ and several major soil gradation parameters such as fines content, F_c, and mean grain diameter, D_{50}. With a help of this study, the relations between SPT *N*-value and Swedish W_{sw} and N_{sw} values can be obtained as shown in Fig. 5. It is to note here that the void ratio range, $e_{max} - e_{min}$, typically takes a value of 0.4 to 0.45 for clean sands, and becomes larger for non-plastic silty sands with more fines content. For comparison, the correlations proposed by Ueda (1957) and Inada (1960) are also shown in Fig. 5.

Fig. 6 View from *top*

Subsurface Soil Profiling

Swedish weight sounding tests can give almost 10 m deep penetration to relatively loose sand deposits. By making use of its penetrability, a dozen of Swedish weight sounding tests were performed at the reclaimed soil deposit, where extensive fluidization and subsidence were observed in Tanno area of Kitami in Hokkaido during 2003 Tokachi-oki Earthquake, which occurred at 4: 50 am on September 26, 2003, as shown in Fig. 6, (Tsukamoto et al. 2009c). The subsurface soil profiling by means of Swedish weight sounding tests gave rise to the cross section of the loose reclaimed deposits as shown in Figs. 7 and 8, and revealed the progressive sequence of fluidization and subsidence, where the ground surface of 35 m wide and 150 m long subsided, and the subsurface layer of volcanic soil beneath the subsided farming field experienced extensive liquefaction and erupted onto the surface until all the subsurface deposits flowed downstream. The details of the reconnaissance investigations are described by Tsukamoto et al. (2009c).

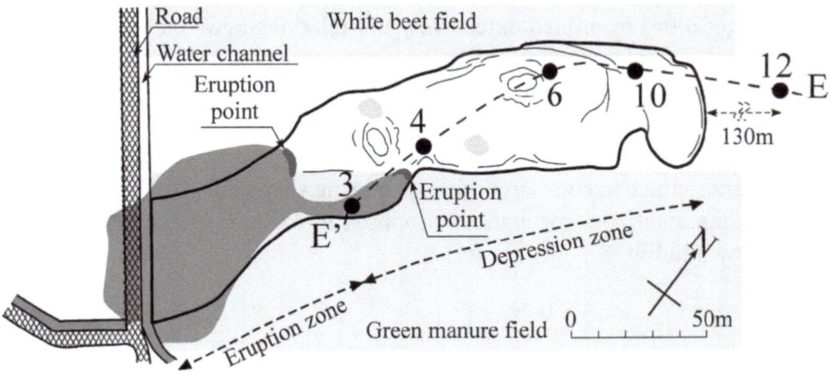

Fig. 7 Positions of field tests

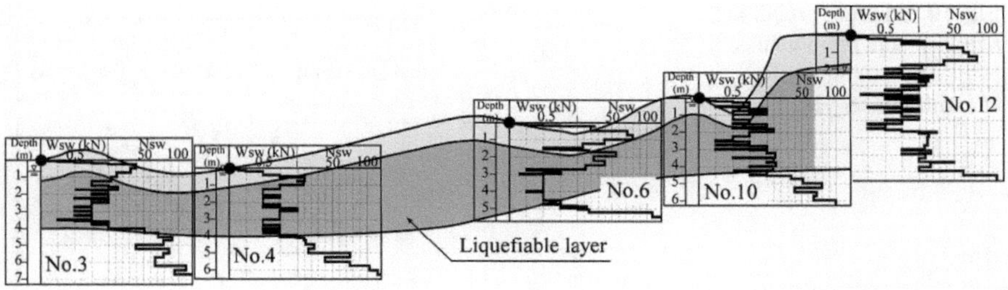

Fig. 8 Soil profile estimated from field tests at cross section E-E'. (After Tsukamoto et al. 2009c)

Undrained Shear Strength for Flow Failure

The undrained shear strength of non-plastic silty sands has been extensively examined by means of saturated and consolidated undrained triaxial compression and extension tests, (Tsukamoto et al. 2009b). The values of undrained shear strength were evaluated within the framework of a steady state concept, and the average values observed in triaxial compression and extension tests were plotted against the relative density D_r. The typical relations between the undrained shear strength ratio S_{us}/σ'_o and relative density D_r are illustrated in Fig. 9, for fine clean sand, fine silty sand with F_c less than 10% and fine silty sand with F_c more than 10%. Herein, the undrained shear strength ratio S_{us}/σ'_o was defined as follows,

$$\frac{S_{us}}{\sigma'_o} = \frac{q_s \cos \phi_s}{2\sigma'_o} \tag{5}$$

where q_s and φ_s are the deviatoric stress, $\sigma'_v - \sigma'_h$, and the internal friction angle observed at states of phase transformation. By using the typical relations between N'_{sw1} and D_r shown in Fig. 4 and the relations between S_{us}/σ'_o and D_r shown in Fig. 9, the typical relations between N'_{sw1}-values and the undrained shear strength

Fig. 9 Summary relations between undrained shear strength ratio S_{us}/σ'_o and relative density D_r from triaxial tests

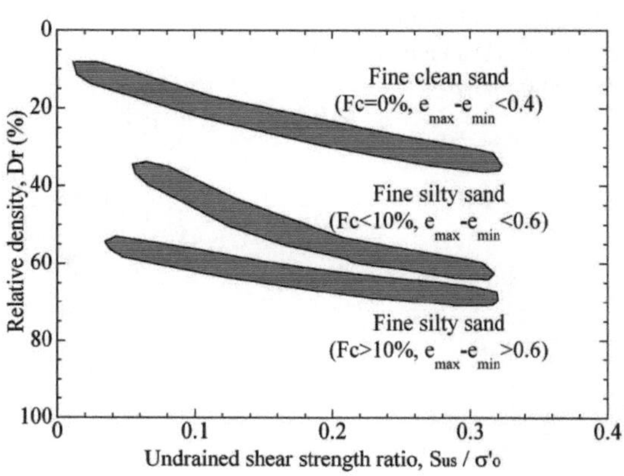

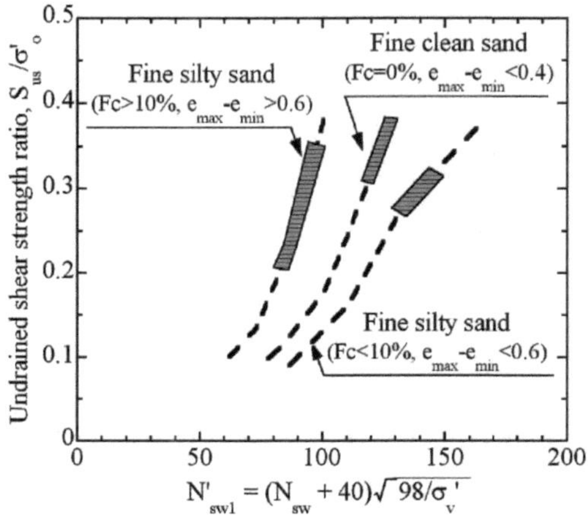

Fig. 10 Summary relations between converted values of N'_{sw1} and undrained shear strength ratio S_{us}/σ'_o

ratio S_{us}/σ'_o can be obtained as shown in Fig. 10. The relations between N'_{sw1}-values and S_{us}/σ'_o shown in Fig. 10 are somewhat found to be non-linear. However, the evaluation of the undrained shear strength ratio S_{us}/σ'_o from N'_{sw1}-values can be more conveniently done as described by Tsukamoto et al. (2009a) and Tsukamoto (2009), with one soil type -dependent parameter C_{sw} as follows,

$$\frac{S_{us}}{\sigma'_o} = \frac{N'_{swl}}{C_{sw}} = \frac{N_{sw}+40}{C_{sw}}\sqrt{\frac{98}{\sigma'_v}} \tag{6}$$

The dependency of the parameter C_{sw} on the soil gradation was examined by Tsukamoto et al. (2009a), by plotting the values of C_{sw} against the void ratio range $e_{max} - e_{min}$, as shown in Fig. 11, where the plot for Inagi sand obtained from the present study is also added.

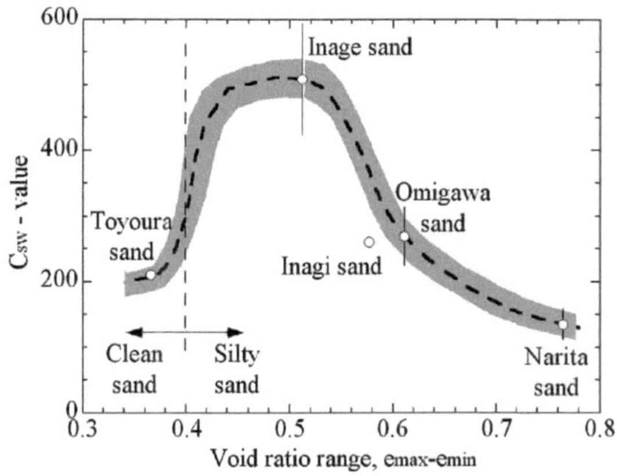

Fig. 11 Plots of values of C_{sw} against void ratio range $e_{max} - e_{min}$. (Data mostly after Tsukamoto et al. 2009a)

Liquefaction Strength

The methods for determining the liquefaction (cyclic) resistance of soils from the blow count of SPT and the tip resistance of CPT have been extensively examined based on the vast amount of field performance data obtained during the past earthquakes, as summarized by Ishihara (1996). However, there have been few studies of the same kind for Swedish weight sounding tests. Nevertheless, it would be possible to estimate the possibility of occurrence of soil liquefaction by means of Swedish weight sounding tests as follows, (Tsukamoto and Ishihara 2011). The SPT *N*-values can be estimated from Eq. (4), and the values of the relative density D_r can be estimated first by using the following empirical expression, (Tsukamoto et al. 2004a),

$$D_r = \sqrt{\frac{(N_{sw}+40)(e_{max}-e_{min})^{2.2}}{90}\sqrt{\frac{98}{\sigma'_v}}} \tag{7}$$

It is then possible to estimate the liquefaction resistance R_l from the SPT *N*-values and the values of D_r, by using the empirical formula proposed in the past literature, such as Tatsuoka et al. (1980). By assuming the value of peak ground acceleration at a particular site, it is possible to calculate the depth-wise profile of the factor of safety against liquefaction, F_l, by employing the conventional procedure used in the liquefaction analysis. In the case of 2000 Tottori-ken Seibu Earthquake, the soil liquefaction observed at the reclaimed soil deposits of Takenouchi Industrial Complex has been successfully evaluated from the results of depth-wise profile of F_l thus calculated.

On the other hand, it is also possible to establish the relation between the relative density D_r and liquefaction resistance R_l, based on the results of saturated and consolidated undrained cyclic triaxial tests. In the present study, the data on Toyoura clean sand and Inagi silty sand are examined, where the soil properties of the two soils are indicated in Table 1. The data on Toyoura clean sand were reported by

Table 1 Soil properties and specimen preparation for undrained cyclic triaxial tests

Location	Toyoura sand	Inagi sand
Specific gravity G_s	2.657	2.726
D_{50} (mm)	0.19	0.16
Atterberg limits	NP	NP
Fines content F_c (%)	0	26.9
Maximum void ratio e_{max}	0.973	1.803
Minimum void ratio e_{min}	0.607	1.023
e_{max}-e_{min}	0.366	0.78
Specimen preparation method	Air pluviation	Wet tamping

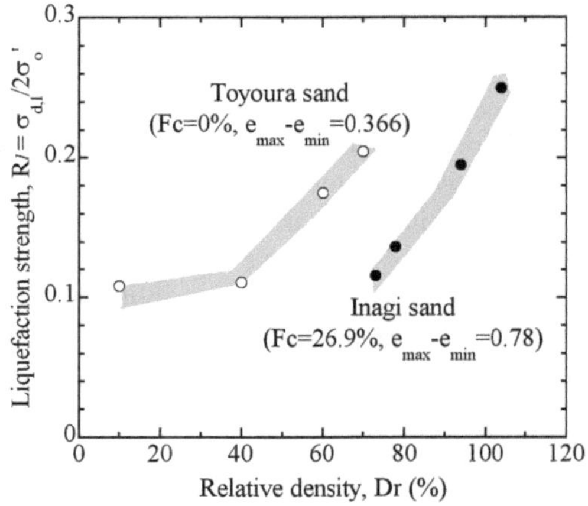

Fig. 12 Plots of liquefaction strength R_l against relative density D_r from triaxial tests

Tsukamoto et al. (2002) and Ishihara and Tsukamoto (2004), and the data on Inagi silty sand are obtained in the present study. It is to note that the soil sample of Inagi sand used for triaxial tests has different soil properties from that used for calibration chamber tests described above. Herein, the liquefaction strength R_l is defined as $\sigma_{d,l}/2\sigma'_o$, where $\sigma_{d,l}$ is the cyclic stress, $\sigma_d = \sigma_v - \sigma_h$, to cause the double amplitude axial strain $DA\varepsilon_a$ of 5% in 20 cycles, and σ'_o is the effective confining stress at consolidation. Figure 12 shows the plots of the liquefaction strength R_l against the relative density D_r thus obtained from triaxial tests. On the other hand, by using Eq. (7), the value of N'_{sw1} can be obtained for each value of D_r shown in Fig. 12. The plots of R_l against N'_{sw1} thus obtained can then be derived as shown in Fig. 13. For the sake of comparison, the relations derived by using the empirical expression proposed by Tatsuoka et al. (1980) are also indicated. The empirical expression proposed by Tatsuoka et al. (1980) requires the value of N_1 and the mean diameters of soils D_{50}. In Fig. 13, the values of N_1 are inferred from Eq. (4), and the values of

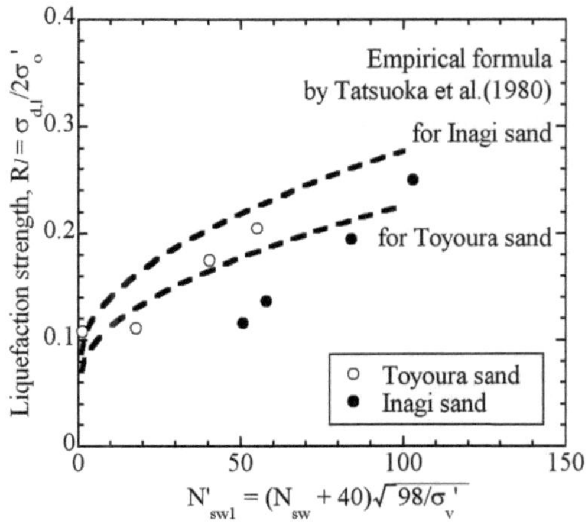

Fig. 13 Summary relations between converted values of N'_{sw1} and liquefaction strength R_l

D_{50} are assumed as shown in Table 1. It is noteworthy that the empirical expression tends to overestimate the liquefaction resistance of Inagi fine silty sand, compared with the results of triaxial tests.

Post-Liquefaction Settlement

One of the major consequences of soil liquefaction during earthquakes is the post-liquefaction settlement of soil deposits, which would sometimes lead to fatal damages on lifeline facilities and important structures such as gas storage tanks and power houses. The procedure for estimating the post-liquefaction settlement has recently been proposed by Tsukamoto and Ishihara (2010), based on the previous study by Tsukamoto et al. (2004b). At a couple of sites examined by Tsukamoto and Ishihara (2010), a series of Swedish weight sounding tests were conducted and the results were used for evaluating post-liquefaction settlement. The N_{sw}-values were first converted to the SPT *N*-values and the relative density D_r, based on Eqs. (4) and (7). The values of the liquefaction resistance were then evaluated from the empirical expression proposed by Tatsuoka et al. (1980), based on the calculated SPT *N*-values and some soil properties. The factor of safety against liquefaction F_l was then evaluated by using the procedure in usual practice, based on the observed acceleration records near the sites. The depth-wise profile of F_l was then estimated. The value of $\varepsilon_v/\varepsilon_{vmax}$ at each depth was then obtained from Fig. 14. Herein, the value of ε_{vmax} was evaluated from Fig. 15. Figure 15 shows the following unique relation for clean sands and non-plastic silty sands,

$$\varepsilon_{v\,\max} = \frac{3}{10}\varepsilon^*{}_{vr}(1 - D_r{}^*) \tag{8}$$

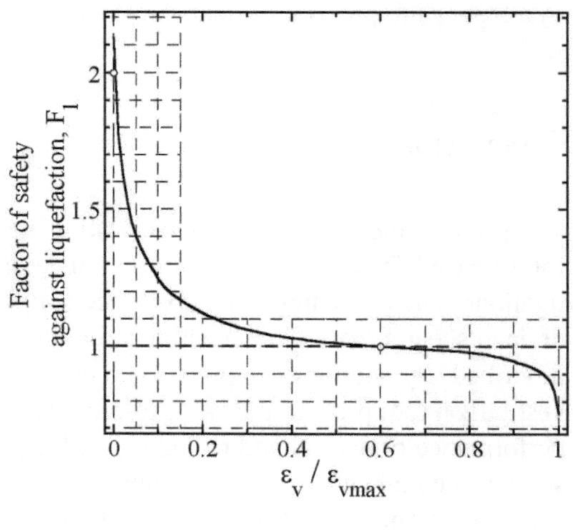

Fig. 14 Suggested relation between factor of safety against liquefaction F_l and $\varepsilon_v/\varepsilon_{vmax}$. (After Tsukamoto et al. 2004b)

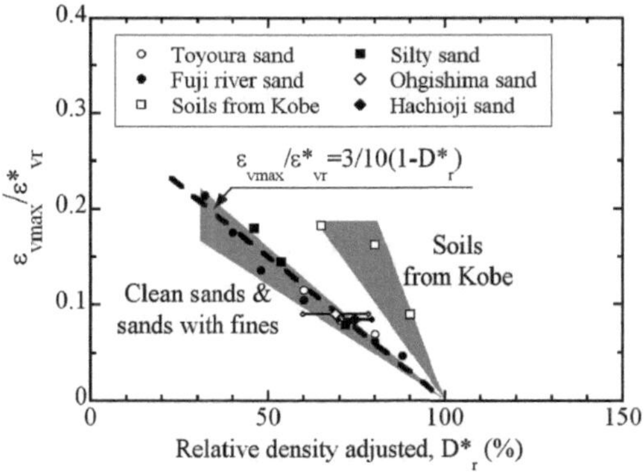

Fig. 15 Plots of $\varepsilon_{vmax}/\varepsilon^*_{vr}$ and relative density D_r^*. (After Tsukamoto and Ishihara 2010)

$$\varepsilon^*_{vr} = \frac{e_{max} - e^*_{min}}{1 + e_{max}}, \quad D_r^* = \frac{e_{max} - e}{e_{max} - e^*_{min}} \tag{9}$$

where the conceptual minimum void ratio was assumed as a fixed value of $e^*_{min}=0.6$. It is to note here that fine clean sand such as Toyoura sand indicates the minimum void ratio e_{min} almost equal to 0.6, while non-plastic fine silty sands indicate the minimum void ratio e_{min} larger than 0.6 and almost equal to 1. However, since fine non-plastic silty sands exhibit volume compressibility, they easily show the void ratio lower than the laboratory-determined minimum void ratio e_{min} as low as 0.6, when subjected to confining stress. This is the reasonable consequence of assuming the conceptual minimum void ratio e^*_{min}, when non-plastic fine silty sands are concerned. When fine clean sands with no fines are concerned, which exhibit negligible compressibility with this account, it is reasonable to assume that $e^*_{min}=e_{min}$. The value of ε^*_{vr} corresponds to the volumetric strain occurring due to the change in the void ratio from the maximum void ratio e_{max} to the conceptual minimum void ratio e^*_{min}. The post-liquefaction settlement can eventually be obtained as integration of the volumetric strain ε_v against depth from the bottom to the top of a liquefiable layer. Three case history studies were conducted by adopting this procedure, (Tsukamoto and Ishihara 2010), and it proved to work well.

Conclusions

The present study was aimed at facilitating and integrating the use of portable and robust Swedish weight sounding tests for future earthquake reconnaissance investigations. The procedures for subsurface soil profiling and for determining the undrained shear strength, liquefaction strength and post-liquefaction settlement were described. By adopting the procedures described in the present study, this field investigation tool proved to be able to provide good assistance in analyzing the field performance of natural and reclaimed soil deposits during earthquakes. The future study is recommended on establishing the evaluation of liqeufaction-induced lateral spreading by means of Swedish weight sounding tests.

Acknowledgement The author would like to acknowledge the advices and guidance from Professor Kenji Ishihara throughout the present study. He also acknowledges that the laboratory tests and field tests described in the present study were conducted with a help of a number of the past students at Tokyo University of Science.

References

Cubrinovski M, Ishihara K (2004) Maximum and minimum void ratio characteristics of sands. Soils Found 42(6):65–78

Inada T (1960) On the use of Swedish weight sounding test results. Domest J Jpn Geotech Soc Tsuchi-to-Kiso 8(1):13–18 (In Japanese)

Ishihara K (1996) Soil behaviour in earthquake geotechnics, The Oxford engineering science series. Clarendon Press, Oxford

Ishihara K, Tsukamoto Y (2004) Cyclic strength of imperfectly saturated sands and analysis of liquefaction. Proc Jpn Acad Ser B 80(8):372–391

Orense R, Vargas-Monge W, Cepeda J (2002) Geotechnical aspects of the January 13, 2001 El salvador earthquake. Soils Found 42(4):57–68

Tatsuoka F, Iwasaki T, Tokida K, Yasuda S, Hirose M, Imai T, Konno M (1980) Standard penetration tests and soil liquefaction potential evaluation. Soils Found 20(4):95–111

Towhata I, Prasad SK, Honda T, Chandradhara GP (2002) Geotechnical reconnaissance study on damage caused by 2001 Gujarat Earthquake, India. Soils Found 42(4):77–88

Tsukamoto Y (2009) Determining the undrained residual strength of soils from field penetration tests with some case history studies. In: Kokusho T, Tsukamoto Y, Yoshimine M (eds) Performance-based design in earthquake geotechnical engineering—from case history to practice. CRC Press, Balkema (Theme Lecture, Proceedings of IS-Tokyo2009 Tsukuba, Japan, June 15–18, 2009), pp 179–192

Tsukamoto Y, Ishihara K (2010) Analysis on settlement of soil deposits following liquefaction during earthquakes. Soils Found 50(3):399–411

Tsukamoto Y, Ishihara K (2011) Use of field penetration tests in evaluating occurrence of soil liquefaction at reclaimed deposits during 2000 Tottori-ken Seibu Earthquake. Proceedings of 5th International Conference on Earthquake Geotechnical Engineering, Santiago, Chile, January 10–13, 2011, CD Paper No. UOFTS

Tsukamoto Y, Ishihara K, Nakazawa H, Kamada K, Huang Y (2002) Resistance of partly saturated sand to liquefaction with reference to longitudinal and shear wave velocities. Soils Found 42(6):93–104

Tsukamoto Y, Ishihara K, Sawada S (2004a) Correlation between penetration resistance of Swedish weight sounding tests and SPT blow counts in sandy soils. Soils Found 44(3):13–24

Tsukamoto Y, Ishihara K, Sawada S (2004b) Settlement of silty sand deposits following liquefaction during earthquakes. Soils Found 44(5):135–148

Tsukamoto Y, Ishihara K, Harada K (2009a) Evaluation of undrained shear strength of soils from field penetration tests. Soils Found 49(1):11–23

Tsukamoto Y, Ishihara K, Kamata T (2009b) Undrained shear strength of soils under flow deformation. Geotechnique 59(5):483–486

Tsukamoto Y, Ishihara K, Kokusho T, Hara T, Tsutsumi Y (2009c) Fluidisation and subsidence of gently sloped farming fields reclaimed with volcanic soils during 2003 Tokachi-oki earthquake in Japan. In: Kokusho T (ed) Earthquake geotechnical case histories for performance-based design. CRC Press, Balkema, pp 109–118

Ueda Y (1957) On Swedish weight sounding rod. Domest J Jpn Geotech Soc Tsuchi-to-Kiso. Tokyo 5(5):9–12 (In Japanese)

MIX
Papier aus verantwortungsvollen Quellen
Paper from responsible sources
FSC® C105338

If you have any concerns about our products,
you can contact us on
ProductSafety@springernature.com

In case Publisher is established outside the EU,
the EU authorized representative is:
Springer Nature Customer Service Center GmbH
Europaplatz 3, 69115 Heidelberg, Germany

Printed by Libri Plureos GmbH
in Hamburg, Germany